Urban and Landscape Perspectives

Volume 18

Aims and Scope

Urban and Landscape Perspectives is a series which aims at nurturing theoretic refl ection on the city and the territory and working out and applying methods and techniques for improving our physical and social landscapes.

The main issue in the series is developed around the projectual dimension, with the objective of visualising both the city and the territory from a particular viewpoint, which singles out the territorial dimension as the city's space of communication and negotiation.

The series will face emerging problems that characterise the dynamics of city development, like the new, fresh relations between urban societies and physical space, the right to the city, urban equity, the project for the physical city as a means to reveal civitas, signs of new social cohesiveness, the sense of contemporary public space and the sustainability of urban development.

Concerned with advancing theories on the city, the series resolves to welcome articles that feature a pluralism of disciplinary contributions studying formal and informal practices on the project for the city and seeking conceptual and operative categories capable of understanding and facing the problems inherent in the profound transformations of contemporary urban landscapes.

More information about this series at
http://www.springer.com/series/7906

Roberto Gambino • Attilia Peano†
Editors

Nature Policies and Landscape Policies

Towards an Alliance

Springer

Editors
Roberto Gambino
Attilia Peano†
Interuniversity Department of Regional and Urban Studies and Planning (DIST)
Politecnico and Università di Torino
Turin
Italy

ISBN 978-3-319-05409-4 ISBN 978-3-319-05410-0 (eBook)
DOI 10.1007/978-3-319-05410-0
Springer Cham Heidelberg New York Dordrecht London

Library of Congress Control Number: 2014955441

Cover image: Po River Regional Park from Crescentino's bridge, near Turin (Italy).
Photo by Ippolito Ostellino, 2010.

Printed on acid-free paper

Springer is part of Springer Science+Business Media (www.springer.com)

With great affection we remember Attilia Peano – former Full Professor in Town and Regional Planning at the Politecnico di Torino (DIST) and CEN PPN Director – and the precious contributions she has given during the course of her life in the field of urban planning, landscape planning, and nature and cultural heritage conservation, being a protagonist in the debate on these topics at national and international level.
Her death (18th August 2013) interrupted her participation in several research activities which are still ongoing, and, in particular, in the international research, that has been carried on by the CED PPN since 2010, concerning the relationship between Landscape policies and Nature Conservation policies.
This book is the outcome of this CED PPN research, and we would like to dedicate it to our friend and colleague Attilia, hoping in this way to remember her passion and her valuable guide in facing the subject here presented.

Roberto, Gabriella, Emma, Luigi

Foreword

Since the beginning of the 1990s, the European Documentation Centre on Nature Park Planning (CED PPN, DIST, Politecnico and Università di Torino) has conducted ongoing research into nature and heritage conservation policies and their relationship with urban and regional planning. Particular attention has been given both to European parks and other protected area[1] policies and to European landscape[2] policies.

In 2008, CED PPN launched an innovative research programme concerning the *connections* between nature conservation policies and landscape policies. This is a major subject since the risks related to global change and the continuous worsening of environmental conditions challenge the effectiveness of area-based nature conservation policies, demanding that they be 'territorialized', thereby overcoming the traditional separation of protected areas with respect to the wider context. This demand is at the basis of the so-called new conservation paradigm (5th IUCN World Parks Congress, Durban 2003). Landscape, as a bridge between nature and culture, could play a crucial role in this direction, helping conservation policies to open up to the territorial, social and economic context, extending their scope and improving their effectiveness. Protected areas, in turn, could act as extraordinary learning laboratories for landscape policies, giving them the regulatory capacity generally gained over the course of a long history of policies and planning. Since 2008, this research programme has been discussed at several international meetings such as the 4th IUCN World Conservation Congress (Barcelona 2008) and the 5th IUCN World Conservation Congress (Jeju 2012).

[1] "A protected area is a clearly-defined geographical space, recognized, dedicated and managed, through legal or other effective means, to achieve the long-term conservation of nature with associated ecosystem services and cultural values" (Dudley, N., Ed., 2008. Guidelines for Applying Protected Area Management Categories. Gland, Switzerland, p. 8).

[2] Landscape means "an area, as perceived by people, whose character is the result of the action and interaction of natural and/or human factors" (Council of Europe, European Landscape Convention, 2000, art. 1).

In the context of this activity, in 2010 CED PPN, in order to take stock of considerations, research and experiences at the international level in relation to the above-mentioned theme, invited a wide group of experts in different disciplines (architects, planners, geographers, biologists, ecologists, historians, jurists, economists, sociologists) from various institutional bodies (universities, administrative authorities, international organisations such as IUCN, EUROPARC, UNESCO, UNISCAPE and others) to participate in a 'research book' project. The experts were asked to deal, starting from their disciplinary background, with the main research thesis – concerning the possibility and opportunity of an alliance between nature conservation policies and landscape policies – and with its political, social, scientific and cultural implications. The idea behind the project was not to draw up a comprehensive, unified proposal on the issue to be subsequently published and discussed but rather to start – on the basis of a book – a process of dialogue stimulating the critical enrichment of existing knowledge on the matter and inspiring further debate.

The most interesting considerations gathered since 2010 are presented in this book, which is a 'research book' in the true sense in that it collects and makes it possible to compare a set of varied contributions on the subject, highlighting agreements and convergences as well as disagreements and divergences about the proposed nature-landscape alliance.

The contributions, after an introduction (Gambino) which presents an overall line of reasoning, are divided into three main parts.

Part I – New Paradigms. In this first part, some general and theoretical considerations about the current and potential relationships between nature conservation policies and landscape policies are reported. Experts discuss the new cultural paradigms that might form the basis of the envisaged alliance. This topic is approached with reference to

- Diverse geographical contexts: the global context (Phillips, Brown, with reference to some international tools such as IUCN Category V, Protected Landscapes), the European context (Ritchie, starting from the EUROPARC Federation experience), the US context (Bray, with reference to the role of a large landscape conservation approach in protected area policies)
- Diverse disciplinary backgrounds such as ecology (Gibellli and Santolini, who analyse the role of ecological functionality for landscape conservation; Guarino et al., who deal with the conflicts between human activities and ecosystem conservation), law (Desideri, who focuses on the legal framework for a comprehensive approach to nature conservation and landscape protection), architecture (Buyck and Vales, who analyse the concept of landscape, also in relation to its design), geography (Raffestin, who explores the role of landscape image in understanding territorial reality)

Part II – From Nature to Landscape and Back. In this second part of the book, experts discuss the mutual interactions between nature policies and landscape policies, focusing on three main topics: (i) regulations and institutional frameworks, (ii) policies, (iii) actions and tools.

With reference to *regulations and institutional frameworks* (i), the book gathers contributions concerning different geographical contexts: the global context (Andrian and Tufano, with reference to the relationship between Biosphere Reserves and protected areas), the European context (Angelini, with specific reference to the Alpine Convention; Romano and Zullo, who focus on the relationship between protected areas, EU Natura 2000 sites and landscapes) and some specific national contexts (such as the Netherlands and the United Kingdom in Voghera or Italy in Moschini and in Besio).

As far as *policies* (ii) are concerned, contributions deal with

- Policies concerning specific landscape types (the urban landscape in La Riccia), particular territorial or institutional contexts (the Andalucian vegas and delta areas in Miguel and Perèz Campaña, the Switzerland institutional system in Hammer and Leng), protected area types (the IUCN category V protected areas – Protected Landscapes – in Salizzoni)
- Policies concerning wider issues such as the strategic role of tourism in protected areas (Danelutti et al., Coda Zabetta), governance processes and community participation for nature conservation (Weizlbaumer et al., Barbera et al., Salvatore, Brunetta), the 'sense of limit' in landscape planning and design (Mazzino), biodiversity policies for landscape conservation (Ferroni et al., Seardo), the concept of protected areas as 'nodes' of networks extended beyond their boundaries (Pigliacelli and Teofili), the role of cooperation policies for landscape management (Nicoletti)

With reference to *actions and tools* (iii), experts present and discuss methodologies and instruments concerning landscape planning and nature conservation in their relationship (Paolinelli, Castelnovi, Sargolini, Tosini, Dudley and Stolton, Laven et al.), also focusing on specific aspects such as visual perception analysis of landscape (Franchini and Greco), management and planning of landscape scenic values (Cassatella) and cultural heritage enhancement (Beltramo). This section also includes contributions on environmental and landscape assessment (Bravi and Gasca, Bottero et al.) and on financing of nature conservation policies (Cetara).

Part III – Experiences and practices. In the third part of the book, a number of case studies, mainly regarding protected areas or special landscape and institutional contexts, are presented. They provide interesting examples of the integration of aspects and policies related both to nature and landscape (Godone et al., Vinardi, Deambrogio and Zocco, Balletti and Soppa, Ghersi, Corsani and Morelli, Martinelli and Simone, Pinzello, Matoda). Some of the case studies allow also to discuss specific issues such as the relationship between energy production and nature-landscape conservation (Natali and Silvestri, Mininni and Rizzi) or the connection between tourism and nature-landscape conservation (Calcagno Maniglio and Simone, Valle and Dongiovanni).

As a 'research book', this publication does not set out to exhaust the topic of the relationship between nature conservation policies and landscape policies. Rather, it aims at opening up some lines of enquiry into the matter, orienting research efforts towards new possible directions. These lines of enquiry should take stock of the following 'lessons' learned from the various contributions, concerning

- The dynamic dimension of current problems related to the joint conservation of nature and landscape and consequently the need to always look to the future
- The strategic role of the diversification and integration of knowledge, visions and competences on the matter, adopting a multi-, inter- and trans disciplinary approach
- The need for trans-scale approaches (global, national, regional and local) in planning and managing nature and cultural heritage
- The crucial role played by the 'project' for a 'good governance' of the regional realities and thus the importance of identifying the values, aims, tools and, most importantly, actors to be involved in this challenge

Gabriella Negrini
Interuniversity Department of Regional and Urban Studies and Planning (DIST)
Politecnico and Università di Torino
Viale Mattioli 39, 10125, Turin, Italy

Emma Salizzoni
Interuniversity Department of Regional and Urban Studies and Planning (DIST)
Politecnico and Università di Torino
Viale Mattioli 39, 10125, Turin, Italy

Acknowledgments

I would like to thank the CED PPN Editorial Board and Research Group, that has supported the drawing up of this book: Gabriella Negrini (research and editorial project coordination), Emma Salizzoni (research project), Luigi La Riccia (editorial project).

CED PPN

European Documentation Centre
on Nature Park Planning

CED PPN (European Documentation Centre on Nature Park Planning) Interuniversity Department of Regional and Urban Studies and Planning (DIST), Politecnico and Università di Torino. www.cedppn.polito.it

Roberto Gambino

Abbreviations

AECID	(Spanish) Agency for International Cooperation and Development
ANCSA	Associazione Nazionale Centri Storici e Artistici/Italian National Association of Historic and Artistic Centres
ANPIL	Aree Naturali Protette di Interesse Locale/Natural Protected Areas of Local Interest
APE	Apennines Park of Europe
APIs	Areas of Particular Importance for biodiversity and ecosystem services
BRs	Biosphere Reserves
CA	Conjoint Analysis methodology
CAP	Common Agricultural Policy of the European Union
CBA	Cost-Benefit Analysis
CBD	Convention on Biological Diversity
CCAs	Community Conserved Areas
CDCULT	Steering Committee for Culture (Council of Europe)
CDPATEP	Steering Committee for Cultural Heritage and Landscape (Council of Europe)
CE	Choice Experiments
CED PPN	European Documentation Centre on Natural Park Planning (DIST, Politecnico di Torino)
CEV	Corporate Ecosystem Valuation (WBCSD)
CITIES	Convention on International Trade in Endangered Species of Wild Fauna and Flora (1973)
CLLD	Community-Led Local Development (ENRD)
CMS	Convention on the Conservation of Migratory Species of Wild Animals (or Bonn Convention 1979)

CIVILSCAPE	Non-governmental Organizations for European Landscape Convention
CNR-IRPI	Italian National Research Council, Research Institute for Geo-hydrological Protection of Turin
CoE	Council of Europe
CoM	Covenant of Mayors (EU)
COMPACT	Community Management of Protected Areas for Conservation program
COP	Conference of the Parties
COPI	Cost of Policy Inaction
CVM	Contingent Valuation Method
ECC	European Economic Community
ECTS	European Charter for Sustainable Ecotourism
EEA	European Environment Agency
EESC	European Economic and Social Committee (EU)
EGTC	European Grouping of Territorial Cooperation
EIA	Environmental Impact Assessment
ELC	European Landscape Convention
ENELC	European Network of Local and Regional Authorities for the Implementation of the European Landscape Convention
EPA	Enlarged Partial Agreement on Cultural Routes (Council of Europe 2010)
EPE	Environmental Protection Expenditure
ESPON	European Spatial Planning Observation Network (EU)
ESs	Ecosystem Services
EUAP	Elenco Ufficiale delle Aree Protette/Official list of Italian Protected Areas
EC	European Commission
EU	European Union
EUROSTAT	Statistical Office of European Union
FAI	Fondo Ambiente Italiano/Italian Environment Fund
GIS	Geographic Information System
GYE	Greater Yellowstone Ecosystem
HABAP	English Highway Agency (Biodiversity Action Plan)
HLF	Heritage Lottery Fund
HPM	Hedonic Pricing Method
IALE	Italian Society of Landscape Ecology
ICCAs	Indigenous peoples' and Community Conserved Areas and territories
ICOMOS	International Council on Monuments and Sites
IGMI	Istituto Geografico Militare Italiano/Italian Military Geographic Institute
ILNM	Federal Inventory of Landscapes and Natural Monuments of National Importance

INU	Istituto Nazionale di Urbanistica/Italian National Institute of Urban Planners
IPA	Indigenous Protected Area (Australia)
IPAL	Integrated Programme on Arid Lands of UNEP
IPBES	Intergovernmental Science-Policy Platform on Biodiversity and Ecosystem Services
ISOS	Federal Inventory of Swiss Heritage Sites
ISPRA	Istituto Superiore per la protezione e la ricerca ambientale/ Institute for Environmental Protection and Research (Italy)
IUCN	International Union for Conservation of Nature
IUCN-Med	IUCN Centre for Mediterranean
LCA	Landscape Character Assessment
LC	Landscape Capacity
LCC	Landscape Conservation Cooperative
LVIA	Landscape and Visual Impact Assessment
MaB	Man and Biosphere UNESCO Program
MAP	Madrid Action Plan
MEA	Millennium Ecosystem Assessment (2005)
MATTM	Italian Ministry for Environment and Territory and Sea
MiBAC	Italian Ministry on Cultural Heritage and Activities
NE	Natural England
NIAs	Nature Improvement Areas
NCAs	National Character Areas
NGO	Non-governmental Organization
NHA	National Heritage Area
NIAs	Nature Improvement Areas
NBSAPs	National Biodiversity and Action Plans
OECD	Organization for Economic Cooperation and Development
OEP	Arco Latino, Osservatorio del Paesaggio/Arco Latino European Landscape Observatory
PA/PAs	Protected Area/Protected Areas
PES	Payments for Ecosystem Services
PIT	Piano di Indirizzo Territoriale/Regional Design Plan
PEBLDS	Pan-European Biological and Landscape Diversity Strategy
PEER	Partnership for European Environmental Research (EU)
PoWPA	Programme of Work on Protected Areas (CBD)
PPGIS	Public Participation Geographic Information Systems
PPR	Piano Paesaggistico Regionale/Regional Landscape Plan
PPTR	Piano Paesaggistico Territoriale Regionale/Regional Territorial Landscape Plan (Apulia Region)
PSR	Piano di Sviluppo Rurale/Rural Development Plan
PTCP	Piano di Coordinamento Provinciale/Province Coordination Plan
SCBD	Secretariat of the Convention on Biological Diversity
SCIs	Sites of Community Importance

SCZs	Special Conservation Zones
SEA	Strategic Environmental Assessment
SEAP	Sustainable Energy Action Plan
SNB	National Strategy for Biodiversity
SPAs	Special Protection Areas
SUME	Sustainable Urban Metabolism for Europe project
TEEB	The Economics of Ecosystems and Biodiversity
TCM	Travel Cost Method
UN	United Nations
UNDRIP	United Nations Declaration on the Rights of Indigenous Peoples (2008)
UNEP	United Nations Environment Programme
UNESCO	United Nations Educational, Scientific and Cultural Organization
UNISCAPE	European Network of Universities for the Implementation of European Landscape Convention
UNWTO	United Nations World Tourism Organization
USNPS	U.S. National Park Service
VC	Venture Capital
WBCSD	World Business Council for Sustainable Development
WCPA	IUCN World Commission on Protected Areas
WDPA	World Database Protected Areas
WHS	World Heritage Sites
WNBR	World Network of Biosphere Reserves
WWF	World Wide Fund for Nature
ZNIEFF	Zone naturelle d'intérêt écologique, faunistique et floristique/ Inventory of natural zones of ecological, faun and floristic interest
ZPPAUP	Zone de Protection du Patrimoine Architectural, Urbain et Paysager/Zone of Protection for Architectural, Urban and Landscape Patrimony

Contents

Contributors

Giorgio Andrian Formerly UNESCO office, Padua, Italy

Paolo Angelini Italian Ministry for the Environment, Land and Sea, Rome, Italy

Franca Balletti Department of Sciences for Architecture, Polytechnic School, University of Genoa, Genoa, Italy

Federica Barbera Protected Areas Department, Legambiente Onlus, Rome, Italy

Brenda Barrett Living Landscape Observer, Harrisburg, PA, USA

Silvia Beltramo Interuniversity Department of Regional and Urban Studies and Planning (DIST), Politecnico and Università di Torino, Turin, Italy

Mariolina Besio Department of Sciences for Architecture, Polytechnic School, University of Genoa, Genoa, Italy

Sergio Bongiovanni GIS Consultant, Turin, Italy

Marta Bottero Interuniversity Department of Regional and Urban Studies and Planning (DIST), Politecnico and Università di Torino, Turin, Italy

Marina Bravi Interuniversity Department of Regional and Urban Studies and Planning (DIST), Politecnico and Università di Torino, Turin, Italy

Paul M. Bray Albany University (Retired), Albany, NY, USA

Jessica Brown New England Biolabs Foundation, Ipswich, MA, USA

IUCN-WCPA Specialist Group on Protected Landscapes, Ipswich, MA, USA

Grazia Brunetta Interuniversity Department of Regional and Urban Studies and Planning (DIST), Politecnico and Università di Torino, Turin, Italy

Jennifer Buyck Institute d'Urbanisme de Grenoble, Grenoble, France

Claudia Cassatella Interuniversity Department of Regional and Urban Studies and Planning (DIST), Politecnico and Università di Torino, Turin, Italy

Paolo Castelnovi ex-Politecnico di Torino, Turin, Italy

Luca Cetara European Academy of Bolzano, Representing Office, Rome, Italy

Giulio Cioffi CREDIA WWF, Fiuminata, MC, Italy

Gabriele Corsani Department of Architecture, University of Florence, Florence, Italy

Carla Danelutti IUCN Center for Mediterranean Cooperation, Málaga, Spain

Ángeles De Andrés Caramés ECOTONO, Equipo Consultor Turismo y Desarrollo, S.L., Madrid, Spain

Almudena De Velasco Menéndez ECOTONO, Equipo Consultor Turismo y Desarrollo, S.L., Madrid, Spain

Franca Deambrogio Sportello INFOFIUME, Parco Fluviale del Po e dell'Orba, Casale Monferrato, Italy

Carlo Desideri Institute for the Study of Regionalism Federalism and Self-Government, ISSiRFA-CNR, Rome, Italy

Maria Giovanna Dongiovanni SiTI Istituto Superiore sui Sistemi Territoriali per l'Innovazione, Turin, Italy

Nigel Dudley School of Geography, Planning and Environmental Management, University of Queensland, Brisbane, QLD, Australia

IUCN World Commission on Protected Areas and Equilibrium Research, Bristol, UK

Valeria Ferretti Interuniversity Department of Regional and Urban Studies and Planning (DIST), Politecnico and Università di Torino, Turin, Italy

Franco Ferroni Biodiversity, Protected Areas and Agricultural Policies, WWF Italy, Rome, Italy

Monica Foglia INEA Research Grant at Agricultural Society La Quercia della Memoria, CREDIA WWF, San Ginesio, MC, Italy

Caterina Franchini Interuniversity Department of Regional and Urban Studies and Planning (DIST), Politecnico and Università di Torino, Turin, Italy

Roberto Gambino Interuniversity Department of Regional and Urban Studies and Planning (DIST), Politecnico and Università di Torino, Turin, Italy

Matteo Garbarino Department of D3A, Università Politecnica delle Marche, Ancona, Italy

Gabriele Garnero Interuniversity Department of Regional and Urban Studies and Planning (DIST), Politecnico and Università di Torino, Turin, Italy

Emanuela Gasca Interuniversity Department of Regional and Urban Studies and Planning (DIST), Politecnico and Università di Torino and SiTI Istituto Superiore sui Sistemi Territoriali per l'Innovazione, Turin, Italy

Adriana Ghersi Department of Sciences for Architecture, Polytechnic School, University of Genoa, Genoa, Italy

Gioia Gibelli SIEP-IALE Italian Society of Landscape Ecology, Milan, Italy

Franco Godone CNR – IRPI, Torino, TO, Italy

Danilo Godone Department DISAFA and NatRisk, Università degli Studi di Torino, Grugliasco, TO, Italy

Elena Greco Interuniversity Department of Regional and Urban Studies and Planning (DIST), Politecnico and Università di Torino, Turin, Italy

Riccardo Guarino Dipartimento STEBICEF, Università di Palermo, Palermo, Italy

Thomas Hammer Centre for Development and Environment CDE, University of Bern/Switzerland, Bern, Switzerland

Jennifer Jewiss Department of Leadership and Developmental Sciences, University of Vermont, Burlington, VT, USA

Luigi La Riccia Interuniversity Department of Regional and Urban Studies and Planning (DIST), Politecnico and Università di Torino, Turin, Italy

Daniel Laven Department of Tourism Studies and Geography/European Tourism Research Institute, Mid Sweden University, Östersund, Sweden

Marion Leng Centre for Development and Environment CDE, University of Bern/Switzerland, Bern, Switzerland

Annalisa Calcagno Maniglio Department of Sciences for Architecture, Polytechnic School, University of Genoa, Genoa, Italy

Nicola Martinelli A6 Lama San Giorgio Protected Area Plan, Bari, Italy

Marzio Marzorati Project Land Stewardship, Legambiente Lombardy, Milan, Italy

Claudia Matoda Interuniversity Department of Regional and Urban Studies and Planning (DIST), Politecnico and Università di Torino, Turin, Italy

Francesca Mazzino Department of Sciences for Architecture, Polytechnic School, University of Genoa, Genoa, Italy

Patrizia Menegoni ENEA, Unità Tecnica AGRI-ECO, Rome, Italy

Maria Valeria Mininni DiCEM UNIBAS, Potenza, Italy

Nora Mitchell Rubenstein School of Environment and Natural Resources, University of Vermont, Burlington, VT, USA

Giulio Mondini Interuniversity Department of Regional and Urban Studies and Planning (DIST), Politecnico and Università di Torino, Turin, Italy

Emanuela Morelli Department of Architecture, University of Florence, Florence, Italy

Renzo Moschini Gruppo San Rossore, San Rossore Park, Pisa, Italy

Ingo Mose ZENARiO – Center for Sustainable Spatial Development, Applied Geography and Environmental Planning Research Group, Carl von Ossietzky University Oldenburg, Oldenburg, Germany

Anna Natali eco&eco, Economia ed Ecologia, Ltd, Bologna, Italy

Gabriella Negrini Interuniversity Department of Regional and Urban Studies and Planning (DIST), Politecnico and Università di Torino, Turin, Italy

Domenico Nicoletti European Landscape Observatory of Arco Latino, Padula, Italy

Antonio Nicoletti Protected Areas Department, Legambiente, Rome, Italy

Concha Olmeda ATECMA, Villalba, Spain

Gabriele Paolinelli Department of Architecture, University of Florence, Florence, Italy

Rocío Perez-Campaña Environmental Planning Laboratory (LABPLAM), Department of Urbanism and Spatial Planning, University of Granada, Granada, Spain

Adrian Phillips UK Countryside Commission, IUCN's World Commission, Cheltenham, UK

Paolo Pigliacelli Federparchi-Europarc Italia (Italian Federation of Parks and Natural Reserves), Rome, Italy

Sandro Pignatti Forum Plinianum, International Association for Biodiversity and System Ecology, Rome, Italy

Ignazia Pinzello University of Palermo, Palermo, Italy

Claude Raffestin Université de Genève, Genève, Switzerland

Carol Ritchie EUROPARC Federation, Regensburg, Germany

Rosanna Rizzi DiCEM UNIBAS, Potenza, Italy

Bernardino Romano University of L'Aquila, L'Aquila, Italy

Emma Salizzoni Interuniversity Department of Regional and Urban Studies and Planning (DIST), Politecnico and Università di Torino, Turin, Italy

Rita Salvatore University of Teramo, Teramo, Italy

Riccardo Santolini Department of Earth, Life and Environmental Sciences (DiSTeVA), Carlo Bo University of Urbino, scientific campus "Enrico Mattei", Urbino, Italy

Massimo Sargolini School of Architecture and Design, University of Camerino, Camerino, Italy

Bianca M. Seardo Interuniversity Department of Regional and Urban Studies and Planning (DIST), Politecnico and Università di Torino, Turin, Italy

Emanuele Sibona Department of DISAFA and NatRisk, Università degli Studi di Torino, Grugliasco, TO, Italy

Dominik Siegrist Institute for Landscape and Open Space, HSR University of Applied Sciences, Rapperswil, Switzerland

Francesco Silvestri eco&eco, Economia ed Ecologia, Ltd., Bologna, Italy

Marianna Simone Master IUAV, Venice, Italy

Silvia Soppa Department of Sciences for Architecture, Polytechnic School, University of Genoa, Genoa, Italy

Sue Stolton IUCN World Commission on Protected Areas and Equilibrium Research, Bristol, UK

Corrado Teofili Federparchi-Europarc Italia (Italian Federation of Parks and Natural Reserves), Rome, Italy

Alessandro Tosini Politecnico di Torino, Spinetta Marengo, AL, Italy

Massimo Tufano The Regional Agency for Protected Areas in Lazio, Rome, Italy

Simone Tulumello Instituto de Ciências Sociais, Universidade de Lisboa, Lisbon, Portugal

Luis Miguel Valenzuela-Montes Environmental Planning Laboratory (LABPLAM), Department of Urbanism and Spatial Planning, University of Granada, Granada, Spain

Teodoro C. Vales Institute d'Urbanisme de Grenoble, Grenoble, France

Marco Valle SiTI Istituto Superiore sui Sistemi Territoriali per l'Innovazione, Turin, Italy

Maria Grazia Vinardi Interuniversity Department of Regional and Urban Studies and Planning (DIST), Politecnico and Università di Torino, Turin, , Italy

Angioletta Voghera Interuniversity Department of Regional and Urban Studies and Planning (DIST), Politecnico and Università di Torino, Turin, Italy

Norbert Weixlbaumer Department of Geography and Regional Research, University of Vienna, Vienna, Austria

Massimiliano Coda Zabetta SiTI Istituto Superiore sui Sistemi Territoriali per l'Innovazione, Turin, Italy

Dario Zocco Parco Fluviale del Po e dell'Orba, Casale Monferrato, Italy

Francesco Zullo University of L'Aquila, L'Aquila, Italy

About the Editors

Prof. Roberto Gambino, Full Professor of Urban and Regional Planning at the Politecnico and Università di Torino – DIST (Interuniversity Department of Regional and Urban Studies and Planning, ex DITer, Interuniversity Department of Urban and Regional Planning) from 1980 to 2010 and Emeritus Professor since 2012; Director of the ex DITer from 1995 to 1999; President of the Degree Course on Territorial, Urban and Environmental Planning, II School of Architecture, Politecnico di Torino, from 2007 to 2010; Director from 1994 to 2010 of the CED PPN (European Documentation Centre on Nature Park Planning) – DIST, member of IUCN – International Union for Conservation of Nature; since 2000, member of the IUCN-CEESP (Commission on Environmental, Economic and Social Policy); Vice-Rector of the Politecnico di Torino from 2006 to 2009; President from 1991 to 1996 of the Scientific Committee for the Nature Parks of Piemonte Region; since 1998, member of the Legambiente Scientific Committee; since 1976, President of the ANCSA (National Association for Artistic and Historic Centers) for Piemonte and Valle d'Aosta Regions; author and coordinator of a wide activity of scientific research and planning including urban plans of Turin and other cities, landscape and regional plans for some Regions and Provinces, plans for several National and Regional Nature Parks; author of a great number of publications, books, reports, and papers presented at national and international conferences, seminars, Ph.D. and Master's degree programs, graduate schools, and other educational institutions concerning particularly Protected Areas and nature conservation, environment, and landscape policies. E-mail: roberto.gambino@polito.it

Prof. Attilia Peano†, Full Professor of Urban and Regional Planning since 1994 and Senior Professor since 2012 at the Politecnico and Università di Torino – DIST (Interuniversity Department of Regional and Urban Studies and Planning, ex DITer, Interuniversity Department of Urban and Regional Planning); Director of the ex DITer from 1999 to 2007 and vice-director from 2009 to 2011; member of the Administration Board of the Politecnico di Torino since 2004; coordinator of the Ph.D. course in Environment and Territory at the Politecnico di Torino and member of the faculty committee of the Ph.D. programs in Landscape Planning at the Università di Firenze; since 2010, Director of the CED PPN (European Documentation Centre on Nature Park Planning) – DIST, member of IUCN – the International Union for Conservation of Nature; since 1997, Chairperson of the Italian Urban Planning Institute's National Landscape and Environment Committee (INU); since 1997, member of the IUCN-WCPA (World Commission on Protected Areas); since 1999, member of the Legambiente Scientific Committee; President, from 2004 to 2010, of the Council of the Gran Paradiso National Park (Italy); member, since 2011, of the Commissione Tecnica Urbanistica (C.T.U.) and of the Commissione Vincoli Paesaggistici of the Piemonte Region; author and coordinator of a wide activity of research and application works regarding regional and landscape plans for some Regions and Provinces, environmental and assessment studies, several National and Regional Nature Parks plans; author of a large number of publications, books, reports and contributions presented at national and international conferences, seminars, Ph.D. and master's degree programs, graduate schools, and other educational institutions concerning particularly environmental planning and design, landscape, and Protected Areas policies.

Chapter 1
Introduction: Reasoning on Parks and Landscapes

Roberto Gambino

Abstract This chapter introduces the question of the relationship between nature conservation policies and parks policies for an active conservation of the natural and cultural heritage, facing effectively the incessant worsening of ecological conditions and the growing risks raising from global changes. Reasoning on the role that both the "protected areas" (IUCN 1994) and the "landscapes" (European Landscape Convention 2000) can play in the new frontiers of conservation, it tries to build a new conceptual frame to drive territorial policies. To this end, it gathers reflections and suggestions coming from an international set of experts and experiences related to diverse contexts and disciplinary backgrounds. Bridging nature and culture, local and global, the landscape paradigm helps to understand, plan and manage how policies concerning parks and areas deserving special protection are to be integrated in urban and territorial strategies and responsive regulations. Despite the traditional and present separation between nature conservation policies and landscape policies, their alliance is a powerful key for strengthening the strategies aiming at the enhancement of the inclusive quality of the entire territory.

Keywords Nature conservation • Landscape protection • Landscape policy • Territorial policy • Natural-cultural heritage • Nature park • Protected area • Territorial governance

1.1 Parks and Landscapes Relationship: An Open Question

When, in the early 2000, Adrian Phillips, then president of the WCPA of IUCN (the International Union for Conservation of Nature), proposed the "new paradigms" (Phillips 2003) for the institution and the management of parks and other protected areas, the question of their relations with the landscape had already been put up for discussion. The conceptions of natural parks were changing in relation to the spectacular growth of their number and surface areas, which over a decade had multiplied by more than ten in Europe and significantly increased throughout the

R. Gambino (✉)
Interuniversity Department of Regional and Urban Studies and Planning (DIST), Politecnico and Università di Torino, Viale Mattioli 39, 10125, Turin, Italy
e-mail: roberto.gambino@polito.it

R. Gambino, A. Peano (eds.), *Nature Policies and Landscape Policies*,
Urban and Landscape Perspectives 18, DOI 10.1007/978-3-319-05410-0_1

world, determining an important increase of their ecological, economic, social and cultural impact. A much more relevant development was accompanied by a growing typology and qualitative diversification. As a consequence, National Parks are now part of a richer mosaic of "protected areas", not only the areas strictly ascribable to the six category classification proposed by IUCN in 1994 as a reference grid for national designation (see Box at the end of this Introduction), but also other areas identified at the international level, such as the biosphere reserves (BR), the MaB areas, the world heritage sites by UNESCO, the wetland areas referred to in the Ramsar Convention, the coastal zones subject of the ICZM and, within the European Union, the sites recognised of community interest. Furthermore, states, regions and local authorities and even private organisations have often identified other areas to be protected, without a reference to the IUCN definitions.

These words, "protected areas", cover, hence, a wide and rather heterogeneous and confused set of resources, whose main common character is the institutional recognition of their outstanding interest for the "in situ" conservation of natural heritage and more precisely of biodiversity. And it is just this common interest that makes them potentially complementary.

But it was this same growing interest – as the Convention for Biological Diversity (SCBD 2004) stresses – which highlighted the limits of the protected area policies, both for their spatial delimitation (as affirmed by IUCN in Durban Congress of 2003: "no park is large enough to be effectively defended within its boundaries") and the separation of the defence of their specific values from that of the values present in the surrounding territories. From here, the need to extend public interest from protected areas to the "landscapes" and the involved territories is increased. "Benefits beyond Boundaries" was the title of the Durban Congress, but it was only in the last decades that the conservation's theory and practice accepted the idea that park policies couldn't leave aside expectations and needs of the concerned communities. This idea have already characterised the experieces of the Regional Parks in Italy, France, Germany, Spain and other European countries. Such idea implies a relevant shift of park visions, from an "inward" to an "outward" vision, strengthening the tendency towards the large-scale planning for nature conservation.

In turn, landscape was subjected, the last century, to an irreversible change in direction. Ripped from reductive aesthetic conceptions, often nourished by impressionist interpretations that had guided and legitimated the traditional action of protection in Italy and other European countries (not without appreciable success), landscape takes on a central role in the "environmentalist shift" that was profiled in the 1960s and 1970s, especially in Anglo-Saxon culture. New conceptions restore the importance to earth sciences, giving space to the anxiety of scientific objectification and in some measure to the "new determinism" (McHarg 1966) that, in those years, characterised the evolution of environmentalist thought. However, they also draw attention to the social and cultural significance of landscape, anticipating the reflections and debates that then found a systematic match in the European Landscape Convention launched by the Council of Europe in 2000.

These changes, together with those mentioned above concerning parks and protected areas, make up the "landscape question", especially from the second half of the last century, like a crucial ground of socio-environmental battle and conflict for contemporary territoriality. It is this ground, thick in snares and promises, on which we should aim current and potential relations among the parks and landscapes. The landscape question has a general complex meaning, in that it concerns conflicts and interaction between nature and culture, but it can be usefully addressed moving from two basic concepts (whose definitions are reported in the Foreword of this book):

- The *natural protected areas* (PAs), according to IUCN (1994)
- The *landscapes,* according to the ELC (CoE 2000)

Such concepts may be compared with other spatial articulations, both at the national and international level (such as the above ones) and at the regional or local level (such as the PAs not reflecting the above IUCN definition, and particularly the Sites of Natura 2000 European interest). Obviously, the comparison must take into account not only the differences of meaning and objectives but also the differences of scale and spatial diffusion: for example, while there were 177,000 PAs in 2012 worldwide (Bertzky et al. 2012), less than 1,000 were the sites in the UNESCO List.

1.2 The Context: Emergency and Long-Lasting Crisis

Though the image of parks is often associated with that of "paradise" and that of landscape often evoke the idea of balance and pacification, their relations cannot but be located in a context of crisis. This observation is not necessarily negative, if viewed from the historical perspective, in light then of the millennial event that, before giving us the landscape in which we live, has cyclically alternated the creative phases (structuration) with destructive ones (de-structuration), through moments of crisis – including dramatic and violent ones – which broke the old equilibriums while looking for new ones (re-structuration). The history of landscape is in this sense a story of crises that cannot be understood except in dynamic terms.

This statement, while quite obvious, opens up some useful considerations regarding the parks/landscapes relationship. The first sheds some light on the structural interpretation of territories that, in various experiences of landscape and territorial planning, has attempted to give a rational basis to management policies. It is particularly using the concept of *invariance*, often applied in that interpretation, that help distinguish the elements and dynamics of long duration and prominent importance, to be respected for the purposes of active conservation of the value systems incorporated in the territory (Gambino 2008, 2011).

A second consideration regards the concept of *emergency*, largely employed to designate the most worrying character of the current concept of crisis, its rapid or explosive, dramatic and largely unpredictable or unforeseen display. Increasingly often, public action in the matter of environment and landscape has to do with emergencies, threats or risks that go out from the ordinary and the everyday and must compete with extraordinary or exceptional events, such as those that are connected to global changes. Paradoxically, it is in the ordinary management of the territory that we should look for the most effective answers to the logic of emergency. It is the very scarcity in the daily "care" of the territory that favours the increase of crisis factors or catastrophic effects, such as typically great floods, tsunami or seismic events. Often true "planned disasters", in the measure in which the urban and territorial planning and ordinary management ignore or do not consider adequately the risk factors and the reasons of environmental stability and security. These effects are aggravated by the fact that the logic of emergency tends to polarise the public funding and the administrative processes.

A third consideration concerns the convergence of crisis factors. The evolution of international agreements in the wake of the CBD has shown the need to incorporate measures of different characters (geophysical, agronomical, economical, political and cultural) and to learn from the past, paying attention to traditional practices and "local knowledge". This attention is matched in the statement by the ELC, which sanctions the founding of policies of landscape on the identity expectations and perceptions of the populations involved.

The above considerations reflect a crisis situation that only in part can find an answer in the policies of the protected areas. We may also fear (see Guarino et al. in this volume) that the very logic of concentrating care for the territory to a minority part of the overall territory prefigures the rising of a fundamental "territorial contradiction", which pays the effectiveness of the protection of some areas with the intensification of the economic and productive exploitation of the rest of the territory. In substance, it is an "insular" conception that regards "nature protected areas" (PA) (similarly to historical centres, historical-cultural properties and UNESCO recognised WH Sites) as besieged islands in increasingly hostile contexts.

1.3 Reaction to the Crisis: From Islands to Networks

In the new conceptions that inspire parks and landscapes policies, insular logic has made way to the "reticular" logics, developed from the 1960s in various disciplinary domains. For a society that functions increasingly by networks, under the pressure of technological development and the great economic and social revolutions of post-modernity, the territory tends to configure itself such as "network of networks", in which different relation systems cross over and interact: biological, ecological, social and anthropological, cultural and landscape in a strict sense. Networks that have densified over the centuries and millennia from the development of communications and "territorial capital", but also been fragmented, lacerated, mutilated by the settlement and infrastructure developments that are alluded to with the terms of "overbuilding", settlement dispersion, urban sprawl, etc. The defence or reconstruction of systems of bio-cultural connectivity constitutes one of the central challenges of contemporary environmental politics.

In this framework, the answers that have been profiled over the last decades to the state of crisis can be referred to two main lines.

The first regards the new relations with cultural heritage, in terms of re-elaboration, re-signification and re-use, often destructive, of the tangible and intangible sediments, of memorial deposits and identity legacy, of the city and the infrastructures and of cultural, scientific and technological legacy, and even the apparatus of government and control. The landscape has taken on, in this line, an emblematic role. It increasingly refers to the scale of the visions and policies, extending conservation so that it involves many more actors, such as landowners, organisations and different government bodies, and requires new forms of governance (as J. Brown and others remind in this book).

The second line of reaction regards the new relations with natural heritage, and with agricultural and forest activities, to be re-launched after the devastating marginalisation produced by the "industrial capitalism", in view of the new needs and new hopes fed by emerging economies at the global level. In the "return to the earth", which is profiled in answer to the traumas of unrestrained globalisation, we can ask which role can be carried out by the parks and protected areas, if and under which conditions they can take once again the symbolic and celebrative role which connoted their founding image halfway through the nineteenth century. The new paradigms launched by IUCN attempt to answer this question.

For both the above lines, natural-cultural heritage seems to be the weight-bearing structure of a territorial-economic development, able to give effective and long-lasting answers to the demands of contemporary society. However, this implies, beyond the ambiguity of "sustainable development", that the heritage conservation is linked to *change* and innovation; and symmetrically, the innovative processes and the production of new territorial values are to be founded on the conservation of existing values (Gambino 1997; ANCSA 1990). If we are looking to a real conservation of natural and cultural heritage, the protection measures (such as strict safeguard, maintenance and restoration of "what, where and how" it pre-exists) are not enough. We need to go far beyond the traditional protection policies towards the concept of the "active conservation". This implies important innovation in the frontiers of conservation.

1.4 New Frontiers of Conservation

In light of the above, we can put forward the hypothesis that the global crisis in which we are submerged is destined to unhinge the principles and models consolidated in Western hegemonic culture regarding the relationships between nature, culture and society. *Western vision* seems increasingly less able to offer solutions to the problems in question, aggravated with dramatic relevance by the global changes regarding climate, energy, population growth, use and access of primary resources such as water. That changes force us to remember the lessons coming, almost unexpectedly, from the south of the world. A paradoxical overturning stands out: the west that learns from the south to re-read and re-legitimise that relationship between nature and culture on which has historically founded its power and wealth and to consequently re-design its own expectations and dreams.

A survey carried out by the CED PPN in 2010–2011 a group of internationally famous experts and scholars has highlighted the changes that characterise the *new frontiers* of the active conservation of the natural-cultural legacy, drawing attention to some emerging themes (CED PPN 2011; Peano et al. 2013):

- *Global climate change* and the consequent need for the integration and adaptation of eco-management in multi-sectorial planning and for mitigation of the harmful effects and the incumbent risks.
- *Reconnection* of bio-cultural relations (through space, time and society), reducing fragmentation and "insularisation" of ecosystems, with networks and bio-regional planning.
- *Landscape* protection and enhancement, strengthening territorial identities and local cultures, through landscape-scale planning, caring for the territory and re-launching agro-forest activities.
- *Governance* inclusive and comprehensive based on trans-scale approaches, the empowerment of local communities and measures of management and planning suitable to effectively protect the extra-local values and the common goods. These should involve the ownership of lands, especially in countries such as Italy in which it, even within protected areas, is mainly private.

The new frontiers of active conservation could have important implications on tools and strategies, in particular on the area-based conservation policies. As the international debate has shown, it induces to make clear the distinction between the "protection areas" and the "conservation areas" (Borrini-Feyerabend et al. 2010, 2013; SCBD 2004, 2010). The latter can include not only part of the protected areas (PAs) corresponding to IUCN categories but also a large range of areas not corresponding to such categories, which may contribute to the active conservation of the whole concerned territories, within and outside the PAs, not always submitted to rigid regulation. More generally, this draws the attention on the governance, conceived as a complex interaction of diverse actors and institutions, means and responsibilities leading the processes of territorial transformation. It is not something new ("Since people have been interacting with nature, someone, somewhere has been taking decisions about what to do", as observed by Borrini-Feyerabend G et al. (2010, reprinted 2012)), but it has gained a growing importance for two main reasons: first, the relevance and intertwining of social, economic, political and cultural factors in such processes, and secondly their spatial enlargement outside the protected areas, as a consequence of the scaling up and the reticular complexification of the territorial dynamics.

1.5 Changes in Sense and Philosophy

The new frontiers of active conservation stress the relevance of interaction between parks and landscape policies, suggesting new models, new ways of thinking and – shortly – new paradigms. In the change of direction that assails the conservation policies, the first paradigm to be considered is the *ecological* one, supported by CBD and directly implied in the re-definition of the role of parks and protected areas for contemporary society. In theory, we can consider obsolete the distinction between nature and culture that for a long time has constituted one of the most awkward legacies of modernity. But we must recognise that there are still uncertainties and interpretative gaps around the concept of bio-diversity, especially when we try to apply it in contexts widely and deeply elaborated by the historical processes of anthropisation. Even more uncertain could be the recognition of the specific eco-systemic contribution of parks and other protected areas, especially if we push aside every "insular" idea of separatism (e.g. thinking of the parks as core areas of ecological networks) or if we accept the idea that protected areas, while destined in priority to defending biodiversity, may or should host "even" indispensable cultural values (in particular for the category of *Protected Landscapes*). An idea that is coherent with the historical conceptual roots of parks, in the position of Olmsted (Fein 1972), or of Geddes in "*Microcosmos naturae, sedes hominum, theatrum historiae, eutopia futuris*" (Mcdonald 2009, Ritchie in this book), or in the epistemological appeals of Pignatti (Guarino et al. in this book).

The ecological paradigm cannot avoid crossing over with the *landscape paradigm*, as defined in the ELC. The latter is in fact characterised by three aspects that in some ways are complementary (Gambino 2004):

1. The extension of landscape policies *to the entire territory*, including urban and rural areas, those of an exceptional feature as well as degraded or everyday life ones (abandoning the traditional image of "natural beauty" or extra-urban spaces).
2. The idea that landscape is not only the result of the incessant interaction between natural and human factors, but also the context of life of the populations, an expression of the diversity of their common cultural and natural legacy and a foundation of their *identity*.
3. The need to recognize landscape policies, right from the recognition of the values and the implied problems, on the expectations and the perceptions of interested *populations*.

The innovative charge of the landscape paradigm suggested by the ELC is nonetheless conditioned by the appeal to the necessary integration of landscape in the multi-sectorial policies of territory planning, which can have a direct or indirect impact on the landscape. This is the reason for which, in many of the experiences of landscape planning and design which are carried out in the wake of the ELC, we can observe the reference (besides the ecological and the landscape paradigm) to a third paradigm, the *territorial* one (Gambino 2005). Not only because it is in the territory that, historically speaking, the processes, problems and conflicts that have created or transformed the contemporary landscapes have taken shape, but also because the crucial themes – the contradictions – that outline the new frontiers of natural and landscape conservation seem hard to be faced outside a re-composition of the relationship between human and earth.

At this regard, the evolution of UNESCO conception of World Heritage Sites is interesting. On the one hand, the 1992 convention opens the list of the sites to the "cultural landscapes", recognising their complex natural-cultural values (particularly underlined in the 2012 Recommendation on the Historic Urban Landscapes) (ANCSA 2012); but on the other hand, UNESCO reaffirms the selective character of the list, where only sites of "outstanding universal value" can be admitted. This last concept is in fact closer to the above "island vision" antithetic to the ELC vision, as the relatively small number of the sites (less than 1,000 at the world level) suggests.

1.6 From Landscape Gardening to Nature Conservation

In the territorial perspective, the relationship between parks and landscapes may be specifically clarified, starting from the above concept of active conservation and avoiding any generic references to nature. Having abandoned any simplistic distinction between nature and culture, we can ask ourselves which meaning the conservation of the biodiversity of protected areas may take on, having in mind that they have always been thought of as reservoirs of naturalness at the benefit of even wider territories. In the attempt to answer this question, we may take into consideration the "natural capital" present in the interested territories, "not so much for what it is, rather for what it does" (Gibelli and Santolini in this book), in order to preserve and possibly improve the fundamental functions that ensure the subsistence, quality, fitness and resilience of said territories. More precisely, we can refer to the *ecosystem services* that the natural capital can supply to the above purpose. Moreover, it should not be forgotten that it regards a capital that is still in some way anthropized, historically manipulated and "acculturated" by humans, never completely "new" or truly virginal (Borrini-Feyerabend et al. 2013; Schama 1997), and so exposed to dynamics that are inevitably influenced by the previous successions.

In the PAs, paradoxically, though destined to defend nature, the "cultural" character is in a certain sense even more evident, as far as the objective of protecting and celebrating nature requires regulatory interventions that intercept the development processes and push towards forms of "simulation" or of true "staging" (at most, the image of the Yellowstone National Park planned, constrained and equipped by the enjoyment network). A certain ideal continuity appears evident from the *landscape gardening* of Capability Brown and the great contemporary landscape architects (Turner 1987) to the active current management of a less "natural" nature, increasingly exposed to the temptations of territorial gardening, of urban greening programmes. And it is just this continuity that allows to find in the contemporary culture the roots of the projectual dimension of landscape. Revealing the links connecting the perception and interpretation with the design of landscape, beyond the silent dialogue between the pre-existent reality and the landscape invention.

1.7 Responsive Regulation of Natural-Cultural Processes

In light of these considerations, it seems opportune to pose the problem of *responsive regulation* of natural dynamics, in the full awareness of the ambiguity and contradictions above. A regulation that must originate from the "lessons" of the past, but which cannot in any way leave aside the choices for the future, the strategies and the policies that we intend to develop. From here the importance to put at the base of any choice the fundamental principles that the evolution of scientific thought and international cultural debate has focussed and continually re-proposes: suffice here to allude to the principles of precaution, limit, co-evolution and savings of primary resources and common goods. Over the past decades, the framework of principles shared at the international level has known, as we said before, significant advances. For example, the concept of protection has increasingly made way to that of "care for" the territory and the attention has moved to agro-forest activities, food supply chains, water management and, in general, everything that directly has to do with "mother earth".

These principles are gaining attention and respect worldwide, despite the strong differences of the national backgrounds, in terms of constitutional framework (see, for instance, the basic statement of Art. 9 of the Italian Constitution about the absolute priority of protection against any other option), provisions of the ordinary legislation, administration patterns and, more generally, cultural attitudes and education.

Nevertheless, it is very evident that these principles and these orientations cannot find adequate consideration within the PAs. Other strategic frameworks, other tools and other policies are needed. These needs may find an answer in the "regional" policies of the landscape, as far as they are able to coordinate different forms of environmental infrastructure, to influence the trans-scale planning of the territory (building appropriate legal frames on a local level) and to trigger realising processes of high symbolic value, such as the public acquisition of coastal strips in France and Great Britain.

A particular role should be recognised to the participated procedures of evaluation, as a tool of "explicit" comparison between different systems of values, interests and preferences, For many reasons (such as the growing interferences between urban sprawl and environmental changes and risks, the competition in ecosystem services exploitation, the needs for financial supporting the care for the territory), conflicts and divergences are raising, between private and public interests and even between diverse sectors of the public administration. This poses problems of democracy in decision processes, of priority in protection and management of the common goods and of acknowledgement and defence of the rights and responsibility of the local communities. The shift from an evaluation strictly related to single nature features to a "landscape evaluation" including the economical dimension, with an outward vision, could be a strategic step towards a more responsive and effective multi-scale regulation.

1.8 Image and Representation as a Tool for Knowledge and Action

Faced with the incumbent risks and pressures, the contribution of "hard sciences" to the comprehension and evaluation of the landscape-environmental reality should certainly not be underestimated. Moreover, said contribution can never be reduced to an objective, neutral and unquestionable "dataset", which may leave aside the decisive role of representations in knowledge and management of the territories of our "landscape society". As it has been noted (Raffestin in this book), in order to know the territory, we need to imagine representations, in a creative effort of invention, which puts the responsibility of choice (starting from the appreciation of the values and the issues at stake) in the hands of the various subjects involved. For this, the question of representation is closely linked to the question of delegation, whose interests and rights are represented, and how, a question that is still open.. the ELC has only partially resolved, imposing a reference to the "interested populations" (not necessarily only the "inhabitants").

If it is true that "man does not dominate the imagination but is the place where images are produced", the anthropological and semiologic dimension of the landscape (particularly in parks, according to the vision of the founding fathers such as Olmsted in the mid-nineteenth century) should not be underestimated. Perception and imagination follow each other, between the inalienable subjectivity of the enjoyment experiences and the "common sense" of the landscape. If every landscape, insofar as it is a "paradise substitute", aspires to perfection, how, within ELC logic, can we appreciate its quality? The traditional emphasis on beauty as an implicit attribution of the landscape quality (emphasis that still turns to the vision of protection apparatus, especially in Italy) seems to have given back to landscape the relevance of beauty as a flag well perceived by people against the degradation and ugliness of the environment. But, as ELC reminds us, landscape's question and policies do not concern only the "natural beauty" traditionally protected, but rather every landscape, included the degraded ones, taking into account the whole set of values involved. Even more, this new conception avoids the risk to overshadow the consideration of the "staging" of the landscape representations (the "landscape as a theatre": Turri 1998) and of the ambiguous meaning of the spectacle of nature within the landscape policies.

In this sense, "the landscape is not a tangible reality but a picture of a tangible reality" (Raffestin, in this book); and we can talk of landscape building as "a production of landscapes by means of landscapes", as it is particularly evident in the circular dynamics concerning touristic areas and flows. This ambiguity between reality and representation, reminding the "hermeneutic circle" of Gadamer (1975) and reflected in ELC landscape definition focusing on the perception, may be seen as a fruitful conception insofar it puts in question the false objectivity of landscape project and territorial planning and recalls their responsibility in face of the incessant processes of change.

1.9 Specialisation and Integration in Knowledge and Planning Approaches

The recognition of the symbolic function of landscape is essentially founded on its holistic and integrated vision, fruit of the inter- and trans-disciplinary interaction of diversified cognitive and project approaches. The structural interpretation of the territory and other readings of the landscape reality referred to in various recent experiences of landscape planning, move from the integration of these approaches. But the process of integration is hindered not only by the separation of the administrative competences attributed to the different institutional bodies but also by the rigidity of the closures and the divisions between the different disciplinary domains, partly deriving from academic traditions. Also the scientific and technological developments make difficult such integration responding to the needs of regulation and control of contemporary society. Therefore, integration interferes with specialisation at all levels, as noted also by the ELC recommendations regarding training and education. Further integrations will derive from the activities of observation, communication and debate developed in charge of the three organisations provided by the Council of Europe: ENELC, UNISCAPE, CIVILSCAPE.

One of the most worrying effects of this interference concerns the *divergence* of the policies and the strategies of action, even in the presence of substantial convergence in values to be protected. An emblematic example is the contrast in Italy between the Framework Law (1991) on Protected areas and the Italian Cultural Heritage and Landscape Code (Repubblica Italiana 2004) in the matter of landscape planning, which attributes to both of them a sort of primacy towards any other plan. A no less worrying effect concerns the difficulty to give place to more or less complex forms of comprehensive planning process, of compact planning and of transnational aggregation and cooperation, such as the European Union Directive of "Natura 2000" Network, the Euro-Mediterranean initiatives of the "Latin Arch", the Alps and Carpathian Conventions. Attempts, those, that seem destined to fail, if adequate reforms, continue to lack.

1.10 Different Governance for Different Conservation Goals

Drawing the attention from the inside to the outside of protected areas suggested by the new paradigms of management, space enlargement, scaling up and the complexification of the conservation policies requires important changes in the models for governance. The traditional models, based on a few rules – mainly prohibitions – managed by an externally ordered authority, generally distinguished by the authorities of local and regional government, are shown to be increasingly inadequate for the purposes of an effective and efficient public regulation of the current processes. It is particularly true in front of the market failure and the consequent need for mitigation and adaptation to contrast the most undesirable effects of global change.

From here arises the need to realise articulated forms of multilateral governance, aimed at stimulating co-responsible and cooperative action of the involved subjects (stakeholders, rightsholders, local government institutions, private operators and owners), based on flexible and shared strategic frameworks. IUCN and CBD have identified four types of governance (Borrini-Feyerabend et al. 2013):

A. Governance by government acting at various levels
B. Governance by various rightsholders and stakeholders
C. Governance by private individuals and organisations
D. Governance by indigenous peoples and/or local communities

In order to guide the definition of appropriate forms of governance, taking into account both the objectives of management (as defined with the six IUCN categories) and the political, institutional and environmental conditions, a matrix has been suggested combining the six categories with the four types. The need for new forms of governance is firstly disclosed in some categories of protected areas such as "Protected Landscapes" (IUCN Category V) or areas managed by local communities (Category VI) (see Box at the end of this introduction). But it is increasingly evident that it is a need regarding – albeit in different ways – all protected areas, including every tangible area in which the objectives of protection and the safeguarding of eco-systemic services cross over with the economic and social ones or clash with the private ownership of the land.

In the IUCN ambit, it has been hoped that the diversification of governance models reflects the diversification of the objectives of protected areas categories. Reasonable requirement if we accept the idea that governance "learns" from the lessons of the past and adheres coherently to the distinctive characters and values of each area. However, perhaps it should be stressed that the realisation of *good governance* is deep down none other than a salient aspect of that drawing of attention from parks to landscapes that make up the subject of this reflection. In this sense, the adoption of interactive and cooperative processes concurs to move the focus of regulation from the local level to the level of eco-region and large landscape, contributing to re-launch local power, its autonomy and solidarity.

1.11 Community-Oriented Policies Versus Competitiveness Strategies

The political and cultural connection of parks to landscapes allows us to confront the crucial dilemma of policies of active conservation more effectively: the choice between cohesion policies and territorial competitiveness, between identity valorisation and opening to the global networks. A dilemma that may find partial response in the perspectives of integration here evoked. But what can, on the contrary, present itself in all the sourness of the current practices: an identity to be defended for the consolidation of local cultures, or to be sold to the highest bidder or to be brandished as a weapon in the disputes with others? And is it an opening up to the global networks in terms of cooperation or abuse or conflict?

If "living is building", the landscape policies cannot avoid referring to the inhabitants; therefore to the local communities hinged on the territory and to their activities, mainly agriculture and forestry have largely shaped our landscape heritage. Also, this close relationship must bring into account impressive phenomena of mobility, such as the great migrations from less developed countries to more developed ones, the rural and agricultural exodus from the country to the industries and to the cities and the commuters' movements in the wide urban area. No less relevant is the role of tourism in all its forms: weekly or holiday touristic flows from urban areas to places of leisure, proliferation of second homes and the supporting infrastructures and services. All of these break the foundational relationship between territory and settled communities; moreover, letting new "care taker" figures emerge, like the "new peasants" who return to the abandoned or marginalised mountain, or like the young entrepreneurs who invent a trade for themselves in the "urban countryside". In general, tourism represents a risky ambivalence between the benefits brought or promised and the environmental and cultural costs generated – firstly at the disadvantage of the landscape. But in any case costs and benefits, do not compete within the protected areas (with a reductively company-oriented vision), but at the level of the interested territories. At this level, interests and "rights to nature" of the urban communities cross the "rights to the city" of the rural communities.

1.12 Towards a New Alliance Between Nature Policies and Landscape Policies

The reasons exposed in the contributions gathered in this book and summarised in the previous paragraphs of the Introduction seem, briefly, to confirm the strategic role that parks and protected area policies are required to play in actively conserving natural and cultural heritage aimed at re-qualifying contemporary territories. But these also indicate that the efficiency and effectiveness of the said policies depend crucially on the possibility of connecting them with landscape policies or more precisely with the territorial policies of the different sectors that may find in landscape, widely speaking, a strategic framework of reference. In this framework, the focal attention should be dedicated to peri-urban spaces, in which the transformative dynamics that regard the urban asset, the context of life and its values meet significant risks and problems such as the instability of the hydro-geological structure, the land and primary resources consumption, the ecological un-balance and fragmentation, the recovering and re-use of the urban heritage, the consolidation and expansion of public spaces and the common properties. This is the main point of contention between collective interests and rights, on one hand, and individual privileges and interests, on the other. And it is especially – but not only – here that the policies of parks and areas deserving special protection are to be integrated in urban, territorial and landscape planning, to face the challenges arising in the new frontiers of conservation. The landscape approach is not to be seen as a mere spatial enlargement of the conservation horizons but as a key for strengthening the strategies aiming at enhancement of the inclusive quality of the territory.

The CED PPN Research: A Focus on the Protected Landscapes in the European Protected Areas System
Gabriella Negrini and Sergio Bongiovanni

The CED PPN Research

The European Documentation Centre on Nature Park Planning (CED PPN) carries out research activities focussed on planning and policies of European nature parks and protected areas and the relationship between nature conservation policies and landscape policies.

Since its institution (1994), the Centre has worked to constantly broaden its field of research, by increasing the number of European countries and protected areas considered. The main qualitative and quantitative features of protected areas have been also analysed with reference to the IUCN protected area classification system. The main research steps undertaken by the Centre during this period refer to three surveys: the first two related to studies carried out and published in 2003 (AP, 2001–2003, which included 33 EU Countries[1]) and 2008 respectively (Parks for Europe, 2007–2008, which included 39 EU Countries[2]), the third one – currently being carried out at the Centre –

(continued)

[1] Ministero dell'Ambiente e della Tutela del Territorio e del Mare-Direzione Generale per la Protezione della Natura, CED PPN – Politecnico di Torino (eds) (2003) AP. Il sistema nazionale delle aree protette nel quadro europeo: classificazione, pianificazione e gestione. Alinea, Florence. Research developed in partnership with Ministero dell'Ambiente (2000–2002).

[2] Gambino R, Talamo D, Thomasset F (eds) (2008) Parchi d'Europa. Verso una politica europea per le Aree Protette. ETS Edizioni, Pisa. Research "Parks for Europe. Towards a European Policy for Protected Areas", developed in partnership with Federparchi (The Italian Federation of Parks and Nature reserves) e AIDAP (The Italian Association of Parks Managers) 2007–2008.

G. Negrini (✉)
Interuniversity Department of Regional and Urban Studies and Planning (DIST), Politecnico and Università di Torino, Turin, Italy
e-mail: gabriella.negrini@polito.it

S. Bongiovanni
GIS Consultant, Via Peyron 12, 10143 Turin, Italy
e-mail: sergio.bongiovanni@fastwebnet.it

concerning the review and update of the 2008 survey. Such ongoing study, starting from an analysis of the data provided by the European Environment Agency as of 2013,[3] intends to broaden the scope of the research of the previous studies, considering 41 European countries and over 86,495 Protected Areas (PAs) (Tables 1.1, 1.2, and 1.3).

It is worth emphasizing that the IUCN protected area classification, based on management objectives, is not just a formal procedure. On the contrary, it is a useful instrument to guide planning and management processes of protected areas, aiming, as advocated by IUCN itself, at creating "a common language" (Bishop et al. 2004; Dudley and Stolton 2008)[4] and at promoting "system policies" for protected areas. CED PPN has focused on this issue part of its research on natural park planning in Europe, attempting to achieve a systemic comparison of all the categories defined in the national legislation of each country and gathering them into five groups, highlighting a possible correspondence with the IUCN 1994 classification (Fig. 1.1 and Table 1.2).

The European Protected Areas

The research carried out by CED PPN highlighted a number of relevant issues on protected areas in Europe:

1. A high territorial impact, affecting a relevant part of the European Countries (in 2008: over 75,000 PAs covering over 90,000,000 ha, nearly 18 % of their territories and approximately 25 % of the population involved); as observed by EEA (2012)[5] Europe results, among the world regions, with the higher incidence of protected areas.
2. A spectacular growth (up 23 % over the 1996–2006 decade, in terms of protected surface), compared with world growth; this trend seems to continue, underlining the cultural, social and economic role of protected areas.

(continued)

[3] Nationally designated areas (CDDA – Common Database on Designated Areas) – status 2013, European Environment Agency.

[4] Speaking a Common Language, Research Project developed by the Cardiff University and Equilibrium Consultants, in collaboration with the IUCN and World Conservation Monitoring Centre, WCMC. See: Bishop K, Dudley N, Phillips A and Stolton S (2004) Speaking a Common Language: the Uses and Performance of the IUCN System of Management Categories for Protected Areas. IUCN and UNEP-WCMC; Dudley N and Stolton S (eds) (2008) Defining protected areas: an international conference in Almeria, Spain. IUCN, Gland.

[5] EEA (2012) Protected areas in Europe – an overview. EEA Report No 5/2012, European Environment Agency, Copenhagen.

Table 1.1 Number, protected surface and territorial incidence in the European countries 2003, 2008, 2013

	n	ha	% EU territorial surface
EU (33) (2003)	32,479	70,495,804	14.1
Source: IUCN 1997, CED PPN 1999–2002			
EU (39) (2008)	75,388	90,452,545	18.0
Source: EEA 2007, CED PPN 2007–2008			
EU (41) (2013)[a]	86,495	102,333,075	20.3
Source: EEA 2013, CED PPN 2013			

[a]Provisional data. Compared to 2008, data have taken into account Montenegro and Kosovo, while Turkey is not being considered yet. Furthermore, in both cases, the considered protected areas categories set have been selected according to CED PPN criteria defined in 2008

Table 1.2 Incidence (%) of the protected surface by the CED PPN Groups (See Fig. 1.1)

	NP	*RP*	*PL*	*R*	*NM*	*Tot.*
EU (33) (2003)	**14.1**	**24.5**	**44.6**	**16.4**	**0.3**	**100.0**
Source: IUCN 1997, CED PPN 1999-2002						

Table 1.3 Incidence (%) of the protected surface by the IUCN categories

	Ia	*Ib*	*II*	*III*	*IV*	*V*	*VI*	*NC*	*Tot.*
EU (39) (2008)	**1.7**	**4.0**	**14.1**	**0.5**	**10.9**	**52.1**	**2.6**	**14.1**	**100.0**
Source: EEA 2007, CED PPN 2007-2008									
EU (41) (2013)*	**1.8**	**4.4**	**13.4**	**0.4**	**13.3**	**49.9**	**5.7**	**11.0**	**100.0**
Source: EEA 2013, CED PPN 2013									

[a]Provisional data

3. A significant diversification of growth trends among various countries, mainly concentrated in "old Europe" – 15 EU Countries (in 2008) along with a promising raising trend of the Eastern countries.
4. A large number of environment, landscape and protection different categories (over 100 national protected areas types), as well as different models and approaches in planning, management and conservation.

(continued)

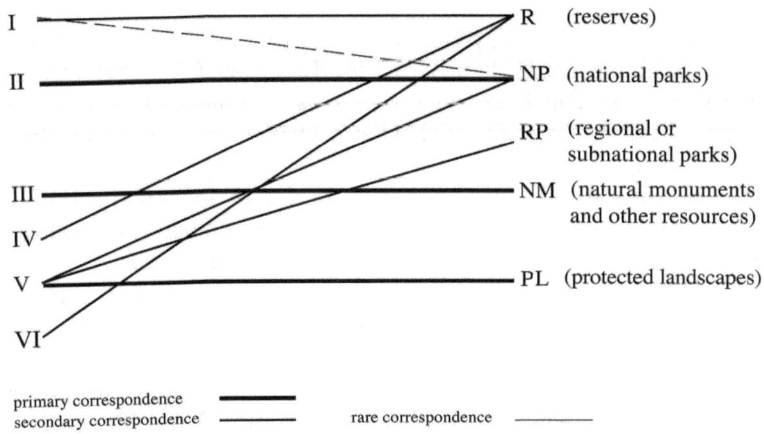

Fig. 1.1 CED PPN categories groups corresponding to IUCN categories (Source: CED PPN 1996)

The European Protected Landscapes (Category V IUCN)

Within this research book, it seems appropriate to give a specific account of the characters of Protected Landscapes (category IUCN V) within the European protected area system. It is, in fact, the most representative, among the six IUCN protected area categories, in relation to the potential and desirable alliance between nature policies and landscape policies. This is clear from the definition of the category itself (see Phillips, Dudley, Brown, Salizzoni and others in this book) and the objective assigned to it: "to protect and sustain important landscapes/seascapes and the associated nature conservation and other values created by interactions with humans through traditional management practices" (Dudley 2008).[6] Category V protected areas, whose aim is "to help people protect and sustain important landscapes and their associated values" (Phillips and Brown 2008),[7] is also particularly representative of the great changes undergone by nature conservation policies, with specific reference to the increasing role of local people and communities in protected areas management (e.g. see Community Conserved Areas – CCAs or ICCAs – shared management and other governance models) and the growing need to integrate conservation and development policies (within and outside protected areas).

- In 2008, Protected Landscapes amount over 9,200 areas, covering a surface of 47,000,000 ha, namely, 52 % of the European protected surface

(continued)

[6] Dudley N (ed) (2008) Guidelines for Applying Protected Area Management Categories. IUCN, Gland.

[7] Phillips A, Brown J (2008) Category V. In: Dudley N and Stolton S (eds), op. cit.

(3 % of the global protected surface), confirming the intertwine between men and nature which characterize the European PAs. Protected Landscapes may include large areas, such as in France, Czech Republic, Romania, United Kingdom, etc., but also smaller areas, such as in Sweden, Greece, Denmark, etc.

- They include extremely heterogeneous protected areas national designations, a really diversified reality as for features, objectives, management and planning tools. Among the main protected area, national categories that are Protected Landscapes, the following can be mentioned: Protected Landscape, Landscape Protection Area/Section, National Park, National Forest Park, Nature Park, Country Park, Regional Park, Area of Outstanding Natural Beauty, National Scenic Areas, Heritage Coast, Aesthetic Forest, Nature Reserve and National Park-Buffer zone. We can observe that the category of Protected Landscape includes, in 2008, 61 % of PAs identified by national legislation as "Parks" (National, Regional, Provincial Parks, Nature Parks, Rural Parks, National Forest Parks, etc.) that cover, in surface terms, 45 % of the category.
- Protected Landscapes grow more than all other categories and are mainly located in Central Europe (29 % in 2008), mostly in France, Germany, Portugal, United Kingdom, etc., despite their absence (in national legislations) in a number of central European countries, such as Italy, Belgium, France and United Kingdom; they are present, above all, in the Continental Biogeographic Region (over 50 % in 2008).
- As for their environmental contexts (in 2008), they are mainly found in mountain areas (over 50 %) and river systems (57 %). Unexpectedly, their presence is lower along the coasts (21 %), generally highly inhabited, a fact already observed by the 2008 IUCN Guidelines (Dudley 2008).
- Considering the relationship with local contexts,[8] Protected Landscapes are situated mainly in contexts characterized by agro-forestry activities (50 % in 2008) and they are mainly located in densely populated areas.
- They are often close to each other or to other PAs, working thus as buffer zones compared to other areas characterized by higher naturalness degree.

The ongoing analysis of data provided by EEA 2013 shows that Protected Landscapes continue to play an essential role in the European PAs system as well as in biodiversity conservation and sustainable development policies (see Table 1.3 and Fig. 1.2).

We can say, indeed, that, up to now, category IUCN V continues to be the most representative compared to others, including about 12,000 protected

(continued)

[8] Concerning the local contexts analysis developed in 2008, data referring to the NUTS3 (Territorial Statistical Units) data, in the framework of the ESPON Programme.

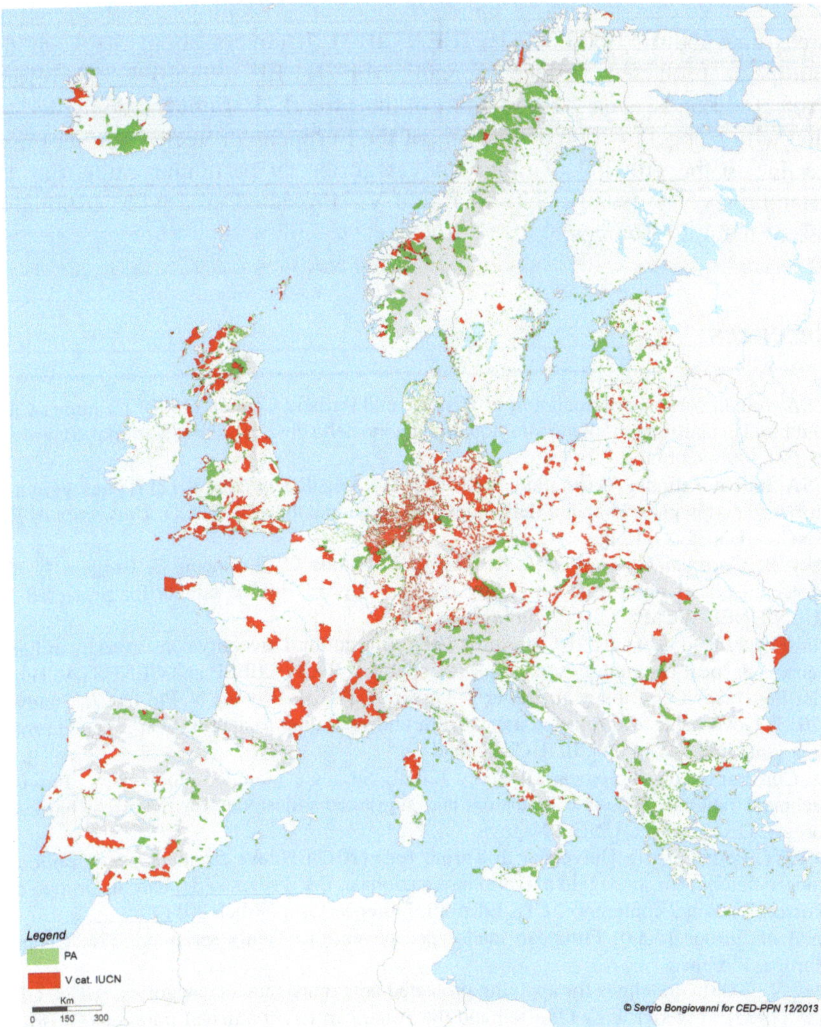

Fig. 1.2 Protected landscape in the European protected areas system (Source: EEA 2013, CED PPN 2008–2012)

areas and about 51,000,000 ha (EEA 2013). It covers about 50 % of the European protected surface, spread over 41 EU countries. Therefore, Protected Landscapes seem to confirm their trend. A slight incidence reduction with respect to 2008 data (52 % of the European protected surface) may be due to the growing of other PA categories, in particular categories IV (Habitat/species management area) and VI (Protected area with sustainable use of natural resources).[9]

References

ANCSA, Italian National Association of Historic and Artistic Centers (1990) La nuova Carta di Gubbio. Un contributo italiano alla riqualificazione della città esistente. National congress, 26–28 Oct 1990, Gubbio

ANCSA, Italian National Association of Historic and Artistic Centers (2012) Il paesaggio urbano storico: le strategie e le azioni della nuova raccomandazione UNESCO. Università di Roma/ANCSA/UNESCO, Rome

Bertzky B, Corrigan C, Kemsey J, Kenney S, Ravilious C, Besançon C, Burgess N (2012) Protected planet report 2012: tracking progress towards global targets for protected areas. IUCN/UNEP-WCMC, Gland/Cambridge

Borrini-Feyerabend G et al (2010, reprinted 2012) Bio-cultural diversity conserved by indigenous peoples & local communities: examples & analysis. IUCN/CEESP and CENESTA, Tehran

Borrini-Feyerabend G, Dudley N, Jaeger T, Lassen B, Pathak Broome N, Phillips A, Sandwith T (2013) Governance of protected areas: from understanding to action, Best practice protected area guidelines series no. 20. IUCN, Gland

CBD, Convention on Biological Diversity (2010) Strategic plan for biodiversity. Provisional technical rationale, Possible indicators and suggested milestones for the Aichi biodiversity targets. Decision X/2. UNEP, Nagoya

CED PPN, Politecnico e Università di Torino (ed) (2011) Nuove frontiere per le politiche di conservazione. Un sondaggio a livello internazionale. Per il rilancio dei Parchi, Gruppo di San Rossore National conference. ETS Edizioni, Florence/Pisa, 28 Feb 2011

Council of Europe (2000) European landscape convention, Treaty series no. 176. Council of Europe, Florence

Dudley N (2008) Guidelines for applying protected area management categories. IUCN, Gland

Fein A (1972) Frederick Law Olmsted and the American environmental tradition. G. Braziller, New York

Gadamer H-G (1975) Hermeneutics and social science. Philos Soc Critic Cult Hermeneut 2 (4):307–316, D. Reidel Publishing Company, Dordrecht

Gambino R (1997) Conservare Innovare: paesaggio, ambiente, territorio. UTET, Turin

Gambino R (2004) Il paesaggio come chiave interpretativa. CRU 16:26

Gambino R (2005) Paysage et territoire en Europe et en Italie. International congress: Paysage question. Faculté de philosophie. Universitè Jean Moulin, Ecole d'architecture, Lyon

Gambino R (2008) Landscape analysis and planning. Master world heritage at work: planning and management for world heritage sites. UNESCO, ILO, SITI, Venaria Reale, Turin

[9] Category IUCN IV primary management objective is "To maintain, conserve and restore species and habitats"; Category VI primary objective is "To protect natural ecosystems and to promote the sustainable use of natural resources, when the conservation and the sustainable use are mutually beneficial" (Dudley 2008).

Gambino R (2011) L'interpretazione strutturale del territorio. XXVII INU Congress, La città oltre la crisi, Livorno
Gambino R, Talamo D, Thomasset F (eds) (2008) Parchi d'Europa. Verso una politica europea per le aree protette. ETS, Pisa
IUCN (2003) Benefits beyond boundaries. Fifth IUCN World Parks Congress, Durban
IUCN/WCMC (1994) Guidelines for protected area management categories. IUCN, Gland/Cambridge
Macdonald M (2009) Sir Patrick Geddes and the Scottish Generalist tradition. The Sir Patrick Geddes lecture, 20 May 2009. Royal Society of Edinburgh, Edinburgh
McHarg J (1966) Ecological determinism. In: Darling FF, Milton JP (eds) Future environments of North America. The National History Press, Garden City
Peano A, Gambino R, Negrini G, Salizzoni E (2013) Parks & landscapes in Europe: towards an integration of policies and management. Fifth symposium for research in protected areas, 10–12 Jun 2013. Hohe Tauern National Park Center, Mittersill/Salzburg
Phillips A (2003) Turning ideas on their head the new paradigm for protected areas. J George Wright Soc 20(2), The George Wright Forum, Hancock
Repubblica Italiana (2004) Codice dei Beni Culturali e del Paesaggio. Italian cultural heritage and landscape code. Legislative decree no. 42/2004, Rome
SCBD, Secretariat of the Convention on Biological Diversity (2004) Biodiversity issues for consideration in the planning, establishment and management of protected area sites and networks, CBD technical series no. 15. SCBD, Montreal
Schama S (1997) Paesaggio e memoria. Mondadori, Milan
Turner T (1987) Landscape planning. An introduction (living environment). Nelson Thornes Publisher Ltd, Cheltenham
Turri E (1998) Il paesaggio come teatro. Dal territorio vissuto al territorio rappresentato. Marsilio, Venice

Part I
New Paradigms

Chapter 2
Nature Conservation and Landscapes: An Introduction to the Issues

Adrian Phillips

Abstract The dialogue between landscape protection and nature conservation is often hampered by conceptual difficulties, but recent developments in our understanding of landscape, as in the European Landscape Convention, have shown how nature, in all its forms, is a key element in landscape. Similarly, recent developments in nature conservation show how landscapes can be made more resilient. Nature conservation and landscape protection converge around the idea of working at the scale of distinctive landscape units. This convergence is explored first through the example of IUCN's Category V "protected areas" (Protected Landscapes/seascapes), which have been shown to be effective instruments for nature conservation and for the protection of agro-biodiversity. Three complementary national programmes in the UK are then described: National Character Areas which identify 159 areas of England which are distinguished by their nature conservation, landscape and other factors; Nature Improvement Areas which are designed to create, improve, extend and connect nature areas across broad tracts of England; and the Landscape Partnerships programme by which lottery funding is made available throughout the UK to support such large-scale initiatives. In all cases nature conservation is helped by being addressed through a landscape context.

Keywords Landscape protection • Nature conservation • European landscape convention • Resilience • Large scale • Distinctive landscape areas • Category V protected areas • IUCN • National character areas • Natural England • Nature improvement areas • Landscape partnerships • Heritage lottery fund

It can be difficult to reconcile nature conservation and landscape protection. Too often those who espouse one of these causes have negative views about the other; indeed there is still a conceptual gulf between many devoted nature conservationists and those who have a passion for landscape protection. But this need not be so.

The first part of this short chapter shows that there is in fact much synergy between the conservation of nature and protection of landscape. That relationship is

---

A. Phillips (✉)
30 Painswick Road, Cheltenham GL50 2HA, UK
e-mail: adrian.phillips@gmx.com

R. Gambino, A. Peano (eds.), *Nature Policies and Landscape Policies*,
Urban and Landscape Perspectives 18, DOI 10.1007/978-3-319-05410-0_2

examined through various perspectives, by looking at the different ways in which the word 'landscape' is used, investigating the idea of a resilient landscape that is good for nature conservation and examining the landscape as a forum where conflicts that affect nature can be resolved. The second part of the chapter will look at several ways in which nature conservation and landscape protection have been brought together in practice, globally and nationally.

2.1 Looking at Landscape Through Various Perspectives

'Landscape' is a slippery notion and can be linguistically confusing. It does not translate easily. For example, neither the French word *paysage*, with its rural overtones (Girot 1999), nor the German one *landschaft*, with its territorial ones (Cosgrove 2004), are exact translations. It also means different things in different areas of policy: notably it is used by conservation biologists as a scalar adjective, meaning a larger area than a site (as in 'landscape scale conservation'), but by some geographers and others in a more comprehensive and integrated way. This latter meaning has found its way into the world's first international treaty on landscape: the European Landscape Convention (ELC) which defines landscape thus:

> An area, as perceived by people, whose character is the result of the action and interaction of natural and/or human factors (Article 1a).

The ELC view of landscape is therefore all-embracing. Nature, in all its forms, is part of it, but so too are people; landscape contains all of the evidence of the past that remains and of the present; and it is rich in social, cultural, economic and environmental values. In this sense landscape extends beyond aesthetics, scenery and geography to include ecological processes and human well-being (Selman 2012). So landscape is a holistic concept; and *a* landscape is a distinctive geographic expression of that concept. Though nothing will stop debate about the meaning of landscape, the existence of this internationally agreed definition provides a reference point, and the ELC definition is used here in discussing how landscape protection relates to nature conservation. Since nature is an essential element within landscape thus defined, the conservation of nature becomes one way in which landscapes can be protected.

Landscapes are not just the passive outcome of people's impact upon the environment: they also do something *for* people (Selman 2012), because they have functions, structure and meaning. The functions of landscape are associated with biophysical processes that it contains and the way that human use it; the structure is represented by natural components (mountains, rivers, forests, the sea, etc.), and land uses and buildings created by people, all of which are visible elements in landscape; meaning is about the various values that we attach to landscapes (Piorr 2003). Landscapes are always subject to change, but the pace of change over the past hundred or more years is without historical precedent. Where rapid change takes place, often driven by global forces, the biophysical functions of landscapes are undermined, natural and historic components are damaged or destroyed, and the meaning that the landscape can convey is lost. As this happens, landscapes lose their distinctive character and their diversity; a nice French word for this process, where everywhere tends towards looking the same, is "banalisation".

Can landscapes be made resilient in the face of change so that their nature conservation and other values are sustained, so that they remain diverse and distinctive? Resilience implies an ability to recover from perturbations, even evolve and thereby retain or recover lost qualities. It is interesting that theories about resilience use the same terminology as we apply to landscapes (Edwards 2009):

> Resilience is the capacity of an individual, society or system to adapt in order to maintain an acceptable level of function, structure and identity.

At least as far as nature conservation is concerned, there is now broad agreement about how to make a landscape resilient. In response to the fragmentation of habitats, the loss of species and a range of threats, notably that from climate change, conservation strategies should seek to protect areas with high natural values, buffer and extend them, link them up (improving connectivity) and restore areas that have become degraded. Initiatives of this kind have been taken in many countries, encouraged by recommendations from the Parties to the Convention on Biological Diversity (CBD)[1]; a recent project in England is described below.

If we are to manage change so that landscape values continue to thrive, we need not only to understand ecological history and the interaction of "culture" and "nature" but also to engage in the diverse governance and management arrangements by which conflicts can be resolved. This is often best done within distinctive landscape units; such a strategy might be called the "landscape approach" (Brown et al. 2005; Phillips and Borrini-Feyerabend 2009). In some countries, formal structures exist which are focused on certain landscape areas, most obviously where these are of special heritage landscape quality and may be recognised as IUCN Category V protected areas (see below). However, other landscape areas may also be identified for concerted action – such as degraded areas around cities, areas of economic decline chosen for environmentally based regeneration or areas which have distinctive character worthy of protection and management but no formal recognition as a protected area. These all provide the political context within which policies of nature protection can be applied. Examples of the various kinds of areas where landscape and nature conservation issues can be resolved are explored below.

From this discussion, four broad principles emerge:

- Though landscape protection and nature conservation are rooted in different disciplines and use different language, they converge around the idea of working at large scale and across disciplinary boundaries.

[1] See, for example, Target 11 of the Strategic Plan for Biodiversity 2011–2020 (COP 10 Decision X/2).

- The broad concept of landscape (e.g. as defined by the ELC) embraces the natural elements; and landscape protection, management and planning must therefore contain a strong nature conservation element.
- An important aspect of a resilient landscape is that nature is able to adapt to change.
- By focusing on distinctive landscape areas, it is possible to reconcile the objectives of landscape protection and nature conservation.

The second part of this chapter explores how these principles play out in practice through one international mechanism and three national ones. They all show how nature conservation and landscape protection can be brought together.

IUCN's system of categorising protected areas by their management objectives is now widely known (Dudley 2008; Bishop et al. 2004). The system's purposes are to facilitate the planning and management of protected areas, improve information about their management and help regulate activities in protected areas. By providing international standards for protected areas management, the system acts as a global framework, recognised by the CBD, for classifying the variety of protected area types around the world.

At the core of the category system is the definition of a "protected area":

> A clearly defined geographical space, recognised, dedicated and managed, through legal or other effective means, to achieve the long-term conservation of nature with associated ecosystem services and cultural values.

IUCN explains that the definition should be applied in the context of certain principles. In the context of the relationship between nature and landscape the most important is (Dudley 2008):

> For IUCN, only those areas where the main objective is conserving nature can be considered protected areas; this can include many areas with other goals as well, at the same level, but in the case of conflict, nature conservation will be the priority.

If the definition is met, then a protected area can be assigned one of six management categories as follows:

- Category Ia: Strict nature reserve; Category Ib: Wilderness area
- Category II: National park
- Category III: Natural monument or feature
- Category IV: Habitats/species management area
- Category V: Protected landscape/seascape
- Category VI: Protected Area with sustainable use of natural resources

One of these categories is specifically focused on landscape: Category V or Protected Landscapes/Seascapes, for which the detailed definition is:

> A protected area where the interaction of people and nature over time has produced an area of distinct character with significant ecological, biological, cultural and scenic value; and where safeguarding the integrity of this interaction is vital to protecting and sustaining the area and its associated nature conservation and other values.

So looking just at these two definitions, it is clear that Protected Landscapes provide the conditions for reconciling nature conservation and landscape protection. They cannot be recognised as protected areas in the first place unless priority is given to nature conservation, but, at the same time, this kind of protected area allows for the protection of a range of values that goes way beyond nature conservation.

A recent IUCN study that has looked of the part played by Category V protected areas in conserving wild biodiversity (Dudley and Stolton 2012) tends to support this theoretical analysis. The authors examined Category V case studies from around the world, asking the question "how valuable are such areas for nature conservation?" Whilst the conclusions were not always clear cut, many examples were found where protected landscape managers had used the approach to increase the protection given to nature and to do so within a context of also supporting local communities and economies, and taking into account the full range of landscape values. Because nature conservation in a protected landscape is often about working through and with local communities, and acknowledging their aspirations for change, "management, including management for biodiversity, is seldom simply about keeping things as they are" (ibid. p. 100). Management for change is indeed the unique challenge of nature conservation in the context of landscape protection.

Another study in the same series looked at the role that protected landscapes play in safeguarding the rare varieties of domesticated crops, livestock, etc. that can often be found in such areas (Amend et al. 2008). This review of agro-biodiversity values showed the importance of these resources within many protected landscapes and the dependence of local communities (and their landscapes) upon their survival. This finding has since been reinforced by work undertaken as part of the Satoyama Initiative (http://satoyama-initiative.org/en). Whilst the protection of agro-biodiversity is not regarded by some as nature conservation (Locke and Dearden 2005), in fact, safeguarding such forms of biodiversity is an important component of the CBD. Protecting agro-biodiversity within Category V protected areas can help sustain the community, the economy and the landscape itself – and maintain a genetic storehouse for future generations of humankind.

Three national examples from the UK also show how protected areas, landscape protection and nature conservation can be approached in an integrated way.

England is a relatively small country with a diverse landscape that reflects both its complex geology and long history of human occupation. Historical geographers and others have attempted to capture the character of its landscape in a range of studies (Dower 1945; Hawkes 1951; Hoskins 1955), some of which were used in designating Protected Landscapes. Similar studies were made of ecological values (Tansley 1945; Huxley 1947) as a basis for nature conservation policies. However, for over 50 years, the two strands – landscape protection and nature conservation – developed separately with separate legislation, institutions and designations. Towards the end of the last century, these movements came together, a process that was given official recognition when the government decided that England should follow Wales and Scotland in merging its previously separate landscape and nature conservation bodies to create Natural England (NE). NE was given a range of responsibilities for nature conservation, landscape protection and public enjoyment of nature and the countryside. As well as bringing together nature and landscape, NE also embraced the idea of a holistic approach to the entire environment, urban as well as rural, marine as well as terrestrial. It took on board the message of the ELC that "all landscapes matter" (not just the "best") and also recognised the value of the ecosystem services that nature provides. One result of this process of consolidation and integration is the 159 National Character Areas (NCAs), which together provide a comprehensive analysis of the English landscape character. Each area is a distinct natural unit, "defined by a unique combination of landscape, biodiversity, geodiversity and cultural and economic activity". Since their boundaries follow natural lines in the landscape, not administrative ones, they are a good decision-making framework for the natural environment. NCAs are being promoted for use in land use planning and land management (see http://www.naturalengland.org.uk/publications/nca/default.aspx).

The second example from within England emerged from a major government-sponsored study, which examined the threats to wildlife which have caused species to be lost and habitats to be degraded (Lawton et al. 2010). In response to these trends, the Lawton report called for a national ecological network made up of more, bigger, better and more joined up natural areas, set within a wider landscape where nature is managed sympathetically. This strategy has been adopted by the government which has encouraged the development of so-called Nature Improvement Areas (NIAs) to provide large-scale connectivity across the countryside. Each NIA aims to create more and better-connected habitats at a landscape scale, providing space for wildlife to thrive and adapt to climate change. Twelve NIAs were approved in 2011 with a modest initial funding of £7.5 m, but well over five times as much match funding has already been secured for implementation. The NCAs provide a framework within which NIAs and similar projects can be implemented (see http://www.naturalengland.org.uk/ourwork/conservation/biodiv-ersity/funding/nia/default.aspx).

Funding for schemes for nature conservation and for landscape protection, management and planning is always in short supply, and budgets for this kind of work in most European countries are being cut. However, within the UK National Lottery, funding for heritage work of all kinds is bearing up well. These funds are disbursed by the Heritage Lottery Fund (HLF). HLF funds a Landscape Partnership (LP) programme with about £30 m annually. Each LP is led by a partnership of local, regional and national interests with the aim of conserving areas of distinctive landscape character. In practice, this is done through a package of integrated projects designed to protect and restore natural and historic heritage features within the landscape, engage communities and encourage volunteering, increase public access to the natural and historic heritage and improve the skills needed for landscape management. HLF funding for LPs is made available for up to 5 years. The areas that are chosen for LP schemes are usually in the range from 20 to 200 km^2. Some are within Protected Landscapes; some will coincide with NIAs; all must take account of the NCA guidance. The LP programme has been very successful, both in terms of conserving heritage assets and engaging communities and volunteers. It offers a practical way to achieve nature conservation, historic building conservation and landscape protection with strong community support (see also http://www.hlf.org.uk).

This short review shows that conceptually nature conservation and landscape protection are entirely compatible and indeed mutually reinforcing. Whilst there may be a few cases where a trade-off will need to be made between the conservation of natural habitats and some historic feature in the landscape, the landscape protection and the nature conservation communities have nothing to fear from working more closely together – and potentially a lot to gain.

References

Amend T, Brown J, Kothari A, Phillips A, Stolton S (2008) Protected landscapes and agrobiodiversity values. IUCN, Gland

Bishop K, Dudley N, Phillips A, Stolton S (2004) Speaking a common language: uses and performance of the IUCN system management categories. IUCN, Gland

Brown J et al (2005) The protected landscape approach: linking nature culture or community. IUCN, Gland

Cosgrove D (2004) Landscape and landshaft. GHI. 35(2):57–71, Washington, DC

Dower J (1945) National parks in England and Wales. HMSO, London

Dudley N (2008) Guidelines for applying protected area management categories. IUCN, Gland

Dudley N, Stolton S (2012) Protected landscapes and wild biodiversity. IUCN, Gland

Edwards C (2009) Resilient nation – how communities respond to systemic breakdown. Demos, London

Girot C (1999) Four trace concepts in landscape architecture. In: Corner J (ed) Recovering landscape. Princeton Architectural Press, New York

Hawkes J (1951, reprinted 1991) A land. Pelican Books, London

Hoskins WG (1955, reprinted 1988) The making of the English landscape. Hodder and Stoughton, London

Huxley J (1947) Report on the conservation of nature in England and Wales. HMSO, London

Lawton J et al (2010) Making space for nature: a review of England's wildlife sites and ecological network. Department of environment, food and rural affairs, London

Locke H, Dearden P (2005) Rethinking protected area categories and the new paradigm. Environ Conserv 32(1):1–10

Phillips A, Borrini-Feyerabend G (2009) Politiche per il paesaggio per la conservazione della natura – Diversità, equità e cambiamento. Urb 139:52–56

Piorr HP (2003) Environmental policy, agri-environmental indicators and landscape indicators. AGR Ecosyst Environ 98:17–33

Selman P (2012) Landscapes as integrating frameworks for human, environmental and policy processes. In: Plieninger T, Bieling C (eds) Resilience and the cultural landscape. Cambridge University Press, Cambridge

Tansley AG (1945) Our heritage of wild nature. Cambridge University Press, Cambridge

Chapter 3
Bringing Together Nature and Culture: Integrating a Landscape Approach in Protected Areas Policy and Practice

Jessica Brown

Abstract Taking a landscape approach to conservation requires us to embrace the complex diversity inherent in these places: recognizing natural and cultural values, tangible and intangible heritage, and history and present-day uses. Such an approach is interdisciplinary and inclusive, linking nature and culture and engaging people in stewardship. This contribution explores emerging trends that support the landscape approach in protected areas policy and practice. Reflecting on various milestone events over the past decade, it briefly reviews key conceptual and policy developments that reinforce collaborative and community governance of protected areas and of the broader landscape/seascape. It discusses the recent review of the IUCN protected area management categories and emergence of the governance framework in protected areas. It notes the growing adoption at national and provincial levels of protected landscapes designations based on IUCN Category V and, in parallel, the use of the World Heritage cultural landscape designation.

Keywords Protected areas • Landscape • Seascape • Bio-cultural diversity • Communities • Governance

3.1 Introduction

Protected areas are key elements in any strategy to conserve and sustain bio-cultural diversity[1] in the landscape and seascape. However, in the world of conservation practice, a perceived divide between nature and culture persists, expressed in how we value different resources, assign institutional responsibilities, and communicate

[1] Bio-cultural diversity can be defined as "the true web of life: diversity in both nature and culture. It's a living network made up of the millions of species of plants and animals that have evolved on Earth, and of the thousands of human cultures and languages that have developed over time. Languages, cultures, and ecosystems are interdependent. They're bound together through the

J. Brown (✉)
New England Biolabs Foundation, IUCN-WCPA Specialist Group, 240 County Road, Ipswich, MA 01938, USA
e-mail: jbrown@oldtownhill.org

R. Gambino, A. Peano (eds.), *Nature Policies and Landscape Policies*, Urban and Landscape Perspectives 18, DOI 10.1007/978-3-319-05410-0_3

priorities. Bridging the nature-culture gap is critical to making protected areas of all kinds relevant to people and meeting future conservation challenges. We risk ignoring the "full value of parks" (Harmon and Putney 2003), unless we embrace a diverse array of values – natural, as well as cultural, tangible, and intangible – in protected areas planning, designation, and management (Phillips 2005). Equally important, the persistent nature-culture divide is at odds with the world views of many indigenous peoples. Why, then, are integrative approaches not the norm?

Landscape offers a bridge between nature and culture but also between protected areas and the broader territorial context. Taking a landscape approach to conservation requires us to embrace the complex diversity inherent in these places: recognizing their natural and cultural values, tangible and intangible heritage, and history and present-day uses. Such an approach requires that we recognize the central role of indigenous and local communities in shaping these landscapes and ensuring their stewardship over time (Brown et al. 2005). Rich in bio-cultural diversity, these places are with us today because of this long and complex relationship, and because of the care provided by the communities who live in and near them.

Worldwide, the growing recognition and adoption of diverse governance regimes for protected areas (Dudley 2008; Borrini-Feyerabend et al. 2013), in particular Indigenous and Community Conserved Areas (ICCAs), are broadening nature conservation beyond a dominant "western" model and helping to bridge the nature-culture divide, by offering a more holistic view of the landscape and seascape (Kothari et al. 2013).

Reflecting on emerging trends in protected areas globally, this contribution will explore a few key developments that support the landscape approach in protected areas policy and practice. Considering the decade between the last World Parks Congress (Durban, South Africa, 2003) and the upcoming 6th World Parks Congress (Sydney, Australia, 2014), it briefly reviews conceptual and policy developments that reinforce collaborative and community governance of protected areas and the broader landscape/seascape. It discusses the recent revisions to the IUCN protected area management categories and notes the growing adoption at national and provincial levels of protected landscapee designations based on IUCN Category V and, in parallel, the use of the World Heritage cultural landscape designation.

3.2 The Protected Landscape Approach

Protected landscapes are cultural landscapes that have coevolved with the human societies inhabiting them. They are protected areas based on the interactions of people and nature over time, rich in bio-cultural diversity not in spite of, but rather *because,* of the presence of people. The *protected landscape approach* links

myriad ways in which people have interacted with the natural environment" (http://www.terralingua.org/, accessed on 12 October 2013).

conservation of nature and culture and fosters stewardship by people living in the landscape. Such an approach is interdisciplinary and inclusive, and recognizes the central role of indigenous and local communities in shaping and caring for the landscape. While grounded in experience with Category V protected areas, this approach is broader than a single protected area category or designation. Rather, it relies on different tools and designations to achieve protection and on an array of governance models and processes to sustain people's relationship to the land (Brown et al. 2005).

At the same time, the term "landscape approach" increasingly is being used to refer to scale, recognizing that conservation is most effective at the level of ecosystems and large landscapes. Landscape-scale conservation relies on achieving connectivity among habitats and ecosystems, as well as bio-cultural features. A growing emphasis on connectivity (Worboys et al. 2010) has pushed conservation planning to reach beyond the boundaries of protected areas and to encompass all the different management categories and governance regimes. In fact, this approach is congruent with the protected landscape concept, as was discussed in a recent meeting (International Academy for Nature Conservation, Isle of Vilm, Germany, April 2013) that explored the role of Category V protected areas in achieving connectivity conservation through case-study experience from all over the world.

In "taking conservation to scale," it is essential to work with a broad array of stakeholders across a mosaic of land uses and ownership patterns, engaging people in stewardship (Brown et al. 2005). In this model, areas under community and private governance play a crucial role, alongside government-protected areas, in achieving connectivity across the broader landscape. As Laven et al. write in this volume, extending conservation to the wider landscape involves many more actors, such as landowners, organizations, and different government bodies, and requires new forms of governance, including those based on coordinating and facilitating networks of partners (Laven et al. 2014).

3.3 Setting the Stage for new Policies and Practice

The World Parks Congress Convened by IUCN once a decade, the World Parks Congress[2] is a premier global convening on protected areas, reviewing current

[2] The debate over the past few decades on the role of protected areas in society has been spirited, challenging, and ultimately constructive. The 5th World Parks Congress in 2003 (Durban, South Africa) was a watershed event in this debate, producing the Durban Accord, which enshrined the rights and responsibilities of indigenous and local communities and raised the profile of diverse governance regimes. Other major global gatherings include subsequent meetings of the Conference of the Parties (COP) to the Convention on Biological Diversity (CBD), in particular COP 7, which adopted the Programme of Work on Protected Areas. Key gatherings of the International Union for Nature Conservation (IUCN) included the World Conservation Congresses of 2004 (Bangkok, Thailand), 2008 (Barcelona, Spain), and 2012 (Jeju, Korea), respectively. Also

status and setting the agenda for protected areas conservation for the decade to come. At the 5th World Parks Congress in 2003 the role of communities in creating and managing protected areas was, for the first time, a central part of the debate, launching significant work on the theme of governance. The topic of protected landscapes and seascapes also featured prominently at the Congress, explored in a series of sessions bringing out case-study experience from diverse regions. Noting the importance of places where the interactions of humans and nature over time have contributed to biodiversity, participants argued for a greater understanding of the link between cultural diversity and biodiversity. They recognized the need for linkages between protected areas and the broader landscape in order to achieve conservation goals, while calling for more holistic approaches that incorporate social and cultural dimensions. Significantly, workshop participants advocated a "protected landscape approach" and began to articulate its elements (Brown et al. 2005). The Congress gave impetus to the work of the WCPA protected landscapes Specialist Group, whose mission is to advance the protected Landscape approach and which has produced a number of publications over the past decade.[3] In the Durban Accord, the 3,000 participants in the 5th World Parks Congress affirmed the value of protected areas as:

> Those places most inspirational and spiritual, most critical to the survival of species and ecosystems, most crucial in safeguarding food, air and water, most essential in stabilizing climate, most unique in cultural and natural heritage and therefore most deserving of humankind's special care.

Alongside the Durban Accord and Action Plan, the 5th World Parks Congress produced a "Message to the Convention on Biological Diversity," with numerous specific recommendations related to the involvement of indigenous and local communities and the need for rights-based approaches to conservation. Importantly, these points were subsequently taken up in the CBD Programme of Work on Protected Areas and thus have helped to shape policy in the countries that are signatories to the Convention (Kothari et al. 2013). Looking ahead, the next World Parks Congress, which will take place in Sydney, Australia, in late 2014, will have as one of its three priority objectives "to position protected areas within goals of economic and community wellbeing." Among the eight thematic streams of the 6th World Parks Congress will be one on "Enhancing the Diversity and Quality of Governance" and another on "Respecting Indigenous and Traditional Knowledge and Culture."[4]

---

during this period, in 2007, the UN General Assembly adopted the United Nations Declaration on the Rights of Indigenous Peoples (UNDRIP), which provides a strong basis for the involvement of indigenous peoples in all forms of conservation and development (Kothari et al. 2013).

[3] http://www.iucn.org/about/work/programmes/gpap_home/gpap_biodiversity/gpap_wcpabiodiv/gpap_landscapes/

[4] http://worldparkscongress.org/streams.html, accessed 15 October 2013

The Convention on Biological Diversity and Programme of Work on Protected Areas Immediately following on the 2003 World Parks Congress, the Programme of Work on Protected Areas (POWPA) was adopted at the 7th Conference of Parties to the Convention on Biological Diversity[5] (COP 7) in 2004, incorporating a major element related to governance, participation, equity, and benefit sharing (Kothari et al. 2013). This element of the POWPA sets targets for equitable sharing of costs and benefits arising from protected areas, and for ensuring full and effective participation by indigenous and local communities in establishment and management of protected areas.[6] Also at COP 7 a decision was adopted encouraging governments to mainstream agricultural biodiversity in their conservation plans and strategies and to recognize and support the efforts of local and indigenous communities in conserving agro-biodiversity.[7] Work on advancing goals set in Article 8(j), the portion of the CBD concerned with traditional knowledge systems and practices, has been steadily advancing through the subsequent COPs and through regular meetings of an ad hoc working group. In its most recent meeting (October 2013), the working group framed recommendations for "work that will build knowledge networks, support capacity development and integrate the traditional knowledge and customary practices of indigenous and local communities into the science-based work of the Convention on Biological Diversity."[8]

The World Heritage Convention and Communities In parallel, with the 2007 adoption of *Community* as the "fifth C" in its strategic objectives (complementing objectives related to credibility, conservation, capacity building, and communication), the World Heritage Convention has highlighted the important role of indigenous and local communities in conservation of World Heritage sites (Te Heuheu et al. 2012). Truly meeting this challenge will require enabling meaningful participation by local communities "upstream" in the process of nomination, fostering their active involvement in conservation at site level, and ensuring that sustainable development near World Heritage sites brings benefits to local communities. UNESCO is now developing an indigenous peoples policy and has highlighted these issues in recent publications (Galla 2012; Te Heuheu et al. 2012). However, progress to date has been uneven and many challenges remain to ensuring meaningful participation in the nomination process and management (Larsen 2012). At site level, initiatives concerned with fostering community engagement in World Heritage are helping to lay the groundwork for future progress. One excellent example is the Community Management of Protected Areas for Conservation program (COMPACT), which has been working in eight World Heritage

[5] The Convention on Biological Diversity is an international treaty for the conservation of biodiversity, the sustainable use of components of biodiversity, and the equitable sharing of the benefits derived from the use of genetic resources. With 193 parties, the Convention has near universal participation among countries.

[6] COP 7 Decision VII/28

[7] COP 7 Decision VII/3

[8] Press Release, CBD, Montreal 12/10/2013

sites over the past decade, demonstrating that community-based approaches to management and governance can increase the effectiveness of biodiversity conservation while bringing benefits to local communities (Brown and Hay-Edie 2013).

3.4 Conceptual Developments

Review of the IUCN Protected Area Management Categories In 2008 the IUCN World Commission on Protected Areas (WCPA) brought out a revised set of guidelines for its protected area management categories, following a year-long process of review and debate that included a "Summit" bringing together practitioners from diverse regions and perspectives (Dudley and Stolton 2008). This resulted in a more precise and arguably more inclusive definition of protected areas[9] and an updated schema of the six management categories which now – in a significant conceptual development – are set in the context of crosscutting governance types (Dudley 2008).

The 2008 Summit provided a forum for robust discussion and debate about emerging new paradigms for protected areas, generally, and about the value of Category V (and VI) protected areas in achieving conservation goals (Locke and Deardon 2005; Mallarach et al. 2008), ultimately reinforcing the place of Categories V and VI among the protected area management categories. An analytical paper commissioned for the meeting reviewed the management objectives, challenges, distinguishing features, and role in the landscape/seascape of Category V and presented a revised definition for discussion (Phillips and Brown 2008). As a result, the following updated definition for Category V is included in the current version of the IUCN guidance for protected area management categories:

> A protected area where the interaction of people and nature over time has produced an area of distinct character with significant ecological, biological, cultural and scenic value; and where safeguarding the integrity of this interaction is vital to protecting and sustaining the area and its associated nature conservation and other values. (Dudley 2008)

The Governance Framework The emergence of a framework for understanding governance has opened the door to recognizing protected areas created and cared for by a diverse array of stewards. While the debate about specific directions is far from over, these and other events reinforced an important shift in thinking: from the conventional view that protected areas are places created and managed by governments to one that recognizes that they are also places that are created and managed by communities or individuals in diverse arrangements. It is no longer questioned

[9] "A clearly defined geographical space, recognized, dedicated and managed, through legal or other effective means, to achieve the long-term conservation of nature with associated ecosystem services and cultural values." (Dudley 2008)

that human societies have been devising conservation regimes for millennia, long before governments created the first national park (Borrini-Feyerabend 2002).

Hence, the growing use of the protected area "matrix" as a tool for planning and, importantly, the recognition that robust national systems of protected areas can and should draw on all of the different management categories and all of the different governance options (Borrini-Feyerabend et al. 2013; Dudley 2008). The conceptual breakthrough that came about in the decade since has been the understanding that these very different governance regimes are crosscutting. In other words, any kind of protected area – from a strict nature reserve to an extractive reserve – can be found under any of the four governance regimes. Countries are now encouraged to expand their national systems by incorporating the full range of governance types (Kothari et al. 2013).

3.5 Sample Developments "On the Ground"

New Category V Designations Over the past decade or so, the Category V model, long associated with the protected area systems of Europe, has increasingly been taken up and applied in diverse regions of the world, in places like Andean South America, East Africa, and Oceania (Brown et al. 2005; Amend et al. 2008; Dudley and Stolton 2012). These new protected landscape/seascape designations typically draw on the Category V definition as presented in the IUCN guidelines (see earlier section), adopting it to a specific national or provincial context.

Recent examples from diverse regions are illustrative. The province of Quebec has created a designation called *paysage humanisé* (or "living landscape") in keeping with Category V and modeled after the regional nature Parks of France and Belgium. The province introduced the designation as a means of increasing biodiversity conservation, particularly on private lands, while encouraging sustainable rural development (Blattel et al. 2008). Brazil's system of protected areas includes the *Area de Proteçao Ambiental* (Environmental Protection Area), a designation similar to Category V (Lino and Britto de Moraes 2005). In Ecuador, with the introduction of a new Law of Culture, the potential creation of an Ecuadorian Heritage Cultural Landscape designation is being explored. Such a designation would be based on values of Ecuadorian identity, sustaining biological as well as cultural diversity and declaring heritage in the Andean sense of "patrimony" worth protecting (Sarmiento and Viteri STET). Earlier this year, the expanded Dhimurru Indigenous Protected Area (IPA) in Australia was formally recognized by the Australian and Northern Territory Governments as a Category V Protected Area. The IPA now comprises some 550,000 ha incorporating extensive areas of land and sea – consistent with coastal Aboriginal people's holistic view of land and sea as indivisible components of their traditional "country" (Dermot Smyth, pers. com 2013). It is a signal development, as its expansion and recognition has been achieved through a collaborative planning and partnership approach led by the Dhimurru aboriginal communities who are the traditional owners of the territory. The Dhimurru IPA management plan spells out specifically its correspondence with the Category V definition and guidelines.[10]

World Heritage cultural landscapes When the category of Cultural Landscapes was included within the framework of the World Heritage Convention in 1992, recognizing outstanding examples of the "combined works of nature and man," it created a new opportunity to inscribe sites that embody the interactions between humans and nature and contain diverse tangible and intangible values (Rossler 2003; Finke 2013). Recent studies have documented the considerable overlap between Category V protected areas and World Heritage cultural landscapes (Phillips 2003; Rossler 2005; Finke 2013).

The 1992 revision of the WH Operating Guidelines was an important milestone, allowing for recognition of indigenous values as they relate to the landscape and bringing better balance to the World Heritage List (Te Heuheu et al. 2012). At the same time, as discussed in a recent review of communities and rights within World Heritage, evaluation of cultural landscapes poses distinct challenges. Increasingly, these new designations must seek to bridge the separation between cultural and natural values, and also between Outstanding Universal Values and those values that are locally held by contemporary communities. As well, new standards are evolving for the *linkages* between the cultural and natural (Larsen 2012).

Over the past two decades, 82 cultural landscapes worldwide have been inscribed as World Heritage.[11] Recent inscriptions highlight the growing role of indigenous and local communities in the nomination process. For example, the Konso people of Ethiopia played an active role in the nomination and 2012 inscription of the Konso Cultural Landscape, which embodies agroecological practices, rituals, and sacred sites and is described by UNESCO as "a spectacular example of a living cultural tradition….[that] demonstrates the shared values, social cohesion and engineering knowledge of its communities."[12] Similarly, indigenous communities advocated for the nomination and designation of World Heritage cultural landscapes such as the Sacred Mijikenda Kaya Forests of Kenya (inscribed in 2008) and Papahānaumokuākea in the Hawaiian Archipelago of the United States (inscribed in 2010).[13]

[10] http://www.dhimurru.com.au/sea-country-ipa-management-plan-launch.html. Accessed 22 October 2013

[11] http://whc.unesco.org/en/culturallandscape/. Accessed 21 October 2013

[12] http://whc.unesco.org/en/list/1333/, accessed 21 October 2013, and http://www.christensenfund.org/2012/08/01/ethiopia-konso-people-celebrate-unesco-world-heritage-support/, accessed 01 Oct 2013

[13] http://whc.unesco.org/en/list/1231;whc.unesco.org/en/list/1326. Accessed 01 October 2013

3.6 Discussion: Toward a Framework for Integrating Nature and Culture in Landscape Conservation

The creation of national and provincial level designations based on Category V and continuing progress in designation of World Heritage cultural landscapes in diverse regions, particularly in nominations led by indigenous and local communities, are two important trends. With global instruments like the CBD incorporating concerns related to governance, participation, equity, and traditional knowledge among others, and with the recent approval of UNDRIP, the global policy arena increasingly requires that nature conservation strategies address the rights and responsibilities of indigenous and local communities. Importantly, the emergence of the governance framework has been a major conceptual breakthrough in our understanding of protected areas, making it possible to engage a broader array of actors in stewardship and work at larger landscape scale.

These developments and others over the past decade support taking a landscape approach in protected areas policy and practice, reaching beyond existing boundaries in ways that encompass diverse governance regimes and engage communities in stewardship. They lay the groundwork for strategies that bring together more closely "nature conservation policies" with those of territorial planning policies affecting the broader landscape. Slowly, the nature-culture divide is being bridged. A key challenge remains in understanding the complex array of linkages between the two and – per the Category V definition – "safeguarding the integrity of this interaction," recognizing that protected landscapes involve *process*, as well as *place*, and that sustaining a relationship between people and the land is basic to their future.

References

Amend T et al (eds) (2008) Protected landscapes and agro-biodiversity values, Volume I in the series, values of protected landscapes and seascapes. IUCN World Commission on Protected Areas and GTZ/Kaperek Verlag, Heidelberg

Blattel A, Gagnon G, Côté J, Brown J (2008) Conserving agro-biodiversity on the Gaspé Peninsula of Québec, Canada. A potential role for Paysage Humanisé Designation. In: Amend T et al (eds) Protected landscapes and agro-biodiversity values, Volume I in the series, values of protected landscapes and seascapes. IUCN World Commission on Protected Areas and GTZ/Kaperek Verlag, Heidelberg

Borrini-Feyerabend G (2002) Interview article in special issue of PARKS on local communities and protected areas. Parks 12(2). IUCN, Gland and Cambridge

Borrini-Feyerabend G, Dudley N, Jaeger T, Lassen B, Pathak N, Phillips A, Sandwith T (2013) Governance of protected areas: from understanding to action, Best practice protected area guidelines series no. 20. IUCN, Gland

Brown J, Hay-Edie T (2013) COMPACT: engaging communities in the stewardship of world heritage. UNDP, New York

Brown J, Mitchell N, Beresford M (eds) (2005) The protected landscape approach: linking nature, culture and community. IUCN and the World Commission on Protected Areas, Gland

Dudley N (ed) (2008) Guidelines for applying protected area management categories. IUCN, Gland

Dudley N, Stolton S (eds) (2008) Defining protected areas: an international conference in Almeria, Spain. IUCN, Gland

Dudley N, Stolton S (eds) (2012) Protected landscapes and wild biodiversity, Volume III in the series volume I in the series, values of protected landscapes and seascapes. IUCN World Commission on Protected Areas, Gland

Finke G (2013) Landscape interfaces: world heritage cultural landscapes and IUCN protected areas. Thematic study published by the IUCN world heritage programme. International Union for Nature Conservation (IUCN), Gland

Galla A (2012) World heritage: benefits beyond borders. UNESCO Publishing and Cambridge University Press, Paris/Cambridge

Harmon D, Putney A (eds) (2003) The full value of parks: from economics to the intangible. Rowman and Littlefield Publishers, New York

Kothari A, Camill P, Brown J (2013) Special issue: policy and practice of community-based conservation. Conservation and Society, 11(1):1–15. Medknow Publications, Bangalore

Larsen PB (2012) IUCN, world heritage, and evaluation processes related to communities and rights. Independent review prepared for the IUCN world heritage programme. IUCN, Gland

Laven D, Mitchell N, Jewiss J, Barret B (2014). Lessons learned from US experience with regional landscape governance: implications for conservation and protected areas In: Gambino R, Peano A (eds), Nature Policies and Landscape Polices, Urban and Landscape Perspectives 18, Springer International Publishing, Switzerland

Lino C, Britto de Moraes M (2005). Protecting landscapes and seascapes: experience from coastal regions of Brazil. In: in Brown J, Mitchell N, Beresford M (eds), The Protected Landscape Approach: Linking Nature, Culture and Community. IUCN, Gland.

Locke H, Deardon P (2005) Rethinking protected area categories and the new paradigm. Environmental Conservation 32(1):1–10

Mallarach JM, Morrison J, Kothari A, Sarmiento F, Atauri JA, Wishitemi B (2008) In defense of protected landscapes: a reply to some criticisms of category V protected areas and suggestions for improvement. In: Dudley N, Stolton S (eds) Defining protected areas: an international conference in Almeria, Spain. IUCN, Gland

Phillips A (2003) Cultural landscapes: IUCN's changing vision of protected areas. In: Rössler M (ed) Cultural landscapes: the challenges of conservation, World heritage papers no. 7. UNESCO World Heritage Centre, Paris

Phillips A (2005) Landscape as meeting ground. In: Brown J, Mitchell N, Beresford M (eds) The protected landscape approach: linking nature, culture and community. IUCN, Gland

Phillips A, Brown J (2008) Category V. In: Dudley N, Stolton S (eds) Defining protected areas: an international conference in Almeria, Spain. IUCN, Gland

Rössler M (2005) World heritage cultural landscapes: a global perspective. In: Brown J, Mitchell N, Beresford M (eds) The protected landscape approach: linking nature, culture and community. IUCN, Gland

Rössler M (2003) Linking nature and culture: world heritage cultural landscapes. In: Rössler M (ed) Cultural landscapes: the challenges of conservation, World heritage papers no. 7. UNESCO World Heritage Centre, Paris

Sarmiento F, Viteri X (2014) Discursive heritage: sustaining Andean cultural landscapes amidst environmental change. In: Taylor K, Mitchell N, St Clair A (eds) Conserving cultural landscapes: challenges and new directions. Routledge Press, New York

Te Heuheu T, Kawharu M, Tuheiva R (2012) World heritage and indigeneity, World heritage, no. 62. UNESCO World Heritage Centre, Paris

Worboys G, Francis W, Lockwood M (eds) (2010) Connectivity conservation management: a global guide. Earthscan, London/Washington

Chapter 4
The Place of Protected Areas in the European Landscape: A EUROPARC Federation Perspective

Carol Ritchie

Abstract Approaches to nature conservation have expanded from mainly species approach in the mid-eighteenth century to landscape-scale and ecosystem services-based approaches today. The EUROPARC Federation has played a significant role in the support of protected areas in developing this landscape conservation approach. Utilising the words of Patrick Geddes, an eighteenth-century Scottish biologist and planner and pioneer in linking man and the environment, when he described 'The Typical Region' as *Microcosmos naturae, sedes hominum, theatrum historiae, eutopia futuris*, he described his landscape concept as first and foremost ecology: a 'microcosm of nature', but it is also the *sedes hominum*, the seat of humanity, the place where human beings make their lives as part of that ecology. And linked to this is the dramatic *theatrum historiae*, the theatre of history, the past experience that should inform the future. Finally, it is the *eutopia* or 'good place' of the future, a place that Geddes believed could be achieved through local and international cooperation and adoption of sustainable technologies (Macdonald M (2009), *Sir Patrick Geddes and the Scottish Generalist Tradition. The Sir Patrick Geddes lecture,* Royal Society of Edinburgh, Edinburgh, 20 May 2009). In this article EUROPARC revisits Geddes original description, for 2013, looking particularly at the role protected areas play in the European landscape.

Keywords Protected areas • Biodiversity • Cultural landscape • Europe • Sustainable tourism

4.1 Microcosmos Naturae

Humans have long attached reverence and special significance to particular places in the natural world. Whether for ancient spiritual significance, locations for food and water or seemingly limitless landscapes to be feared and admired, the people of Europe have imbued implicit values to parts of their surroundings.

C. Ritchie (✉)
EUROPARC Federation, Waffnergasse 6, Regensburg 93047, Germany
e-mail: c.ritchie@europarc.org

R. Gambino, A. Peano (eds.), *Nature Policies and Landscape Policies*,
Urban and Landscape Perspectives 18, DOI 10.1007/978-3-319-05410-0_4

Our parks and protected areas today retain their original universal values and yet deliver many vital benefits to the European society. They remain the reservoirs of biodiversity, sources of water, cultural and economic assets, places of recreation, health and well-being and spiritual replenishment. The creation of Europe's protected areas was built on that shared inheritance of valuing special places; they are the expression of faith in the future, a pact between generations and a promise from the past to the future.

Despite decades of legislation and action, however, Europe's biodiversity remains vulnerable. 15 % of terrestrial mammals are classified threatened with extinction and our ecosystems continue to be fragmented (EEA 2010). Yet despite clear evidence indicating the loss of biodiversity in Europe, there appears no societal consensus of its occurrence or of the connection with human activity in affecting biodiversity loss. If we are to realise the promise envisaged by the early initiators of protected areas in Europe and secure a healthy natural landscape, then a more coordinated approach both within the network of protected areas and out with their boundaries must be envisaged.

The EUROPARC Federation's mission is to increase effectiveness in conserving and enhancing natural and cultural heritage, for the well-being and benefit of future generations. The founders of the Federation committed themselves, at the birth of the network in 1973, to further the conservation of Europe's nature through international cooperation. It was this belief in our fragmented European landscape, with increasing pressure on our natural resources, that opportunities to make the environment more resilient to systemic risks and change are needed to be found.

EUROPARC members understood that nature knows no boundaries; therefore, EUROPARC is founded on the principle that the future protection and conservation of nature is achieved through international cooperation. International cooperation works best through personal contact. That through personal contact comes mutual understanding, shared experiences, knowledge and innovation. This delivers better support and management of protected areas, which ensures the future protection and conservation of nature, because nature knows no boundaries.

Europe's protected areas, including its national and nature parks, biosphere reserves and other designations, play a key role in protecting biodiversity within landscapes. They are vital to the continent's economy, contributing over EUR 15 billion a year in jobs, food and other services for the people of Europe (EEA 2012). However, for effective biodiversity conservation, protected areas must be seen, within a wider landscape, which would include the contribution of fisheries, agricultural and forestry policies and coordinated land use planning with a greater shift towards sustainable development.

The European Commission recently launched its new vision for 2050 and its states. By 2050, European Union biodiversity and the ecosystem services it provides – its natural capital – are protected, valued and appropriately restored for biodiversity's intrinsic value and for their essential contribution to human well-being and economic prosperity, and so catastrophic changes caused by the loss of biodiversity are avoided.

It will take a concerted effort by all actors working together to realise that vision. The existing Natura 2000 and Emerald Networks including nationally and regionally protected areas, as well as the recognition and need for large areas of wilderness, are essential assets to build future natural, cultural and economic prosperity. A landscape devoid of nature is not only empty of life, but fundamentally unsustainable.

Humans have been part of the European landscape for millennia, and in many regions, these long-established human and nature codependent relationships have generated cultural landscape with high nature values. It is clear therefore, that one of the major assets to realise the vision for European biodiversity, is people. The natural and cultural landscapes identified primarily by the protected area network in Europe are a catalyst to engage people in the process of conservation.

4.2 Sedes Hominum

Europe is endowed with a variety of landscapes, embedded in cultural identity. Often sculpted by nature and shaped by people, these landscapes have influenced the course of history, inspired artists and thinkers and enriched lives. They are our natural life support system. But these protected areas represent the land that is our inheritance – we hold it in trust for those who come after us.

In Europe especially, people are part of the landscape and should not be apart from it (Council of Europe 2000). The landscape therefore is a synthesis of the physical and natural physical features of a place, as well as the human use of the resources creating, over time, a fusion of interdependence from which emerges a cultural identity. Within a physical landscape will coexist a social distinctiveness, defining the self-image of a region and its sense of place, which differentiates it from other regions. Traditionally, a physical place although evidenced in nature and natural resources is often the covert dynamic backdrop to people's habits, food and architecture, indeed permeating every part of their lives.

Landscapes have often been viewed primarily as visions of beauty, serenity and scenes from nature, but laterally have included the splendour of pastoral and human modified views. A modernist scene could include the majesty of urban scenes. Clearly, human relationships with the landscape and what we think of as a landscape continue to change.

As the population in Europe grew over millennia, valued places were marginalised, to the very sacred, the less useful and those areas available to an exclusive elite. All that now remains of the once vast natural treasures of European society are often held in protected areas. Many of these protected areas remain as the physical and natural testaments to regions where the cultural landscape has now in the present day virtually disintegrated due to societal changes towards urbanisation and globalisation.

Within Europe's cultural landscape however lies the store of our collective understanding of people living and utilising natural resources, which had sustained communities over millennia. As European society risks an increasing disconnection from nature and the landscapes that people need, a renewed interest in investing in cultural landscapes through European protected areas has emerged.

The need for a more sustainable use and development of resources, in order to address the loss of nature in cultural landscapes, was one of the drivers behind the creation of EUROPARC's Charter for Sustainable Tourism.

This innovative tool for protected areas seeks to assist in 'expanding' the boundaries of participation and involvement whilst seeking to balance different perspectives in the future management of a region. Embedded in the 'Charter system', applied by protected areas and their surrounding region, is the renewal of the area's cultural identity as the key to sustainable regeneration. The underlying concept is a return of a sense of place, achieved through identification with the area's natural and cultural assets. Participation by local people is the key to the success of sustainable tourism, with protected areas acting as the catalyst to drive environmental, social and economic benefits, in other words, finding again the connection between nature and culture that sustains communities and landscapes.

With, in 2013, 119 parks in 13 countries, certified under the Charter for Sustainable Tourism in protected areas, many areas are on their way to rediscovering and even redefining their cultural identity. Operating more sustainably has resulted in over € 441 M invested in sustainable management in these areas. With 700 products supported through these sustainable tourism partnerships bringing local products and culture to the marketplace, ensuring the livelihood of many and keeping local traditions alive, and with over 3,000 organisations throughout the network committed to working together to deliver sustainable development and conservation, reviving local engagement and bringing greater social cohesion all of which have sustained and revived cultural landscapes in differing parts of Europe, restoring and preserving knowledge, skills, and identity for future generations.

As European society lurches from financial to ecological crisis and back again, the need to appreciate our natural assets means supporting the structures that underpin protected areas, supporting communities and visitors and encouraging scientists who monitor effects and impacts and farmers and fishermen who manage sustainably our cultural and living landscapes.

4.3 Theatrum Historiae

The modern concept of European national parks and protected areas culminated a century ago in the designation of the continent's first, in Sweden, in 1909. Those that followed started with high mountains, often inaccessible, yet, iconic landscapes in the national psyche. As the century progressed, early values of national parks were enhanced with further concepts of species protection and ecological integrity and the need to ensure exemplars of Europe's great diversity of habitats, landscapes and seascapes, to be preserved in perpetuity. In more recent years, public recreation, sustainable development and community involvement models of protected area governance have become the norm, and the European Union's network of Natura 2000 sites has been added to the earlier foundations, ensuring that 18 % of European land has some kind of nature protection.

The first real global acceptance of human impacts of the environment came from the 1972 Stockholm Conference on the Human Environment. Further the IUCN took the idea of sustainable development, then in its infancy, and combined it with nature conservation in their forward-looking World Conservation Strategy of 1980. The seminal report of the Brundtland Commission in 1987, 'Our Common Future', mapped out the need for and gave a universally accepted definition of sustainable development as:

> development that meets the needs of the present without compromising the ability of future generations to meet their own needs. (UN 1987)

The report indicated that sustainability is needed to be achieved at the environmental, social and economic levels.

The Rio Earth Summit in 1992 spawned the Convention on Biological Diversity and Local Agenda 21, with climate change as an increasing concern on the agenda. The IUCN published 'Parks for Life' (1994), to which the EUROPARC Federation was a steering group member and contributor. This publication advocated a European network of protected areas, working in a participatory manner with local communities where unique landscape qualities should be recognised and managed. The report contained a recommendation to create trans-frontier parks. As a direct result, EUROPARC created the Transboundary programme 'Following Nature's Design' with, in 2014, 10 transboundary areas, formed by 23 protected areas in 13 countries.

These events and reports, amongst many others, presented critical shifts in thinking, coupled with the expressions of concerns and challenges from the nature conservation community that permeated policy and practice in protected area and landscape management.

The Council of Europe's European Landscape Convention introduced a Europe-wide concept focused on the quality of landscape protection, management and planning and covering entire territories. It very much advocated a people-centred approach, betraying very much the influence of the thinking of the previous three decades.

Despite continuing challenges towards culture and nature, all is not lost, the report 'A Vision for a Wilder Europe' (Sylvén and Widstrand 2013) indicates that the 'last three decades have seen significant improvements of Europe's fauna and flora, with four international agreements playing a particularly important role as impetus for conservation action: the Berne Convention (1979) and the European Union's Birds Directive (1979), Habitats Directive (1992) and Water Framework Directive (2000)'. Further that there has been something of a wildlife comeback in Europe. According to the 'Living Planet Report' (2012), the period 1970–2008 saw an average increase in animal population size of 6 % in the Palearctic realm (which mostly includes data from Europe). Better environmental protection is one explanation put forward to be a contributing factor to this increase. However, recent changes in land use with abandonment of farmland, reduced hunting pressure and higher productivity of many ecosystems (due to more nutritional input from human activities) play a part.

4.4 Eutopia Futuris

Landscapes provide the context of daily life and bear witness to our forebears. Yet today, landscapes and the protected areas within them face challenges and pressures unforeseen by our enlightened predecessors. Protected areas are at the frontline of climate change, where effects on ecosystems are already evident and mitigation to retain Europe's natural heritage critical. They are our early warning system, and monitoring and enlightened management are needed to secure the very systems that support us. Concerns over multiple uses of protected areas do also loom large, and the call to retain large areas as wilderness as well as to re-wild more managed urbanised areas grows.

Europe's landscapes, together with the network of protected areas, need to regain the values of our early ancestors and realise fully the worth of sustainable management of our natural and cultural heritage.

Importantly, to achieve environmental sustainability, reduce biodiversity loss and find again a renewed sense of place, the Natura 2000 sites, with national protected areas, must be more fully supported and managed. European policy, including the European Landscape Convention, should be fully implemented. These actions alone would bring a substantial improvement and lighten the impact on Europe's stressed environment.

Due to rapid urbanisation in Europe, there has been a dramatic change in land use and abandonment of previously cultivated or grazed land over the last decades. This trend is projected to continue although influenced by population changes and demand for resources particularly in the face of climate change. Land abandonment is a current and future issue in Europe with significant opportunities for nature conservation and implications for cultural identity and landscapes.

Further, through collaboration between national and regional protected areas, Natura 2000 sites, and private landowners, the management experience gained through long-established protected areas could be better shared. EUROPARC has strived to achieve this through a collective approach to seeking solutions and international cooperation in raising standards, in all areas of parks and landscape management.

If the future of the cultural identity, as well as our natural heritage, is to be secured, then the voice of and active participation of young people must be heard. Young people are at the heart of the local communities living in and around protected areas. They are also the future advocates, guardians and potential employees of these areas. By providing young people with opportunities to participate in their work, protected areas fulfil their educational role whilst enhancing individual understanding of their aims and appreciation of the natural and cultural resources they protect. EUROPARC has operated the Junior Ranger Programme since 2002 and now boasts a total of almost 50 partners with thousands of young people involved. In 2013, the Federation undertook its first youth conference from 18- to 21-year-olds, graduates of the Junior Ranger Programme, and now seeks to create an advocacy programme to ensure that young people have a say in their future.

The EUROPARC Federation was founded on the need to protect and enhance Europe's natural and cultural protected areas. These principles are embedded in the work of the Federation, bringing together those who care for Europe's protected areas. EUROPARC facilitates and stimulates collaboration and partnerships. Such a partnership must include the private sector whose use of and influence on natural and cultural landscapes is the foundation of their business opportunities.

It is also important if protected areas are to retain their valuable role in landscapes that they are encouraged to build strong teams of professional protected area staff. Working alongside local landowners businesses and communities, an ethos, to encourage innovative, forward thinking, outward-looking inclusive approaches that will deliver complex integrated management is needed for resilient landscapes and protected areas of the future.

We all have a twenty-first-century responsibility of great importance. It is to proclaim anew the meaning and value of protected areas and their key role in sustaining European landscapes. Protected area managers, politicians and policy makers need to look beyond the park boundaries and transfer knowledge and experience to become important players in the social and economic development of regions and countries. This translates into development based on careful planning and use of natural resources and takes special care of the landscape and all its components. Sustainable development where protected areas are present means securing human well-being by maintaining nature, the very source of life on earth. EUROPARC would support the call for the exploration of a new European Wilderness Convention under the auspices of the Council of Europe with increased commitment from states for the protection of wilderness landscapes, transboundary cooperation and strengthened links to the Convention on Biological Diversity.

This collective work would restore the important third pillar of a human society, following that of social and economic, the nature and natural resources. Thus, protected areas and conservation movements as a whole can contribute to a real equilibrium for the future of our shared inheritance.

Today, the network of protected areas represents what we as humans treasure about our environment. In its richness and beauty, it is a living example of the human capacity to connect with the life of our planet and represents a deep well of hope for the future.

References

Council of Europe (1979) Convention on the conservation of European wildlife and natural habitats. 19.IX.1979, Bern

Council of Europe (2000) European landscape convention. Treaty series no 176. Available via http://conventions.coe.int/Treaty/en/Treaties/Html/176.htm. Accessed 15 Nov 2013

EUROPARC (2010) Following nature's design. Promoting cross-boarder cooperation in nature conservation. Available via http://www.europarc.org/uploaded/documents/455.pdf. Accessed 15 Nov 2013

European Economic Community (EEC) (1979) Birds Directive. 79/409/EEC

European Environment Agency (EEA) (2010) Biodiversity baseline. Technical report no 12/2010. Available via http://www.eea.europa.eu/publications/eu-2010-biodiversity-baseline. Accessed 15 Nov 2013

European Environment Agency (EEA) (2012) Annual report 2012 and environmental statement 2013. Available via http://www.eea.europa.eu/publications/eea-annual-report-2012/at_download/file. Accessed 15 Nov 2013

European Union (EU) (1992) Habitats Directive. 92/43/EEC

European Union (EU) (2000) Water Framework Directive. 2000/60/EC

IUCN (1994) Parks for life. Future challenges and directions for protected areas in South Asia. Proceedings of the commission on national parks and protected areas, 42nd working session, Islamabad, 21–23 Sept 1994

Macdonald M (2009) Sir Patrick Geddes and the Scottish Generalist tradition. The Sir Patrick Geddes lecture, Royal Society of Edinburgh, Edinburgh, 20 May 2009

Sylvén M, Widstrand S (eds) (2013) A vision of a Wilder Europe. WILD 10, The Wild Foundation

UN, Brundtland Commission (1987) Report of the World Commission on Environment and Development: our common future. Available via http://www.un-documents.net/wced-ocf.htm. Accessed 15 Nov 2013

WWF (2012) Living Planet Report 2012: biodiversity, biocapacity and development, Gland

Chapter 5
From Park-Centric Conservation to Whole-Landscape Conservation in the USA

Paul M. Bray

Abstract America's traditional parks like its national parks were a creation of a time when today's environmental problems like climate change did not exist. It allowed for parks to be separate and apart from neighboring land holdings and inward looking in their management. Now park managers must be outward looking in order to protect park resources and to realize the ecological, social, and economic roles parks play in their entire ecosystem, nation, and world are parts of their responsibilities. This chapter traces the evolution of conservation from focusing on individual parks and features to addressing whole or large landscape conservation.

Keywords National parks • Conservation • Environmental movement • Large and whole landscapes • Public trust doctrine • Heritage areas • Place

5.1 Introduction

The creation, maintenance, and enjoyment of national, state, and local nature parks are a great achievement in the USA.

The national park system established in 1916 now comprises 401 areas covering more than 84 million acres in every state (except the state of Delaware). Although the best known areas are the great scenic national parks like Yellowstone and the Grand Canyon, over half the areas of the national park system preserve sites that commemorate persons, events, and activities important in the nation's history. They include national parks, monuments, battlefields, military parks, historical parks, historic sites, lakeshores, seashores, recreation areas, scenic rivers and trails, and the White House.

Most areas in the national park system are fully owned by the government, and they are inward looking. Valley Forge National Historical Park, for example, is located very near a large shopping mall. Many national parks have adjoining

P.M. Bray (✉)
University at Albany (Retired), SUNY. 1400 Washington Ave, Albany, NY 12222, USA
e-mail: pmbray@aol.com

R. Gambino, A. Peano (eds.), *Nature Policies and Landscape Policies*,
Urban and Landscape Perspectives 18, DOI 10.1007/978-3-319-05410-0_5

developed communities with hotel and motel accommodations, retail shopping, and restaurant and food establishments. These "gateway" communities generally have little connection with the resources of the park other than their role in mining for money from the park visitors.

In other words, American parks are generally distinguished both by their unique natural and cultural resources and the fact that they are separate and apart from inconsistent surrounding land uses. Some national parks are large enough to have a variety of resources and encompass large land or water areas to help provide adequate protection of the resources. But little consideration has been given by the managers of National and State Parks for the negative externalities generated by parks themselves like air pollution from vehicles associated with park visitors.

On the other hand, park managers pay attention to pollution impacting their parks like the Great Smoky Mountains National Park from outside areas like air pollution shrinking scenic views, damaging plants, and degrading high-elevation streams and soils. Park authorities point out that even human health is at risk when pollution directly affects their parks. Most pollution originates outside the park and is created by power plants, industry, and automobiles. Great Smoky Mountains National Park has an array of air quality initiatives underway, including research and monitoring.

The inward-looking, park-centric approach to national parks developed out of conditions in the nineteenth and twentieth century.

The birth of the environmental movement in the USA is attributed to the emergence of Earth Day as an annual day to rally the public for environmental goals beginning in the 1970s. This is well after the establishment of many of the national parks. Earth Day was soon followed by the enactment of comprehensive air and water pollution laws and a wide variety of other environmental protection initiatives. National parks and state park systems had their gates up before the age of environmentalism, and therefore, some park management could more easily isolate itself from the national and global ecological issues like air and water quality, wildlife protection, invasive species, and climate issues.

At the same time that national parks were emerging and expanding, the national economy was growing at breakneck speed. Industrialization began in the nineteenth century, and the automobile and sprawl dominated the twentieth century. Conservationists viewed parks as a refuge from the effects of commerce and economic development and did not want parks associated with economic activity. When the National Park Service was created in 1916,

> (...) parks' legal boundaries themselves seemed the most important protectors – fending off timber baron, mining companies, and an army of hucksters ever ready to surround a national treasure with a sprawl of shoddy tourist attractions. (Vv.Aa. 2009)

Inherent conflicts within the park mission have also been a force for keeping parks separate and apart. There has been a 100-year-old debate started between preservationist John Muir and forest manager Gifford Pinchot. This debate was about choosing between protecting nature for its intrinsic value as we have done in national and state parks versus being utilitarian and practicing sustained harvesting of forests in resource management areas. Some parks like New York State's Adirondack Park have examples of both sides of the debate. Half of the park is protected by the constitutional forever wild clause, which prohibits cutting of trees, while a large portion of privately owned forest land in the park is used for sustained yield of forest products.

The conflict between preservation of park resources and beneficial enjoyment by the public was also a force for keeping parks separate. Many forms of recreation like snowmobiling are not consistent with the preservation and protection goals of parks and reinforced the desire to keep parks separate and apart from whole landscapes. These kinds of conflicts keep park management attention focused inward.

In addition, vertical political configurations between levels of the government and horizontal configurations between neighboring municipalities also work to keep parks separate and apart.

5.2 Meeting New Challenges and Taking Advantage of New Opportunities

In the last few decades, there has been increasing awareness that natural resource issues like biodiversity protection, water management, invasive species, and climate adaptation need to be addressed at the scale of large landscapes or ecosystems for effective management.

The threats to Yellowstone National Park, for example, were coming from land holdings of the federal government which oversees most of the million acres of the Yellowstone ecosystem. Federal holdings include two national parks, two national wildlife refuges, parts of six national forests, plus land held by the Bureau of Land Management. However, when the park was delineated in 1872, few of any understood that the science of preservation of Yellowstone called for a much broader stewardship approach.

Notwithstanding increased recognition of the threats to Yellowstone National Park in the 1980s, the conservative Reagan administration did not take the problem seriously. Yet, as Paul Schullery points out in his book, *Searching for Yellowstone: Ecology and Wonder in the Last Wilderness*,

> Now, less than a decade later, the National Park Service is routinely involved in regional planning. The park's connections to the rest of the GYE (Greater Yellowstone Ecosystem), made clear by the wanderings of grizzly bears, the migrations of elk and bison, and the geothermal aquifers that cross park boundaries in many directions, are now seen as giving Yellowstone superintendents a strong mandate to speak out on issues affecting the GYE. Management of the park, for so many decades a fairly contained assignment, now involves paying attention to a minimum of 20 million acres of land, 90 percent of which is beyond the boundaries. (Schullery 1997)

Private conservation organizations have adopted the approach of working at a whole-landscape scale. The Nature Conservancy that at one time focused narrowly at specific habitats of endangered species has changed its focus to whole landscapes or what it now calls "whole systems" as necessary to achieve its conservation goals. Disconnected pieces of natural systems often do not survive nor do natural features thrive just by being in traditional gated parks separated from the remainder of its ecosystem.

The Wilderness Society with the mission to protect wilderness and inspire Americans to care about wild places has also adopted a landscape approach. It recognized that islands of wilderness even if legally protected are not viable over the long term. The science of conservation biology makes clear that even large wilderness areas must be surrounded by buffer zones and connected with other wild places if their ecosystems are to function and be sustained over time.

5.3 Emerging Trends in Whole-Landscape Conservation

Heritage areas have been the wild card in whole-landscape conservation movement. They have increased in number and significance while still not being fully understood or accepted by traditional conservation interests and institutions. They are a model for integrated conservation of natural and cultural resources and of whole landscapes, yet frequently without the full support of state and federal park agencies and other conservation interests.

Since the mid-1970s, heritage areas have emerged from the ground up first as the whole city as an urban cultural park and then as regions as heritage areas. New York State has 20 state-designated heritage areas, and there are now 49 national heritage areas that include some vast areas like the whole state of Tennessee and the 542-mile Erie Canalway National Heritage Corridor. Most heritage areas are generated by local activists who reach out to their state or national legislators and seek state legislative or congressional designation. This designation is the basis for their creation. As the heritage area program law in New York State declares, heritage areas are "an amalgam of natural and cultural resources." They are theme based associated with the history of their territory and have the intersecting goals of conservation, recreation, education, and sustainable economic development addressed in their state or federally approved management plans. While they encompass entire landscapes, most of the land is privately owned or, if publicly owned, not necessarily for park purposes.

National heritage areas are not accepted as part of the national park system. State heritage areas although they may be legally under the wing of state park agencies are at best stepchildren of their state park agencies. On the one hand, the theme of a heritage area and intersecting goals can serve to be a connector of the diverse elements of a landscape, but separate interests and priorities of the component elements of a heritage area makes it difficult to have the cohesion that comes with sole ownership by one entity. The resulting push and pull and lack of overall leadership makes heritage areas an excellent testing ground for a maturation into whole-landscape conservation, but it is difficult.

5.4 Other Federal Initiatives

Despite the failures at the federal level of initiatives like the Yellowstone ecosystem initiative, efforts continue to protect North America's natural and cultural resources and landscapes to sustain our quality of life and our economy. In 2009 now retired US Interior Secretary Ken Salazar signed Secretarial Order No. 3289, directing the Department of Interior bureaus to stimulate the development of the Landscape Conservation Cooperative (LCC) network as a response to landscape-scale stressors, including climate change. The cooperatives are intended to provide scientific support for conservation activities that address a variety of broad-scale land-use pressures and landscape-scale stressors including, but not limited to climate change that affect wildlife, water, land, and cultural resources.

LCCs are intended to look at whole landscapes and engage a diverse community of conservation partners working on a given landscape. This fosters conservation agencies and organizations working on complex and interconnected conservation issues to work together across jurisdictional lines to inform management of natural and cultural resources of a whole landscape.

LCCs are less formal and more process oriented than heritage areas and parks. Therefore, they are less threatening to the many diverse entities in a landscape that may fear a whole-landscape entity take over its individual jurisdictions. Currently, there are 22 LCCs including those in the Pacific Islands and Caribbean. By building working relationships including through steering committees and having successes from these relationships, whole landscapes may establish ongoing institutional frameworks and management systems to achieve the highest order of whole-landscape management.

5.5 Private Sector Initiatives

Private sector efforts for policy and action on the level of whole landscapes (taking conservation to scale) are also taking shape. Recently, the Lincoln Institute of Land Policy did a Policy Focus Report on Large Landscape Conservation: A Strategic Framework for Policy and Action (www.lincolninst.edu/subcenters/regional-collaboration). It was based on the recognition of growing challenges from land and water issues from local to global including land-use patterns, water management, biodiversity protection, and climate adaptation that require new approaches. Large landscapes were deemed to be the scale most desirable for meeting the challenges of today. The report identified three criteria for large landscape conservation: (1) multijurisdictional; (2) multipurpose including but not limited to mixes of issues like environment, economy, and community; and (3) multi-stakeholder like public, private, and governmental actors.

The dialogue that started at the Lincoln Institute continued with meetings in the west and with a northeast program on management of large landscapes in the northeast led by the Regional Plan Association.

5.6 Challenge of Meeting the Challenges of Territorial and Large Landscape Conservation

Traditional parks like the American national parks are intended to both provide beneficial enjoyment for the public and to protect park resources for future generations. To some degree public expectation will be for territories and large landscapes to also provide recreation and public enjoyment as well as protection for natural and cultural resources. Diversity in political jurisdictions therein, land ownership, and land uses in territories or large landscapes provide a much higher challenge than the traditional public estate park.

Success with territorial and large landscape conservation is going to depend on public acceptance, which will require in large part the public's recognition that it is benefiting there from. Otherwise, you have what Garrett Hardin called "The Tragedy of the Commons" where the commons suffer because the public is not invested in the values of the commons, which can include water, wilderness, and air and does not believe it has stewardship responsibility unless the commons directly affects them.

The environmental movement, which has included enactment of clean air and clear water laws, has been a significant step towards caring for the commons. Yet, for example, the American public's general indifference to addressing global warming in a meaningful way is a sign of the difficulties inherent in stewardship of the commons.

A positive sign in the USA is a growing recognition and appreciation of qualities of place and the interconnection of natural and cultural resources. Heritage areas are based on management of intersecting goals of conservation, recreation, education, and sustainable economic development. Not long ago one often heard that we need to make a choice between conservation and economic development. Now there is increasing recognition that development should be sustainable and, in fact, pursuing conservation and development are mutually compatible.

In addition, along with a public understanding and commitment to stewardship as part of territorial and large landscape conservation, a legal framework is needed as an underpinning for stewardship.

The public trust doctrine with its origins in Roman and English civil law offers the underpinning and means to protect the territorial and large landscape natural and cultural resources. The doctrine historically provided that title to tidal and navigable waters and the lands beneath are vested in the sovereign for the benefit of the people. What is included under the doctrine has changed from time to time in different cultures consistent with what is deemed to be in the public interest.

Application of the public trust doctrine depends upon clear findings regarding the values of resources and goals, comprehensive plans defining appropriate and inappropriate uses, and delineation of local versus regional and state jurisdictions. Legislation and management plans should provide for area-wide balancing of public trust uses and other uses that benefit the public and for substantial public review.

5.7 Conclusion

The promises of regional or territorial resources management/large landscape conservation to solve ecological, social, and economic problems and better serve human and societal needs are great. That we are inextricably moving in both public and private sectors towards territorial/large landscape conservation approaches even in the face of resistance like what happened with the Yellowstone ecosystem is evidence of their validation. Despite the challenges that come with the promises, farsightedness, new initiatives, and many layers of collaboration from the public and private sectors will make for their ultimate realization. Prospects for new application, for example, of the historic public trust doctrine, can provide the means for management including resource protection for territory or large landscapes.

References

Vv.Aa (1981) New York urban cultural park system: summary plan. New York State Office of Parks and Recreation, Albany

Vv.Aa (1985) National parks for a new generation: visions, realities, prospects. The Conservation Foundation, Washington, DC

Vv.Aa (1991) Partnerships in parks & preservation: proceedings and bibliography. Cosponsored by the National Park Service, The New York State Office of Parks, Recreation and Historic Preservation and the National Parks and Conservation Association and its New York Chapter, Albany, New York

Vv.Aa (2009) Advancing the National Park Idea. National Parks Second Century Commission report. National Parks Conservation Association, Washington, DC

Bray MP (1988) Possibility of parks unbounded, environment. University of Waterloo, Canada

Bray MP (1994) The role of the public trust doctrine in planning and management of the New York State Canal System. Albany Law J Sci Technol 4(1):47–56

Cranz G (1982) The politics of park design: a history of urban parks in America. The Massachusetts Institute of Technology, Cambridge, MA

Eugster JG (2003) Evolution of the heritage areas movement. GWF 20(2):50–59

McKinney M et al (2010) Large landscape conservation: a strategic framework for policy and action. Lincoln Institute of Land Policy, Cambridge, MA

Schullery P (1997) Searching for Yellowstone: ecology and wonder in the last wilderness. Houghton Mifflin Company, Boston/New York

Slade DC (ed) (1990) Putting Trust Doctrine to work: the application of the Public Trust Doctrine to the management of lands, waters and living resources of the coastal states. Connecticut Department of Environmental Protection, Coastal Resources Management Division, Connecticut

Stein WS, Bray MP (eds) (1989) Parks: New Directions in Resource Planning: proceedings from a symposium at Cornell University. Sponsored by Cornell University's Department of City and Regional Planning and its Graduate Program in Landscape Architecture, the New York Parks and Conservation Association and the New York Regional Plan Association, Cornell University, Ithaca, New York

Chapter 6
Ecological Functions, Biodiversity and Landscape Conservation

Gioia Gibelli and Riccardo Santolini

Abstract The Italian strategy for biodiversity underlines the necessity to integrate nature conservation policies with the active participation of stakeholders and of the government at any level. Moreover, there is a need to combine ecology and economy in a framework where natural conservation is an added value of social and economic development. Policies based on ecosystem services (ESs) and sustainable activities are able to reduce the marginalization of the local productions and to improve new economies fitted on the local resources. In this framework, the park's role is not only linked to the natural conservation aims but it is becoming critical for an innovative development, based on the huge nature role for the overall collective wealth. The Natural Capital value includes all benefits provided by nature to ecosystems and human habitats. It is a new concept that has to be considered developing green economy. Biodiversity and ecosystem status are critical to underpin the landscape resilience and to adapt the territorial systems to climate change. It is urgent to develop a policy framework that recognizes the interdependence of climate change, biodiversity, and ecosystem service and that implements concrete actions at any levels: global, national, and regional, based on the quality improvement of ecosystems and the ESs. Suitable indicators assessing the ecological quality of a landscape and the policy effectiveness should support the policy framework provided in order to restore ecological function to the overall system.

Keywords Ecological functions • Ecosystem services • Biodiversity • Landscape conservation

---

G. Gibelli (✉)
SIEP-IALE Italian Society of Landscape Ecology, Via Senato 45, Milan 20121, Italy
e-mail: gioiagibelli@gmail.com

R. Santolini
Department of Earth, Life and Environmental Sciences (DiSTeVA), Carlo Bo University of Urbino, scientific campus "Enrico Mattei", Urbino 61029, Italy
e-mail: riccardo.santolini@uniurb.it

R. Gambino, A. Peano (eds.), *Nature Policies and Landscape Policies*,
Urban and Landscape Perspectives 18, DOI 10.1007/978-3-319-05410-0_6

6.1 The Park: How?

Nowadays, the issue of managing protected areas is strategic and pivotal in the valorization and conservation of landscapes. The evident failure of several summits on climate change underlines the incapability of more powerful and more industrialized countries to solve with viable and sustainable actions a dramatic situation, which has not only economic but also social and environmental consequences.

> An economic development which does not consider the environment, not only runs the risk of not being able to be maintained, but fails in terms of quality and loses value. (Musu 2008)

The European level, at least, is concerned by some economic and territorial policies: the first OECD World Forum on "Statistics, Knowledge and Politics," in Palermo (2004), promotes the "Global Project on Measuring the Progress of Societies." The UN, the World Bank, and the European Commission also share such matters of the OECD. The Istanbul Declaration (2007) is committed in going "beyond the GDP." The document, "Not only GDP. Measuring progress in an evolving world," COM (2009) 433, Brussels, proposes some operational paths towards an integrated GDP with environmental and social indicators and their introduction into national accounts. This document was an answer to the European Commission, which had launched initiatives for the implementation and the assessment of community policies in 2007.

The national strategy for biodiversity underlines the necessity to integrate all the conservation policies with the active participation of all institutional entities and all stakeholders. Conservation and sustainable use of biodiversity is, indeed, a primary need to guarantee humanity a future and to preserve economic prosperity and wealth.

Consequently, there is a need to combine ecology and economy in a framework where natural conservation is an added value of local economic development. Then it is important to recognize ecosystem services (ESs) and sustainable activities, which face and reduce land marginalization and improve new economies fitted on the local resources.

In this framework, the park is designed not only as a natural conservation tool but also as a tool for social and economic development, by recognizing what nature does for the collective wealth. Developing green economy is an opportunity to consider the environmental consequences of a new production system in terms of ecosystem functioning and ecological and Natural Capital value alteration, in which biodiversity is a component that structures and conditions the whole landscape.

6.2 Biodiversity and Ecological Functionality

During the last years, many initiatives have been implemented in order to bring the biodiversity and the ecological functions assessment at the core of conservation and management strategies, the aim being the next planning choices (TEEB, The Economics of Ecosystems and Biodiversity; COPI, Cost of Policy Inaction; IPBES, Intergovernmental Science-Policy Platform on Biodiversity and Ecosystem Services).[1]

In this paradigm, the relationship between ecosystems and biodiversity is critical but evident in its effect on the ecosystem functioning (Balvanera et al. 2006). In an ecosystem the species diversity corresponds to the complexity of interactions among them and to the way that energy has to cross a community. Complexity, complementary behavior between the species and energy fluxes are linked to the ecosystems stability.

So the alteration of biodiversity (the change in the population and community parameters, determined by both direct and indirect factors and by changing landscapes) causes changes in the stability of ecosystem processes. According to Hooper et al. (2005), given species combinations are complementary when using resources and can increase the average productivity and retention rates of nutrients. Complementarity is one of the emerging properties needed by a community: the relationships among different species allow the reusing of a much wider set of resources than usual. This concept is valid also in landscape: the presence of some complementary different landscape elements inside a land unit is one of the landscape factors of resilience. Diversity at different level of scale is one of the ecosystem's properties key controls.

So, the specific richness and the functional role of a species are important elements in maintaining the ecosystem functionalities. Since species participate in different ecosystem functions and have an influence on their quality, more than one species are needed to keep an ecosystem fully active in a multifunctional way.

This concept is very useful in order to understand deeply the biodiversity levels needed to maintain a multifunctional ecosystem decreasing the ecosystems vulnerability dealing, for example, to climate change (Santolini 2010). If species of a given functional group operate at different scales in the same landscape, they contribute to increase the landscape resilience and to reduce competitiveness within the species of the same functions: in this way, the resilience effect due to the complementarity among species completes the resilience effects within the different scale levels. This is due to the partial overlapping of the ecological niches of the species (redundancy), which live at the same scale level (Peterson et al. 1998). As a consequence, by combining the functional redundancies of different scales, the ecological resilience is strengthened. De Bello et al. (2008) define a functional peculiarity as

[1] TEEB: www.teebweb.org; COPI: http://ec.europa.eu/environment/nature/biodiversity/economics/teeb_en.htm; IPBES: http://ipbes.net

a feature of an organism which has demonstrable links to the organism's function", that is related to its role in the ecosystem or its performance.

Even though it is easier to think of functional peculiarity at the species level, the notion can be also extended to groups of organisms with similar peculiarities and with similar effects and/or characteristics and also to landscape elements or groups of them. Functional diversity is considered as the type, the range, and the relative abundance of the functional characteristics of a community (or a landscape unit) with important consequences for ecosystem (and landscape) processes (De Bello 2008). This brings to consider the opportunity of using focus species (Lambeck 1997) or focus guilds. These are assessed even by the expert-based approach (Amici and Battisti 2009), efficient to determine green ecological infrastructures by means of a group or of groups of species playing a given ecological role. Such guilds are characterized by spatial and functional ecosystem needs and are able to successfully describe an ecological system with a given quality level, able to supply ESs. In fact, the functions that support ESs often depend on key species, guilds, or kinds of habitat (Kremen 2005). In this way it is possible to overcome the species-specific ecological net methodologies with a great contribute to the territorial governance.

More change it the spatial features of a land that provides ecosystems services, the more important is that those services are recognized. In fact, some services can be categorized in relation with their spatial and chorological characteristics (Costanza 2008) that could be related to a patch of the eco-mosaic (e.g., wood raw material) or to an area of geographical reference (e.g., water catchment). As a consequence, each patch of the mosaic has a potential in the ESs supplying capacity depending on its proper spatial characteristics; hence, variations in soil use can imply decreasing performances in the ESs providing (Burkhard et al. 2012; Scolozzi et al. 2012), thus making an ecological system, also a landscape unit, dystrophic and vulnerable.

The relationship between biodiversity and ecosystems functionality (Haines-Young and Potschin 2011) gives the opportunity to finding new monitoring ways of the landscapes characterized by multifunctional green infrastructures. These infrastructures are not only more convenient and suitable for natural life but they also could join either the land users' interests (farmers, foresters, tour operators, etc.) and the overall society ones by providing high-value ESs such as soil and water conservation, landscape quality, heritage services, and a lot of other services to the human habitat. This kind of services is growing a lot in importance also inside the metropolitan areas, where protected areas could be the resilience answers of the technological ecosystems.

6.3 Biodiversity and Landscape

An intense effort to integrate cultures and principles was made with the drafting of the European Landscape Convention (ELC), which suggests a common definition of landscape, which is an important conceptual and operational reference. Such a definition considers the idea that landscapes evolve by the time because of natural forces and of the action of human beings. Moreover, it points out the idea that the landscape is part of a whole, whose natural and cultural elements are simultaneously considered.

The strict relationship between landscape and biodiversity is quite clear in North European cultures. Otherwise in Italy, as far as the contents of the convention are concerned, it is generally necessary to highlight how innovative is to recall ecological and economic values, which are usually ignored both in theory and in practice (Gibelli 2008).

This cultural gap is partially due to the fact that, in Italy, inside the territorial governance processes, the landscape deserved very little importance. In general it comes into being at the end of the plan and project processes as a final "finishing touch" with the aim of improving the aesthetic quality of a given object. These strong aesthetic and cultural visions often drive landscape in the dimension of the triviality. This happens, especially, in a country where culture is hardly considered as a founding condition of civilization and development. On the contrary, the awareness of the interrelation between natural resources and landscape highlights the fundamental role that landscape plays towards sustainability, concept often recalled by the ELC, and that it should play in plans and programs. Within the latter, it should be considered as the departure point of a systemic knowledge and as the final quality objective of programming tools, the aim being that of guaranteeing adequate environmental quality levels and high standards of living for human populations.

Hence, nature and culture equally contribute in the evolution of landscapes. The natural components which are at the basis of ecosystems and of resources (rocks, waters, soils, vegetation, air, etc.) and the processes (water treatment, CO_2 fixation, soil preservation, etc.) are the matrix, though not perceived. Every landscape and its values (not the price) evolve from the relationships between that unperceived elements and from their status.

6.4 Landscape Conservation

Landscape revitalization should not just derive from the desire of restoring the landscape historical value and memory, but it should derive from a principle: landscape design must move from the aspects that can reduce its vulnerability and improve sustainability.

Disappearing habitats and fragmentation are considered all over the world as a key question, in relation with biological diversity conservation (IUCN 1980). The fragmentation process of natural environment is even considered as the first cause of biodiversity loss. Consequently, an operational response to such a problem was found in the paradigm of ecological networks. Ecological networks or green infrastructures are more and more enriched with multiple functions of landscape revitalization, conservation of stock resources belonging to Natural Capital, ESs providing, and other functions different from those of "mere" conservation of connectivity, for which they were created (Santolini 2012).

The main causes of the alterations are linear infrastructures and urban sprawl. The effects of urban sprawl on biodiversity are immediately evident. It is well known that urban sprawl causes radical changes of the landscape even if no important alterations of the territory occur. The issue of urban sprawl is an example of how in the long run the sum of phenomena at the local scale produces extremely problematic and unpredictable effects at higher scales if hierarchic relations among scales are ignored (see the first paragraph). That is why landscape overall structure plays such a vital role even dealing with sectorial and limited phenomena. Urban sprawl and land infrastructures break up the landscape structure, which, in turn, causes further and more complex effects on biodiversity (Gibelli 2011).

The extent of the phenomenon depends on numerous variables, which are linked to many structural aspects of the landscape, among which the landscape and environmental context, the environmental typologies, the extent and configuration of residual habitats, the degree of connectivity among them, the distance from other environmental typologies, and the time elapsed from the beginning of the process (Battisti 2004). However, urban sprawl is considered as another primary factor of biodiversity loss at the world scale, because of both direct effects (soil consumption) and indirect effects, with important consequences on the loss of ecological functions and ESs and on vulnerability increase.

Because of the abovementioned reasons nowadays, in many parts of Italy, protected areas are more and more isolated in an unsuitable matrix, which is often coupled by a progressive artificialization of functions (e.g., wet areas) and a consequent loss of functional links between them and with key biomes such as the sea. Moreover, the speed up of the territorial changes due to the current technology and huge availability of energy creates adaptive problems to natural systems and to the human ones, both adapted on biological rhythms (Gibelli 2003). At the same time, some mechanisms slow down or even stop, mainly because of fragmentation of management, which tends to crystallize spatial planning and to protect ecosystems at predefined successional levels (wet areas, falling woods, etc.). The resulting effect is the loss of the ecological functionality of ecosystems and of landscapes. Most of ESs are, indeed, threatened and with negative trends for the next 50 years, despite availability of ESs, is considered to be as an essential basis for human wealth and as a key factor for reducing poverty (MEA 2005).

Influencing factors of ecosystems and of their functionality are, indeed, a series of pressures, deriving from factors which are related with policies and technological development and which depend on expectations, consumption choices, and consequent behaviors. The field of intervention and control of such pressures is mainly regional and local. This means that government sectors and territorial planning are mutually responsible for developing actions aiming at:

1. Maintaining and/or restoring ecological and evolutionary processes, which stimulate or support biodiversity in the whole territory and not only in protected areas.
2. Protecting clusters of habitats and ecosystems large enough to maintain an adequate resilience of the system at a larger scale. This implies a good ability in reacting to disturbances and changes and event on the long run. This is possible thanks to clusters of diversified and interacting ecosystems, which reduce the vulnerability of the landscape and environmental system, by maintaining its functional effectiveness.

As a consequence, in order to protect the biodiversity of a territory, it is necessary to work at different scales and with an integrated approach. What was said about landscape vulnerability, spatial scales, speed modification, and adaptation times plays a remarkable role in relation with the effects of climate changes, which the tools for territorial governance need to start considering. Currently, there are two main approaches to diminishing change-related risks: reducing greenhouse gases and implementing adaptation strategies based on the reduction of the vulnerability of systems. The two strategies can be implemented at two levels: the first one at the very large level of both national and supranational policies and individual behaviors and the second one at the intermediate levels of territorial governance policies. These correspond to the possibilities offered by planning and by territorial governance tools in general, where protected areas play a key role in maintaining the critical Natural Capital (Gibelli 2011).

Clearly, results can be reached only through an overall commitment. This means that every stakeholder has to collaborate so as to build new environmental balances. Exploring the meaning of vulnerability of landscape systems is apparently a key sector in determining strategies and valuable adaptive criteria.

To this extent, it is worth recalling that one of the ten messages for 2010 of EEA (2009) is called "Global change and biodiversity." Its "key message" is the following:

> The variety of life underpins our social and economic wellbeing and will be increasingly an indispensable resource in the battle against climate change. However, our consumption and production patterns are depriving ecosystems of their capacity to withstand climate change and deliver the services we need from them. As we understand more about the ways that climate change is impacting biodiversity, it becomes clear that we cannot tackle the two crises separately. Their interdependence requires us to address them together.

The conservation of a landscape therefore calls the protection of the ecological functionality of its ecosystems, and today, it is essential that the interdependence among climate change, biodiversity conservation, and landscape quality is translated into concrete actions at all levels: global, national, and regional. According to EU WG (2009), maximizing synergies in different fields require to develop three types of actions:

- Maintaining and restoring the biodiversity and ecosystems those underpin resilience and the capability to mitigate and adapt to climate change. This includes building up our "green infrastructure."
- Developing a policy framework that recognizes the interdependence of climate change, biodiversity, and ecosystem services. Such a framework should facilitate cross-sectorial interaction, drawing in areas such as agriculture, forestry, and business, and also support further research.
- Use this cross-sectorial framework to design and implement concrete actions based on ecosystems and the ESs provided. Examples include developing soft coastal defenses and maintaining and restoring floodplains, vegetation cover, and green infrastructure in order to restore ecological function of the system, optimal indicator to assess the ecological quality of a landscape, and the effectiveness management of actions.

References

Amici V, Battisti C (2009) Selecting focal species in ecological network planning following an expert-based approach: a case study and a conceptual framework. Landsc Res 34:545–561

Balvanera P, Pfisterer AB, Buchmann N, He J-S, Nakashizuka T, Raffaelli D, Schmidt B (2006) Quantifying the evidence for biodiversity effects on ecosystem functioning and services. Ecol Lett 9:1146–1156

Battisti C (2004) Frammentazione ambientale, connettività, reti ecologiche. Un contributo teorico e metodologico con particolare riferimento alla fauna selvatica. Provincia di Roma, Assessorato alle Politiche agricole, ambientali e Protezione civile, Rome

Burkhard B, Kroll F, Nedkov S, Müller F (2012) Mapping ecosystem service supply, demand and budgets. Ecol Indic 21:17–29

Costanza R (2008) Ecosystem services: multiple classification systems are needed. Biol Conserv 141:350–352

De Bello F, Lavorel S, Díaz S, Harrington R, Bardgett R, Berg M, Cipriotti P, Cornelissen H, Feld C, Hering C, Martins da Silva P, Potts S, Sandin L, Sousa JS, Storkey J, Wardle D (2008) Functional traits underlie the delivery of ecosystem services across different trophic levels. Available at http://www.rubicode.net/rubicode/RUBICODE_Review_on_Traits.pdf. Accessed 7 Oct 2013

EEA (2009) 10 messages for 2010 Climate change and biodiversity, Copenhagen. Available at http://www.eea.europa.eu/publications/10-messages-for-2010/message-1-biodiversity-and-climate-change. Accessed 4 Nov 2013

EU WG (2009) Towards a strategy on climate change, ecosystem services and biodiversity. European Union Ad Hoc Expert Working Group on Biodiversity and Climate Change. Available at http://ec.europa.eu/environment/nature/pdf/discussion_paper_climate_change.pdf. Accessed 7 Dec 2009

Gibelli MG (2003) Il Paesaggio delle frange urbane. Franco Angeli, Milan

Gibelli MG (2008) Paesaggio e Paesaggi: tante definizioni per una parola sola. In: Teofili C, Clarino R (eds) Riconquistare il Paesaggio. La Convenzione Europea del Paesaggio e la conservazione della Biodiversità in Italia. WWF Italia ONG ONLUS, Ministero dell'Istruzione, dell'Università e della Ricerca, Rome, pp 108–123

Gibelli MG (2011) Paesaggio e biodiversità. Ri-Vista Ricerche per la progettazione del paesaggio, Firenze University Press, 15–16:5–21

Haines-Young R, Potschin M (2011) The links between biodiversity, ecosystem services and human well-being. In: Raffaelli DG, Frid CLJ (eds) Ecosystem ecology: a new synthesis. Cambridge University Press, Cambridge, UK, pp 110–139

Hooper DU, Chapin FS III, Ewel JJ, Hector A, Inchausti P, Lavorel S, Lawton JH, Lodge DM, Loreau M, Naeem S, Schmid B, Setälä H, Symstad AJ, Vandermeer J, Wardle DA (2005) Effects of biodiversity on ecosystem functioning: a consensus of current knowledge. Ecol Monogr 75:3–35

IUCN (1980) The world conservation strategy. IUCN/UNEP/WWF, Gland

Kremen C (2005) Managing ecosystem services: what do we need to know about their ecology? Ecol Lett 8:468–479

Lambeck RJ (1997) Focal species: a multi-species umbrella for nature conservation. Conserv Biol 11:849–856

Millennium Ecosystem Assessment (MEA) (2005) Ecosystem and human well-being: a framework for assessment. Island Press, Washington, DC

Musu I (2008) L'economia e la natura. In: Cadelo E (ed) Idea di Natura. Marsilio, Venezia, pp 69–88

Peterson G, Allen CR, Holling CS (1998) Ecological resilience, biodiversity, and scale. Ecosystems 1:6–18

Santolini R (2010) Biodiversità, servizi ecosistemici e prospettive nella gestione interamministrativa. In: Ferroni F, Romano B (eds) Biodiversità, consumo di suolo e reti ecologiche. La conservazione della natura nel governo del territorio. WWF Italia, Ministero dell'Università e della Ricerca Scientifica, Cogecstre Ed., pp 76–84

Santolini, R (2012) The indicators for the ecological network: indicators for monitoring of durable development in the Alpine regions. Platform "Ecological Network" of the Alpine Convention. Available at http://www.alpconv.org/en/organization/groups/WGEcologicalNetwork/Documents/20120628_Ecol_Network_Indicators_syntesis.pdf. Accessed 4 Nov 2013

Scolozzi R, Morri E, Santolini R (2012) Delphi-based change assessment in eco-system service values to support strategic spatial planning in Italian landscapes. Ecol Indic 21:134–144

Chapter 7
A Territorial Contradiction

Riccardo Guarino, Patrizia Menegoni, Sandro Pignatti, and Simone Tulumello

Abstract Spatial planning and environmental restoration are essential corollaries to the management of protected natural areas; however, without a sound awareness of the evolutionary consistency of biocoenoses, the harmonious integration between human activities and ecosystem preservation remains an unattainable utopia. The theorisation of a balanced welfare, inspired by the universal tendency of ecosystems to reach a steady state, has to go along with the defection from any economic greed.

Keywords Parks • Protected areas • Sustainability • Spatial planning • Human behaviour

The ever-increasing importance given to nature conservation in Europe in recent decades has brought to the setting up of a system of protected natural areas extended to 18 % of the EU territory, mostly thanks to the transposition and implementation of Directives 79/409/EEC and 92/43 EEC.[1]

Frequently, European protected areas have limited extension and are close to densely populated areas characterised by pervasive urbanisation and infrastructures. What is under protection in Europe is not a primordial nature, of which very

[1] http://ec.europa.eu/environment/nature/index_en.htm, http://natura2000.eea.europa.eu/

R. Guarino (✉)
Dipartimento STEBICEF, Università di Palermo, Via Archirafi 38 I, 90123 Palermo, Italy
e-mail: guarinotro@hotmail.com

P. Menegoni
ENEA, Unità Tecnica AGRI-ECO, Via Anguillarese 301, 00123 Rome, Italy
e-mail: patrizia.menegoni@casaccia.enea.it

S. Pignatti
Forum Plinianum, International Association for Biodiversity and System Ecology, Via Lavinio 22, Rome 00183, Italy
e-mail: sandro.pignatti@gmail.com

S. Tulumello
Instituto de Ciências Sociais, Universidade de Lisboa, Avenida Professor Aníbal Bettencourt 9, Lisboa 1600-189, Portugal
e-mail: simone.tulumello@ics.ul.pt

R. Gambino, A. Peano (eds.), *Nature Policies and Landscape Policies*, Urban and Landscape Perspectives 18, DOI 10.1007/978-3-319-05410-0_7

few traces remain, but the still-surviving elements of a traditional cultural landscape, rich in natural features of which the establishment of protected areas try to salvage the most significant relicts.

The new EU policies consider natural areas as a resource to be managed through measures and initiatives aiming not only to preserve biodiversity but also to meet the demands of local people, in order to ensure the best compromise between ecosystem integrity and socio-economic development (Petermann and Ssymank 2007). The new managerial paradigm is therefore based on a collaborative approach, agreed and shared by local communities along with all the other stakeholders.

Unfortunately, it is very difficult to find an optimal balance in the expectations of those who propose, who use and who manage protected areas. The risk is to invest resources in protecting and perpetuating what we like most, making a sort of 'large-scale gardening' – gardening at the scale of landscapes – often at odds with all natural processes and dynamics, such as the shrub encroachment in abandoned rangelands, a frequently observed natural process throughout Europe, that is causing the rarefaction of orchids very dear to man.

As often happens, it is necessary to establish priorities and to make choices. Nonetheless, about biodiversity, real and perceived, as well as about advisability and effectiveness of actions taken to protect it, there is a great variety of opinions that makes difficult the implementation of programs and the evaluation of results, also due to some confusion of roles between ecologists and planners (Guarino et al. 2011).

Knowledge about vegetation, ecoregions and ecosystem relationships is an essential element in planning and land management (Pignatti 1994, 1995; Blasi and Paolella 1995; Biondi 2007). It is necessary to understand where and, more importantly, how much it costs (in terms of 'environmental sustainability') to invest energy and resources to counter the natural dynamics.

It is not always given due importance to this knowledge base, and it often happens that, in choosing the management strategies, the impact on employment of the 'interventionist' approach is preferred without considering that nature, in order to remain such, should not be excessively subject to the deterministic control by man.

However, outside protected areas, weeding of roadsides and cultivations is practiced without hesitation; the continuity between trophic ecosystems and agro-systems is compromised in order to promote all that is functional to the production system in the global market. Not even the management and conservation of protected areas escape the market rules and require, thus, the availability of resources to invest. This brings us to the obvious contradiction that to safeguard very limited portions of the planet, the remaining areas are exploited with increasing intensity (Guarino and Pignatti 2010).

As previously mentioned, European protected areas have a strongly 'urban' character; this stimulates a constant search for innovative solutions in the management of such complex areas. The will to protect not mediated by a thorough and dispassionate understanding of ecosystems can easily run into errors or end up to mainly accommodate the requests of those who look at nature protection primarily in an economic-productive way, demanding guarantees, benefits and services.

7.1 Protected Areas or Theme Parks?

As we have seen, the protection of nature in Europe is inextricably linked to the preservation of surviving fragments of our collective past, an ancestry from which we freed thanks to the recent technological and socio-economic development. Like the historical city centres, which are protected and restored to last over time, even the protected areas are often subject to maintenance and conservative restoration. Significant differences exist between pre- and post-industrial cities, based on the juxtaposition city/nature and on the concept of the urbanised area itself. In the past, cities were a closed entity opposed to the res nullius of the outer territory (Salzano 1998). When cities were surrounded by walls, the unknown, the unknowable and the unpredictable were kept outside. In more recent times, urban expansion and population growth have gradually blurred the city boundaries, until, in the post-industrial urban sprawl, the res nullius has come to penetrate the city itself, along with a functional complexity that has made us accustomed to use, but not to know, and much less to control, many items and spaces of our daily lives.

While in the past we were frightened by what was outside the city, currently, it is the specialisation – and functional segregation – of the modern urban space that intimidate us (Ellin 1996): a progressive occupation of physical space, unable to build the city. Among the areas that – at least in appearance – are still relatively immune to such contamination, there are the natural spaces, which can be seen as a belated acceptance of the devastation wrought by the territorial city: the wildlife reserves of modern Europe, however great, can be well interpreted as recreational appendages of urban spaces. They are used by most of the people to relax, to do a little exercise, to visit unusual places, to buy local products and to imagine how it was in the past.

Oddly, the establishment of protected natural areas, which occurred over the past two decades at an unprecedented rate, is contemporary to the need of creating 'newly urban' spaces. Think of the malls, entities that assert themselves as public venues similar to cities but without their flaws: safe, reassuring and with an easily recognisable spatial and functional organisation. Consider the reaction of the urban centres to the processes of gentrification that replicate the characteristics of a mall through a 'renewal' based on urban marketing and surveillance systems.

Sometimes, these spaces are built from scratch: City Walk is a pedestrian and commercial area built in the 1990s in Los Angeles, 'an urban area painstakingly reproduced (even to the extreme of wedging candy wrappers into the pavement [. . .]) and idealized because it wants to be the best essence of the city, completely free from the violence of Los Angeles' (Codeluppi 2000, our translation)'.

In other cases, are the historical centres to be modified according to profit-oriented models. Thus, many Italian towns have seen cleared their social fabric,

Fig. 7.1 Palazzo del Gran Cancelliere, Palermo

replaced by a space tailored to tourism requirements. Connected to these processes is the falsification of historical spaces, pushed towards aesthetic stereotypes consistent with their commercial role. It is an example – perhaps unintended – the PPE (detailed executive plan) for the historic centre of Palermo released in 1989, which, in reaction to some types of urban speculation, requires the accurate reconstruction of entire blocks and is populating the city of architectures that are historically fake like the Palace of the Grand Chancellor in the homonymous square (Fig. 7.1), which seems a restoration but it is an almost entirely new building.

Even more complex is the situation of Venice: the city was not developed 'against' the surrounding environment, which for a millennium has been maintained as a necessary enclosure for the city, and provided food resources (fisheries) and safety from external attacks. Between the city and the lagoon has remained an interactive relationship (just think of the importance of tides) that man has changed over the centuries with the diverting of rivers flowing into the lagoon and the consolidation of the lidos. This has allowed the development of Venice as a political and commercial centre, the development of the first industrial complex (the Arsenal) and a thriving culture. Over the past two centuries, the city has lost these features and in recent years much of the population has migrated to the mainland, while the lagoon has been progressively depleted by erosion and pollution. In this way, the old balance between the town and the lagoon is lost: both are now (for various reasons) protected areas, but the cultural and commercial meaning of the first and the natural one of the second are being upset.

The metaphor we have built seems to reveal a sad fate: protected areas, whether they are natural parks, historical centres or quaint villages, are pushed – unknowingly? – towards a 'productive' function; the object to be protected becomes a valuable frame within which to develop employment and investment, tourism and territorial marketing. In this context, visitors become users/consumers: they usually reserve to the frame a rather superficial aesthetic/contemplative evaluation and assess their experience mainly on the quality of services offered by the administrators. This new realm is a city of simulations, television city and the city as a theme park (Sorkin 1992).

The 'sanctuary' (in Russian: *Zapovednik*) is an exception to this general trend and, as a natural environment protected *erga omnes*, should be considered a positive example, although elitist and expensive, because it requires a difficult management (control of herbivores, biodiversity monitoring, etc.), which often clashes with the reluctance of administrators and public opinion to accept the non-usability of areas that, to remain such, require maintenance patrolling and monitoring costs (Sessions 1995; Boreiko et al. 2013).

7.2 Pandemic Park Foundations

So many parks have been recently founded all over the world! Some example are (Google search: 'park' 2013): national, regional, pelagic, river, mountain, valley, wildlife, urban, public, cultural, school, college, music, literary, research, technological, archaeological, Jurassic, safari, amusement, recreational, commercial, private, pocket, wind, solar, car and even sushi park! Despite their diverse nature, all these areas share an implicit 'need' for protection, fence, boundary and sectoriality. According to Diez (1853), the term 'park' derives from the Latin word *parcere* (i.e. to impede): the place where wild animals of every kind are locked up, in order to take delight in hunting at any time. According to others, the term derives from the ancient German word *berkan* (modern: *bergen*): to cover, to save and to defend. In fact, the word *perku* already existed in Akkadian, with the meaning of defence, frontier and barrage. In connection with these concepts is the root 'pork' (in Latin: *porcus*), originally indicating the enclosure, the courtyard where the domestic pig (in Latin: *sus*) was kept and later designating the beast itself. The *porcus* stands clearly out from *aper* (i.e. the wild boar), which is the same beast but lives in open spaces, in freedom.

Our history of supporters or detractors of parks is largely based on the metaphorical contrast between a pig living in a closed, fenced and protected place and a wild boar routing in the forest without supervision. The pig, symbol of the rational use of animal breeding, has originated from the clever domestication of a wild boar. Similarly, the park, a protected place, is the outcome of a metaphorical domestication of Dante's forest 'savage, rough, and stern/which in the very thought renews the fear'.[2] The pristine nature, reduced to a paltry fragments, does not more induce awe but inspires a protective instinct. In the modern city, men undergo an inexorable fascination towards nature, and the greater the fascination, the stronger the process of alienation against it. The Italian writer Calvino (1963) has masterfully represented such fascination in the short stories of Marcovaldo:

[2] http://www.worldofdante.org

The Marcovaldo's love for nature can only be felt by a city man (...) Dad -the children said- are the cows like trams? Do they make stops? Where is the terminus of the cows? (Calvino 1963, our translation).

7.3 Towards a Participated Landscape

Beauty and harmony of nature, together with its efficiency, have inspired most of speculative thinking and art forms that have marked the human history. Human nature and its technical and cultural expressions mirror the complexity of the phenomenon of life. Through the centuries, rural communities have managed their environment and farmed the land in their own natural way, creating a rich diversity of landscapes, choral representation of historical identity of the territory and cultural human heritage (Fig. 7.2). We now tend to recognise in that model of development the precursor of 'sustainability'.

In the past, even the human welfare was associated with a balanced and durable state of satisfaction, inspired to the ecological concept of climax. The ἀταραξία of the Greeks and the *otium* of the Latins are expressions of a pleasure to be enjoyed noting wisely the satisfaction not of one's own desires, but of his own needs. Modern man has redefined the perception of welfare and simplified its semantic breadth: all parameters are set on the purchasing power of goods, products and services that in many cases are necessary just because they are depicted as such by the new global socio-economic order. Paradigm for this change is the gradual shift from the theorisation of a balanced welfare, inspired by the universal tendency of ecosystems to reach a steady state, towards an incremental and bulimic welfare, no longer inspired by nature, but fuelled by its devastation. In doing so, the speculative power of analytical thinking has been equally simplified and increasingly bound to the binary logic of cost/benefit analyses (Menegoni et al. 2011). Cheap and pervasive information services broadcast this new concept of welfare, emphasising in the popular imagination the gap between the 'polluted' places of our everyday life and the 'intact' places of protected areas.

Fig. 7.2 Cultural landscapes of Campagna Iblea (*left*) and Valle d'Itria (*right*).

To overcome this contradiction, it is necessary to design new logistic networks, integrated on a local scale. We urgently need a planning that links the man to his territory and not the restorer to his object. These aims are achievable only if we are able to put every single man in a new position of awareness and responsibility.

The spaces to be (re)planned will no longer be, as they were in pre-industrial times, the result of unconscious, choral, attempts to best use land, resources and local materials. They will be, instead, the result of a planning well integrated to the social context and to the strategic sharing of ideals and models alternative to those of consumerism and of the global market. So, not the return to an edenic, pre-industrial world, but the evolution from a world centralised by the global economy towards a world where global technologies and knowledge will be used to boost local economies, to emphasise the local diversities and to encourage the decongestion of the trade routes that underpin the current human habits, linked to products and services standardised on a national and, increasingly, continental scale. To do this, the planner cannot ignore the political value of acting on behalf of an ethical necessity, imposed by the not sustainable environmental and social costs of current consumption patterns.

Under this perspective, even the 'sanctuary' takes on a new meaning: it does not only matter for the rarity or the particular aspect of species and vegetation layers but also for its value as an ethical model, a physical space where an efficient and optimal balance is established between the external factors (climate and soil) and the local communities (bacteria, plants, animals), a living example of self-organised order, able to maintain and preserve in a steady state all the ecosystem functions which are needed also by the human species. The tools to convey this message are the virtual channels of the web and the mass media that the new planners should learn to use with skill at least equal to that of those who use them as catalysts of global consumption patterns. The physical elements of the new landscapes will be much stronger the greater the number of people who believes in and supports a re-localisation of consumption habits and particularly of those related to the human nutrition. The new landscapes will be more durable the greater the number of people who will use their free time to set up the network of collaboration and proactive interaction that is functional to the development and maintenance of a participated, unmediated and alive cultural landscape (Guarino and Menegoni 2010). If most of us will keep on spending our free time in malls, spas and television, land protection in an integrated and systemic view risks being perceived as yet another action to share passively, to be supported by providing a small contribution money, without changing our habits too. In this way, we will not go very far.

The new way of planning should be social and territorial at the same time: if the aim is to promote, not just for aesthetic reasons, more sustainable landscapes, we should be able to recognise in the parsimony of our ancestors the precursor of the moral and personal commitment of modern innovators. A parsimony no longer imposed, as in the past, by poverty and limited resources, but by the awareness of how gross – and inefficient from environmental and thermodynamic standpoint – is to consume products whose packaging and transport costs outweigh the production costs (Patel 2009).

The challenge goes far beyond the ability to redesign the territory: it lies in making desirable a sober lifestyle, aware of the environmental consequences of all our actions; it lies in making choices oriented to the re-territorialisation, i.e. the downsizing and the localisation of production districts, in close proximity to trading posts and disposal places; and it lies in favouring the most direct relationship between production and consumption.

References

Biondi E (2007) Paesaggio, biodiversità e sviluppo sostenibile: il case delle praterie appenniniche. In: Finco A (ed) Ambiente, Paesaggio e Biodiversità nelle politiche di sviluppo rurale. Aracne, Rome

Blasi C, Paolella A (1995) Progettazione ambientale. La Nuova Italia Scientifica, Rome

Boreiko V, Parnikoza I, Burkovskiy A (2013) Absolute "zapovednost'" – a concept of wildlife protection for the 21st century. Bull Eur Grassland Group 19–20:25–30

Calvino I (1963) Marcovaldo, ovvero Le stagioni in città. Einaudi, Turin

Codeluppi V (2000) Lo spettacolo della merce. I luoghi del consumo dai passages a Disney World. Bompiani, Milan

Diez CF (1853) Etymologisches Wörterbuch der romanischen Sprachen. Scheeler, Bonn

Ellin N (1996) Postmodern urbanism. Blackwell, Cambridge

Guarino R, Menegoni P (2010) Paesaggi marginali e paesaggi mediati. Ecosc 3:32–33

Guarino R, Pignatti S (2010) Diversitas and biodiversity: the roots of a 21st century myth. Rend Lincei – Sci Fis Nat 20(4):351–357

Guarino R, Bazan G, Marino P (2011) La sindrome delle aree protette. In: Pignatti S (ed) Aree Protette e Ricerca Scientifica. ETS, Pisa, pp 143–158

Menegoni P, Guarino R, Pignatti S (2011) Economia, ecologia e tecnologia: riflessioni su una convivenza difficile. Naturalmente – Fatti e Trame delle Sci 24(2):8–12

Patel R (2009) The value of nothing: how to reshape market society and redefine democracy. Picador, London. Italian edition: Patel R (2010) Il valore delle cose e le illusioni del capitalismo (trans. Oliveri A). Feltrinelli, Milan

Petermann J, Ssymank A (2007) Natura 2000 and its implications for the protection of plant syntaxa in Germany, with a case-study on grasslands. Ann Bot (Rome) 7:5–18

Pignatti S (1994) Ecologia del paesaggio. UTET, Turin

Pignatti S (1995) L'ecosistema urbano. Accad. Naz. Lincei XXI Semin. Evoluz. Biol. pp 137–167

Salzano E (1998) Fondamenti di urbanistica. Laterza, Rome-Bari

Sessions G (1995) Deep ecology for the 21st century. Shambhala, Boston

Sorkin M (ed) (1992) Variations on a Theme Park: the New American City and the end of the public space. Hill and Wang, New York

Chapter 8
Legal Frameworks for Nature Conservation and Landscape Protection

Carlo Desideri

Abstract In a brief period, a new paradigm for a comprehensive approach to nature conservation and landscape protection has been put on the agenda, mainly through developments of international and European law. The implementation of the new paradigm by national states requires a common language but not a uniformity of policy patterns. Existing legislative differences between European states not only show different degrees of evolution but also a wealth of cultures and experiences that can be turned 'into strength'. Two elements of a comprehensive approach are appearing to emerge. The first one concerns the need for national governance both for assuring the guiding and monitoring of policies to be undertaken and for the definition of efficient nationwide principles and standards. The second element concerns the need for the largest possible sharing of responsibility. To this end the people's awareness and participation, environmental agreements, new rights and duties and the use of property to assure protection can all provide great help. In any case, the task to be accomplished is very hard. The development of the modern environmental law is a very short-term experience with respect to power structures and legal schemes, which are often very ancient and deeply rooted but unsound in regard to the needs of the new paradigm.

Keywords Biodiversity conservation • Landscape protection • Nature conservation • National governance • Legal framework • Environmental agreements • Environmental rights • Environmental duties

8.1 The New Paradigms: An introduction

Modern policies for nature conservation and landscape protection began to be set up recently, after the Second World War. Previously some important initiatives had also taken place in European countries, but aesthetic concerns had been predominant and the protection had been mainly concentrated on delimited scenic landscapes and areas of natural beauty.

C. Desideri (✉)
Institute for the Study of Regionalism Federalism and Self-Government, ISSiRFA-CNR,
Via dei Taurini 19, Rome 00185, Italy
e-mail: desideri.carlo@gmail.com

R. Gambino, A. Peano (eds.), *Nature Policies and Landscape Policies*,
Urban and Landscape Perspectives 18, DOI 10.1007/978-3-319-05410-0_8

Even more recently there have been further innovations in international law. In particular the Convention on Biological Diversity (1992) introduced new ideas and new principles intended to stimulate a deep rethinking of the policies on nature conservation. In Europe, this innovation has proceeded in a significant way since the setting up in 1992 of Network Nature 2000 (Bromley 1997); this happened while the European environmental law was also improving and some important principles – such as prevention, precaution, integration and subsidiary principle – were established (Maastricht Treaty 1992). A first essential feature of the new nature conservation policy is that it is now based on the aim of conservation of biodiversity which is a more precise concept than the previous one of generic 'nature'. This also means that the new policy must find its legitimisation and guidelines in a scientific context. The other essential feature is that a systematic approach is needed. This doesn't eliminate the rationale for having specific protected areas. But the construction of a network of sites with attention to the 'ecological connectivity' between them must be the key rationale of the conservation policy. It is also worth noting that, according to the Habitat Directive (2000), all factors which could have a negative impact on the conservation of species and habitats must be considered, even if they are external to the perimeter of classified sites. In the same direction, the EU has recently proposed the launching of 'green infrastructures' (European Commission 2013), conceived as natural or quasi natural areas capable of offering 'ecosystemic services' (biodiversity is one of them). Moreover, environmental objectives, and biodiversity conservation in particular, are now seen as the strongest if not the only reason which in the immediate future can justify the European agricultural policy.

In short, biodiversity conservation requires an approach which must make reference to the entire territory and no longer to specific delimited areas (European Commission 2011). Here this new approach – that necessarily also involves landscape – appears to be consistent with the new vision and strategy proposed by the European Landscape Convention (2000). The aim of the Convention in fact is to definitely overcome the traditional idea of landscape viewed only as a matter of aesthetics, consequently protecting only delimited scenic landscapes. Its aim is instead to promote policies capable of taking care of all types of landscapes: those of special value, those of ordinary life and even those which are degraded and must be restored. The Convention doesn't give any directions about nature conservation in particular. But that a new paradigm capable of integrating nature and landscape is needed is already clear in the definition of landscape.

As consequences of the above-mentioned innovations, two other essential elements must be considered. The first one is that the integration of policies, which were conceived and built in the past as if they were different sectors, is now an essential feature of the new paradigm. This is the case of the relationships between nature conservation and landscape protection policies, and in turn their connection with land planning and management, as well as policies for agriculture, forestry, fishery and others if they are relevant. The second element is that legal provisions for policy settings and implementation can't be limited to the traditional 'command and control' scheme. In fact the raising of public awareness and consent and the sharing of responsibility between all the actors involved – not only public authorities, but individuals, groups and private organisations – are now considered essential factors for success. This also requires a variety of tools, initiatives and measures which should be applied in an elastic way in accordance with places and situations.

In conclusion, in a very few years, a rapid change of paradigm has been put on the agenda mainly through developments in international and European law. In the face of this rapid change, there are however national states and legislative systems which were built and consolidated before the environmental issues emerged. Moreover, the first legislative schemes for nature conservation and landscape protection adopted by the individual states are often different and also have different degrees of development and of effectiveness. Differences also exist of course between the individual countries in terms of public sensitivity and attitudes towards the care for nature and landscape.

8.2 Legal and Policy Patterns in European States: Some Examples

Here it is possible only to give a rapid glance at the differences between some European states. This can help us to detect some difficulties and problems that the new paradigm could encounter, as well as to see that there are various types of solutions and tools already at work, which could be useful. Legislative differences, as well as territorial and cultural ones, don't necessarily represent a critical point. They may be useful and turned "into strength" (European Commission 2008).

First of all, differences can be seen in the relevant provisions of the constitutions of the individual states. So, for example, in the Italian Constitution, there is a fundamental principle (article 9) which still belongs to the original text of 1948 and which establishes that the Republic must protect 'the landscape and the historic and artistic heritage of the Nation'. Instead 'the safeguarding of the environment [and] of the ecosystem' appears only as an item of the list of the reserved legislative powers of the state, since the constitutional reform of 2001. In Germany, in the Basic Law, there is the fundamental principle (article 20a introduced in 1994 and modified in 2002) according to which 'mindful also of its responsibility towards future generations the state shall protect the natural foundations of life and animals'. The example of France is also different in that the *Charte de l'environnement*, which was integrated into the Constitution in 2005, establishes the right of everyone to live in a balanced environment, which respects health conditions. In addition the *Charte* places great emphasis on duties towards the environment charged both to public authorities and to individuals. These three examples show different degrees of evolution of the constitutional provisions and that different concepts are used, perhaps suggesting different points of view in particular about the relevance given to nature and its components.

In the second place, differences between European states also emerge in the provisions of the current legislation. In this respect, the legislative development in Germany appears to be an important example. Here already the Federal Nature Conservation Act of 1976 established the legal basis for a comprehensive approach to nature conservation and landscape protection, as well as mechanisms directed towards its integration in land planning. It also contains provisions on the classification and management of protected areas. After a revision in 2002, which also provided for integrating into the Act some provisions on agriculture and forestry, a large reform was recently introduced with the new Act in 2010. In France, several laws were adopted with regard to nature conservation (1976), safeguarding of the coast (1986) and landscape protection (1993). A recent law of 2009 establishes new measures for biodiversity conservation with the creation of 'green infrastructures' to ensure 'ecological connectivity'. These laws are characterised in large part by provisions conceived as being integrative of the legislation on land planning. As for protected areas, after a 1960 law, new provisions on national parks were established by the more recent law of 2006. A comprehensive legislative approach to nature conservation and landscape protection is assured by the *Code de l'environnement*, developed since 2000. But some relevant aspects are also in the *Code de l'urbanisme*. Both *Codes* are constantly brought up to date. By and large the first *Code* contains the general concepts and principles and some specific provisions on protected areas; the second one contains detailed provisions directed to integrate nature conservation and landscape protection in the system of land planning and management. Italy offers an example of legislation which still strongly appears tied to individual sectors. A law on land planning and management was adopted in 1942 and has never been revised. At its side legislation that referred to landscape was developed, which in a first period considered only specific areas and later, since 1985, has been extended to larger parts of the national territory, but without succeeding in changing the regulatory tools, which in practice remained those originally intended for limited areas, and without a real integration with land planning. A general law on nature conservation doesn't exist. However, an important law on protected areas was adopted in 1991 (with minor modifications introduced later).

In the third place, differences exist with regard to the administrative pattern in the concerned matters. It is easy to notice the parallelism with the evolution of the legislative systems. In fact the states which have been most engaged in the development of a comprehensive legislation in general have also adopted (Germany, France) a pattern of unified ministerial responsibility on such matters as nature conservation, landscape protection and land management (also considering the development of infrastructures). In Italy, instead individual ministerial responsibilities still exist for the environment (where nature conservation is also considered) on one hand and the landscape, together with cultural heritage, on the other hand. The administrative reorganisation in Great Britain appears to be an important example of evolution in response to the emergence of the new paradigm. In fact a new agency, Natural England, was created for English territory by the Natural Environment and Rural Communities Act of 2006, which unifies tasks which were before divided between the Countryside Commission (Agency since 1999) responsible for the landscape and English Nature responsible for the nature conservation. In Germany, since 1993, the Federal Agency for Nature Conservation has also been responsible for a large range of activities about nature conservation and landscape protection and management. Both the English and the German Agencies, which have an independent status, have important tasks of scientific relevance.

8.3 A Tentative Legal Framework

At this point it is possible to sketch the profile of a comprehensive legal framework for nature conservation and landscape protection, even if in a very synthetic way. This profile can be carried out on the basis of the new paradigms developed in international and European laws as well as of some developments which occurred in domestic legislative systems of individual European states, with two warnings.

First. It is obvious that the new paradigm can, and must, work as a stimulus to renovate national legislations and to improve effectiveness whenever necessary. Moreover, it is also very important that a common language be adopted. This is necessary for the consistency of individual state policies with respect to the shared ends and for a real assessment of implementation processes and their results. However, this doesn't also mean that a uniform policy pattern should be adopted by the states. Against this there are not only the existence of different legislative traditions and, not least, of different cultural identities and public sensitivities. There is mainly the fact that the new paradigm itself must be applied in an elastic way, giving space to adaptable solutions based on experiences and values present in the various countries and places.

Second. The task to be accomplished is very hard. It is true that especially in some European countries a more rapid evolution can be noticed. But the problems to be tackled are still enormous: for instance, Europe didn't reach its 2010 objective for biodiversity conservation; soil consumption still appears unstoppable. The development of modern environmental law is very recent, and on the other hand many current power structures, legal schemes and connected behaviours which are unsound in respect to nature and landscape, and which must be modified and corrected, are often very ancient and deeply rooted, having been consolidated in the course of centuries in a world which was very different from the present one. Just think of the idea of the existence of unlimited resources disposable for human beings. To tackle such problems as the loss of biodiversity and climate change, which have accelerated in a very short space of time, now the search for new legal solutions and new policies is a very urgent need. If this urgency does not rapidly become a question of common sense, we could ask ourselves if the needed changes will arrive on time.

Looking now at the fundamental elements of a comprehensive legal framework for nature conservation and landscape protection, they appear to be two apparently contradictory aspects; but in reality they are complementary.

The first element is the need for a policy governance in the individual states at the national level, other than obviously at the international and European levels. The conservation of biodiversity is a global issue with a scientific rationale, and states are asked to take responsibility for it. Two different examples illustrate the importance of the national governance. The first one concerns the fact that a policy which invokes a scientific basis and is looking for a systematic approach needs a national dedicated and eminent structure which is independent in respect to short-term political considerations. This structure must be capable of giving strong support to decisions at every level and of assuring an objective review of policy implementation and of its results. In this perspective qualified agencies such as those operating in some European countries in regard to nature conservation and landscape protection appear to be an example of a positive solution concerning the issues to be confronted. The second example concerns the need to limit the still strong consumption of soil and the growing landscape fragmentation caused by urban and infrastructural development, even if at different levels of intensity in individual European countries. To face this issue through the application of national principles and standards is unavoidable when a new paradigm involving the entire territory is adopted. The experiences of some states already offer examples: in France the legal principles establishing that every development must happen only "in continuity" with the existing urban settlements and in Germany the principle that priority must be given to the "reuse" of already developed areas and, since 1998, the strict standard fixing a quantitative objective of territory consumption. There is currently a debate in Italy as well on the need to introduce a national standard.

The second element is the need to have as much sharing as possible of awareness and responsibility, together with the mobilisation of all the forces and experiences which can be appreciated as being valid and useful. This element is complementary to the first one, in the light of the subsidiary principle. In this regard, the powers and tasks of the regions or of the federal member states, as well as those of local governments, must be considered. As is well known, history shows that there is always an oscillation between the devolution of powers in some periods and centralisation in others. For instance, centralisation appears to be prevalent now in Germany and in Italy, with special regard to environmental matters. In any case regions, federal member states and local governments still remain very important for the implementation of the new paradigm, taking into consideration the powers they have in respect to land planning and often in such relevant sectors as agriculture, tourism, infrastructures, etc. However, it is also very important to mobilise individuals, groups, farmers, nature users, communities, private organisations and in general all relevant stakeholders. In this regard all new strategies insist first of all on the increase of public awareness, on training and education and on participation of people. In second place, they want to stimulate and support sound forms of behaviour, which not only respect nature and landscape but – and here is the main novelty – take affirmative action for their maintenance. This appears the fundamental aim, for instance, of the environmental agreements, which are suggested by the Habitat Directive and are already widespread in such countries as Germany,

Austria and France, but not in Italy (EEA 2012, pp. 96–99). And it is also the aim of the agreements contemplated for the implementation of agro-environmental measures by the European agricultural policy. In addition the recent German Federal Nature Conservation Act of 2010 gives priority, whenever possible, to 'contractual agreements'. 'Command and control' don't disappear, but their role is downsized with respect to a new approach directed at enabling individuals to behave as true agents of nature conservation and landscape protection. The rights and duties towards the environment contemplated by the French constitution as well as the statement of the quoted 2010 German Act that every individual should contribute "to the realization of the purposes of nature conservation and landscape management" can also be seen in the same perspective. A growing relevance of the approaches based on consensus and the sharing of responsibility is also evident in the evolution of legal schemes for protected natural areas. This is the case, for instance, of the Italian legislation, and more recently of the new 2006 French law on national parks, which introduced the *Charte du parc* and some connected agreements for identifying the area of the park, for establishing the shared objectives and the management guidelines. Finally, it is worth noting the importance of protection experiences on the part of specialised subjects that manage sites of natural and cultural relevance which they purchased on the market or received in legacies or as gift (Desideri and Imparato 2005). The most famous example is the National Trust which now manages more than 270,000 ha in England, Wales and North Ireland (a similar Trust does exist for Scotland). In addiction this Trust has more than three million supporters and operates thanks to the help of thousands of volunteers. Moreover many farms exist in the land managed by the Trust operating on the basis of environmental agreements signed with the Trust itself. The Trust is a private organisation (a registered charity) to which since 1907 the law has given the privilege to declare with binding legal effect which of its properties cannot be sold or expropriated. Some other private organisations, such as the *Conservatoires d'espaces naturels* in France and the *Fondo Ambiente Italiano (FAI)* and the WWF in Italy, act in ways which are analogous but are not as large or relevant as the National Trust. Instead the *Conservatoire de l'espace littoral* in France is legally a public body which has been in existence since 1976 with the mission of managing natural coastal areas that are its property and were mainly purchased in the market, even if the Conservatoire also has the power to expropriate private properties and preemption rights. It is worth noting that the Conservatoire normally operates by giving the management of its sites to local authorities, farmers and some private organisations on the basis of agreements.

In conclusion the national governance of the system and the sharing of responsibility, mobilising all the valid and available forces, should go hand in hand for the real accomplishment of the hard undertaking that is now on the agenda.

References

Bromley P (1997) Nature conservation in Europe. Policy and practice. E & FN Spon, London

Desideri C, Imparato E (2005) Beni ambientali e proprietà. I casi del National Trust e del Conservatoire de l'espace littoral. Giuffrè Editore, Milan

European Commission (2008) Green paper on territorial cohesion, 6.10.2008 Com (2008) 616 final

European Commission (2011) Our life insurance, our natural capital: EU biodiversity strategy to 2020, 3.5.2011 Com (2011) 244 final

European Commission (2013) Green infrastructure-enhancing Europe's natural capital, 6.5.2013 Com (2013) 249 final

European Environmental Agency (EEA) (2012) Protected areas in Europe-an overview. Report n.5/2012

Chapter 9
Beyond Gardens and Nature Reserves: Contemporaneous Landscapes

Jennifer Buyck and Teodoro C. Vales

Abstract There have been many attempts at providing definitions of landscape, and many points of view have been expressed. The word *paysage* is not based on a latin one, which explains the lack of consensus as regards its meaning. The word appears to be based on the term *pays*, which refers to a physical environment. The *Littré* dictionary mentions that *paysage* (landscape) is both "an area of flat land which can be observed" and "a country-style type of painting." The word therefore has several meanings and is very widely used. But what landscape are we talking about at a time where everything is part of the landscape? The word is overused and this results in confusion. What role did the *pays* play in the invention of *paysage*? What is the place of nature in it? What is the relationship between man and landscape?

Keywords Landscape • Contemporaneous landscape • Landscape concept • Cultural heritage • Nature

9.1 About Landscapes

The word and notion *paysage* is believed to have originated in the Netherlands, around 1415. The term was then used in Italy, where it was influenced by works related to perspective. Until then, landscape was only used as a scene. It later became the main object of paintings. The status of landscape progressively changed (Mérot 2009). Originally, landscapes were visible through windows and were part of indoor scenes. Windows and views then absorbed the rest of the paintings to the point where they actually became the paintings.

As a framing device, the window helps to understand the origin of the word *paysage*. *Pagus* is used to delimitate, to fix, to anchor. As a consequence, it is doubly inscribed in reality through the observed environment and through the spectator who is watching.

Natural – and not so natural – landscapes. The Grand Robert dictionary gives three definitions of the word *paysage*. The first one reads: "part of a country, area of

J. Buyck (✉) • T.C. Vales
Institute d'Urbanisme de Grenoble, 14, avenue Marie Reynoard, 38100 Grenoble, France
e-mail: jennifer.buyck@upmf-grenoble.fr; teovales@yahoo.co.uk

R. Gambino, A. Peano (eds.), *Nature Policies and Landscape Policies*, Urban and Landscape Perspectives 18, DOI 10.1007/978-3-319-05410-0_9

land offered by nature to an observer" (Rey 1992). This definition is somewhat ambiguous because of the use of the term "nature" which should be defined. Looking around us, we can wonder what part of nature is not artificial. Is there such a thing as a natural space? Parks and gardens have been planted by man for man. Rural areas, with their farms, are also spaces of culture. The first definition which the Grand Robert dictionary gives therefore reflects a vision of landscape as "site," "view," or "décor." The second definition is related to the pictorial origin of the word: "pictorial or graphic figuration of an area of land in which nature is central whereas figures (men or animals) and buildings are secondary" (Rey 1992). The last definition is figurative. *Paysage* thus becomes a "general aspect," a "situation," and allows to establish the state of the art in various fields.

It is the first definition, which is of most interest to us here, the two others being relative to painting and state of the art. Many dictionaries give three such definitions of the word *paysage*. The first definition given by *Dictionnaire de la langue française d'Emile Littré* is of particular interest: "area of land that can be seen as a whole" (Littré 1994). This definition appears to be separate from "nature" – the definition of which could be even more difficult to reach than that of landscape – but it does not highlight the subjectivity of the individual facing the landscape.

Landscape and its vocabulary. Vocabulary related to landscape does not provide much help in understanding this definition confused. Let us take the word *paysagiste* (landscape architect). According to the *Grand Robert* dictionary, a *paysagiste* is both an artist who paints landscapes and someone who draws garden plans, and who designs green spaces in cities (Rey 1992). The same word is therefore used to talk about different disciplines, whether artistic or not. It can indeed refer to a painting that aims at representing land, trees, or flowers, with various degrees of artistic subjectivity. Similarly, the word *paysagisme*, as "art and technique of landscape" (Rey 1992), is just as slippery. The words *paysagé* or *paysager*, however, are much clearer although used more rarely. They are defined as follows: "arranged in a manner that creates a natural landscape effect" (Rey 1992).

Landscape outside France. The English language offers many definitions of landscape, just like French does. The Oxford English Dictionary (Simpson and Weiner 1998) states that landscape is a "picture representing natural inland scenery," or "a view or prospect of natural inland scenery, such as can be taken in at a glance from one point of view," or "the depiction or description of something in words." The vocabulary of landscape is however very clear: a landscape architect works on "the planning of parks of gardens to form an attractive landscape, often in association with the design of buildings, roads." An artist who paints landscapes is referred to as a landscape painter or landscapist. Confusion is impossible.

The word *pays* is the basis of many Western words meaning landscape: *landschaft* in Germany, *landschap* in the Netherlands. Similarly, the word "land" is used to refer to *pays* in both cases. In Italy, the word *paesaggio* is used to talk about landscape and is based on *paese*. In Spain, the words *paisaje* and *pais* are used. In Greece, *topoi* means landscape and *topos* refers to *pays*.

9.2 Nature at the Center of Contemporaneous Landscapes

Landscape as ready-made. Land is an important notion in landscape, but these should not be confused: it is important to distinguish landscape and its material and physical element features. It is for this reason that Pascal Aubry in *Mouvance II: soixante-dix mots pour le paysage* chose to define "the invention of landscapes" (Berque 2006) rather than landscape itself. This allows to highlight the fact that landscape is a recognition rather than a creation: "as a landscape, the concrete and pre-existing space can be invented" (Aubry 2006). The invention of landscape is thus a type of ready-made (Duchamp 1998) in which the observed space (in this case, *pays*) is transformed through emotions, along a footpath, for example.

The relation we establish with our environment is not necessarily related to landscape. Pascal Aubry suggests several criteria in the same paper. Firstly, the observation should be made on site and should require all senses. The second point is related to the perception angle, in other words of the visual field. The horizon should be visible. It can be a fictitious line where land and sky meet, but it can also be cultural. The spatial arrangement of the area of land should be such that it can be described. The last point is related to nature. The territory should indeed be depending on this notion.

The concept of nature has been divided into three parts since antiquity – from Cicero to Petrarch. Jean-Pierre Le Dantec's *Jardins et paysages, textes essentiels* (Le Dantec 1996) points to the meaning of this subdivision. The site, or wilderness, refers to untouched nature, independent from human activity. Cultivated nature, in other words the countryside, refers to the second dimension of nature and includes all human interventions on nature. The third dimension is related to the association and art and nature for the purpose of aesthetics, philosophy, or leisure. These three definitions show that the invention of landscape relies on nature which itself is a complex and multifaceted notion.

Thinking about landscape. Let us now take a closer look at landscape theory. For it to exist, one has to able to represent landscape and use a word to refer to it. There is a difference between landscape theory (*pensée du paysage*) and *pensée paysagère*, which does not necessarily require words.

Landscape theory thus entails the following transformation: a collection of physical objects, as previously defined, is transcribed into a visual or mental image. This process generates action: landscape theory leads to action in favor of landscape. Pierre Donadieu considers that today's society is a *société paysagiste* (Donadieu 2002), in other words a landscape architect society.

> Arguing that the Western society is a landscape architect means that is pursues two simultaneous goals in building its living environment. On the one hand, it seeks to materialize philosophical, ethical or aesthetics ideals, on the other hand, it wants to control the soft forms of the world to live in, whether as a source of comfort and dreams, or as a testimony of the advent of an ideal world. (Donadieu 2002)

This means that the relationship between man and the outside world is guided by two quests: the material satisfaction of physical and biological needs and the immaterial satisfaction of psychological and mental aspirations. Today's society refuses to be part of the world it has been given and seeks to build another world. It is for this reason that this society does not limit itself to existing landscapes, which are the result of overwhelming trends and forces. It craves to be someplace else (Hervieu and Viard 2005) and has an unfulfilled desire of purity and serenity. This ideal varies in time and everyone does not share it. But everyone knows where he or she is and where he or she would like to be. This collective dream constantly tries to materialize, in either private or public space. Gardens are material symbols of this desire for landscape.

The garden world. Jean-Pierre Le Dantec defines the word garden (*jardin*) as "a portion of territory designed in a unique way and (generally) planted for food production and/or aesthetic purposes" (Le Dantec 2006). Its owners concentrate their efforts on it. They can grow edible or ornamental plants and reveal the symbolic space of their cosmos (Nys 1999). It is partly in this place, which has become symbolic, that the idealized link between nature, home, and civilization can be materialized (Baridon 1998). Mowing the lawn, adding garden gnomes, pruning trees, or allowing plants to grow are some of the choices made by garden aficionados. These choices appear to be trivial. But they are also a reflection of the attention we pay to landscape and to out constant attempt at transforming it.

We previously showed that landscape has two definitions. It can be pictorial, by staging part or all of an area. It can also refer to an area of land, which is being observed. Based on this, the various specialists of landscape – whatever their field of study may be – constantly contribute to the production of research literature through new definitions. These definitions are diverse, but they are all related to *pays*. As ready-made, landscape creates distance vis-à-vis the land, which is being observed. This distance can consist in the production of visual or mental images and can therefore be said to belong to the field of representation. These representations can be strongly desired or, on the contrary, clearly refused, and they generate an ongoing action from society on landscape.

9.3 Contemporaneous Landscapes and Design

Landscape design by landscape designers (*concepteurs de paysage*) also follows this route since it aims at formulating an idea or an intention in order to physically and immaterially transform the relationship between the country and the people who live in it. Landscape design can therefore be defined as the "intention to design space taking into account the material and immaterial future of an area in order to improve its habitability" (Donadieu 2006). It is therefore about anticipating the spatial and social future of the area being dealt with.

The image-landscape. Landscape, as a possible way of looking at space, is one of many interpretations of an area being observed. The way we look at land is a major aspect of landscape. More generally, this look is inherent in any relation between object and subject.

> The world is not an outside reality which is perceived, for it is an object which establishes it as a world and as an exteriority. Any object is above all an object for someone, just as a subject can be defined in contrast with an object. . .the world is the relationship between a subject and an object, and any segregation between one and the other makes no sense. (Staszak 1998)

Geographer and orientalist Augustin Berque suggest we use the concepts of *trajection* (Berque 2000) and *médiance* (Berque 1990) to talk about the relations between objects and subjects. Landscape is a reality, which is interior to the subject, who takes ownership of it through images which lie at the basis of any action on landscape. These images are highly distorting, as suggested by Serge Morin in *Progrès, paysages et identities dans les hautes terres camerounaises:*

> Landscape is the (conscious or unconscious) polysemous display of projects, heritage, rules, limits and dynamics of a society in a given space or place. It allows to delimitate and to identify a territory. This way, it is an expression of a certain socio-spatial system [. . .], it is a built appearance [. . .] it does not show everything, dissimulates facts and can even distort reality. It is not a mere mirror. (Morin 2001)

As image and reality, this landscape changes all the time. These changes sometimes cause strong social reactions, whatever their scope. These changes in the landscape are produced by changes in the area but also by the evolution of the relationship between society and the landscape it finds in the area. Major trends (Hervieu and Viard 2001) can be identified when it comes to the relationship between landscape and society. But this dialectic is heterogeneous for “each person, based on his own ethics, has an idea of the priorities that should be retained and of the solutions which should be applied” (Donadieu 2002).

Landscape that makes sense. By contemplating and judging the spectacle of the world it lives in, the landscape-architect society builds landscapes, which are most often materialized as photographic images. It turns the world into landscapes that then cause intimate or shared emotions, when in contact with ancient or threatened societies. Landscapes often stage heritage and reflect the difficulty of transmitting cultural goods since these are both material and immaterial. Our society pays great heed to cultural heritage as an “active measure of the past” (Mohen 1999). It is therefore forbidden to forget and memory becomes a duty. Landscape can therefore become heritage. In a similar way, landscape can be considered as a space of nature. At a time when everyone knows natural resources are under major threat (Rio de Janeiro Summit 1992), public authorities are opting for the protection of nature. This way, each area has its own charter setting out the good way to use the place. Everything is naturalized. The landscape is seen to be a resource reservoir. Landscape can also be seen as a territory (Montricher 1995), in other words as an invention of the State (in the past) and of societies which live in it (today). At present, choosing where to live is a real question. The place where we are born is rarely the place where we live in. Areas are therefore becoming increasingly standardized, in order to become places where one will inevitably find the same feature: the “search for the sense of his living environment” (Donadieu 2002).

Landscape transmission. For centuries, man has used his technical power on nature. He is thus liable for what is done. Today, the European landscape convention, published in 2000, is the framework in this field. Nature's timescale that of cycles and seasons has been replaced by a historical timescale. Choices made in this field refer to the assumptions that compose our vision of the world and our project as a society. The perception of landscape by political institutions and by citizens depends on its value. It is assessed and criticized. This value serves as a basis for decisions related to the alteration or preservation (Nora 2001) of landscape. By questioning the value of everyday landscapes, we question the practices of spatial anticipation in the face of the "increasing control which we seek to have on terrestrial and sidereal space to design it and live in it" (Boutinet 2005). The question of value is a complex one. It has various foundations. Value can be considered as a price, as a value which is added to the cost, such as an aesthetic, heritage-related, or environmental interest. These values are contradictory, relative, and arbitrary. The question of landscape transmission is therefore directly linked to the practices of spatial anticipation which always attempt at preempting in order to make the desirable occur. Several attitudes are possible to reach this goal: regulation, knowledge, and imagination of the future. Foresight allows to protect material or immaterial, public or private goods. Landscape is no exception. It is in this context that the French State has taken charge of the protection of listed buildings, natural sites, and landscape. French protection of listed buildings was created in 1913 by chapter II, book IV, of the *Code du patrimoine* and by *décret* 2007–487, issued on March 30, 2007. They define the protection of listed buildings as a recognition of public interest for monuments and sites which are more specifically related to art and history. The protection of listed buildings was created in 1913 by chapter II, book IV, of the *Code du patrimoine* and by *décret* 2007–487, issued on March 30, 2007. They define the protection of listed buildings as a recognition of public interest for monuments and sites which are more specifically related to art and history. France has a special set up referred to as *Zone de Protection du Patrimoine Architectural, Urbain et Paysager* (*ZPPAUP*). It was created on January 7, 1983 by the decentralization laws and was later extended through *loi paysage* on January 8, 1993. Since February 24, it has been the first article of *Code du Patrimoine*. It aims at ensuring the protection of landscape and urban heritage, at enhancing areas and sites that should be protected for historic or aesthetic reasons.

Every day, the numerous inventories currently being carried out for listed buildings, natural heritage, landscapes of outstanding beauty and natural zones with an ecological, faunistic, or floristic interest show what is under threat in order for the State to anticipate the spatial future through regulation. The inventory of natural zones of ecological, faunistic, and floristic interest (*ZNIEFF*) is a program launched by Loi Bouchardeau, dates July 12, 1983. But the current quest for landscape anticipation does not stop here. Attempts at predicting the evolution of the land that surrounds us are numerous. They belong to the field of prospective. The anticipation race reaches its climax with our will to imagine our future. According to Pierre Donadieu, it is the favorite type of anticipation for our landscape architect society (Donadieu 2002). Inventing the future of our landscapes means having an influence either on land itself or on the way we look at it in order for it to appear as landscape. Imagining the future can take many forms, from ideas to materialization.

Lastly, projects (Donadieu 2006), and in particular those related to landscape, are operatory anticipations of the desirable future of our land. They can follow current trends or can be at odds with them. In any case, the main goal of a project is to favor land's habitability (Lévy 2002). The question of habitability is central here. Is landscape anticipation a guarantee of a better life? This question is still raised. In any case, it is clear that our current practices make our society liable in terms of landscape. As Javier Maderuelo said, we believe that "si el paisaje que estamos construyendo, no es satisfactorio, entonces es que nos estamos equivocando" (Maderuelo 2005). In other words, "if the landscape we are building does not satisfy us, we are going down the wrong road."

References

Aubry P (2006) Mouvance: du jardin au territoire II, soixante-dix mots pour le paysage. Ed. de la Villette, Paris, pp 55–56

Baridon M (1998) Les jardins: paysagistes, jardiniers, poètes. Robert Laffont, Paris, p 1239

Berque A (1990) Médiance de milieux en paysages. Reclus, Montpellier, p 162

Berque A (2000) Ecoumène: introduction à l'étude des milieux humains. Belin, Paris, p 271

Berque A (2006) Mouvance: du jardin au territoire II, soixante-dix mots pour le paysage. Ed. de la Villette, Paris, p 119 (A)

Boutinet JP (2005) Anthropologie du projet. Presses Universitaires de France, Paris, p 405

Donadieu P (2002) La Société paysagiste. Actes Sud, Arles, p 150

Donadieu P (2006) "Projet de paysage", Mouvance: soixante-dix mots pour le paysage. Ed. de la Villette, Paris, pp 85–86 (B)

Duchamp M (1998) Marcel Duchamp parle des ready-made. Phillipe Collin et L'Echoppe, Paris, p 22

Hervieu B, Viard J (2001) Au bonheur des campagnes. Éd. de l'Aube, La Tour-d'Aigues, p 155

Hervieu B, Viard J (2005) Au bonheur des campagnes et des provinces. Ed. de l'Aube, La Tour d'Aigues, p 153

Le Dantec JP (1996) Jardins et paysage. Textes critiques de l'antiquité à nos jours. Larousse, Paris, p 634

Le Dantec JP (2006) Mouvance: du jardin au territoire II, soixante-dix mots pour le paysage. Ed. de la Villette, Paris

Lévy J (2002) Sur les conditions d'habitabilité de l'espace. Annales de géographie 111 (626):395–405

Littré PE (1994) Littré. Encyclopedia Britannica Inc., tome V, Chicago

Maderuelo J (2005) El paisaje: génesis de un concepto. Abada Editores, Madrid

Mérot A (2009) Du paysage en peinture dans l'Occident moderne. Gallimard, Paris, p 443

Mohen JP (1999) Les sciences du patrimoine. Odile Jacob, Paris, pp 17–18

Montricher N (1995) L'aménagement du territoire. La Découverte, Paris, p 124

Morin S (2001) Progrès, paysages et identités dans les hautes terres camerounaises. Les montagne tropicales, 16

Nora P (2001) "Introduction", Tri, Sélection, Conservation, Quel patrimoine pour l'avenir?. Ed. du patrimoine, Paris, pp 15–17

Nys P (1999) Le jardin exploré: une herméneutique du lieu. Les Ed. de l'Imprimeur, Besançon, p 247

Rey A (1992) Grand Robert de la langue française (2ème édition). Tome VII, Dictionnaire Le Robert, Paris

Rio de Janeiro Summit Conference (1992) The United Nations conference on environment and development report, 3–14 June 1992

Simpson JA, Weiner ESC (eds) (1998) The Oxford English dictionary, vol VIII. Clarendon, Oxford, pp 628–629

Staszak JF (1998) Dans quel monde vivons-nous? Les discours du géographe. L'Harmattan, Paris

Chapter 10
From the Territory to the Landscape: The Image as a Tool for Discovery

Claude Raffestin

Abstract Until the nineteenth century, the Western society made a clear difference between territory and landscape, between reality and image. Now, there is a confusion between reality and its image. The landscape as image is an instrument to have an idea of reality, to know a particular character of reality. We want to analyze the reality of different representations, which constitute the fount of our knowledge and also the fount of the sociocultural perceptions of the members of our societies. The images inform on the vision of the society. The people create landscapes through the different languages and give possibility to understand the sociocultural meanings of the landscapes.

Keywords Landscape • Territory • Image • Reality • Geostructure • Geogram

10.1 The Idiocy of Reality

If we look at the *Encyclopédie* of Diderot and d'Alembert, the landscape is described as

> (...) le genre de peinture qui représente les campagnes et les objets qui s'y rencontrent. (Diderot and d'Alembert 1751)

For any contemporary dictionary, the landscape is a portion of territory considered from a perspective and a descriptive point of view. This latter concept has been adopted, with variants, by the European Landscape Convention (2000).

Hence, for the eighteenth century, the landscape is a portrayal as opposed to a tangible reality; it is the "product" of human observation, staged through any language. The pictorial connotation is evident, but there are other possible portrayals that depend on the languages available. If we talk about images, we do not just consider pictorial images, but every other kind of image too (Manguel 2001).

"The land where lemons blossom" Goethe (1795): the Italian landscape between nature and invention

C. Raffestin (✉)
Université de Genève, 7, rue Calvin 1204, Genève, Switzerland
e-mail: craffestin@hotmail.com

R. Gambino, A. Peano (eds.), *Nature Policies and Landscape Policies*,
Urban and Landscape Perspectives 18, DOI 10.1007/978-3-319-05410-0_10

For the twentieth century, on the other hand, the tangible image and the picture which portrays it are synonyms and, in everyday language, almost exactly the same thing. Between the eighteenth and the twentieth centuries, fission gave way to fusion. We can mourn its loss because we were gradually deprived of a tool of outstanding value.

This "tangible reality" that we live in, the territory, is the result of constant alterations and is subject to ongoing change because it is an immediate expression of everyday life and history. The territory is, in short, the fruit of a 1/1 scale diachronic production process, in continuous evolution.

The landscape, to the contrary, is the result of a mental production process, which originates in human observation, mediated by different languages: natural, pictorial, sculptural, logic-formal, or mathematic. The landscape or picture of the territory is always a two-dimensional historic document (apart from scale models) – unlike the three-dimensionality of the territory, which develops synchronously – and is the expression of a pause in time: a snapshot. Insofar as it is the result of the passage from reality to portrayal, it constitutes the intellectual journey necessary to produce the tools needed to analyze reality and becomes a tool of knowledge. The arrangement in succession of several synchronic images allows the construction of a diachronic axis, which can help read the evolution of the territory.

It is hard to establish when man was able to make the intellectual leap that enabled his progress beyond the mere presentation of things, to reach their portrayal, feeling the need to "double" the tangible world with that of images.

Images or portrayals of reality are ways not only of describing tangibility but also of explaining it from a scientific, literary, and artistic point of view, depending on the language adopted. Consequently, they take on very different functions. The process that lies behind the creation of a portrayal is always based on a program or theory of observation, sometimes declared but more often implicit. This is why, theoretically, an infinity of images of reality is possible, because they can be an infinity of viewpoints, but in short, there is actually a limited number in the same way that there is a limited number of metaphors of reality: however poorly defined, the reality of "territory" has become the inescapable subject of contemporary geographic portrayal.

Observation is of great interest, because it presumes that in order to know the territory, it is necessary to "imagine" portrayals that are nothing more than different points of view from which to look at the subject portrayed. To quote Rosset, "reality is described as idiotic because it is unique." This means that it has the ontological privilege of being inimitable, without being able to imitate anything else (Rosset 1977).

When we look at the territory from an exclusively geographic point of view, it becomes a more or less well-defined geostructure, starting from which we can build up geograms or images (Fig. 10.1): when faced with the portrayal, which uses words or pictures to immortalize reality, we find something real "brilliance." Reality is like the sun for Icarus, when we get too close, we can get burned. Consequently, taking into consideration reality means not really possessing reality, but something under a light less bright (Rosset 1977).

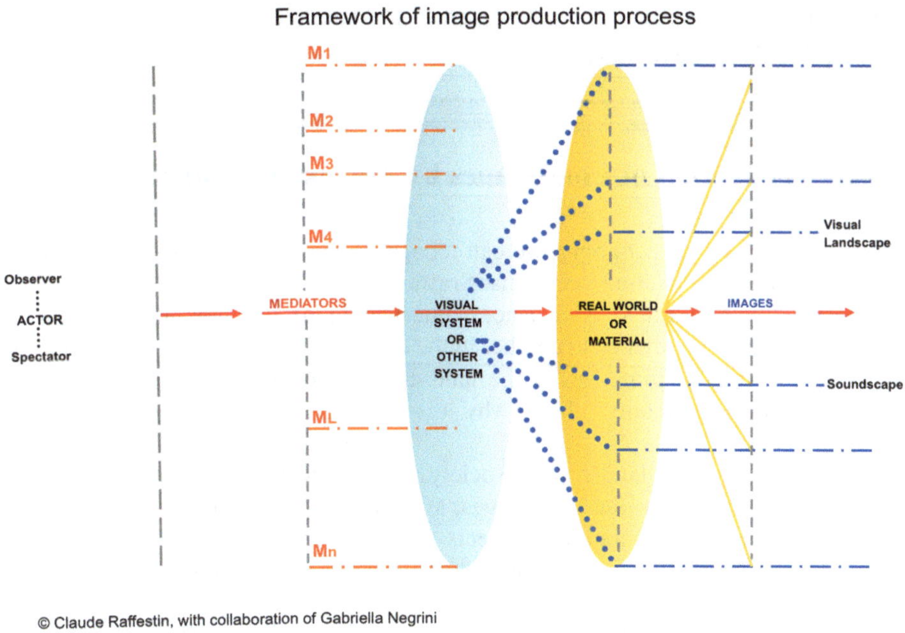

Fig. 10.1 Framework of image production process (Source: author's elaboration)

At the end of the portrayal process, reality vanishes and is replaced by one or more more-or-less banal images: reality often precedes its portrayal, the function of which is not to evoke a contemporary reality of perception but to reveal a prior reality. The most convincing portrayal usually indicates a prior reality or one which had at least started to become real before being recognized as such. This means that it only became known the second time around, because the first was that of the appearance of this reality (Rosset 1977).

This latter observation is good at showing how a reality can often be recognized a long time after its first tangible realization. Contemporary history shows how the industrial landscape, for example, meaning the image of industry, has contributed to publicizing that reality when it first began to decline. This delay coincides with the moment in which the industrial companies were able to critically observe and create images of the reality around them. This coincided with the start of the crisis within the system.

To better identify the role of portrayal as a tool of knowledge, we can use the metaphor (Dematteis 1985) and its formative elements, significance, and meaning. It is, however, necessary to point out that there is no real similarity because the relationship between significance is, in the first case, arbitrary, while it is not when we look at the relationship between tangible figure and image. The geogram, for example, which we mentioned above, is the fruit of a specific point of view (geographic); is created to present something which is considered relevant, interesting, and conventional with regard to the tangible reality observed; and is not, therefore, arbitrary.

10.2 Production of Landscapes by Means of Landscapes

With reference to the economy, we can recognize the transposition of the famous formula of the great Italian economist Sraffa (1963): "production of commodities by means of commodities." This is not plagiarism on my part, but use of a "model" transferred from one science to another in the continuity of a homology. Contrary to what people usually think, using the landscape is not a tangible reality but the picture of a tangible reality. This is why it is fair to think that we produce pictures using pictures.

Nowadays, in our "landscapist" society, we like to take old territories, particularly in cities, and try out new images, awaiting the possibility to adapt them to the new territorialities. Modern architects-town planners have to deal with puzzle territories, the elements of which are waiting to be rethought through new images or portrayals to tangibly give them another life in an almost baroque situation of difference and styles that belong to different moments in history. The contemporary architect-town planner is more a creator and a planner of images than a builder in the classic sense of the term.

The landscape can be defined in this sense as an image, a "geogram" of the tangible reality or geostructure. Hence, we are looking at a double system of relationships between the developing tangible reality and the different possible portrayals of this reality.

When we talk about a product of history and of historical document, we sustain that there is a reality produced, in an uninterrupted diachronic way, and that, synchronically, images are produced to enable the discovery of this reality. At this point, a distinction is essential: reality is produced according to a continuous method, but the image is produced according to a discontinuous method because it would not be interesting, even if it were possible, to conserve uninterrupted images.

To understand this double system, we can quote the approach of the painters of the 1780s and their landscapes of Rome, abstraction, and rationalism to reveal the implicit geometric forms of reality through the "Cercle de David" which made Umberto Eco say:

> I felt like these Frenchmen copied from artists who came later: De Chirico, Metaphysics. (Ottani Cavina 1994)

The drawn and/or painted picture is a tool for discovery. Discovery of what is hidden in reality, but also, and especially, a discovery of "oneself in a social environment" via the elimination or highlighting of particular elements: the influence of David on Saint-Ours can be seen in the elimination to portray the Geneva of the Alps, the lake, etc., to portray Geneva as though it were Athens, an ancient city with its Acropolis, its temple, and its walls; an ideal city which corresponds to good political management, in which the perfection of forms reflects the democratic system according to the principles of revolution (Ottani Cavina 1994).

Once again we find ourselves faced with a double system of relationships: the picture as a tool to discover what is hidden in the various places that can be painted and the picture as a historical tool to make it possible to understand the different ideologies promoted by representatives of society of want to give their observations an identity (Dardel 1986).

Consequently, geographic reality is a pretext for the production of images, images that can teach two things at the same time, on the place and on the mediated observation of the person offering the portrayal.

In any case, every form of portrayal is a tool to break down reality. The image, and probably it would be better to use the plural sense, is always necessary to take hold of reality. There is no knowledge of reality without the help of an image. The knowledge we produce in relation to reality is contained, partly, in the portrayals we use to create it. Reality is an illusion, in that we never have such a direct relationship. At every level of reality, there is a social construction, and this is why we always work, often without even realizing it, on portrayal rather than matter.

10.3 The Mechanism of Portrayal Representation

According to Céline Masson (2004), the work is a double picture: that of the portrayal (dream, joke, and symptom) and that of the presence of the other to whom images, words, and actions are referred. In short, the work of portrayal is used to find one's own identity and the origin of the objects of one's personal desire. The landscape is the perfect product of desire. The landscape is also perfect as portrayal is a way of stopping time for the object portrayed. Time is shut down at that instant and the subject becomes eternal, as though it were suspended in time:

> The portrayal of things is formed through the invasion of the mnemonic images of the object or of its traces, in other words, portrayal is what brings to life the mnemonic trace of the object, meaning the localised inscription of the event. (Céline Masson 2004)

Aristotle had already theorized on how the image could be the indicator of something absent. There is no doubt that there is also a way of taking control of space and the forms that surround it, reproducing and transporting a form into a different material, and imprinting it onto another support; man becomes the author who controls its fabrication or even the creator if he reworks and reinvents it. It is the work of man to replicate the world in pictures, this being a way of standing back from reality. But the landscape is always the search for perfection that reality is incapable of providing. Man seeks another image behind everything he sees, there is always the desire for something else; the landscape is the fruit of dissatisfaction, because there is always something missing. Obviously portrayal is not a copy:

> The truth is that, for a painting to portray a subject, it has to be a symbol of it, standing for it, referring back to it; and no degree of similarity is enough to create the relationship of reference required. A painting that portrays – or a passage that describes – a subject refers to it, denoting it. Denotation is the crux of portrayal and is independent of similarity.

This means that the theory of portrayal as a copy is stopped right from the start, for the very fact that it is impossible to specify what has to be copied because, to quote Kant, the innocent eye is blind and the virgin mind is empty. On this topic, E. H. Gombrich wrote some essential things in "Art and Illusion":

> A portrayal or a description is suitable, effective, illuminating, subtle and suggestive to the extent by which the painter or writer grasps unprecedented and significant relationships and devises means to make them manifest. (Gombrich 1960)

However, this idea poses an evident problem in relation to realism. A painting is realistic insofar as it is a clever illusion, inducing the viewer to presume that it is what it portrays or that it has the same features. The measure of realism proposed coincides, in other words, with the probability of exchanging the portrayal for what is portrayed.

This explanation is plausible, but, in the case of the landscape, it is not completely convincing, because realism is not absolutely what we have to measure. If, every time, the landscape seeks a perfection in the image that cannot be found in the tangible reality, then it is the measure of probability that replaces the portrayal for that portrayed.

As far as the territory is concerned, the search for perfection has been present throughout civilization, in all the myths and religions through the notion of paradise: either we have lost it and have to find it again or we have to enter in after death. I see the landscape as being a substitute for paradise, a golden age of the territory, and naturally, in the same way, that paradise has been conceived differently by different people, and at different times throughout history, the landscape has gone through a series of different conceptions. Perfection is manifest within and not outside of a historical context. While it is true that "utopia" corresponds to "ucronia," people talk more easily about the perfection of the territory than that of time. If we can imagine a perfect place, a paradise, we cannot imagine a perfect history, other than the extension of perfection into eternity, and this rules out history, replacing it with a time without suffering, characterized solely by pleasures, as in the case of the paradise created by the various religions. This is none other than the end of history, not in the sense used by Fukuyama but in the sense of the nonrealization of the world. How can I be aware of perfection if not through emotions and feelings? Portrayal encourages the gift of memory to preserve it. This is a relative perfection. We all have nostalgia of perfection, of paradise. Talking about realism is another way of going back to the problem of similarity. Portrayal is unstable, just like expression: "As much in the case of portrayal as that of expression, certain relationships are established stably by force of habit; but in neither of the two cases is there such a thing as an absolute, universal or unchangeable relationship."

René Char (1948) succeeded in approaching the idea of perfection when he wrote:

> Poetry is love made of desire which remains desire (*le poème est l'amour réalisé du désir demeuré désir*).

Now this is not quite the same, but it is similar: we can write that the picture of the landscape is the landscape of the perfect territory that has remained a desire. The apparition of the landscape as the desire for perfection or the perfect territory is absolutely vital. In actual fact, it is the apparition of images, of living through or inside images. It is always more obvious that portrayal is manifest through a terrible phenomenon of transformation, of atrophy, or of hypertrophy. Deformation is one of the important mechanisms of portrayal, but it has still to be fully clarified. In order to do this, we would have to work on the matter of caricature.

Portrayal obeys an inner visual space:

> Giacometti had grasped a very clear idea of what depended on vision and what depended on the viewpoint, in the sense of the "scopic" pulse mentioned by Freud. This invasive disturbance translates the collision between vision (the object perceived) and the viewpoint (the thing invested by the pulse and makes it hard to create form. [...] This triggers the formation of an inner vision, which opposes the desire to copy reality and present the perceived view. A conflict is created between two needs: the eye's perception of the view as it is and as he wishes to reproduce it; the transformation made between this vision and an inner reality, between the eye and the view. It is the adjustment that has to be made that causes this crisis of portrayal.

This is why the Angélus de Millet and Breughel's Fall of Icarus are not landscapes: "It is because, quite rightly, the landscape is the opposite of a background: the «land» has to emerge completely on the surface, the land alone, everywhere." This means that the landscape contains no presence, it itself being the entire presence.

The portrayal of the geostructure (tangible reality) is a "photograph" formulated suing one or many languages that make it possible to obtain different geograms (image of reality). The geostructure is always in time, but the geogram stops time, making it a historical document. The geostructure has a value in terms of use, while the geogram has a value in terms of exchange. The geogram, meaning the landscape, as a means of accessing knowledge of reality, has all the characteristics of the fiduciary currency. The value of this currency does not lie in the similarity between reality and portrayal but in the nature of the observation, which imposes a particular way of seeing. The geogram takes on value because it is able to identify the nature of the observation. In a manner of speaking, the geogram is the sign of the wealth of reality, but not in the sense of similarity, more in the sense of creating a different way of understanding. This currency has three features: the crystallization of values of a historical subject, a means of seeing reality and gaining knowledge of it, and, lastly, a means of treasuring and accumulation. On the matter of this problem, Georg Simmel expressed himself strongly:

(...) Our observation can join the elements of a landscape, grouping them together in one way or another. It can move the accents in numerous ways, even changing the centre and the limits. (George Simmel 1988)

This unity is the product of *Stimmung*, the German word for "mood." What contributes best to defining portrayal, as a fiduciary currency, is the correspondence between the mediators used to produce the image and those used to read it. It is clear that as time goes by, the mediators change, and then the images are no longer legible in the same way. Between the seventeenth century and the start of the nineteenth century, the vision of the Alps, for example, changed completely, and the comparisons that can be made between the legible version and the real-life experience in the period between the two centuries are irrelevant. There are at least two mechanisms that work contrarily between the geostructure and the geogram: atrophy and hypertrophy. The "landscape" in this sense is a conception of the world, which is important because man needs it to live within the territory. If man lived the territory like an animal, he would only pay attention to the resources that he needs to satisfy his needs, but, because he possesses culture, he also lives the territory with his memory and his imagination. The appropriation of the territory takes place also, and possibly mainly, through portrayal. It is not possible to live the territorial reality without thinking of the image of this reality. The production of ideal landscapes, utopian landscapes, and imagination is proof of the enormous role that portrayals and symbols play for man.

If archaeology is a fundamental science to decipher the different phases of construction, destruction, and reconstruction of a tangible territory, iconography, via the production of representations, is an essential means to elaborate a historical memory of observation and a memory of the ways of symbolic appropriation of things (Turri 1998). The landscape, as a portrayal, possesses a double character of autonomy, as form and as production. At this point, it is helpful to remember that the landscape has a life which is independent of that of the territory. This observation is valid for all the portrayals of the arts or science. The symbols and metaphors take ambitions into account, and, in this sense, they constitute an authentic fiduciary currency which is extremely valuable when it becomes legal tender in people's minds.

The mechanism might seem paradoxical because, in the same way that currency is not wealth but the symbol of wealth, portrayal is not reality. The more spectacular the portrayal, the higher the value of the landscape, a value of exchange: rather than be a counterpoint of picturesque details, the landscape is a combination: a convergence, a moment of life. An inner link, an "impression," unites all the elements. The landscape is unified around a dominant affective shade, which is perfectly legitimate despite rejecting a purely scientific reduction. It calls into discussion of man as a whole, his existential links with the Earth, or, if you prefer, his original geographic stance: the Earth and the basis and means for his realization. The landscape is not a closed circle but an unfolding shape. It is geographic in its "extensions," in the real or imaginary background that the space opens up for observation. The landscape creates no shadows because, as mentioned earlier, it is a portrayal, but it acts as a shadow for the territory or, more precisely, is the territory that has not generated a landscape, in the sense of creating a portrayal with a certain value: landscape is not merely as, intended by geographers, the physical space built by man to live and work in but also the theater in which everyone plays their part, being actor and spectator at the same time. This is considered in the Greek sense of *theatron*, derived from *thasasthai*, meaning to contemplate, watch as a spectator, which takes us back to the position in which man finds himself when, having emerged from the frenzy of life, after battling, working, and building, he looks out onto the battlefields or, according to the metaphor of Lucretius, he stands on the shore to watch a ship sink.

It is a pleasure to see the ingenuity of Simon Schama when he writes:

> Even the landscapes we believe to be most independent from our culture can, if we look closer, turn out to be a product of it. (Simon Schama 1995)

If I may say so, there is a reason for this slight irony. Schama's "discovery" is, without wishing and without knowing it to be so, a tribute paid to the role of portrayal in culture. Indeed, thanks to culture, man does two things of importance: he creates tangibly and spiritually. The landscape is the spiritual tribute paid by intellect to the territory, made with man's hands and brain.

References

Char R (1948) Seuls demeurent, Fureur et mystère. Gallimard, Paris, p 162
Dardel E (1986) L'uomo e la Terra. Natura della realtà geografica. Unicopli, Milan
Dematteis G (1985) Le Metafore della Terra, la geografia umana tra mito e scienza. Feltrinelli, Milan
Diderot D, D'Alembert JB (1751) L'Encyclopédie, 1re éd. Dix-sept volumes plus neuf volumes de planches, Paris
Goethe JW (1795) Wilhelm Meisters Lehrjahre. Unger, Berlin
Gombrich EH (1960) Art and illusion. A study in the psychology of pictorial representation. Phaidon, London
Manguel A (2001) Le livre d'images. Actes Sud, Lerméa
Masson C (2004) Fonction de l'image dans l'appareil psychique. Éditions Ères. Ramonville Saint-Agne
Ottani Cavina A (1994) I paesaggi della ragione. Einaudi, Turin
Rosset C (1977) Le réel. Traité de l'diotie. Editions de minuit, Paris
Schama S (1995) Paesaggio e Memoria. Mondadori, Milan
Simmel G (1988) La tragédie de la culture. Editions Rivages, Paris
Sraffa P (1963) Production of commodities by means of commodities. Prelude to a critique of economic theory. Vora & Co. Publishers PVT Ltd, Bombay
Turri E (1998) Il paesaggio come teatro. Marsilio, Venice

Part II
From Nature to Landscape and Back

Chapter 11
Biosphere Reserves and Protected Areas: A Liaison Dangereuse or a Mutually Beneficial Relationship?

Giorgio Andrian and Massimo Tufano

Abstract "BR are much more than just protected areas" was remarked in 1995, on the occasion of the Seville Conference, a milestone in the history of the Man and Biosphere (MaB) Programme of UNESCO. Being inspired by this ultimate remark, the authors of the paper decided to analyse the evolution of the relationship between the Biosphere Reserves (BRs) and the Protected Areas (PAs), in terms of conceptual and operational approaches developed by both the respective scientific and practitioners' communities. From being initially identified as sub-portions of pre-existing protected areas – as observed in the early designations dating back to the 1970s and 1980s of the last century – BRs have become larger, in terms of their total extensions, and multifunctional, in terms of their zoning drivers, including the PAs as only one portion of the entire designated territories. Paradoxically, the MaB Programme has never explicitly considered the "landscape" as a specific BR attribute (or nomination criteria), on the contrary to what has happened at UNESCO within the World Heritage Convention, where the concept of *cultural landscape* has become an official category of designated sites since 1992. The analysis concludes by observing how the relationship between BR and PAs may be easily transformed into a *liaison dangereuse* when a clear distinction of the respective primary functions and their territorial implications (the zoning) has not been applied. The two governance regimes can be easily confused with each other with clear difficulties of engineering appropriate management measures and risks of reducing their effectiveness.

Keywords Biosphere reserves • Protected natural areas • Zoning • Governance • Learning laboratories • Landscape

G. Andrian (✉)
Formerly UNESCO office, Via del Santo, 26-35123 Padua, Italy
e-mail: projects@giorgioandrian.eu

M. Tufano
The Regional Agency for Protected Areas in Lazio, Rome, Italy
e-mail: mtufano@regione.lazio.it

R. Gambino, A. Peano (eds.), *Nature Policies and Landscape Policies*,
Urban and Landscape Perspectives 18, DOI 10.1007/978-3-319-05410-0_11

In 1995, on the occasion of the Seville Conference, a milestone in the history of the MaB Programme was remarked: "Biosphere Reserves are much more than just protected areas" (UNESCO 1995).

Why, after 24 year of existence, did the MaB Programme need to "take some distances" from the "traditional" protected areas community? Why do Biosphere Reserves (hereinafter referred as BRs) continue to exist despite their close similarity with other categories of protected areas? Additionally, is the relationship between the *MaB discourse* and the *nature conservation discourse* a mutually reinforcing one or rather a *liaison dangereuse* resulting in creating confusion at the scale of the individual designated territories' management regimes? And, ultimately, how is the landscape dimension treated in the MaB Programme?

These are the curiosities which drove the authors in navigating through the 40 years long history of the MaB Programme and its governance and the evolution of the BRs, the territories designated as "open space laboratories", in relation to their statutory functions.

11.1 Forty Years "Young". A Brief Historical Overview of the MaB Programme

Officially launched in 1970[1] as an intergovernmental initiative by UNESCO, the Man and Biosphere Programme (MaB) dates back its very origins in 1968, when the "Biosphere Conference" was organised, in collaboration with Food and Agriculture Organization of the United Nations [FAO] and International Union for Conservation of Nature [IUCN]. At this conference, over 300 representatives from more than 60 countries and international organisations discussed the world's growing environmental problems – 4 years before the first report of the "Club of Rome" and before the founding of United Nations Environment Programme [UNEP].

The Programme structure was initially designed to primarily serve research purposes. Scientists immediately began discussing the details of the initiative, especially in terms of concrete interdisciplinary strategies and projects for a modern environmental policy combining protection and yield of the biosphere.

In fact, only within 1 of the 14 proposed research projects (viz., the nr.8) the idea of a worldwide network of designated territories emerged. BRs mentioned only under this theme proposed:

> As basic logistical resources for research where experiments can be repeated in the same places over periods of time, as areas for education and training, and as essential components for the study of many projects under the Programme.[2]

[1] On 23 October 1970, the foundation of the MaB programme was adopted by the 16th UNESCO General Conference.

[2] UNESCO (1971), p. 21.

This shows that the original idea of the "reserves" already incorporated different essential elements, which later on become their "statutory roles"; in fact, in addition to the relevant "scientific" function, they were expected to address economic, educational, cultural, and recreational purposes at the same time.

Despite this initial comprehensive concept, emphasis was thus placed on both the conservation and logistical functions, with the development role left rather undefined:

> The idea and the term "biosphere reserves"[3] were officially launched, but in a somewhat hazy manner, with little clarity in terms of their exact role and nature. (UNESCO 2002)

The first 57 BRs were approved in 1976 in 9 countries; a further 61 were added to the network in 1977. By 1981, 208 BRs had been designated in 58 countries. Most of them were largely corresponding to pre-existing protected areas, being the initial interpretation of the MaB primarily focused on *conservation* (mostly of genetic resources and biological diversity).[4] Hence, in the mid-1980s, while the "biosphere reserve" was gaining ground as a conceptual alternative to the "national park", the solidarity that should have bound the international community to consider BRs as priority sites for testing and validating approaches to integrated conservation and development operations was undermined (Hadley 2011).

The political and institutional aspects of the MaB Programme have always been playing a relevant role within the UNESCO activities. In fact, at the early stages, the UN agency offered a unique platform to debate innovative scientific issues and to bring well-known specialists from the various related disciplines; as far as the MaB developed and become an important intergovernmental programme, the geopolitical issues strongly emerged, as a consequence of its institutionalisation.

This widening recognition of BRs during the 1980s was reflected in an observation by the MaB Council, at its 11th session in November 1990, that:

> (...) the general interest in BRs had probably never been greater, even though the quality of the international BR network was highly uneven and lacked credibility as an operational network. (UNESCO 1990)

It was within such a context that the MaB Council requested that an Advisory Committee on Biosphere Reserves be set up, in order to consolidate the work of the

---

[3] This original denomination has been recently subject to criticisms (see, for instance, the Proceedings of the International Conference "Biodiversity and Society", held in New York, Columbia University, May 2001), mostly due to the fact that it evokes images of places, which are almost untouched and remotely located. No consensus has been reached so far at the international level to change the official name; but it is allowed to use local (national) names that may not contain the word "reserve" (e.g., the *Monviso Biosphere Reserve* is called *Area della Biosfera*, in the Italian language).

[4] "The designation of BRs was delegated by the MaB Council to its six-member Bureau. The main criteria to approve BRs was their conservation role, together with the presence of research facilities or a particularly history. In fact, the Bureau adopted a very flexible approach, considering it sufficient for the areas proposed by the MaB Committees to appear of interest for the conservation of ecosystems, possess appropriate legal protection and be the object of a reasonable amount of research work" (Hadley 2011).

international BR network at the time when the overall MaB Programme itself was being reviewed.[5]

In 1995, on the occasion of the Seville Conference, a complete revision of the MaB and its functioning mechanisms was carried out: the adoption of the Seville Strategy and the Statutory Framework of the World Network marked the separation line between the so-called "first" and "second" generations of BRs.

The "paper park"[6] effect affected also the World Network of Biosphere Reserves, especially after the immediate post-Seville period.[7] This leads to call for a specific meeting to assess the first 5 years of implementation of the Seville Strategy. The Seville +5 conference was held in Pamplona (Spain), in 2000, with the main purpose "to analyze the three levels of implementation of the Seville Strategy; namely, the international, the national and the site levels" (UNESCO 2002, p. 27).

As a consequence, more attention has been given to the other two major BR functions: namely, the development and logistic support, resulting in core zones that were less extended when compared with the buffer and transitions ones. Fostering economic and human development – which has to be socioculturally and ecologically sustainable – becomes equally important as identifying areas of exclusive nature conservation scope.

More recently, and in relation with the UN Decade on Education for Sustainable Development (UN DESD 2005–2015), BRs have been identified as learning sites for sustainable development, to remark the importance of the original scientific approach in testing innovative management solutions to reconcile protection and local development. In the light of the more recent development of the *learning laboratory concept* (LLab), BRs are to be seen as "processes as well as settings in which a group (e.g. a management team) can learn together" (Nguyen et al. 2010).

Between the early BR designations and the recent developments of the MaB Programme, about two decades passed, during which significant changes in the nature conservation and sustainable development discourses have occurred at international level.[8] The MaB Programme milestones reflect how BR activities have both influenced and been influenced by the major worldwide conceptual and operational developments. In the 1970s, the main concern was that vast amounts of

[5] The statutes of the Advisory Committee, stipulate that it has the task "to advise the Director-General of UNESCO on the scientific and technical matters concerning the designation, evaluation and management of BRs as well as the development, operation and monitoring of the international BR network".

[6] The term has been used to describe the conditions where protected areas exist "on paper" but not in reality (Phillips 2003).

[7] The 1996–2000 period has registered an average of 12.6 BR nomination per year (UNESCO 2002).

[8] The primeval global issues of the "development" and the "human development" evolved in the "sustainable development", "system of objectives" (Vallega 1995).

data and models were missing on human-environment relations,[9] while in the 1980s much data was at hand but the understanding of managing these relationships was still insufficient. Later on (in the mid-1990s), MaB and BRs addressed more intensively structural political challenges. It became more and more central to understand large-scale rather than small-scale phenomena as well as focusing on joint solutions with local inhabitants rather than maintaining confrontations of interests.

The constantly growing number of BR designations reflects the potentials of the original concept in terms of accommodating new emerging issues of global concern, such as the loss of biodiversity, the climate changes, and the education for sustainable development, while maintaining the ultimate objective of MaB Programme – "aiming to set a scientific basis for the improvement of the relationships between people and their environment globally" – very up to date.

11.2 The Mechanism Makes the Difference. The MaB Programme Governance System

The governance system that has been developed within the MaB Programme reflects, first of all, its intergovernmental constituency; since its very beginning, the decision-taking mechanisms are based on an ICC (currently called International Coordination Council and composed by representatives from the UNESCO Member States) combined with a panel of independent experts (the Advisory Committee).

The idea that the BR has to serve research objectives first leads to the design of an international network, grouping all the designated territories (later on formalised in the World Network of Biosphere Reserves (WNBR), to have the possibility to monitor at a global scale natural phenomena and their interrelations with the human communities' activities, by confronting data collected in the various sites belonging to the same group (e.g., the mountain BRs). The WNBR has been enriched by regional scale groups (e.g., EuroMaB, AfriMaB, IberoMaB) and thematic networks (e.g., island and coastal BRs, mountain BRs).

[9] One example of the scientific work of the MaB programme in the first decade of its existence is the research on desert areas: after long droughts in the Sahel zone from 1968 to 1973, UNESCO began to draw up research plans for the Sahel region of Niamey (Niger), North Africa, and large-scale irrigation systems in the Sahara. UNESCO-MaB's and UNEP's IPAL (Integrated Programme on Arid Lands) project in northern Kenya was funded by Germany until the mid-1980s. Numerous publications resulted from this project. In 1977, the World Map of Arid Regions was published within the framework of the MaB Programme, and a series of case studies was formulated in Pakistan and Chile, e.g., on how to control desertification. In 1978, the desert laboratory in Mapimi (Mexico) was inaugurated. The intensity of these research activities was held up during the 1980s, and new projects were added continuously.

Differently from the Word Heritage sites, no Convention has been created within the MaB: in fact, the Statutory Framework – adopted in combination with the Seville Strategy, in 1995 – has served the purpose to set the guiding principles and the designation criteria for the BRs. An interesting debate has always animated the MaB community in relation to the opportunity of creating a more binding international treaty to regulate the Programme's activities. Some countries adopted a specific legislation, which identifies the category of "biosphere reserve" as an additional type of protected areas. On the other hands, other States are convinced that creating additional legal frameworks to specifically regulate the BR existence would add additional rigidities to the territorial governance and reducing the efficiency of the designation. Ultimately, on the occasion of the third World Congress of BRs (held in Madrid in 2008), the Madrid Action Plan (MAP) was adopted, containing a specific mention to the legal recognition issue; it affirms that BRs would benefit from "an enhanced legal recognition where appropriate" and that "States be encouraged to include BRs in their own legislation" (target 11, action 11.1.).

The process (Fig. 11.1) leading to the designation of the BRs is voluntary and promoted by the UNESCO Member States, which adhere to the MaB Programme. The MaB Committee (or the MaB Focal Point, in case that the Committee doesn't exist) proposes – through the official channel of the National Commission for UNESCO in the country – suitable territories to be designated as BRs; the Nomination Form is the model form provided by the MaB Secretariat to guide the preparation of the official dossier. The local proponent(s) (typically a protected area or a consortium of local governments) coaches the process for the official dossier preparation, in close cooperation with the national authorities in charge

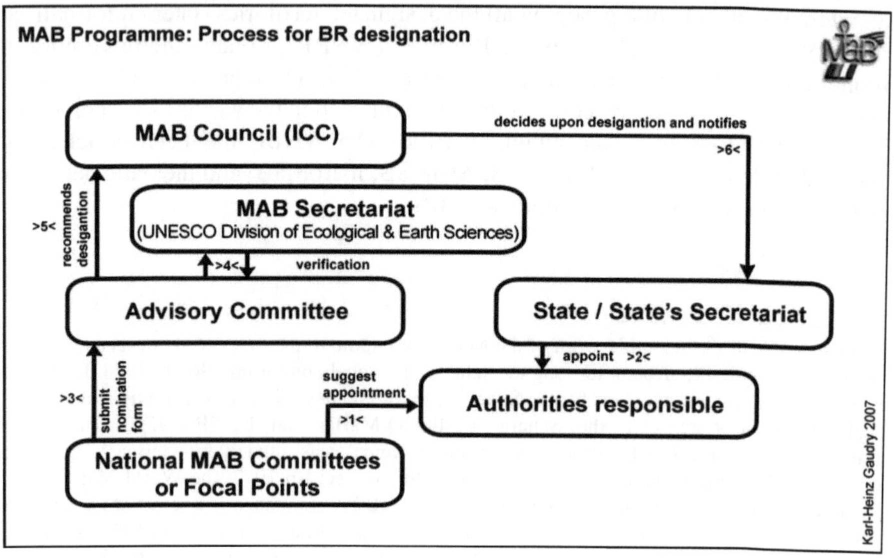

Fig. 11.1 The BRs' designation process

(steps 1 and 2). Once finalised at the national level, the Nomination Form is submitted to the MaB Secretariat, by September 30th of each year (steps 3 and 4).

All the decisions are taken by the main MaB governing body, the ICC, that consists of 34 Member States elected by UNESCO's biennial General Conference. Prior to the annual session, the Advisory Committee of the MaB ICC evaluates all the new proposals coming from the Member States – via the Secretariat (step 4) – by formulating recommendations on the quality of the dossier and advising whether or not to approve their inclusion in the WNBR.[10]

Differently from the UNESCO World Heritage Natural Sites designation process that has foreseen a preliminary evaluation of the dossier by IUCN (the external Advisory Body), the MaB ICC relies exclusively on the opinion of the set of international experts constituting the Advisory Committee.[11] Once the decision is taken, the notification is sent back to the Member States (step 6) through the official UNESCO channels.

The overall process is the result of the interplay of three major scales: (a) the *local*, where the original proposal is formulated; (b) the *national*, where the proposal is firstly validated in relation to the state's priority; and (c) the *international*, where the ultimate decisions are taken.

In this multi-scale governance system, there's a limited influence of the traditional protected areas discourses; in fact, only one (out of three) major BR *function* is clearly related to *conservation,* and the complete evaluation process conducted by the Advisory Committee's experts reflects the multifunction MaB approach.

At the international scale, the MaB Secretariat coordinates the Programme's activities with those of other major Multilateral Environmental Agreements (MEAs) counterparts; in particular, important working relationships have been developed with the other four biodiversity-related conventions.[12]

In analysing its governance system, there are no evidences that the BR designation itself effectively contributes to reinforce the nature conservation policies at the scale of the individual sites (Stoll-Kleeman 2007). The existence of the MaB "logo" on a given territory does not directly imply any additional protection regime per se; only if the country's legislations prescribe a special recognition to BRs – as a separate category of protected areas – the international recognition may result in an added value; otherwise, it can only serve the purpose of alerting the public opinion when some potential (or existing) threats are challenging the site.

[10] The three options used by the Advisory Committee in recommending the ICC are (1) *to be accepted*, when the dossier is complete; (2) *to be deferred*, when the dossier is acceptable but some important elements are missing; and (3) *to be rejected*, when the dossier reflects a proposal that – for some reasons – cannot fulfil the required criteria.

[11] Only recently (on the occasion of the 25th ICC), IUCN was asked by the ICC to provide an official proposal on how to structure a possible external advisor role for the MaB Program.

[12] Namely, the Convention on Biological Diversity (CBD 1992), the Convention on International Trade in Endangered Species of Wild Fauna and Flora (CITES 1973), the Convention on the Conservation of Migratory Species of Wild Animals (CMS or Bonn Convention 1979), and the Convention on Wetlands of International Importance (Ramsar 1971).

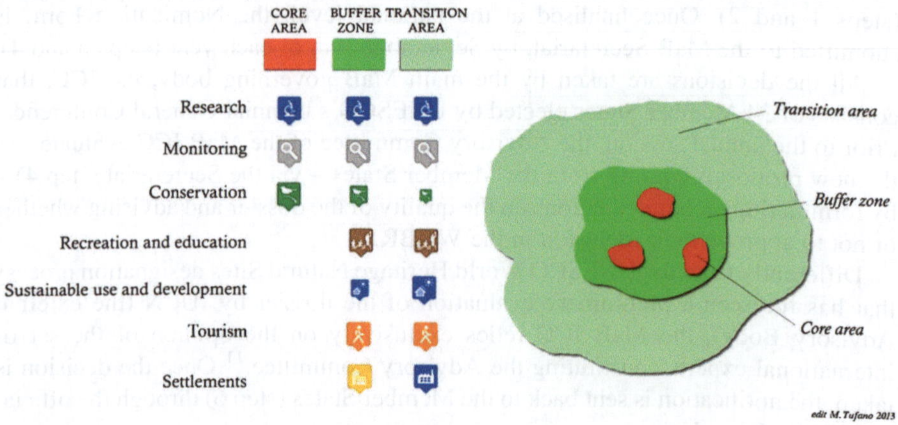

Fig. 11.2 Zoning and functions of BRs

11.3 The BRs Zoning: Spatial Evolution of the Relationship Between Areas and Functions

The *Statutory Framework of the World Network of Biosphere Reserves*[13] defines them as:

> Areas of terrestrial and coastal/marine ecosystems or a combination thereof, which are internationally recognized within the framework of UNESCO's Program on Man and Biosphere (MaB).

Proposed by national governments, they may be designated as BRs only at the condition that they "meet a minimum set of criteria"[14]; the Nomination Form (the last version being introduced by the MaB Secretariat in January 2013) lists them at Chap. 4. In Fig. 11.2, the distribution of the key activities characteristic of each BR is illustrated in relation with the MaB zoning system.

"Although originally envisioned as a series of concentric rings, the three zones have been implemented in many different ways, in order to meet the local needs and conditions" is reported in the "Biosphere Reserve Concept" section of the Statuary Framework. The same "model" has been shaped around very many different designated territories all over the planet, resulting in a very diversified spectrum of BRs; this reflects "one the greatest strengths of the biosphere reserves concept, that has been the flexibility and creativity with which it has been realized in various situations".

[13] The Statutory Framework of the World Network and the Seville Strategy were adopted under 28/C/Resolution 2.4 of the UNESCO General Conference, in November 1995.

[14] The Biosphere Reserve Concept, in "The Seville Strategy and Statutory Framework of the World Network", UNESCO 1995, p. 4.

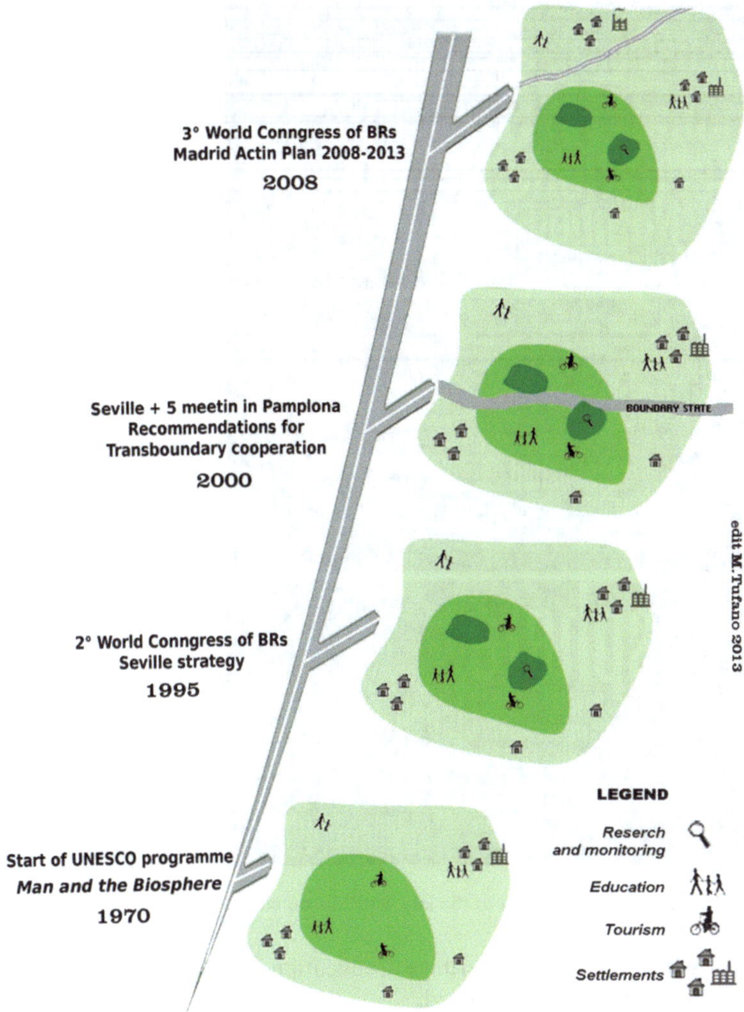

Fig. 11.3 The chronological development of the BRs' zoning and functions (Source: authors' elaboration of the original by Lange 2011)

Figure 11.3 illustrates how the initial BR concept – simple zones and few functions – has been developed over time, in relation with the MaB Programme milestones. Only after 1995, a clear three-zone and three-function territorial scheme has been introduced, which has been well reflected in the so-called "second generation BRs" layout.

Despite the fact that the total surfaces of the BR have increased over time – primarily due to the inclusion of larger transition zones – the MaB Programme has never explicitly considered the landscape as a specific attribute (or a possible nomination criteria); it seems paradoxical while confronting with the activities

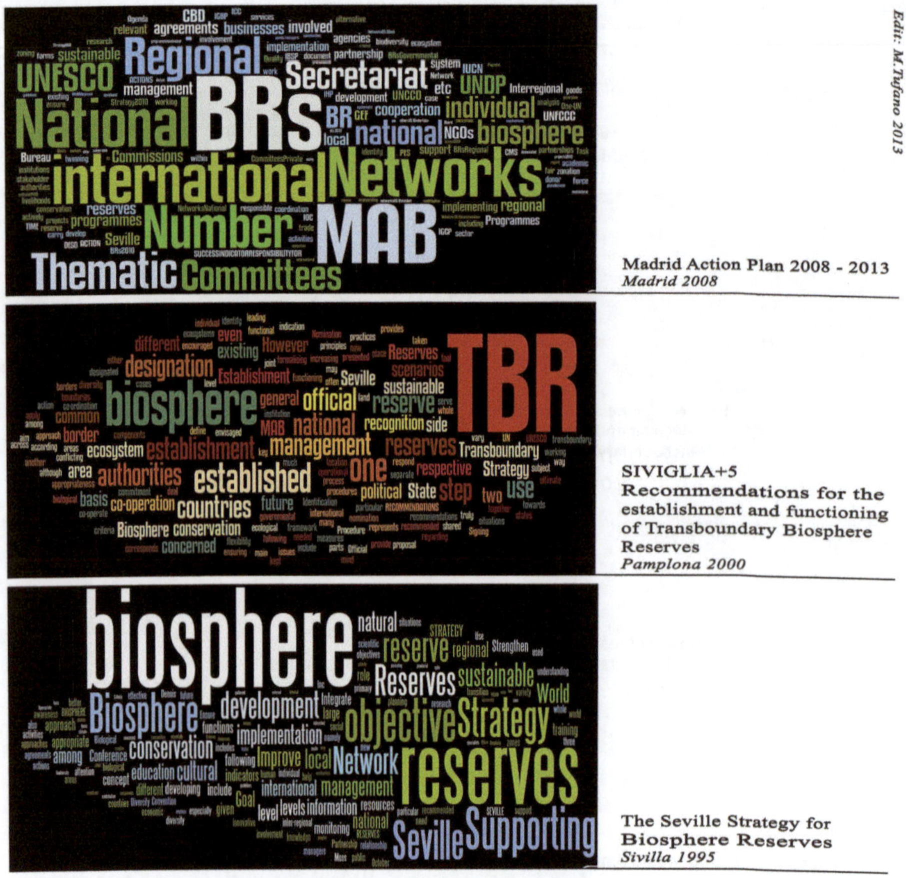

Fig. 11.4 The word cloud applied to the key MaB documents

developed within the World Heritage Convention,[15] where the concept "cultural landscape" was officially introduced.[16]

Ultimately (Fig. 11.4), by analysing the wording used in the recent key MaB strategic documents – namely, the Seville Strategy (1995), the Pamplona Recommendations (2000), and the Madrid Action Plan (2008) – it can be observed that the term landscape doesn't appear amongst the most visible ones.

This reflects the lack of specific discussion at the MaB international scale on the appropriateness of incorporating the landscape conceptual and operational categories in relation with the BRs. In fact, in various occasions, scientists and practitioners reported to the MaB Secretariat about researches dealing with the landscape

[15] UNESCO hosts the World Heritage Centre, which acts as the Secretariat of the World Heritage Convention.

[16] On the occasion of the 16th Session of the World Heritage Committee (1992).

aspects of the BRs[17] but a more structured and integrated discussion on these implications in the MaB operational aspects is still absent.

11.4 Conclusions

Looking for significant differences between BRs and PAs proved to be a misleading approach; in fact, scrutinising both the historical overview of the MaB Programme and the analysis of the trends in the designation process, it emerges that the two regimes have always been closely interlinked, both conceptually and geographically. Certainly, the name Biosphere Reserve:

> May generate some confusion and misunderstanding, given the meaning of the term 'reserve' in the common language, as well as in numerous national legislations or according to IUCN classification, environmentalists and practitioners. (Tamburelli 2012)

Adopting a multi-scale approach, the relationship has been analysed at the three major scales of pertinence, namely, the *local*, the *national* and the *international* (Andrian and Gaudry 2010).

At the scale of the individual BRs, it can be observed that each of the proposed territories had – and still have – a component (at least, the core zone) that is represented by protected areas; the legally constituted core area(s) devoted to long-term protection cannot but coincide with pre-existing PAs. This aspect has led to the creation of functional correlations between the protection regimes and the "rest of the territories", that resulted in the need to identify coordination mechanisms to be devoted to the joint management of the whole BR territories. This strong correlation is reflected also at the national level; typically the MaB activities are coordinated by the ministries of environment and physical planning, and the BRs are – in many cases – considered another category of protected areas.

Ultimately, at the international level, if we search for the MaB Programme in the UNESCO website, we find it located within the theme of "Science for a Sustainable Future", under the "Geology, Ecosystem and Biodiversity" section.

From a conceptual point of view, it can be noted that at the time of the Biosphere Reserve Conference (1968), the underlying philosophy behind national park was that, in order to protect nature, you need to make it off-limits to the human population (Hadley 2011).[18] More recently (officially after 1995), within the MaB Programme, fostering economic and human development – which has to be

[17] One recent example is given by the paper presented by Pablo B. Eyzaguirre, on the occasion of the Expert Planning Workshop, held at UNESCO HQs on 24–25 March 2011, on "Bio-Cultural Mosaic Landscapes: Centres of Crop Domestication and Eco-Agricultural Adaptation and Innovation in the MaB Global Network".

[18] The oldest National Park in the world may be Bogd Khan Uul, established in 1783 in Mongolia. The first to be established in the West was Yellowstone, in the USA (1872).

socioculturally and ecologically sustainable – becomes equally important as identifying areas of exclusive nature conservation.

Despite the fact that BRs have recently been referred as “more than just protected areas”, the MAB Programme has been constantly confronting with its international counterparts devoted to the conservation issues; just as an example, the World Conservation Monitoring Centre (WCMC) and the World Commission on Protected Areas (WCPA) – both important IUCN organs – established a Biosphere Reserves Task Force. As a result, specific “Resolutions on Biosphere Reserves” were adopted, on the occasion of the recent editions of the World Conservation Congresses,[19] stressing that:

> The characteristic that distinguishes BRs is the combination of functions of conservation and development, inside a unique conceptual framework, with individual sites connected through an international network.

The idea that UNESCO BRs are to be seen as “model region with a global reputation”[20] is an interesting vision for the future of the Programme, but at the condition that a real experimentation of new conceptual and operational modalities is conducted using BRs and their networks, with the intention to test the effects on the co-evolution of the human and natural systems. The involvement of all the stakeholders – including the private sector – at large territorial scales, in the evolving management governance system, will be the best way to use the BR as “learning sites”, as “*environments where policy makers, manager, local people and researchers collaborate and learn together to understand and address complex problems of common interest in a systematic way*” (Nguyen et al. 2010).

A specific discussion on the landscape dimension of the BRs and its implication in conceptual and operational terms is still missing in the MaB Programme; this – in our understanding – represents a limit to a postmodern evolution of the BR-related activities that are to be refocused on larger-scale territorial dynamics and the functional relationships between the various constituting zones. Probably, it is the most appropriate way to keep the original “*idea and ideal*”[21] alive.

References

Andrian G, Gaudry KH (2010) The role of UNESCO designated sites in fostering international cooperation and environmental security in SEE. In: Montini M, Bodganovic S (eds) Environmental security in South-Eastern Europe. Springer, Berlin

Hadley M (2011) Forty years of field laboratories in sustainability. In: Vv. Aa. A world of science, UNESCO, 9(4):2–11

---

[19] Occurred in 1996, 2000, and 2003.

[20] This is the title of the UNESCO Today, the journal of the German Commission for UNESCO, issue no. 2/2007.

[21] “Biosphere Reserves are an idea and an ideal” affirmed Peter Bridgewater, former MaB Secretariat Director, on the occasion of the Madrid Conference (February 2008).

Lange S (2011) The development of UNESCO's MaB Programme, with as special focus on mountain aspects. In: Austrian MaB Committee (ed) Biosphere reserves in the mountains of the world. Excellence in the clouds? Austrian Academy of Science, Vienna
Nguyen NC, Ockie Bosh JH, Maani KE (2010) Creating "Learning Laboratories" for sustainable development in biospheres: a system thinking approach. System Research and Behavioural Sciences. Available via Wiley InterScience. http://www.interscience.wiley.com. Accessed 13 Nov 2013
Phillips A (2003) Caring of the assets – the effectiveness of protected area management. Paper presented on the occasion of the 2003 World Conservation Congress. IUCN, Durban
Stoll-Kleeman S (2007) Success factors for biosphere reserve management. UNESCO Today 2:37–39
Tamburelli G (2012) Global Protected Area Programmes – an overview. Environ Policy Law 42 (2):96–101
UNESCO (1971) Rapport final, Conseil international de coordination du Programme sur l'homme et la biosphère (MAB), première session. UNESCO, Paris
UNESCO (1990) International co-ordinating council for the programme on man and the biosphere. Eleventh session. Paris, 12–16 Nov 1990. MAB report series, no. 62. UNESCO, Paris
UNESCO (1995) The biosphere reserve concept, the Seville Strategy and statutory framework of the world network, UNESCO, Paris
UNESCO (2002) Biosphere reserves, special places for people and nature. UNESCO, Paris
UNESCO (2008) The Madrid Action Plan. MaB secretariat. UNESCO, Paris
Vallega A (1995) La regione, sistema territoriale sostenibile. Mursia, Milan

Chapter 12
Connecting the Alpine Protected Areas in a Wide Ecological Infrastructure: Opportunities from a Legal Point of View

Paolo Angelini

Abstract A supporting legal framework is an essential prerequisite for the establishment of an ecological continuum throughout the Alps. The Alps consist of eight different countries, each of which has its own legal framework. Moreover, the individual countries may have federal states or provinces with specific regulations. Different legislations in force at various governance levels potentially affect ecological connectivity. Analyzing the impediments for the establishment of functioning ecological networks among protected alpine areas in order to preserve biodiversity for the region is a primary target. This activity was performed on the basis of national assessments. The survey was aimed at identifying the obstacles to ecological connectivity and the best tools to establish and/or maintain ecological corridors and networks. Furthermore, the global dimension needs to be taken into account, as well as the EU legal tool "European Grouping of Territorial Cooperation" (EGTC) that seems to be a suitable one in specific Alpine cases, as a way to facilitate and enhance cooperation at interregional and international levels that reach across borders. It enables regional and local authorities and other public bodies from different member states to join together in a cooperation grouping obtaining legal personality. The opportunities offered by an EGTC are therefore worth being considered in a policy making perspective. The contribution is expected to build on the results of the legal work package of the EU Alpine Space Programme Project ECONNECT, where the Italian Ministry for the Environment strongly participated.

Keywords Alpine protected areas • Alpine Convention • Ecological networks

---

P. Angelini (✉)
Italian Ministry for the Environment, Land and Sea, Via C. Colombo 44, 00147 Rome, Italy
e-mail: angelini.paolo@minambiente.it

R. Gambino, A. Peano (eds.), *Nature Policies and Landscape Policies*,
Urban and Landscape Perspectives 18, DOI 10.1007/978-3-319-05410-0_12

12.1 The Protected Areas in the Alps

The relevance of the paper for the current research has two main aspects:

- The emblematic importance of the alpine system in landscape policies, facing risks and processes such as the fragmentation of the ecological continuity, or other negative effects related to global change
- The importance of legal frame and particularly the EGTC to ensure the effectiveness of biodiversity conservation at landscape scale in a transborder context

The protected areas are key to safeguard biodiversity in Europe and in the Alps. Moreover, as repositories of natural and cultural values, they contribute to increasing environmental understanding as well as to socioeconomic well being (Bonnin 2007).

Hence, ecological connectivity is a determining factor for the survival, migration, and adaptation potential of all plant and animal species present in a given habitat and an important factor for the preservation of ecosystem services (Alpine Convention 1994). The development of an ecological network for the entire Alpine region would provide a solution to tackling fragmentation of the Alpine space especially as an adaptation strategy to climate change.

The Alpine Arc is the second biodiversity-rich (Sarukhàn 2005) and at the same time one of the most densely populated regions in Europe (Alpine Convention 2012). In this human-dominated landscape, the natural environment is subject to multiple pressures that are apt to bring about habitat destruction and fragmentation, which not only reduces the overall size of natural habitat but also leads to landscape "patchiness," that is, the isolation of natural areas into distinct habitat "islands" that prevent essential ecological processes from taking place. Indeed, ECONNECT Project (2011d) asserts that:

> An un-fragmented ecological continuum in the landscape would ideally consist of a rich variety of interconnected natural habitats hosting a rich variety of species.. (ECONNECT Project 2011d)

Nevertheless, a good ecological *continuum* largely depends on the existence of an integrity and functioning of ecosystems, including the conservation of biodiversity and provision of important ecosystem services (Santolini 2012).

The integrity and functioning of ecosystems, including the conservation of biodiversity and provision of important ecosystem services, largely depend on the existence of an ecological continuum.

Overcoming habitat fragmentation therefore plays a significant role for achieving effective conservation of the biodiversity levels in the Alpine Arc, in compliance with a number of international (United Nations 1992) and regional conventions and agreements, including inter alia the Alpine Convention, the Convention on Biological Diversity, and the European Union "Habitats" Directive of which the NATURA 2000 network is a central pillar (European Union 1992). It is also in line with the targets set out in the new EU 2020 biodiversity strategy (ECONNECT Project 2011a). The "Habitats" Directive intends to create an ecological network through

Europe. The provisions of Article 10 of this Directive contain measures for improving the ecological coherence of the ecological network (ECONNECT Project 2011b).

According to the Guidance on the maintenance of landscape connectivity features of major importance for wild flora and fauna (Kettunen, 2007) elaborated at the European level for improving the coherence of the Natura 2000 Network, it is clear from the texts of the "Habitats" Directive that the interpretation of the concept of "coherence" is a key issue affecting the implementation of directives.

To meet the requirements envisaged by the Directives, Member States have then to apply the legal provisions for the implementation and the management of the Natura 2000 Network, and different Alpine countries adopt various legal frameworks which can be devised on either the national or regional level.

12.2 A Comprehensive Legal Framework in Support of Ecological Connectivity in the Alpine Region

A supporting legal framework is an indispensable prerequisite for the establishment of transnational ecological connectivity measures for preserving and enhancing an ecological continuum throughout the Alpine Arc that in order to be effective, need to be designed taking account three different sets of issues: (1) identification of legal opportunities and obstacles for the feasibility of every project; (2) legal institutions governing private lands, where fragmentation needs to be overcome; (3) connectivity issues must be taken into account in land use planning processes through an integrated legal framework.[1]

The ECONNECT project analyzed and compared the legal frameworks favoring ecological connectivity among the Alpine States. One tool emerged as being especially appropriate for overcoming legal and social barriers: the European legal instrument EGTC (European Grouping of Territorial Cooperation), was designed to facilitate and promote cross-border, transnational, and interregional cooperation by enabling interest groups and institutions, as well as regional and local authorities from different EU Member States, to form cooperative associations within legislation (ECONNECT Project 2011b).

[1] Connectivity is an issue involving very different scales and multiple and diverse stakeholders. It became clear within the ECONNECT project (www.econnectproject.eu/) that the respect of private landowners' rights is a key element for the conservation and improvement of connectivity. It is impossible to realize a sustainable ecological continuum without the participation of private and public landowners and interest groups. See The ECONNECT Project, Policy Recommendations, p 3.

12.3 An European Instrument for the Facilitation of Transborder Cooperation: EGTC

The EGTC (European Grouping of Territorial Cooperation) is a community legal instrument introduced by Regulation (EC) no. 1082/2006 of the European Parliament and the Council. According to Article 2 of the abovementioned Regulation, the EGTC is meant to "facilitate cross-border, transnational and interregional cooperation [. . .] with the exclusive aim of strengthening economic and social cohesion."

To this purpose it rules that the EGTC shall have in each Member State "the most extensive legal capacity accorded to legal persons under that Member State's national law (Angelini 2009)." Unlike other instruments of cooperation, the EGTC has full legal personality in its own right, thus allowing public authorities of different states to associate and deliver joint services without the need for a prior international agreement to be ratified by national parliaments.[2] In particular, an EGCT can be participated by: Member States, regional and local authorities, and bodies governed by public law, a "body governed by public law."[3]

The State, however, has to agree on the participation of a potential member: to this purpose each prospective member is bound by Article 4 of Regulation (EC) no. 1082/2006 to notify the Member State under which it has been formed of its intention to take part in the Group, sending the State a copy of the proposed Convention and Statutes intended to govern the Group. The State shall then, as a general rule, reach its decision within 3 months from the date of receipt.[4]

Indeed, as mentioned, it is also possible for a Member State to become part of an EGCT, although its main objective is to serve as a cooperation tool for local/regional authorities. This constitutes an important change for territorial cooperation, as the possibility for Member States to participate had not been previously considered in the field of cross-border cooperation.[5] Due to these specific features,

---

[2] The EGTC may therefore acquire or dispose of movable and immovable property and employ staff, and it may also be a party to legal proceedings.

[3] A "body governed by public law" means anybody: (a) established for the specific purpose of meeting needs in the general interest, not having an industrial or commercial character;(b) having legal personality; and (c) financed, for the most part, by the State, regional, or local authorities, or other bodies governed by public law or subject to management supervision by those bodies or having an administrative, managerial, or supervisory board, more than half of whose members are appointed by the State, regional, or local authorities or by other bodies governed by public law (Regulation (EC) no.1082/2006, Art.3).

[4] Should the Member State consider the proposed participation not to be in conformity with either Reg. (EC) no. 1082/2006 or its national law, or that the participation would be detrimental to public interest or public policy, it will give a statement of its reasons for withholding approval (Reg. (EC) no. 1082/2006, Art. 4).

[5] As a consequence, this will allow some Member States to participate in such cooperation where no regions exist (e.g., Slovenia, Luxembourg) or where the envisaged theme of cooperation is a competence of the national level (www.interact-eu.net/the_egtc_regulation/68, 26 May 2009).

the EGTC can arguably be considered as a juridical figure that is particularly suited for the execution of cooperation actions or projects involving established partners in different Member States, namely, those with access to European Union co-financing through structural funds.

12.4 A Cross-Country Comparison of the National Legal Frameworks for the EGTC in the Alps: Instrument to Develop and Alpine-Wide Infrastructure for Protected Areas

In the previous paragraph, we argued that EGTC can be considered a suited juridical figure for the execution of cooperation actions or projects involving established partners in different Member States that can support the legal framework for the creation of a functioning ecological network in the Alps apt to preserve its extraordinarily rich biological diversity.

Nevertheless, differences between Alpine States are still located in the national legal order for the implementation of Regulation no. 1082/2006 (ECONNECT Project 2011c), especially with regard to competencies of territorial entities in the field of transborder cooperation and therefore the sharing of competences within the Alpine States. Analyzing the transnational legal impediments for the establishment of functioning ecological networks among protected alpine areas in order to preserve biodiversity for the region is a primary target also to consistently proceed on with the implementation of the current international and EU law and policies. In this paragraph, we will provide a brief overview of the national and/or regional legislation currently in force in four countries which are Parties to the Alpine Convention concerning the application in the respective national territories of EU Regulation no. 1082/2006, with the aim to analyze and compare the current situation as a basis to identify new challenges to be addressed in order to develop concrete transnational ecological connectivity measures to build an Alpine-wide ecological infrastructure.

The provisions for the implementation of the European regulation on the EGTC are integrated in the Community Law 2008 adopted in July 2009. Chapter III of this text is about the EGTC. According to paragraph 2 of Chapter III, then, the EGTC whose bench is in Italy have the legal personality governed by public law (ECONNECT Project 2011c).[6]

[6] The regulation refers to the notion of body governed by public law defined in Directive 2004/18/CE (Article 9, paragraph 9), but the Community Law 2008 does not quote this Directive. Directive 2004/18/EC of the European Parliament and of the Council of 31 March 2004 on the coordination of procedures for the award of public works contracts, public supply contracts and public service contracts (OJ L 134, 30.4.2004, pp 114–240)

The regional and local authorities (paragraph 3) designed in Article 3 of Regulation no. 1082/2006, are, respectively the Regions and the Autonomous Provinces of Trento and Bolzano, and the local entities designed in Article 2, paragraph 1, of the legislative decree no. 267/2000 (Angelini 2002).

According to Chapter III of the Community Law 2008 (paragraph 3), the object and the duties of the EGTC have to be laid down in a statute, whose minimal requirements are more strict in the Italian text compared to in the European Regulation. Different authorizations have to be given for the creation of an EGTC, such as the agreement of all the interested administrations has to be given for the creation of an EGTC.

On the basis of the application of paragraph 16 of the Preamble of Regulation no. 1082/2006, Switzerland – a Contracting Party to the Alpine Convention but not of the EU – could take part in a EGTC, which is quite relevant considering the Alpine scope of our analysis.

One limit to be highlighted for the participation of Switzerland in an EGTC, "[an] EGTC shall be made up of members located on the territory of at least two Member States"(Article 3, paragraph 2), this provision implying that it is not enabled to create an EGTC between only two States including Switzerland (ECONNECT Project 2011c).

There was a debate in Austria in order to clarify which institutions (Länder o Bund) were competent to adopt the legislation for the EGTC. However, as of an application of the so-called Generalklausel integrated in Article 15 about the sharing of competences between the Bund and the Länder of the Austrian Basic Law/Constitution, it is clarified that it is both competence of the Bund and the Länder, depending on the fields covered by the EGTC. As a result, nine laws adopted at the regional (Länder) level and one adopted at the federal level on the EGTC are into force in Austria.

The *Code général des collectivités territoriales* (CGCT) (Territorial Community Code) was modified Law no. 2008–352 (Decocq 2006). This law adapted the French law to the provisions of the European regulation on the EGTC. It allows the French local entities, within the limit of their competences and the respect of the international engagements of France, to create such a grouping with territorial collectivities, statutory bodies of the EU Member States, like with Member States. The law authorizes also the creation of an EGTC with Border State Members of the Council of Europe. In any case, the new opportunities offered in the French legislation by Article L.115-5 of the CGCT are limited to the creation of an EGTC.[7]

---

[7] Indeed the general interdiction for the territorial collectivities to conclude agreements with foreign states remains. This general interdiction generates problems for the collaboration between France and micro-States like Luxembourg or Monaco. This interdiction is based on constitutional considerations as it was explained in a study of the French Council of State published in 2007.

12.5 New Challenges

Protected areas are a key element of ecological networks due to their spatial role in the network and their function for the initiation and support of the process to maintain and restore ecological connectivity.

The institute of "legitimization" to protected areas management has to be conferred by the competent administrative organ in accordance with the political systems of the individual Alpine countries (federal or centralized systems). In a similar context, protected area managers at the pan-Alpine level should be involved in all the decision making process to actively support the functioning of ecological processes beyond the borders of the protected area itself. For this reason, it is necessary that local or regional authorities grant them instruments of formal competence to engage including within the peripheral zone or entire park region. In this sense, close cooperation with the competent administrative authority in questions of ecological connectivity is fundamental.

The EGCT must not only be seen as a legal framework: on the one hand, it represents an important tool to overcome the borders and expand the scope of international relations passing from a manly bilateral dimension to a multilateral dimension therefore creating opportunity for economic, social, and cultural development of the subject involved; on the other, the EGCT provide potential to solve the issues related with the cooperation activity as the attribution of functions, competences, and accountability.

As in many other fields, the EU enlargement with the integration between the Countries require stable frameworks for the cross-border cooperation, also to promote the sustainable development of a territory that, fulfilling the subsidiary principles, must take into account the geographical level most suited to achieving the goals set, therefore, not only to international or national level but also at the subnational and local realities (Fodella 2005).

Ensuring an ecological continuum between territories characterized by the same ecosystems favors an appropriate dimension of the ecosystems themselves which increases the ability of the ecosystems to host a greater number of individuals of the same species and reduces the risk of consanguinity within the same species. This group of mutually connected ecosystems that can activate an *autopoiesis* support the conservation of the level of biodiversity in the ecosystems under analysis.

In the Alpine arc, this strategy especially concerns the realization of ecological connections between protected areas, meaning that the possibility for concrete practical and legal measures to be foreseen, when necessary, as to be taken even outside of the protected areas, should be considered. This new challenge is gradually emerging on the legal stage, affecting not only environmental legislation but also a number of other fields such as spatial planning and agriculture (Worboys 2010).

References

Alpine Convention (1994) Protocol on the implementation of the Alpine Convention of 1991 relating to the conservation of nature and the countryside. Available via http://www.alpconv.org/en/convention/protocols/Documents/protokoll_naturschutzGB.pdf. Accessed 15 Dec 2013

Alpine Convention (2012) Contribution of the Alpine Convention to the process towards a macro-regional strategy for the Alps. An "Input paper", approved by the XII Alpine conference in Poschiavo, 6–7 Sept 2012

Angelini P, Randier C (2009) The European grouping of territorial cooperation (EGTC), EURAC research, Bolzano. Available via http://www.econnectproject.eu/cms/sites/default/files/Analysis%20of%20EGTC.pdf. Accessed 14 Dec 2013

Angelini P, Egerer H, Tommasini D (eds) (2002) Sharing the experience, mountain sustainable development in the Carpathians and the Alps, international meeting and roundtables, EURAC, Bolzano

Bonnin M (2007) The Pan-European ecological network: taking stock, Nature and Environment, n. 146, Council of Europe

Decocq C (2006) Rapport sur la proposition de loi n°2624, adoptée par le Sénat, relative à l'action extérieure des collectivités territoriales et de leurs groupements, Rapport de l'Assemblée nationale n°3610, p 12. Conseil d'Etat, Le cadre juridique de l'action extéri-eure des collectivités territoriales, Les études du Conseil d'État, La documentation française, Paris

European Union (1992) Council directive 92/43/EEC of 21 May 1992 on the conservation of natural habitats and of wild fauna and flora, Annex II and IV. Available via http://eur-lex.europa.eu/LexUriServ/LexUriServ.do?uri=CONSLEG:1992L0043:20070101:EN:pdf. Accessed 15 Dec 2013

Fodella A (ed) (2005) Creazione di nuove forme di cooperazione transfrontaliera a livello sub-statale per lo sviluppo sostenibile del territorio. In: Strumenti giuridici della cooperazione per lo sviluppo sostenibile di un'area montana transfrontaliera. Proceedings, Eurac-Research, Bolzano

Kettunen M, Terry A, Tucker G, Jones A (2007) Guidance on the maintenance of landscape connectivity features of major importance for wild flora and fauna. Guidance on the implementation of Article 3 of the Birds Directive (79/409/EEC) and Article 10 of the Habitats Directive (92/43/EEC). Institute for European Environmental Policy

ECONNECT Project (2011a) Policy recommendations. Available via http://www.econnectproject.eu/cms/sites/default/files/Policy%20Recommendations.pdf. Accessed 15 Dec 2013

ECONNECT Project (2011b) The Econnect project and the work package 6. Executive summary. Available via http://www.econnectproject.eu/cms/sites/default/files/WP6%20executive%20summary.pdf. Accessed 15 Dec 2013

ECONNECT Project (2011c) The Econnect project. Legal barriers compendium. Available via http://www.econnectproject.eu/cms/sites/default/files/WP6%20compendium%20outcomes_0.pdf. Accessed 15 Dec 2013

ECONNECT Project (2011d) Towards ecological connectivity in the Alps. The Econnect project synopsis. Available via http://www.econnectproject.eu/cms/sites/default/files/Synopsis.pdf. Accessed 15 Dec 2013

Santolini R (2012) The indicators for the ecological network: indicators for monitoring of durable development in the Alpine regions. Platform "ecological network" of the Alpine Convention. Available via http://www.alpconv.org/en/organization/groups/WGEcologicalNetwork/Documents/20120628_Ecol_Network_Indicators_syntesis.pdf. Accessed 15 Dec 2013

Sarukhàn J, Whyte A (2005) Millennium ecosystem assessment. Document 293. Covelo London, Washington, DC

United Nations (1992) Convention on biodiversity, CBD. New York. Available via http://www.cbd.int/convention/text/. Accessed 15 Dec 2013

Worboys GL, Francis WL, Lockwood M (eds) (2010) Connectivity, conservation management. A global guide. Earthscan, London

Chapter 13
Protected Areas, Natura 2000 Sites and Landscape: Divergent Policies on Converging Values

Bernardino Romano and Francesco Zullo

Abstract The identification of Natura 2000 sites in Italy has led to a significant change in the geography of environmental protection, by profoundly strengthening the role of ecological and naturalistic values in a country where the collective culture is traditionally more prepared to understand cultural values. In general, the identification of Natura 2000 Sites, carried out according to the guidelines established by Directive 92/43/EEC, was based on a more scientific and less politically "negotiated" process compared to the one followed for the determination of protected areas, by selecting habitats of community interest and not landscape or historical and cultural values. It seems very clear that these are two different types of areas with partially overlapping values that require forms of territorial planning and governance that optimize multiple conservation goals: while Nature 2000 sites protect habitats, protected areas extend their function to cultural landscapes, historical heritage and traditions. The Ecological Network should be a decisive model to classify values and integrate rules, avoiding excessively specialized approaches and applying instead the typical techniques of preservation biology and connectivity conservation, together with routine urban and infrastructure planning techniques.

Keywords Ecological network • Nature 2000 • Landscape urbanization • Protected areas

13.1 Natural Areas: The National Framework

The establishment of Natura 2000 sites in Italy has led to a significant change in the geography of environmental protection, by profoundly strengthening the role of ecological and naturalistic values in a country where the collective culture is

B. Romano (✉) • F. Zullo
University of L'Aquila, Via Giovanni Gronchi 18, 67100 L'Aquila, Italy
e-mail: bernardino.romano@univaq.it

R. Gambino, A. Peano (eds.), *Nature Policies and Landscape Policies*,
Urban and Landscape Perspectives 18, DOI 10.1007/978-3-319-05410-0_13

traditionally more prepared to understand cultural values. Since 1995, year in which the "Bioitaly" programme cofinanced by the European Commission within the framework of LIFE Nature 1994 was implemented, the Italian regions, aided by various naturalistic groups, have identified almost 2,600 sites among SCIs (Sites of Community Interest) and SPAs (Special Protection Areas). SCIs alone, covering 4,530,400 ha, account for almost 15 % of the national territory (Calvario 2010; Maiorano et al. 2007).

Today, terrestrial protected areas (PAs) set up at national and regional levels amount to 3,163,590 ha (Ministry of the Environment 2010), but only 42 % of SCIs is included in PAs. These figures convey the idea of the geographical pervasiveness of the Natura 2000 programme, which has protected slightly less than 2,600,000 ha, in addition to park territories, almost doubling the Italian natural areas that were already protected.

The union of Natura 2000 sites (SPAs + SCIs), covering just less than 60,000 km^2, accounts for as much as 19 % of the national territory. Although high, this figure ranks Italy only in tenth in the EU of 27 behind countries, such as Bulgaria and Slovenia (with over 34 %) or Greece and Estonia with over one fourth of their respective territories identified as Natura 2000 sites (Natura 2000). However, it is also true that in terms of absolute value of Natura 2000 areas, Italy is preceded only by France, Poland and Estonia, and, in any case, our country is part of the group of six (France, Sweden, Germany, Italy, Poland and Estonia) with over 50,000 km^2 of established Natura 2000 sites (Table 13.1). This aspect is clear evidence of the fact that although Italy is broadly affected by severe phenomena, such as environmental degradation, uncontrolled urbanization (on average in excess of 7.5 % today) and ecosystem fragmentation, it still retains a significant expanse of natural habitats as evidenced by comparisons at European level (Mücher et al. 2009). And this is with a population density (203 inhabitants/km^2) that is lower only to that of Germany (228 inhabitants/km^2), but significantly higher than that of the other four countries considered (Poland 123 inhabitants/km^2, France 102, Estonia 28 and Sweden 20).

Besides their geographical expanse, PAs and Natura 2000 sites are marked by two very different approaches in terms of criteria and identification methods, as evidenced by their not uniform distribution in the country.

In general, the identification of Natura 2000 sites, carried out according to the guidelines established by Directive 92/43/EEC (Habitat), was based on a more scientific and less politically "negotiated" process compared to the one followed for the determination of Pas (Alphandéry and Fortier 2001), by selecting ecosystem values only (habitats of community interest) and not landscape or historical and cultural values. As a result, the study of their relationship with PAs in terms of size and geography provides important information on the efficiency of the environmental protection policy pursued in Italy until the end of the 1990s. This policy has perhaps focused too much on more aesthetical and cultural values, rather than purely naturalistic ones. Table 13.2 shows the two types of protected areas, PAs and Natura 2000 sites, and their distribution in Italian regions.

Table 13.1 National surfaces involved in the Natura 2000 programme in Europe updated to 2010 (Source: http://ec.europa.eu/environment/nature/natura2000/db_gis/)

Country	National surface (km^2)	Nature 2000 terrestrial area (km^2)	Natura 2000 national rate (%)
United Kingdom	244,820	17,683.22	0.07
Denmark	43,093	3,849.09	0.09
Latvia	64,589	7,304.53	0.11
Lithuania	65,301	7,879.07	0.12
France	549,192	68,789.94	0.13
Belgium	30,528	3,870.04	0.13
Malta	316	40.93	0.13
Ireland	70,280	9,122.4	0.13
Sweden	414,864	57124.04	0.14
Netherlands	41,526	5,724.52	0.14
Czech Republic	78,866	11,072.12	0.14
Finland	338,145	48,757.52	0.14
Austria	83,859	12,324.19	0.15
Germany	357,031	55,060.92	0.15
Spain	45,226	8,036.87	0.18
Romania	238,391	42,653.97	0.18
Luxembourg	2,597	471.34	0.18
Italy	301,333	57,736.45	0.19
Poland	312,685	60,781.74	0.19
Portugal	91,990	19,202.45	0.21
Hungary	93,030	19,938.72	0.21
Greece	131,940	35,804	0.27
Estonia	504,782	137,316.84	0.27
Cyprus	5,736	1,627.35	0.28
Slovakia	48,845	14,141.07	0.29
Bulgaria	110,910	37,634.08	0.34
Slovenia	20,273	7,202.98	0.36
Total EU	**4,290,148**	**751,150.39**	**0.18**

PAs exceed one quarter of the total area only in two regions (Campania and the Abruzzi), while the same threshold referred to Natura 2000 sites is exceeded in Liguria, Valle d'Aosta and once again Campania.

A significant aspect is the greater homogeneous geographical density of the two types of areas in the various regions. In fact, the standard deviation compared to the median value calculated in the columns of the regional area rates is 0.74 for PAs and 0.46 for Natura 2000 sites: this suggests an objective and more regular distribution of habitats of value which the park policy has been unable to seize, as it has been greatly conditioned by the local sensibility of regional governments.

Table 13.2 Distribution of PAs and Natura 2000 sites (SCIs) per region (Source: author's elaboration on data of Italian Ministry of Environment Decree 27 April 2010)

		Protected areas		Nature 2000 (SCIs)	
Regions	Regional area (ha)	Area (ha)	Regional area rate (%)	Area (ha)	Regional area rate (%)
Abruzzi	1,082,699	318,352	0.29	252,587	0.23
Basilicata	1,007,280	196,181	0.19	59,114	0.06
Calabria	1,522,338	270,248	0.18	85,976	0.06
Campania	1,360,917	349,251	0.26	363,275	0.27
Emilia-Romagna	2,218,437	93,781	0.04	226,481	0.10
Friuli-Venezia Giulia	785,993	52,624	0.07	132,170	0.17
Lazio	1,722,149	225,086	0.13	143,107	0.08
Liguria	540,595	28,056	0.05	145,428	0.27
Lombardia	2,386,119	149,646	0.06	224,201	0.09
Marche	974,954	88,293	0.09	102,608	0.11
Molise	446,103	6,265	0.01	97,750	0.22
Piedmont	2,538,879	179,717	0.07	282,345	0.11
Puglia	1,953,386	249,308	0.13	465,518	0.24
Sardinia	2,392,008	84,341	0.04	426,251	0.18
Sicily	2,555,398	259,841	0.10	384,065	0.15
Tuscany	2,268,096	127,604	0.06	286,839	0.13
Trentino-Alto Adige	1,360,077	268,017	0.20	301,525	0.22
Umbria	846,108	63,039	0.07	109,667	0.13
Valle d'Aosta	326,093	43,425	0.13	71,619	0.22
Veneto	1,842,400	110,516	0.06	369,866	0.20
Total and average	**30,130,029**	**3,163,591**	**0.10**	**4,530,392**	**0.15**

A clear example of this can be found in 15 regions, including Lombardy, Trentino-Alto Adige and Piedmont, where the density of Natura 2000 sites is higher than that of PAs, in some cases with huge differences: in Valle d'Aosta, Veneto, Friuli, Liguria, Emilia-Romagna, Tuscany and Umbria, the Natura 2000 rates are much higher than double those of PAs, but in Sardinia this value is fourfold higher, while in Molise it is 20-fold higher.

Only in five central and southern regions, where large national parks are found, is this ratio reversed, with Natura 2000 density that is lower than PA density: this is the case of Lazio, Campania, Abruzzi, Basilicata and Calabria, with PAs covering almost fourfold Natura 2000 sites in the latter region.

It is also worth mentioning that, despite broad local variations, national parks on average coincide with SCIs for less than half (48 %) of their expanse (Fig. 13.1).

This analysis once again confirms that the policies on protected areas pursued over the past 50 years have not been able to select the more purely eco-biological

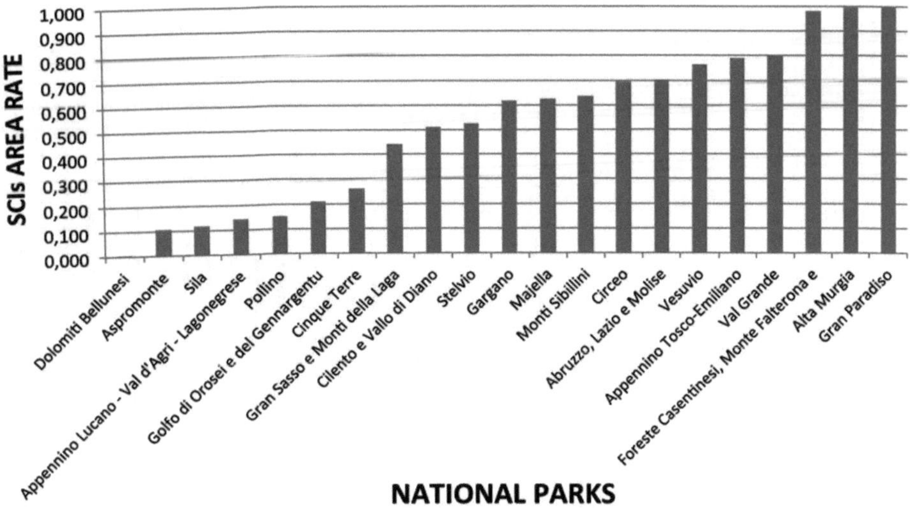

Fig. 13.1 Percentage of SCIs over the overall surface of Italian National Parks (Source: author's elaboration on data of Italian Ministry of Environment Decree 27 April 2010)

aspects of natural environments, but have favoured the greater political and collective sensitivity towards historical and landscape values. On the other hand, Italy, where one of the first laws in the world on the aesthetical value of landscapes was enacted (Law no. 1497 of 1939) has only recently developed a culture capable of fully understanding the ecosystem-related importance of the territory (Settis 2010).

13.2 Landscape Conservation Effects

The foregoing statement is also confirmed by the results in Fig. 13.1 showing incidence rates of PAs and Natura 2000 sites in the different types of landscapes listed by ISPRA in 2004.

PAs have protected most (over 50 %) of the different environmental categories (some volcanic formations, plateau landscapes in mountain areas and the dolomites landscape) but have greatly neglected some systems which the Natura 2000 programme focused on later. In fact, in the case of small islands, wetlands, carbonatic hills, volcanic mountains, granitic hills and lavic flats, the protection provided by Natura 2000 sites covers areas that are eight- to tenfold greater than those considered as PAs.

It may be said that Natura 2000 sites generally protect all Italian landscapes more, with the exception of intermountain basins and plateau landscapes in mountain areas. However, in the latter historical human activities have generally been far more intensive than in adjacent areas and the historical and visual values are normally by far greater than eco-biological ones.

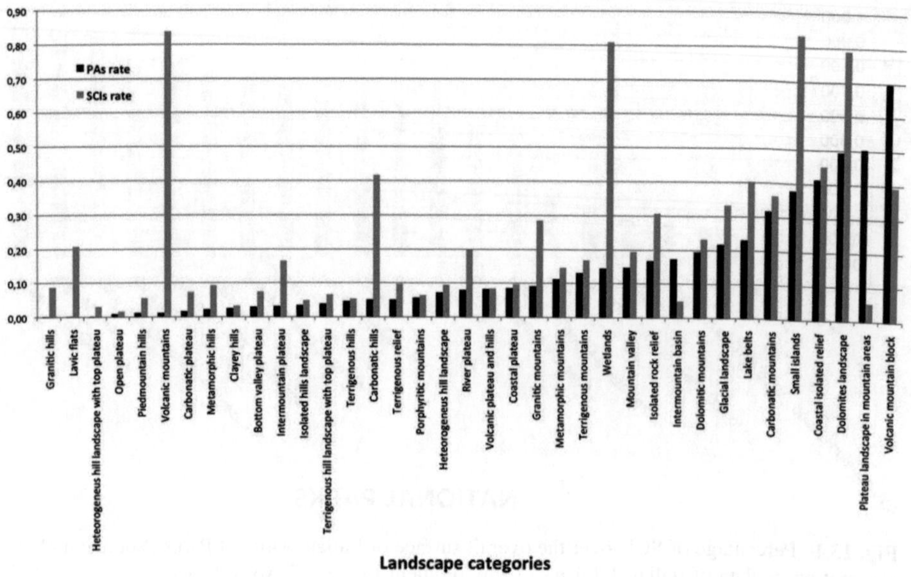

Fig. 13.2 Diagram showing the rate of incidence of the different types of landscapes in PAs and Natura 2000 sites

One aspect that PAs and Natura 2000 sites share is that they both are concentrated massively in medium-high altitudes, thus evidencing that the country's remaining naturalistic areas are found in these morphologically more marginal areas. Italian PAs include over 30 % of belts ranging between 1,000 and 1,800 m above sea-level, but even Natura 2000 sites cover 25 % of these strips. The greater ability of the Natura 2000 programme to seize environmental importance is once again evidenced in the plain and hill belts ranging between sea level and 600 m in altitude. In these areas, Natura 2000 sites, accounting for 42 % coverage, exert a greater pressure than PAs that do not exceed 35 % (Fig. 13.2).

As a result of these morphological features, PAs and Natura 2000 sites are scarcely affected by urbanization. At present, in PAs mean urbanization density is less than 1 % (9 ‰), and in SCI sites it is essentially similar. However, in the latter case, the most significant aspects concern urban transformation in the surrounding areas (Fig. 13.3): considering a buffer of 1 km in width around the sites, in the 1950s urbanization density in these areas was 2.7 %, but since 2000 it has risen to over 14 % (Romano and Zullo 2012a). Although the habitats within Natura 2000 sites have not been altered physically, urbanization causes border disturbances and above all severe consequences in terms of fragmentation between the same habitats, thus reducing or nullifying the potentialities of the ecological network, which the Natura 2000 programme aimed to set up.

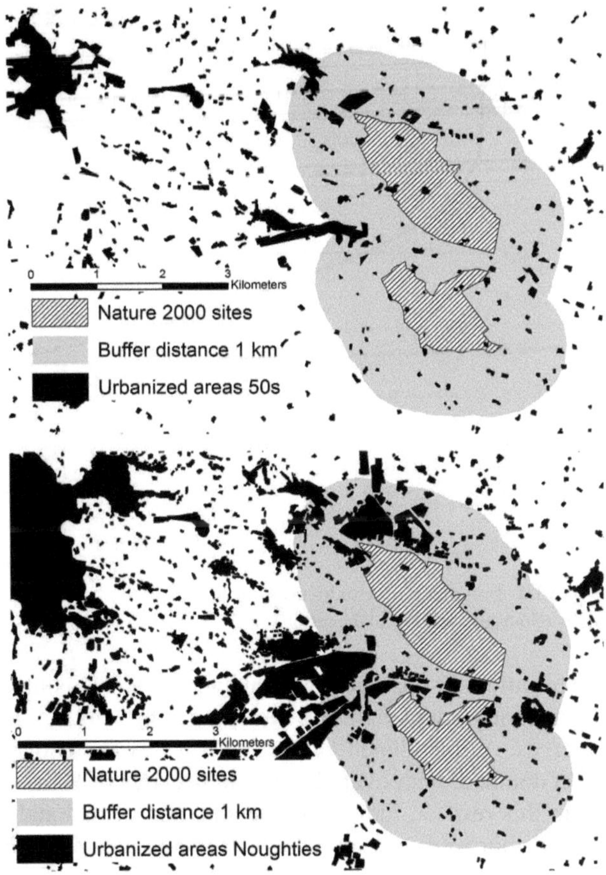

Fig. 13.3 Example of development of urbanized areas in buffer strips of 1 km in width of two Natura 2000 sites (Ansa degli Ornari, IT5210025, and Boschi a farnetto di Collestrada, IT5210077), highlighting that, although newly built-up areas have not affected the inner parts of the sites, they have however isolated and fragmented interstitial areas (Source: authors' elaboration)

13.3 Models for Converging Environmental Policies

In the past 20 years, the introduction of the principles to select habitats of Community interest in the Italian environmental culture has led to forms of political, technical and management disorientations in the traditional approach to conservation. One cause has certainly been the underlying misunderstandings inherent in the Natura 2000 programme itself: pursuing the goal of the European Ecological network through a "set" of isolated areas. In fact, in the planning and management stages, the Natura 2000 sites were often handled as normal protected areas, even if characterized by prevailing ecosystem-related attributes compared to other categories of values.

The same management plans have essentially been unable to have an impact on the changes affecting the surrounding territorial matrices, exactly as has been the case of the plans for conventional protected areas. Given the situation, we can certainly assert that the Natura 2000 programme has offered an important

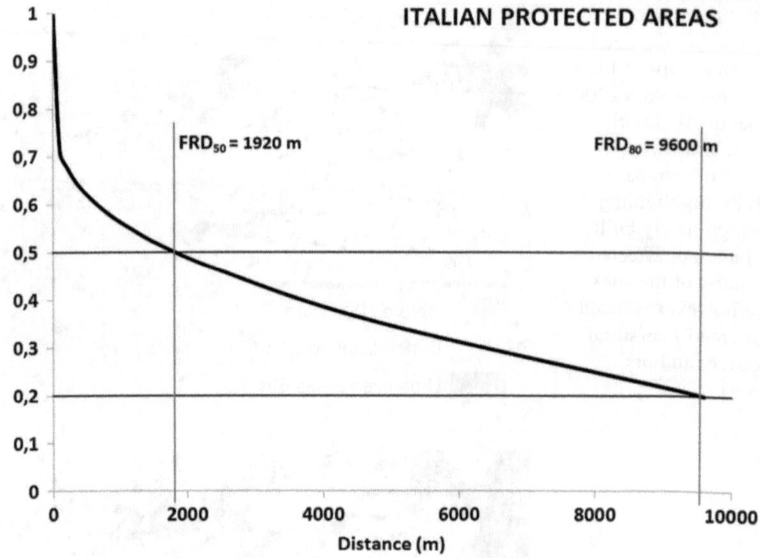

Fig. 13.4 Fragmentation reduction curve and FRD indices for the Italian system of protected areas (Source: authors' elaboration)

contribution to the geographical increment in natural areas to be protected, by significantly supplementing the yet patchy action undertaken through parks and reserves. As mentioned earlier, the various forms of protected areas have almost been doubled, with over six million hectares of PAs + SCIs union.

In this respect, indirectly, the situation of national ecological connectivity has surely improved: a recent study (Romano and Zullo 2012b) shows that the mean distance to be bridged in Italy to reduce fragmentation between PAs by 50 % is of almost 2 km (Fig. 13.4), while this same index applied to the union of PAs and Natura 2000 sites drops to a few hundred metres (566 m). The reduction of these distances between core areas and stepping stones has certainly improved the function of the actual ecological network for some land species, of great importance too from the standpoint of preservation (such as large mammals). However, this system (the Ecological Network) is not recognized formally in Italy, nor legally defined or planned, excluding only two regional cases: Umbria and Lombardy.

On the other hand, the improved efficiency in the preservation of biodiversity achieved through the geographical integration of protected areas is running the risk of being weakened: the creation of Natura 2000 sites seems to have deflated the role of protected areas, even in the eyes of the public opinion. It is most likely a coincidence, due to the economic crisis, that Italian parks are suffering huge cuts in resources over the past years. On the contrary, the Natura 2000 sites are able to draw on various European funds (e.g., ROP-ERDF or LIFE), albeit to a varying extent from region to region, and are expected to attract additional funds once converted into SCZs (Special Conservation Zones) and subjected to PAF (Prioritized Action Frame), as established by the same Habitat Directive.

Based on the above, it seems very clear that these are two different types of areas with partially overlapping values that require forms of territorial planning and governance that optimize multiple conservation goals: while Natura 2000 sites protect habitats, protected areas extend their function to cultural landscapes, historical heritage and traditions.

The planning tools used in these areas and in the territories that contain and link them are numerous: park plans, management plans for Natura 2000 sites, regional landscape plans and municipal land use plans. Only recently, albeit in the absence of a specific systematic national law, have some spontaneous attempts been made to pool the goals or at any rate to avoid contradictions between the various rules pursuing, in part, the same and, in part, different goals (conservation of biodiversity and habitats, retention of historical landscapes and enhancement of cultural heritage).

According to the opinions of many, the Ecological Network should be a decisive model to classify values and integrate rules, avoiding excessively specialized approaches and applying instead the typical techniques of preservation biology and connectivity conservation, together with routine urban and infrastructure planning techniques (Gambino and Romano 2004). This message is having some difficulties in making its way into routine territorial governance policies, although many signs suggest that over the next decade, there could be a significant shift in this direction in Italian culture.

References

Alphandéry P, Fortier A (2001) Can a territorial policy be based on science alone? The system for creating the Natura 2000 network in France. Sociol Ruralis 41(3):311–328

Calvario E (2010) La rete Natura 2000 nel governo del territorio. In: Romano B, Ferroni F (eds) Biodiversità, consumo di suolo e reti ecologiche. Ministero dell'Istruzione, della Università e della Ricerca, Franco Angeli Ed., Rome, pp 57–68

Ministero dell'Ambiente e della Tutela del Territorio e del Mare (2010) Decreto 27 aprile 2010, Approvazione dello schema aggiornato relativo al VI Elenco ufficiale delle Aree Protette. Rome

Gambino R, Romano B (2004) Territorial strategies and environmental continuity in mountain systems, the case of the Apennines (Italy). In: Harmon D, Worboys G (eds) Managing mountain protected areas: challenges and responses for the 21st century. IUCN/Andromeda Ed, Rome

Maiorano L et al (2007) Contribution of the Natura 2000 network to biodiversity conservation in Italy. Conserv Biol 21(6):1433–1444

Mücher CA et al (2009) Modelling the spatial distribution of Natura 2000 habitats across Europe. Landsc Urban Plan 92(2):148–159

Romano B, Zullo F (2012a) Land urbanization in Central Italy: 50 years of evolution. J Land Use Sci. doi:10.1080/1747423X.2012.754963

Romano B, Zullo F (2012b) Landscape fragmentation in Italy. Indices implementation to support territorial policies. In: Campagna M et al (eds) Planning support tools: policy analysis, implementation and evaluation. Franco Angeli Ed, Milano, pp 399–414

Settis S (2010) Paesaggio, costituzione e cemento. Einaudi, Turin

Chapter 14
Regional Planning for Linking Parks and Landscape: Innovative Issues

Angioletta Voghera

Abstract Since 2000, the emerging international indications (new paradigms relating to the protected areas, the European Landscape Convention, the socio-ecological approach of resilience) have expanded the relationship between protected areas and landscape. The paper reflects on conceptual innovations with reference to the methodological approach of some countries: the assessment of the landscape in the United Kingdom as a tool for defining policies and plans capable of integrating and harmonizing the development of human societies with the conservation of ecological and landscape stability and the territorial enhancement policies in the Netherlands that promote territorial development starting from nature and the landscape.

Keywords Landscape • Protected areas • Sustainability • European Landscape Convention • Adaption policies • Mitigation policies

14.1 International Visions

During the 2000s, new conservation paradigms were launched (Phillips 2003), expanding the vision of the relationships between protected areas and landscape.

On one hand, as regards protected areas, the great themes of the importance of populations and local communities for the conservation and management of the natural and anthropic components of the territory emerged, along with the conception of protected areas as tools for socioeconomic and territorial development (IUCN). On the occasion of the most recent IUCN Congresses,[1] this conception has been translated into recommendations, addresses and programmes to integrate the ordinary territory and the protected areas, starting from the consideration of the

[1] World Conservation Congress "People and Nature, Only One World", Bangkok, 2004; World Conservation Congress "For a Diverse and Sustainable World", Barcelona, 2008; World Conservation Congress "Nature +", Jeju, 2012.

A. Voghera (✉)
Interuniversity Department of Regional and Urban Studies and Planning (DIST), Politecnico and Università di Torino, Viale Mattioli 39, 10125 Turin, Italy
e-mail: angioletta.voghera@polito.it

R. Gambino, A. Peano (eds.), *Nature Policies and Landscape Policies*, Urban and Landscape Perspectives 18, DOI 10.1007/978-3-319-05410-0_14

contribution of protected areas to climate control, soil protection, food production and the quality of life within the scope of global challenges (under debate at the IUCN World Parks Congress that will be held in Sydney in 2014).

On the other hand, the European Landscape Convention (2000) and the subsequent Recommendation (Council of Europe, Committee of Ministers 2008) for implementation highlight the need to pay attention to the territory as a whole, linking ecological, archaeological, historical, cultural, perceptive and economic values and incorporating social and economic aspects. The protected sites and landscapes are an important resource for ensuring the "quality of the territory as a whole by defining policies for appraising, planning, developing and managing" the landscape (id. 2008).

So the specific, binding regulations, guidelines or rules developed for protected areas should be made part of general landscape planning and development. This idea sees parks as an engine for experimenting initiatives for the conservation and enhancement of the widespread landscape heritage, bringing all of the values of the territory and the local communities into play in everyday management (natural resources comanagement; Borrini-Feyerabend et al. 2007). Landscape planning and programming become necessary to promote multifunctional development – capable of enhancing the numerous ecological, social and cultural values of the landscape – for the diversification of activities and uses with the aim of controlling the microclimate and guaranteeing soil safety and water quality as well as the conservation and maintenance of the biotopes. The aim is to encourage sustainable development, which places environmental and landscape aspects in close relationships with those of a socioeconomic nature. This kind of development makes reference to the conception of landscape as an ecosystem service (Costanza et al. 1997; Ehrlich and Kennedy 2005; Egoh et al. 2008; Granek et al. 2010), capable of supplying goods and services, contributing directly and indirectly to human well-being.

The close link between policies for nature and the landscape, between conservation and management and between population and territory finds reference in the international cultural debate also in a new development principle, that of resilience, which brings into play the sustainability of the territory (Kates et al. 2001) as an innovation of social-ecological systems (Gallopín et al. 1989; Berkes and Folke 1998), based on the reciprocal interaction and adaptation between man and the environment (Turner et al. 2003).

Gaining consolidation primarily in ecology (e.g. White 1949; Steward 1955), the resilience concept refers to the process of structural change in response to external circumstances. Resilience refers to the capacity to adapt to future changes in the environment of the system concerned, taking on a multitude of meanings, which permeate anthropology, human geography, social science (Folke 2006), risk management (Kasperson et al. 1995), the fight against climate change, and the planning of the territory (Davoudi 2012). In the cultural debate emerges a utopic vision that imagines a future for the city, the territory and the landscape, in its natural and anthropic characteristics, setting a nonlinear transformation process that invests the quality of the performance of the environmental and territorial system. The evolutionary approach linked to resilience is applicable to the landscape systems, strongly characterized by an integration between man and nature, crossed by processes of adaptation to the environment that continue in time (ecological approach; Kauffman 1993) and space (Gunderson et al. 2010) according to multiple balance models, and sustained strongly by the creative capacity of social capital, which is based on the role of the individual and the institutions. A development model, which imagines a territory capable of adapting and innovating through planned and programmed actions and through bottom-up actions and capable of tackling its difficulties and regenerating its memory and its symbolic and landscape system in a sustainable way, has been affirmed.

Adaption processes and transformation of landscape are indissolubly linked; in fact in the landscape the changing process is continuous and affects the single elements that combine within it, conditioning the state of the landscape (Van Eetvelde and Antrop 2004), and the landscape changes can be absorbed by the landscape structure. Landscape resilience integrates the more well-known concept of vulnerability of the landscape (Lyle 1985; Kozlowski 1986; Klausmeyer et al. 2011), highlighting the system components and responding in an integrated and trans-sectorial manner to the transformation processes that touch the landscape. Both of these concepts have already been taken into consideration in the process for the assessment of landscape sensitivity in the United Kingdom,[2] which assesses the stability of the landscape characteristics and values and their ability to recover any losses and/or damages, looking at the innovative actions/pressures generated by the transformation policies (Swanwick 2002) from a dynamic and sustainable viewpoint. The landscape evaluation can support the processes of co-evolution between the anthropic and natural system via techniques capable of measuring and qualifying the choices made in terms of conservation, planning and management of protected and ordinary landscapes.

14.2 Dynamics and Policies of Integration Between Nature and Landscape

In the Netherlands and the United Kingdom, territorial policies try to link conservation and development of nature and the landscape, being strongly anchored to the tradition of planning and programming which is mainly characterized by the combination of strategies and interventions and on the implementative responsibility of the institutional and social stakeholders (Voghera 2011a, b). Moreover, the impacts of policies and projects are assessed in their development through consolidated methods, tested by experts and public administrators, capable of considering the complexity of the values and of the social and collective functions ascribable to nature and landscape (Sijtsma et al. 2013; Swanwick 2002). The effectiveness of landscape enhancement policies lies in society's participation in the definition and

---

[2] Landscape Sensitivity in the Landscape and Visual Impact Assessment (LVIA).

making of decisions as well as in integration at all administrative levels with physical and sectorial planning, creating synergies and maximizing the results.

In the Netherlands, key planning decisions, territorial development strategies relating to the physical planning and quality of the landscape (Agenda Landschap, 2008) – which integrate actions in the sector aimed at safeguarding and enhancing views from the main infrastructures directing the conservation and maintenance of the rural territories ("natural monuments", Monument Act, 1988, art. 6) and of the protected landscapes (The Belvedere Memorandum 1999), representative of the national identity – are defined on a national scale. Moreover, through the different strategies of planning, conservation and project of the territory, anthropic development is controlled, also with setting up of buffer zones between the protected landscape and the ordinary territory, where co-evolutive strategies for ecological and landscape conservation are tested and key enhancement projects are identified. These indications find reference for implementation in the provincial planning that guides the work of the municipalities, monitors the effects of the territorial choices and defines adequate measures for compensation of the projects.

We can highlight some difficulties in this process in Arkemheen-Eemland landscape related to the too weak role of the institutional actors (i.e., province). A significant example is that of Ijsselmeer (a Natura 2000 Site) and of De Hoge Veluwe National Park, where public and private stakeholders, involved in the definition of the local plan, have promoted actions aimed at experimenting a better integration between physical planning and sectorial strategies (water, protected areas, farming, landscape, transport, etc.), leading a bottom-up process of reciprocal adaptation between the natural and the anthropic system. This has led to a plan that is consistent with the single elements/values; territorial and landscape restrictions linked to the park, on one hand, and development – also at economic level – on the other (through the construction of great infrastructures like Hanzelijn, the new railway and motorway link, which, having been assessed and monitored since the project stages, contribute to the enhancement and renewal of the landscape system), are in synergy for the quality of the territory and life.

In integrating sectorial and landscape actions, economy and governance of the territory, the Netherlands is effective at promoting the quality of life and the landscape (verified with the Social Cost/Benefit Analysis of the Landscape, 2007), also for the strategic and financial support of the Dutch Landscape Funding Task Force, which try to guarantee the continuity, the sustainability of the policies and the overall resilience of the system.

Since the 1990s, the United Kingdom has implemented vast landscape strategies[3] and made an overall assessment of the territory, to define actions and impacts. The landscape assessment directs the governance strategies of the territory and the project on different scales (national, regional and district), focusing on the communication of the landscape, using abacuses, manuals and guidelines to outline possible interventions. Assessment is the tool to build co-evolutive strategies for the

[3] Landscape Guidelines, Landscape Convention – A Framework for Implementation, 2007, 2009.

landscape capable, on the different scales of territorial governance, of coordinating the choices of conservation and enhancement of the natural and anthropic components. The aim of the National LCA Guidance (Swanwick 2002) is to contribute, by measuring the landscape capacity and sensitivity, to the definition of responsive solutions and strategies with regard to the evolution of the natural and anthropic components.

The project strategies and indications identified in this way will be compatible with the capacity to support the transformation of different landscape types and values and will be capable of creating a new identity, with opportunities for implementation on a local scale. With different methods (overall landscape sensitivity[4], sensitivity to landscape change,[5] landscape capacity[6]), the sensitivity assessment allows a dynamic viewpoint, through the consideration of physical and visual characteristics, to read the relationship between the pressures generated by the policies for the transformation and development and certain characteristics of the landscape, such as significance, vulnerability, replace ability, stability and resilience with respect to the planning choices (Swanwick 2002). The assessment techniques will have to be different in relation to the scale and the type of planning (strategic, landscape, land use). The same landscape sensitivity index varies, in time and space.

The assessment activity enables the qualification and legitimization of planning choices so that they are able to integrate and harmonize the development of human societies with the conservation of ecological and landscape stability, favouring the passage from one state of balance to another, on the basis of the capacity for adaptation to change, transformation and increase of resilience (Folke et al. 2010). It is possible to highlight and monitor the development of the results of choices and interventions in order to minimize the impact on ecological connectivity in terms of energetic flows, erosive costs, dissemination of pollutants, soil consumption, etc. The aim is to promote, through policies and projects, the

[4] Determination of the "general sensitivity of the landscape", which depends on the characteristic sensitivity relating to the single landscape resources and to visual sensitivity, independent of the type of change envisaged. It is generally used in the preparation of vast regional or subregional strategies or for the LVIA.

[5] Identification of sensitivity with respect to a specific type of change or development. It is achieved by studying the landscape, its characteristics and values and the way in which it is perceived by the population, as well as the nature of the change proposed.

[6] Measuring the landscape capacity (LC), the capacity to respond to the different territorial governance processes. LC depends on the sensitivity to change, visual change and that of the general and specific value of the single landscape elements. LC is used to assess the effects of relevant policies. In general, the assessment categories used are qualitative (high, medium, low), while the quantitative approach finds reference in those assessments that use GIS to organize and integrate the different levels of information, allowing the construction of evaluative matrices and algorithms and the dissemination of information and elaborations among a nonexpert public (through the web, via the principle of transparency) as well as guaranteeing an effective and efficient use of information for the definition of alternative scenarios of action on the landscape (Landscape Guidelines).

conservation of better functionality of the landscape, reducing the potential danger of stochastic events (such as fires, pathologies, periods of drought, etc.) and increasing the stability and resilience of the ecosystems and of the nature-man interactions.

From the assessment to decision-making in the United Kingdom, the landscape action builds up empowerment. Starting with experimentation in natural protected areas and in the countryside, the local stakeholders – involved through techniques like "parish maps", community appraisals, guided visualization, and village design statements – imagine the future of the natural and rural landscape, of the spaces of social aggregation and services and of the buildings, roads and urban furnishings. These involvement actions for the project build up extensive public and private consensus and responsibility on the aims.

14.3 Towards Innovation

The challenge for management of nature and landscape is to develop institutional structures that match the ecological and social processes that operate at different spatial and temporal scales and that address the links between those scales (Folke et al. 2002), casting light on the intricacies of nonlinear dynamics, cross-scale interactions and complex adaptive systems and improving the sustainability and functioning of prevailing social and ecological systems.

The goal is an adaptive comanagement in which a long-term management structure permits stakeholders (as in the Netherlands) to share management responsibility within a specific system of natural resources and to learn from their actions (Ruitenbeek and Cartier 2001). This is the significant direction taken by the experience of the Indigenous Community Conserved Areas[7] which focuses on involving the populations in the management of local ecological resources, supporting the "democratization of the decision-making process", promoting the resolution of conflicts and encouraging participation to enhance social, environmental, cultural and economic diversity (IUCN, Durban 2003). The results lead towards a continuative involvement of the community in management decisions, and efforts lead to the conservation of the territory and associated cultural values.

Protected and sensitive area policies and projects, like in the United Kingdom, allow the orientation and redefinition of the continuous evolution of the landscape, on different scales and from a dynamic viewpoint, through strategies and innovative interventions aimed at sustainability and resilience. Decisional models capable of interacting with the strategic view of policies, plans and projects (of the territorial and landscape governance authorities on different scales) are required, with bottom-up actions, to maximize the benefits deriving from public-private comanagement.

[7] ICCA Consortium website http://www.iccaforum.org/.

References

Berkes F, Folke C (eds) (1998) Linking social and ecological systems: management practices and social mechanisms for building resilience. Cambridge University Press, Cambridge

Borrini-Feyerabend G, Farvar MT, Nguinguiri JC, Ndangang VA (2007) Co-management of natural resources: organising, negotiating and learning-by-doing. GTZ and IUCN/Kasparek Verlag, Heidelberg

Costanza R, d'Arge R, de Groot R, Farber S, Grasso M, Hannon B, Limburg K, Naeem S, O'Neill RV, Paruelo J, Raskin RG, Sutton P, van den Belt M (1997) The value of the world's ecosystem services and natural capital. Nature 387:253–260

Council of Europe, Committee of Ministers (2008) Recommendation CM/Rec (2008)3 of the Committee of Ministers to member states on the guidelines for the implementation of the European Landscape Convention. 6 Feb 2008 -17th meeting of the Ministers' Deputies

Davoudi S (2012) Resilience: a bridging concept or a dead end? Plann Theory Pract 13(2): 299–307

Egoh B, Reyers B, Rouget M, Richardson DM, Le Maitre DC, van Jaarsveld AS (2008) Mapping ecosystem services for planning and management. Agr Ecosyst Environ 127(1–2):135–140

Ehrlich PR, Kennedy D (2005) Millennium assessment of human behavior. Science 309(5734): 562–563

Folke C, Colding J, Berkes F (2002) Building resilience for adaptive capacity in social-ecological systems. In: Berkes F, Colding J, Folke C (eds) Navigating social-ecological systems: building resilience for complexity and change. Cambridge University Press, Cambridge

Folke C (2006) Resilience: the emergence of a perspective for social-ecological systems analyses. Glob Environ Change 16(3):253–267

Folke C, Carpenter SR, Walker B, Scheffer M, Chapin T, Rockström J (2010) Resilience thinking: integrating resilience, adaptability and transformability. Ecol Soc 15(4):20. http://www.ecologyandsociety.org/vol15/iss4/art20/

Gallopín GC, Gutman P, Maletta H (1989) Global impoverishment, sustainable development and the environment: a conceptual approach. Int Soc Sci J 121:375–397

Granek EF, Polasky S, Kappel CV, Reed DJ, Stoms DM, Koch EW, Kennedy CJ, Cramer LA, Hacker SD, Barbier EB, Aswani S, Ruckelshaus M, Perillo GME, Silliman BR, Muthiga N, Bael D, And Wolanski E (2010) Ecosystem services as a common language for coastal ecosystem-based management. Conserv Biol 24(1):207–216

Gunderson LH, Allen GR, Holling CS (eds) (2010) Foundations of ecological resilience. Island Press, Washington, DC

IUCN (2003) Vth IUCN world parks congress. Durban

Kasperson JX, Kasperson RE, Turner BL (1995) Regions at risk: comparisons of threatened environments. United Nations University Press, Tokyo

Kates RW, Clark WC, Corell R, Hall JM, Jaeger CC, Lowe I, McCarthy JJ, Schellnhuber HJ, Bolin B, Dickson NM, Faucheux S, Gallopín GC, Grübler A, Huntley B, Jäger J, Jodha NS, Kasperson RE, Mabogunje A, Matson P, Mooney H, Moore B III, O'Riordan T, Svedin U (2001) Environmental and development: sustainability science. Science 292(5517):641–642

Kauffman S (1993) The origins of order. Oxford University Press, New York

Klausmeyer KR, Shaw MR, MacKenzie JB, Cameron DR (2011) Landscape-scale indicators of biodiversity's vulnerability to climate change. Ecosphere 2(8)

Kozlowski J (1986) Threshold approach in urban regional and environmental planning: theory & practice. University of Queensland Press, St. Lucia

Lyle JT (1985) Design for human ecosystems. Van Nostrand Reinhold, New York

Ministries van VROM & LNV (2008) Agenda landschap. Landschappelijk verantwoord ondernemen voor iedereen. Ministrie LNV, Den Haag

Phillips A (2003) Turning ideas on their head. The new paradigm for protected areas. George Wright FORUM. 20(2):8–32

Ruitenbeek J, Cartier C (2001) The invisible wand: adaptive co-management as an emergent strategy in complex bio-economic systems. Occasional paper number 34. Center for International Forestry Research, Jakarta

Sijtsma FJ, Farjon H, van Tol S, van Hinsberg A, van Kampen P, Buijs A, van Hinsberg A (2013) Chapter 8: Evaluation of landscape changes – enriching the economist's toolbox with the Hotspotindex. In: Heijman W, van der Heide CMJ (eds) The economic value of landscapes. Routledge, London, pp 136–164

Steward J (1955) Theory of culture change. University of Illinois Press, Urbana

Swanwick C (ed) (2002) Judging capacity and sensitivity, landscape character assessment guidance for England and Scotland. The Countryside Agency

Turner BL II, Matson PA, McCarthy JJ, Corell RW, Christensen L, Eckley N, Hovelsrud-Broda GK, Kasperson JX, Kasperson RE, Luers A, Martello ML, Mathiesen S, Naylor R, Polsky C, Pulsipher A, Schiller A, Selin H, Tyler N (2003) Illustrating the coupled human–environment system for vulnerability analysis: three case studies. Proc Natl Acad Sci U S A 100(14): 8080–8085

Van Eetvelde V, Antrop M (2004) Development of European landscapes. Analyzing structural and functional changes of traditional landscapes – two examples from Southern France. Landsc Urban Plann 67:79–95

Voghera A (2011a) Land use indicators for landscape assessment. In: Cassatella C, Peano A (eds) Landscape indicators. Assessing and monitoring landscape quality. Springer, Dordrecht, pp 141–165. doi:10.1007/978-94-007-0366-7

Voghera A (2011b) Dopo la Convenzione europea del paesaggio. Politiche, piani e valutazione/ After the European landscape convention. Policies, plans and evaluation. Alinea Editrice, Florence

White LA (1949) The science of culture: a study of man and civilization. Farrar, Straus, and Giroux, New York

Chapter 15
Landscape and Protected Natural Areas: Laws and Policies in Italy

Renzo Moschini

Abstract The background regarding landscape and protected natural areas in Italy is characterised by a twofold situation: the first concerns the country's long and historical tradition of landscape protection, while the second concerns the significance of parks and protected areas as a consequence of their spectacular growth in terms of numbers and surface area. In this regard, regional authorities have played a crucial role in replacing or accelerating government initiatives. The European context provides a framework for understanding the need for cooperation between stakeholders and institutions both in nature conservation and landscape policies, standing in stark contrast to the current institutional framework in Italy, which is unclear in its separation of such competences.

Keywords Nature conservation • Nature parks • Landscape protection • Legal framework

The document signed by the President of The United States in 1916 to institute the new National Park Service gave it the task of protecting the country's landscape, its natural and historic heritage and wildlife, in order to leave them unaltered and to be enjoyed by future generations. In more recent times, the National Park Service has reconsidered and examined European Parks and their experience, with a particular focus on the subject of landscape, which has evolved differently from the European context. At the international level, today, in accordance with new conservation paradigms, the ecological and symbolic role of protected areas is in need of a coherent integration of nature conservation policies with regional policies, towards a "territorialisation" of policies concerning protected areas which inevitably merges with policies concerning landscapes, thus bridging nature and culture. Italy's Constitution (1947) is unique in Europe in that it specifically refers to landscape.

R. Moschini (✉)
Gruppo San Rossore, San Rossore Park, Località Cascine Vecchie, 56100 Pisa, Italy
e-mail: renzo.moschini@gmail.com

R. Gambino, A. Peano (eds.), *Nature Policies and Landscape Policies*,
Urban and Landscape Perspectives 18, DOI 10.1007/978-3-319-05410-0_15

Italian legislation concerning the conservation of wildlife, in contrast, was implemented in 1991 under the Framework Law no. 394, which refers to the scope and purpose of parks, citing article 9 of the Constitution, regarding landscape, together with article 32, regarding health. Prior to 1991, the belatedly instituted regions used their mandate in the areas of planning, agriculture and environmental protection to establish the first regional parks. As a result, their authority crossed over with that of the *Soprintendenze*[1] with regard to landscape protection. Landscape had traditionally been perceived as an area with a specific aesthetic identity, particularly with regard to its monuments and historical buildings. While the regional and local institutions became involved for the first time in the historical and cultural sphere, which hitherto had been the prerogative of the *Soprintendenze*, similarly governmental decentralisation structures found themselves having to contend with a new interlocutor (i.e. Regions), who were also called upon to operate in the sphere of territorial local planning. Law no. 394 of 1991 granted the regions and local authorities autonomy in planning and establishing new regional parks. Indeed, the then President of the Italian Republic, Oscar Luigi Scalfaro, speaking at the first national Conference of Protected Areas, *Parchi, ricchezza italiana* (Vittoriano, 25–28 Sept. 1997, Rome), declared that the regions, in this regard, held an important role of "constitutional supply" towards the State, which had long deferred assuming its direct responsibilities.

Thus, the "bridge" mentioned in this research between nature conservation and landscape management was established. Several major projects which have come to fruition in recent times – I am thinking not just of the San Rossore Park but many others too, from Ticino Lombardo to the Piedmont Parks – are proof that a completely new direction was taken, although environmentalists have not always completely agreed, for example, with Pierluigi Cervellati's San Rossore project, which somebody deemed to have been too generous with regard to the local landscape and history, to the detriment of the natural dimension. This conception of landscape, linked for the first time to the "entire territory", was subsequently recognised explicitly by the European Landscape Convention (Council of Europe 2000) and in addition has undergone a process at the European level by which each Member State's domestic affairs have met with new responsibilities and supranational standards and have thus suffered a setback as well as a reversal, which must be taken into account today if we want this "bridge" to be effectively put in place. The first serious challenge to this idea of integrating nature and landscape came from the Italian Cultural Heritage and Landscape Code (2004), which removed the "landscape" component from park plans giving priority to landscape plan, just as the ELC was being signed, in 2006. Such regulation was decided upon at a time when planning was still facing severe difficulties, without any institutional involvement or reaction from the parties concerned. It is difficult to say to what extent such acquiescence has been influenced by the argument that the State had been an exclusive competence for landscape. However, its authority was already exclusive

[1] The *Soprintendenze* are decentralised bodies of the Ministry for Cultural Heritage and Activities (Italian: *Ministero dei beni e delle attività culturali e del turismo*).

in the legislation passed in 1991, as the Constitutional Court did not object to include landscape in park planning. Indeed, it was obvious that it did not constitute a transfer of jurisdiction (i.e. removal from the jurisdiction of the State) but was simply the outcome of a need to integrate aspects of landscape which up to that time had been separately managed, with results which were becoming less and less easy to reconcile with the recommendations of the European Landscape Convention. This decision certainly affected Italy's ratification of the European Landscape Convention, to which many have attributed a questionable – if not outright dangerous – concept of landscape, because it concerns the "whole territory", and therefore local communities, who can thus influence the decisions previously taken only by the State. We had returned to dividing and separating what had previously been painstakingly connected and united, at precisely the moment that environmental policies had been weakening, particularly with regard to landscape as well as soil conservation and nature management. The international launch of the "new paradigms" thus comes just in time and, regarding the Italian context, makes it imperative to regain a leading role after an overlong hiatus during which it has been guilty of inaction.

The European scenario may help us to better understand the risks that we are running regarding the management of parks and protected areas, which today are no longer limited to those of national, regional and local institution. Indeed, our national, regional and local parks as our reserves host large part of the Natura 2000 sites (SCIs and SPAs) established by the European Union. The immediate effect of this has been to cause inconvenience and noncompliance, as demonstrated by the (perhaps excessive number of) fines imposed by EU on Italy. Such coexistence is confusing and difficult, not least because of a lack of national policies, which has made it impossible to draw up a reliable classification or to clearly integrate our protected areas, a commitment which, although formally requested and compulsory, has been – as it is well known – thoroughly ignored, with the result that many of our protected areas, especially marine areas, have remained "clandestine". What kind of "bridge" can be established between the landscape and national parks on the Tuscan Archipelago, for instance, if the marine area has not yet been defined?

We must not forget that even on the occasion of the twentieth anniversary of the Framework Law, not only were bizarre views regarding the parks doing the rounds, but interventions which severely affected the national and regional parks were being drawn up. With regard to this crucial aspect, a most disconcerting silence was observed, to the point that it was possible to discern a "brake" on the Framework Law, which had to be removed, while at the same time not even acknowledging that one of the most important pieces of legislation – concerning plans – had been crippled. In order to start off on the right foot, we must therefore take the European context into account, above all because it already has rules and programmes in place to tackle the issue of the abovementioned "bridge" between nature and landscape. One issue concerns the fact that with regard to protecting SCIs and SPAs – including even those located within a national or regional park – European rules and not those of the park that host them must be applied. It is not difficult to understand the contradictory nature of this dual regime when the purposes of the protection of biodiversity are the same. This contradiction applies to all parks, but especially to marine protected areas, where problems of integration appear to have received the least attention (see, e.g. the whale sanctuaries). However, whether we examine the situation in the Alps or the Apennines, the picture remains largely the same.

All of these open up another issue: how do we participate in European decisions and how do we rely also from the financial point of view? It is well known – as I have already pointed out – that Italy "leads" the way, if measured in terms of the penalties and fines which we collect as a result of our failures, while the Basin Authorities and with them the River Parks still do not comply with EU rules. Recent data provided by the Italian Ministry for Cohesion shows that we are barely able to submit viable projects able to obtain EU funds. In July 2013, the European Commission approved the LIFE + funding programme (a fund for the environment established by the European Union in 1992) for a total of 516.5 million Euros, 268.4 million of which will be covered by contributions from EU Member States. 1,078 funding applications were submitted. However, only 202 projects have been selected for co-funding in the three areas concerned: policy and environmental governance, information and communication. Of the 268 proposals received, just 76 projects were approved and selected for funding by partnerships of conservation institutions, government agencies and other bodies, for a total of 241.8 million Euros in funding, 136 million Euros of which was provided by the EU. Of the 607 proposals submitted, the Commission granted its approval to 113 projects, funded with 258.4 million Euros from public organisations, 124.4 million of which will be covered by contributions. The list does not end here. However, I wish to draw attention to these figures as it does not require an expert to realise that such projects are underpinned by a planning process involving multiple institutions and individuals which require an attitude of initiative, having to deal with problems such as oil shortages for vehicles and the fire services, among others. The "bridge" between nature conservation and landscape management requires a systemic approach to protected areas, which must be highly integrated and consistent with European policies. Finally, a remark concerns the role in Europe of Italian protected areas, particularly in the light of recent international developments concerning park policies extending beyond their boundaries. In contrast, Italy still exhibits severe difficulty, as it is shown by the lack of any local or regional policies capable of tackling issues such as urban sprawl, soil consumption, landscape degradation, and the resumption of offshore drilling activities.

References

Council of Europe (2000) European landscape convention. Treaty series n. 176

European Commission (2013) Life + call for proposal. Official Journal n. 2013/C 47/21

Repubblica Italiana (1947) Costituzione della Repubblica Italiana. 27 Dec 1947

Repubblica Italiana (1991) Legge Quadro sulle Aree Protette. Framework Law n. 394/91

Repubblica Italiana (2004) Codice dei Beni Culturali e del Paesaggio. Leg. Decree n. 42/2004

US Department of Interior (1916) National Park Service Organic Act. US Federal Law. 25 Aug 1916

Chapter 16
Evolution of Concepts and Tools for Landscape Protection and Nature Conservation

Mariolina Besio

Abstract Three laws, dealing with landscape, in Italy represent a significant field of study for the analysis of the relationships between landscape policies and nature policies. Although other acts treated the same topic, these are the most important ones since they started the territorial policies on landscape protection (and nature protection) and set landscape planning as an operative tool of protection. These laws, issued in 1939, 1985, and 2004, offered an overall and coherent vision of the subject; this is why I have analyzed them taking different aspects into account: the meaning of the word landscape, the system of values according to which the quality of a landscape is assessed, and the relationship between landscape and nature and the form and structure of the landscape planning tools. Throughout the years, there has been a considerable evolution of the meanings, the relationships, and the operative tools for the protection of the landscape, but at the same time, there has been a relative independence of the meanings and values from landscape planning. Since meanings, values, and relationships are not always expressed explicitly, the laws have been analyzed through methods borrowed from the cognitive sciences, particularly from the theory of conceptualization.

Keywords Landscape • Nature • Meanings • Landscape planning tools

16.1 The Subject and the Method

Landscape and nature exist in a flexible relationship that not only has changed throughout the years but that has changed also according to the person speaking about it. Words refer to general entities that do not have a single meaning. They are used both with the precision of scientific paradigms and with the broad shades of common meaning. Sometimes their meanings can be interchangeable: landscape is nature and nature is landscape; or they can refer to different realities: nature

---

M. Besio (✉)
Department of Sciences for Architecture, Polytechnic School, University of Genoa, Stradone S. Agostino 37, Genoa 16123, Italy
e-mail: besio@arch.unige.it

R. Gambino, A. Peano (eds.), *Nature Policies and Landscape Policies*, Urban and Landscape Perspectives 18, DOI 10.1007/978-3-319-05410-0_16

disregards human presence, landscape is not an objective reality but a cultural mediation between men and what surrounds them (Farina 2001; Forman and Godron 1986; Naveh 1984; Cosgrove 1990; Turri 1990, 2003).

On other occasions, these words are used with meanings that have been coded according to generally acknowledged, but generic, tacit conventions.

Methods, policies, and operative tools of landscape protection and nature conservation cannot exclude explicit references to the meanings of nature and landscape in an action-oriented discipline, such as territorial planning.

In this contribution, the relationship between landscape and nature is analyzed through the laws that implemented landscape planning in Italy. The meaning of these words has changed in the laws that have introduced and then modified the policies and tools of landscape protection from the Second World War to the present (Settis 2010; Antonucci 2009). Since the laws do not always explicitly define what is nature, and what is landscape, a thorough interpretation of the texts has been necessary. The things and objects identified as subjects of the protection policies and tools have been associated with the general ideas they refer to. The concepts develop a reasoning that allows deducing the meanings given to landscape and nature.

The policies have been implemented and have affected landscape and nature through the tools developed in the laws to protect the landscape and the nature protected areas. I have analyzed these texts in order to highlight the evolution of their form and structure, as a consequence of the change of the protection tools. I have evaluated the elements through which they have operated: the objects and things they controlled and the values of nature and landscape that they considered worthy of protection.

Finally, I went through the relationships and connections between meanings and tools, proving how they have not always followed a consistent and explicit path.

16.2 Topics and Concepts

The evolution of the laws emphasizes how the meanings of nature and landscape have changed together with the planning tools. Throughout the three laws that have proposed landscape plans as protection tools, the meanings of nature and landscape have sometimes merged or differed.

The law that introduced the landscape plans in Italy was issued in 1939. Two legislative decrees which were turned into law in 1985 made it compulsory for the regions to design landscape plans. The Cultural Heritage and Landscape Code contains and organizes all the previous laws regarding the landscape, including the European Landscape Convention that, in 2004, defined the criteria according to which landscape plans had to be designed.

Focus on the landscape considered as a cultural and scientific entity, autonomous and independent, started with the interest in nature at the beginning of the nineteenth century, with the Romantic paintings and the geographic studies of the colonial explorations (Romano 1978). The dualism between culture and nature has been continuously discussed ever since. However, it was not until the UNESCO Convention on Cultural Landscapes in 1992 and the European Landscape Convention in 2000 that a clear distinction between the two words was suggested.

In Italy, with regard to landscape, there are two protection tools: the landscape plan and the legal constraint for heritage protection. The landscape plan is an instrument with a broader dimension and a more complex and organic concept of conservation and protection, if compared to the legal constraint for heritage protection. This is why it is better suited for the integration of landscape, nature, and socioeconomic development policies and for the implementation of territorial strategies for sustainable development. The relationship between nature and landscape can appear blurred and sometimes ambiguous in the 1939 and 1985 laws. Nonetheless, a more in-depth approach allows one to deduce the implicit idea of what is landscape and what is nature (Bottani and Penco 2013). The method of categorization helps us to decode the texts, to better understand the definitions and the meanings that the conventions offer, and what kind of representations of nature and landscape they provide (Osherson and Lasnik 1990). Categories are mental acts that consist of thinking about objects, events, and things as elements of a broader set. Concepts are the mental representations and the ideas of the categories. Words, referring to objects, things, and events, partially mirror the structure of the categories, concepts, and ideas (Quine 1970; Putnam 1987). Objects, events, and things can assume different meanings according to the concepts they refer to.

The deconstruction of the text highlights the categories of objects, events, and things that are the reality to which the laws apply to; it goes back to the concepts they refer to and, by inference, to the relationships between the concepts. We can clarify the meaning of the text and terms that are used in it, by reconstructing the implicit reasoning (Frixione 2007).

16.3 The Evolution of Meanings

For Merleau-Ponty "landscape is the world we see," and for Cosgrove "each individual has a partial vision, that perceives only specific elements of the landscape." According to Merleau-Ponty (1980), the landscape is the entirety of what we observe, while according to Cosgrove (1990), not all the observers see the same things. Two elements contribute to the meaning of landscape: the things of the world that we can see as well as the criteria and models that we use to see. Landscape is at the same time both physical and material and cultural and symbolic (Cauquelin 2000).

The visible things that we consider landscape (the objects) as well as the point of view (the criteria and the patterns to define and evaluate the landscape) of the observer (the legislator) have changed in the laws that have dealt with landscape and landscape plans.

In the Law of 1939, the word *landscape* is not used except as an adjective related to planning (art. 12, 25) and with reference to art. 9 of Italian Constitution. Its definition is implicit in the listing of four elements worthy of protection, which belong to limited and defined categories: the first two are singular elements; the other two are groups of elements. Although the items that compose the four categories are different in nature, they have beauty in common. The concept of beauty is associated with landscape: either the landscape is beautiful or it is not. Beauty is a mental category connected with some objects in the territory that stand out for their exception to the normal, insignificant, and neutral surrounding context. Since the law is entitled "Protection of Natural Beauties," we should infer that the meaning of landscape is related to the aspects of nature that are valued as *beautiful*. One of the four elements, "the sets of immovable things that form a characteristic aspect having an aesthetic and traditional value," reveals a broader and somehow ambiguous meaning. The traditional value refers to the manmade structures and to the values of the history of human settlements, although considered under the perspective of an artistic view and a work of art. The 1939 Law does not provide criteria or standards to evaluate beauty or to choose only the elements that are beautiful. Since in a dominant culture it is considered that only the well educated are capable of evaluating beauty, a committee of experts is assigned by law to determine what is worth protecting (Assini 1977).

Even in the 1985 law, the word *landscape* does not appear. The definition of landscape is given through what is listed from (a) to (m) as worthy of being protected, contributing to the four categories of the previous law. Even the items of this list belong to categories of objects scattered unevenly in the territory. Eight categories consist of objects that belong to the world of nature; the rest consists of objects clearly identified on the basis of previous administration acts. This time the objects are clearly identified. Landscape largely coincides with the physiographic features of the country: rivers, orographic systems, glaciers, woods, and humid areas. Otherwise, landscape coincides with areas that have already been considered of natural value (parks) or of historical value (common properties and areas of archaeological interest). The title of the law recalls the "urgent dispositions for areas of particular environmental interest," so we can logically assume that the meaning of landscape is directly associated with nature. Furthermore, the orohydrographic, forestal, and exceptional ecological elements are identified as landscape, that is to say, all the features that are greatly visible, which distinctly characterize the geomorphology of the country. The definition of landscape is matched with singular objects that stand out from the territory for their ecological and environmental relevance.

The new category of the *environment* has triggered a kind of crossover between landscape and nature, and in the decades previous to the issue of the Law, it supplied models of interpretation that well represented the complexity of the dynamics of the natural systems. In the second half of the 80s, the landscape policies are strictly but also ambiguously connected with nature policies, even due to the emerging environmental issues (Di Fidio 1991; TCI 1991).

The European Convention for the landscape sheds light on the semantic ambiguity that in the past caused misunderstandings and lexical ambiguity. In 2004, the Convention was implemented by the Cultural Heritage and Landscape Code, which contains brand new meanings. For the first time, the explicit definition of landscape is given: "a territory expressing identity, whose distinctive character comes from the actions of natural and human factors and from their interconnections." It is a pivotal turning point since the change is not only formal, but substantial. It concerns definitions and meanings but also means of interpretation and assessment. The concept of landscape no longer refers to single categories of territorial objects, each one emerging from a neutral and undistinguished context nor is it connected to the concept of esthetical or environmental quality. It refers to the concept of territory and to one of a specific aspect together with many others: the environmental aspect, the social aspect, the economic aspect, etc. It also pairs with both the concept of identity and the concept of interaction between natural and anthropic factors.

The logical inferences in the concepts ascribe to the landscape the meaning of exterior manifestation of the territorial identity: "aspects and characters that are a material and visible representation of the national identity." This, in turn, is defined upon the relationships that have been established between the dynamics of the natural phenomena (natural factors) and the processes of human settlements (human factors), which are considered as cultural evidence: "as the expression of cultural values." The dimension of the landscape coincides with the whole territory: it includes nature as well as cities, industrial areas, rural areas, etc. Landscape is no longer exclusively represented by natural elements, but it embraces them in a well-structured and complex synthesis. It is no longer bonded to the category of beauty, but it connects to the category of culture and to the capacity of grasping the symbolic value of the relationship between man and nature. The relationship is variable and it can be assessed according to different values of quality and decay. The ethics of lifestyle replaces the aesthetics of the image. The act of evaluating, which is significant in the definition of landscape, becomes much more complex and all-inclusive. It no longer refers to the outstanding levels of quality but to the distinction within a landscape of different manifestations of identity and different levels of quality, including also the lack of quality (Bonesio 2007).

16.4 The Evolution of Landscape Planning

In order to verify the coherence of ideas and concepts with the actions of protection, it is useful to compare the evolution of the meanings of landscape and nature with the evolution of landscape plans. Landscape plans are a complement of urban plans, since they derive their structure (they implement protection by controlling the building industry and the suitability for building) and their language (they govern the transformations through zoning and technical norms).

When general concepts are turned into action, there can often be ambiguity and confusion, due to either the greater inertia of actions in comparison with ideas or to the reduction of general meanings in sectorial procedures. Due to inertia, landscape plans adopt well-established forms; although not oriented towards landscape protection, due to reduction they operate only on some elements of the landscape. The first indication of the technical contents of landscape plans is contained in the ministerial statement that accompanied the law on the "Protection of Natural Beauties." The structure of the plan is similar to the detailed plans that were used between the two World Wars to implement urban plans of expansion and renewal. Only the areas that are protected by legal constriction for heritage protection are taken into account in the landscape plans, which control almost exclusively the building expansion and consider nature only as vegetation in the areas pertinent to the buildings.

The policies of nature are in service of the landscape policies, implemented mainly through the control of building construction (INU 1959). Plans are not compulsory and they have a very limited success: in 1985 only 12 were approved in Italy, not enough to consolidate and develop this practice by means of experimenting practices and actions.

In 1985, the institutional and administrative framework completely changed because the Ministry of Cultural and Environmental Affairs and the Regions were operative, the later with the task to design landscape plans by law.

Neither the 1985 law gave instructions on how new plans should be nor were there solid experiences. It was also ambiguous as it assimilated landscape plans with urban territorial plans that had specific consideration for the environmental and landscape features. Due to the lack of clear direction, each region adopted its own patterns and a confused and contradictory situation derived from all this experimentation. The relationship between nature and landscape was variable and lacked coded foundations (Ciccone 1986, Romani 1994).

The Framework Law no 394 on protected areas was issued in 1991. It dealt with park plans in a systematic and uniform way, with exclusive attention to the conservation of nature. Procedures, competences, and the effectiveness of park plans and landscape plans were completely separated. Landscape and nature policies neither coordinated nor integrated.

In 2004, the Cultural Heritage and Landscape Code deals thoroughly with landscape plans, which become strategic since they have to decide upon the objectives of quality, on the ground of survey and evaluation, and set the policies and actions to achieve them.

The Code was issued almost 10 years ago, but the outcomes are still too limited to evaluate its effects (Priore 2009). In the revision of regional urban plans, some regions such as Apulia and Tuscany are experimenting with integrated procedures of territorial and landscape planning. In these cases, nature does not overlap or identify with landscape, but it integrates as one of the components of sustainability, which is relevant for both ecological and hydrogeological aspects.

16.5 Conclusions

This overview highlights some questions that will have to be addressed to implement territorial policies of real integration between the conservation of nature, landscape protection, and the eco-friendly development of local communities.

1. The effective integration of tools and policies for nature conservation and landscape protection can be fostered with adequate conceptual models and with a clear distinction between the meanings of the words, in order to avoid any lexical ambiguity and overlapping.
2. Landscape, being the visible form of the territory, is a symbolic expression of the relationship between natural and human factors as well as social, civil, and cultural forms of the population that must take part in the implementation of the protection policies.
3. Landscape and nature policies and protection tools cannot disregard the socio-economic development in the pursuance of integrated models of sustainability that are coherent and measured on the local characters of landscape.
4. The integration implies the coordination of protection policies and tools in a unitary territorial dimension.

References

Antonucci D (2009) Codice commentato dei Beni culturali e del paesaggio. Sistemi editoriali, Naples
Assini N (1977) Protezione delle bellezze naturali. Edizione medicea, Florence
Bonesio L (2007) Paesaggio, identità e comunità tra locale e globale. Diabasis, Reggio Emilia
Bottani A, Penco C (2013) Significato e teoria del linguaggio. Angeli, Milan
Cauquelin A (2000) L'invention du paysage. Presse Universitaire de France, Paris
Ciccone F (1986) I piani paesistici: le innovazioni dei sistemi di pianificazione dopo la legge 431. NIS, Rome
Cosgrove D (1990) Realtà sociali e paesaggio simbolico. Unicopli, Milan
Di Fidio M (1991) Tutela dell'ambiente naturale: difesa, gestione e sviluppo della natura e del paesaggio. Pirola, Milan
Farina A (2001) Ecologia del paesaggio: principi, metodi, applicazioni. UTET, Turin
Forman R, Godron M (1986) Landscape ecology. Wiley, New York
Frixione M (2007) Come ragioniamo. Laterza, Rome
Italian National Institute of Urban Planners (INU) (1959) Bilancio dell'urbanistica comunale nel quadro della pianificazione territoriale e paesistica. INU edizioni, Rome
Merleau-Ponty M (1980) Fenomenologia della percezione. Il Saggiatore, Milan
Naveh Z (1984) Landscape ecology: theory and application. Springer, New York
Osherson D, Lasnik H (eds) (1990) An invitation to cognitive science vol 1, language, vol 2, visual cognition and action, vol 3, thinking. The MIT Press, Cambridge, MA
Priore R (2009) No people, no landscape. la Convenzione Europea del Paesaggio: luci e ombre nel processo di attuazione in Italia. Franco Angeli, Milan
Putnam H (1987) Mente, linguaggio e realtà. Adelphi, Milan
Quine WV (1970) Parola e oggetto. Il Saggiatore, Milan
Romani V (1994) Il paesaggio: teoria e pianificazione. Franco Angeli, Milan

Romano G (1978) Studi sul paesaggio. Einaudi, Turin
Settis S (2010) Paesaggio, Costituzione, cemento: la battaglia per l'ambiente contro il degrado civile. Einaudi, Turin
Touring Club Italiano (TCI) (1991) Guida d'Italia: natura, ambiente, paesaggio. Ed TCI, Milan
Turri E (1990) Semiologia del paesaggio italiano. Longanesi, Milan
Turri E (2003) Il paesaggio degli uomini: la natura, la cultura, la storia. Zanichelli, Bologna

Chapter 17
Nature Conservation in the Urban Landscape Planning

Luigi La Riccia

Abstract This contribution briefly reflects on the evolution of nature in the city, according to the existing literature on landscape in the Italian urban planning tradition. New approaches to the urban nature conservation strongly depend on the regulations that planning can give in terms of local ecosystem services, absorption of pollutants in the atmosphere, noise reduction and allocation of places for recreation. The conservation of nature in the city is also part of the global effort to stop the biodiversity decline. In fact, landscape, the urban one, has the ability to introduce the social dimension and is therefore functional to the implementation of urban nature conservation frameworks. Current urbanizations, which are closest to natural areas, often demonstrate at all scales a lack of social and ecological relationships: the risk is a conceptual and physical insularization, which reduces public support to nature conservation, causes a further loss of biodiversity and does not promote the generation of new ecosystem services. One of the main future challenges will therefore be to convert the existing conservation strategies and introduce specific regulations in planning for natural areas that may be better integrated with the urban context: this contribution discusses the fact that the landscape can be the element that may drive this integration.

Keywords Landscape • Urban nature conservation • Sustainable urban development • Green networks

17.1 A New Challenge for Urban Nature Conservation

Nature conservation in the city is one of the biggest challenges for sustainable urban development, as a result of a social and ecological coevolution (Powell et al. 2002). In the studies conducted by Sundseth and Raeymaekers (2006), the value of nature in the city, however, goes far beyond its influence on the inhabitants' quality of life or rather has an intrinsic value: urban areas are surprisingly rich in biodiversity, as demonstrated, e.g., by the presence of Natura 2000 sites in 32 European cities. The

L. La Riccia (✉)
Interuniversity Department of Regional and Urban Studies and Planning (DIST), Politecnico and Università di Torino, Viale Mattioli 39, 10125, Turin, Italy
e-mail: luigi.lariccia@polito.it

R. Gambino, A. Peano (eds.), *Nature Policies and Landscape Policies*, Urban and Landscape Perspectives 18, DOI 10.1007/978-3-319-05410-0_17

conservation and management of nature and biodiversity in urban areas is often very complex: there are more people, stronger development pressures, less space, a multiplicity of actors involved, etc. Often, the analyses reveal that the urban natural reserves are few but large and have a high density. Large natural reserves can be especially important in urban landscapes, as the difference between the urban and natural environment can be high.

It should be noted, however, that the strategies of urban planning and those of nature preservation are generally separated. One possible reason is that the protection of nature has favoured a vision purely "conservative" towards nature outside the city and has made the vision of urban nature conservation trivial and distorted. However, the identification of urban nature is also part of a broader change in perspective within the conservation policies and remains as a necessary point of reference for a sustainable urban development (UNEP 1992; IUCN 2003). In many cities, this change of perspective was manifested through the institution of urban areas for nature conservation, supported by a general concept of "urban landscape". In this sense, Dudley (2008) reminds us that a formally protected nature conservation area may be defined as "a clearly defined geographical space, recognized, dedicated and managed, through legal or other effective means, to achieve the long-term conservation of nature with associated ecosystem services and cultural values".

In the urban context, the establishment of these areas has been started during the twentieth century as a reaction to the rapid degradation of the urban environment due to industrialization and the consequent urban growth. It was therefore seen as a necessary step to keep nature and landscape away from private exploitations. Today, instead, the public interest is more oriented to the preservation of social values, biodiversity of nature and landscape. In the recent decades, in effect, the nature conservation and landscape policies have changed: today, a possible alliance between nature and landscape is assumed to be an essential condition for sustainable development (UNEP 1992; Hooper et al. 2005; IUCN 2012) and lays itself at different scales (Antrop 2001, 2004; Potschin and Haines-Young 2006; Selman 2006; Gambino 2009).

These necessary changes must therefore be addressed by urban planning, taking into account the wide panorama of policies that in many countries are highly formalized and still indissolubly linked to the culture of heritage. This does not mean to consider an idea of heritage linked to a process of museification but to hire the landscape and nature conservation to face not only the risk of biological diversity decline but also the loss of cultural diversity (i.e., the ability to continue to produce diversified cultural values in the future).

For this reason, the maintenance of nature in the city seems to require a strong change of perspective, which has to be mediated by the concept of landscape through specific regulations of urban planning. In this case, it is necessary to overcome the conceptual reductionism, which has been traditionally used to describe nature and landscape components in urban systems, such as open spaces, green spaces, green areas, ecological corridors, greenways and urban parks (Swanwick et al. 2003). The choice of a concept among these almost always depends on the specific issues addressed by the planning, from the point of view of the scale (local or regional), the value of income and ownership (public or private green park), or the spatial configuration (greenway and green belts).

This contribution therefore aims to demonstrate that the conservation of nature in the city is not possible without a broader consideration of the concept of the urban landscape, where the areas for nature conservation may play a central role for the new image and the ecological rehabilitation of the city.

17.2 Evolution of Urban Landscape and Protected Areas in Italian Planning Tradition

The growing awareness of the environmental issue during the 1970s of the twentieth century has laid the foundations for the search for solutions in the field of urban and regional planning. Up to those years, the consideration of nature and landscape in the Italian urban planning tradition has privileged the aesthetic approach, oriented to the historical and cultural heritage of excellence. During those years, when in Italy the debate was focused on the general crisis in planning,[1] at the international level, an important shift on focus could be observed towards *landscape planning* (Turner 1983), a new way of understanding the landscape in the plan, closer to the urgency of reducing ecological problems and supported by an emerging environmentalist currency in the cultural and political scenes. On the one hand, there was a growing need to put an end to environmental disasters; on the other, the issue of landscape emerged forcefully in different disciplinary contexts.

The texts of reference, which have supported this period, are mainly two American books: "Silent Spring" (1962) of the biologist Rachel Carson, about the scourge of pesticides effects and "The Closing Circle" (1971) of the ecologist Barry Commoner. The political and economic relevance of the environmental question was then outlined at the end of the 1960s, thanks to first criticisms put forward by the "Report on Limits to Growth" (Meadows et al. 1972) against the "modern" myth of unlimited growth and the gradual domestication of nature, criticisms that invested the scientific basis of the modern project and the credit provided by the scientific objectification to the false ideas of progress and criticisms that allowed to develop the equation, already understood by Weber (1922), among the progressive "urbanization of the idea of nature" and the "naturalization of the idea of the city". But it is especially with the concept of "sustainable development", introduced in 1987 by the Brundtland Commission, in which the relationship between environmental issues and social issues became crucial, focusing not only on the global environmental emergencies but also on the influence they may have on the local development.

[1] About the crisis of urban planning in Italy or, more specifically, on the "urban plan crisis", and on the centrality of the "ordinary" plans, see Gabrielli (1995).

In the Italian legislative context, the period between the late 1970s and the 1990s, however, saw the enactment of some laws relating to landscape, watersheds and parks.[2] In practices, instead, there was a progressive lack of responsibility in the disciplines of urban planning, which gave way to, firstly, the large-scale territorial studies (priority of the analysis and understanding of the phenomena) and then to the research on architectural quality. The need to seek new rules in a world rapidly changing by size and timelines seemed obvious, but, in fact, the problem was even more clearly given by the separation between the preservation of the landscape, on a large-scale, and the interpretation of changes, within a "localism", that did not seem to provide adequate preservation of the authenticity and integrity.

In 2000, the European Landscape Convention has placed the landscape at the centre of the policies of individual states, introducing significant innovations in concepts and practices, even the ordinary and degraded landscapes, recognizing it as the representative of the identity of people and as an economic as well as ecological and cultural resource, which needs articulated protection, management and planning interventions. The European Landscape Convention has been an important opportunity to emphasize the theme of the ordinary landscape (though not defining in detail the operational rules by which this should be done) and innovate the binomial *nature-city*.

The following experiences have enriched effectively a reflection on the importance of landscape, even if brought back to the environment, which seems to be a common reference with regard to the fields and methods of intervention for the organization of space. In particular, the proliferation of specific sectorial plans (e.g., transport, water resources, parks, businesses, rural development, etc.) represents, in reality, an attempt to face the complexity of the contemporary city.

However, in some Italian experiences, the sustained idea has allowed to experience some attempts to overcome the only restriction-based approach of urban planning as "static" and often not shared by people and institution. In the cases of Reggio Emilia and Bergamo, for example, landscape and nature have been assumed as strategic elements for the identification of scenarios for sustainable urban development and at the same time for testing some initiatives to bring in operational terms a shared sense of urban landscape transformations.

[2] Including the Law no. 431 of 8 August 1985 ("Conversione in legge con modificazioni del decreto legge 27 giugno 1985, n. 312 concernente disposizioni urgenti per la tutela delle zone di particolare interesse ambientale") for the landscape, the Law no. 183 of 18 May 1989 ("Norme per il riassetto organizzativo e funzionale della difesa del suolo"), on soil conservation and the establishment of the Basin Authorities, Law no. 394 of 6 December 1991 ("Legge quadro sulle aree protette"), parks and protected areas.

17.3 Nature Conservation for What?

During the 1980s and 1990s, the debate on the landscape in the urban planning was largely oriented towards the idea of adopting a methodological approach more suitable to describe the increasing complexity of the relations between the new images of the city and territory and the new planning models regarding the emerging challenges of the contemporary city. The subsequent reflection on the landscape that emerged in those years within the disciplinary field of urbanism then revolved around the idea that the instrumentation, particularly the master plan, was to suffer an afterthought, to address the challenges posed by ecologism and sustainable development (La Riccia 2012).

The case of Reggio Emilia is expressive of this new needs: the preliminary design for the PRG of Reggio Emilia (1994) was certainly an innovative tool in the Italian urban planning tradition, as he attempted to demonstrate that the ecological problem can be object of a new form of the urban plan, which is not based exclusively on the cadastral incomes and which allows to obtain the necessary areas for the design of environmental corridors networks. This has been expressed through two considerations: firstly, the need to start from the territory and the environment to "explain" the new problems of the city and, secondly, the idea of a so strong ecologism that seems to place in background the other landscape dimensions, such as the historical as well as the aesthetical. The ecological approach is expressed in the role played by pre-existing environments: urban parks and "green wedges" (i.e., the green areas that continue from the outside towards the city centre) which allow the definition of the city limits and ensure the establishment of a network of environmental and ecological continuity. It also allows assigning a new environmental and landscape value to the "nature in the city", not always explicit in the current urban planning practices, through the simultaneous redesign of rural and natural spaces.

In the case of Bergamo, what is relevant from this point of view is a strong dynamism in the economic field influenced by the territorial system still heavily biased on the metropolitan capital. Among the ideas of the new Plan of Government of the Territory (2010), particular relevance is placed to strategies to avoid the ever-increasing functional specialization: among these, the improvement of urban, environment and landscape quality is taken as a priority for local development and territorial cohesion. The design of green areas represents the main node of intervention: in particular, it is not just a drawing of ecological networks but of *landscapes*, involving both the natural environment and the built environment, aiming, for instance, to improve the visibility of the historic city (the so-called Città Alta) from the plain.

Also in this case, there is a strong consideration to work on urban voids providing for the construction of three large parks (the Park of Trucca, the Park of Martinella and the Park of Porta Sud) and a connecting element, namely, the green belt, which closes the system. This element is designed as the new limit of the consolidated city: a green band affecting the forms and perceptions of the new urban landscape.

The areas in which there are parks and green belts are obtained through the experimentation of new compensatory and equalization measures: these allow, in fact, a better balance between the green areas and new public space design.

The cases of Bergamo and Reggio Emilia remain isolated examples in the Italian panorama. We understand, as the reality of current urban planning still demonstrates, the difficulty in dealing with the problems of the city. Urban design operates by disconnected fragments: excluding almost completely the relationships between tangible and intangible assets. It is, in fact, an approach that appears inappropriate for its simplification: urban planning, carried out "by parts", often through projects in derogation of the master plan, contributes to the inability to agree to a single overall image of the city (La Riccia 2013).

17.4 Conclusions

Since the 1970s, the ecological question has been gradually established in the urban planning debate as one of the main challenges, if not the most important, with which the society was called to confront, gradually acquiring a central role at the local, regional, European and global levels. The new places of the contemporary city appear more and more complex due to the overlap and sedimentation over time of actions mutually incompatible, as described by Bauman (2000) in terms of "liquid modernity". In this sense, the search for a better balance, between proposals of landscape protection and those of sustainability, becomes one of the urgent tasks for the planning and design practices (Nassauer 2007).

Until the 1970s, in Italian urban planning, we could not speak about a real ecological paradigm, but of "urban greening", the distribution of which was generally expected in new districts as well as in historical centres. The creation of urban parks also became one of the focal points of the urban plans. Keeping them indicated a fundamental aspect of environmental continuity in urban space. In the analysed experiences, indeed, the green areas act as a common element for rejoining city and countryside, that is to say, for the redevelopment of the modern districts in order to reduce the pressures on both historic centre and new districts.

The ecological paradigm is therefore a different vision and has guided the practice of urbanism towards a new direction. The environmental provisions now seem to articulate the new practices, coordinating behaviours and reconfiguring the spaces of the city: this means defining new principles and functionalist provisions more ecological.

Recently, many of the project experiences with particular regard to planning and architecture, "sustainable" by declaration, have been shown to be unconvincing from the practical point of view. This is often a problem of scale: much of the work of planners has been targeted to the final outcome of the interventions, but losing the sense of scale, the relationship with the ecological systems, larger, more complex, and the ability to understand the interrelationship with other dynamics, economic, political, social and cultural. The significance of certain issues, the urgency of action for restoring the functionality of ecosystems and urban nature conservation, often led to a sort of "schizophrenia" in relation to the qualitative value of many landscapes, urban and natural. This is due to the fact that the ecological theme, and the comprehension of the problems that it refers to, is placed in the tradition of Italian urban planning as a totally new theme, requiring a new sensitivity and ability to incorporate and address the needs of growth and development. The urban nature conservation requires also new conditions: ecosystems, such as landscape, transcend the scales, beyond just the urban area.

We need to understand, within the rules and plan projects, that green is no longer just a mere architecture of context but contributes, primarily, to create a system, complex, unitary consistent with historical heritage and environmental dynamics. Urban planning is therefore called to consider this aspect, going beyond the mere response to environmental and ecological issues and enabling to understand and appreciate the values of cultural processes underlying the urban and natural landscape, as well as the qualitative effects of choices considered in some way "environmentally sustainable".

References

Antrop M (2001) The language of landscape ecologists and planners – a comparative content analysis of concepts used in landscape ecology. Landsc Urban Plan 55:163–173

Antrop M (2004) Landscape change and the urbanization process in Europe. Landsc Urban Plan 67:9–76

Bauman Z (2000) Liquid modernity. Polity PR, Cambridge. Italian edition: Bauman Z (2002) Modernità liquida (trans: Minucci S). Laterza, Roma-Bari

Carson R (1962) Silent spring. Fawcett Crest, New York

Commoner B (1971) The closing circle: nature, man and technology. Knopf, New York

Council of Europe (2000) European Landscape Convention. Treaty series no. 176, Florence

Dudley N (2008) Guidelines for applying protected area management categories. IUCN, Gland

Gabrielli B (1995) Contro i piani di settore. In: Muscarà C (ed) Piani, parchi, paesaggi. Laterza, Roma, pp 281–287

Gambino R (2009) Lectio Magistralis. Paper presented at the Polytechnic of Turin, Turin, 8 Oct 2009

Hooper DU et al (2005) Effects of biodiversity on ecosystem functioning: a consensus of current knowledge. Ecol Monogr 75:3–35

IUCN (2003) The urban imperative. Urban outreach strategies for protected area agencies. IUCN World Conservation Congress, Durban, 6–17 Sept 2003

IUCN (2012) Nature+. IUCN World Conservation Congress, Jeju, 6–15 Sept 2012

La Riccia L (2012) Cultura del paesaggio e nuove prospettive di sostenibilità nella pianificazione locale. Ambiente Società Territorio 5:37–40

La Riccia L (2013) Paesaggio e pianificazione locale. PhD thesis, Politecnico di Torino (Italy)

Meadows D et al (1972) The limits to growth. Universe Books, New York

Nassauer JI (2007) Cultural sustainability: aligning aesthetics and ecology. In: Carlson A, Lintott S (eds) Nature, aesthetics and environmentalism. Columbia University Press, New York, pp 363–379

Powell J, Selman P, Wragg A (2002) Protected areas: reinforcing the virtuous circle. Plann Pract Res 17:279–295
Potschin MB, Haines-Young RH (2006) Landscapes and sustainability. Landsc Urban Plan 75: 155–161
Selman P (2006) Planning at the landscape scale. Routledge, New York
Sundseth K, Raeymaekers G (2006) Biodiversity and nature 2000 in urban areas. Federnature, Brussels
Swanwick C, Dunnett N, Woolley H (2003) Nature, role and value of green space in towns and cities: an overview. Built Environ 29:94–106
Turner T (1983) Landscape planning: a linguistic and historical analysis of the term's use. Landsc Plan 9:172–192
United Nations Environment Programme (UNEP) (1992) Convention on biological diversity. UNEP, Rio de Janeiro
World Commission on Environment and Development, Brundtland Commission (WCED) (1987) Our common future. Document A/42/427 – Development and International Co-operation: Environment. Oxford University Press, Oxford
Weber M (1922) Wirtshaft und Gesellschaft. Die Stadt. Mohr, Tubingen. Italian edition: Weber M (2003) Economia e Società (trans: Palma M). La città. Donzelli, Roma

Chapter 18
Protection of Peri-urban Agricultural Landscapes: *Vegas* and *Deltas* in Andalucía

Rocío Pérez-Campaña and Luis Miguel Valenzuela-Montes

Abstract In recent years, peri-urban agricultural landscapes are gradually receiving significant support in the context of five main fields. The first is from the European Landscape Convention (ELC), since its Article 2 refers not only to outstanding landscapes but also to everyday or even degraded landscapes. A second support is provided by the field of cultural heritage and cultural landscape, connecting the role of agricultural landscapes to patrimonial values. The understanding of landscape as heritage is quite recent and it is connected to the evolution of the heritage concept itself towards ideas such as the "heritagisation of territories and landscapes". Within this frame, agricultural landscapes are actually being considered as an important part of European heritage. Multifunctionality of agriculture and landscape offers a third approach to strengthen agricultural landscapes based on their role of developing multiple functions and their corresponding economic, ecological and sociocultural services. From the field of protected areas, the advances in its conception and role is leading to an inspirational framework to be applied beyond the traditional boundaries of this body, giving new opportunities to link nature protection and landscape, including agricultural landscapes, which are recognised in some cases as important ecosystems and biodiversity foci. Finally, from a spatial planning viewpoint, different instruments could offer specific solutions to protect and promote *vegas* and *deltas*.

Keywords Peri-urban agricultural landscape • *Vega* • *Delta* • Landscape protection

18.1 Vegas and Deltas in Andalucía (Spain) as Peri-urban Agricultural Landscapes

Vegas and deltas are probably the most characteristic peri-urban, agricultural landscapes in Andalucía. Vegas (agricultural floodplains) are traditional agricultural landscapes linked to historical irrigation systems in the Mediterranean Region

R. Pérez-Campaña (✉) • L.M. Valenzuela-Montes
Environmental Planning Laboratory (LABPLAM), Department of Urbanism and Spatial Planning, University of Granada, Campus de Fuentenueva s/n, Granada 18072, Spain
e-mail: rociopc@ugr.es; lvmontes@ugr.es

R. Gambino, A. Peano (eds.), *Nature Policies and Landscape Policies*,
Urban and Landscape Perspectives 18, DOI 10.1007/978-3-319-05410-0_18

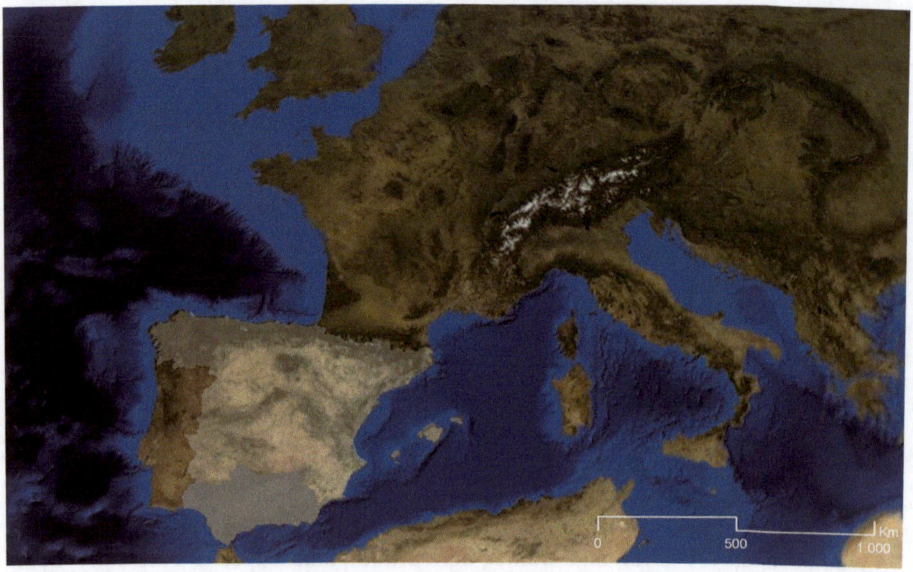

Fig. 18.1 Location of Andalucía (Spain)

(González-Bernáldez 1992; Mata and Fernández 2010), agricultural plains which are known for their high fertility and as a part of the typical landscape trio river-city-*vega* (Regional Government of Andalucía 2012). Delta is a more widely used and better known term to refer to a particular agricultural landscape at the mouth of a river (Meeus et al. 1990; Kruse et al. 2010). Both spaces have been historically occupied (especially the *vegas*) by human use due to, among others, their particular characteristics of geomorphology and topography (river dynamics), edaphology (the base of their fertility), productivity (linked to the agrologic capacity of some soil types and management techniques), water resources, etc.

Figures 18.1 and 18.2 locate *vegas* and deltas in Andalucía (Spain), showing linear and dendritic structures in most of the cases (depending on the fluvial system) and in a few locations more extensive areas, e.g. in the Vega de Granada.

Despite the fact that *vegas* and *deltas* represent just 5 % of the total area in Andalucía, nearly 70 % of urban and infrastructural areas and around 75 % of medium-sized cities are located on or in the vicinity of *vegas* and *deltas*. It implies that these spaces are especially important elements for the territorial structure in the region and represent the main landscape matrix, which composes peri-urban spaces in Andalusian cities and villages.

Fig. 18.2 Location of *vegas* and *deltas* in Andalucía

18.2 Protection of *Vegas* and *Deltas*: Possibilities and Needs

From all the different approaches to protect and promote *vegas* and deltas as agricultural peri-urban landscapes, we have selected five. In some cases the approach provides a general framework or theoretical, academic support and in other cases it may offer specific regulation or planning solutions. With the selected items in the following sections, we try to contribute to a general overview of the current situation for agricultural peri-urban landscapes in general and *vegas* and deltas in particular.

Peri-urban landscapes are considered by the European Landscape Convention (Art. 2) as singular landscapes where different agricultural, urban and natural elements coexist. These landscapes are daily life or everyday landscapes, and for the first time they have been recognised through a European framework marking an important milestone for the future development of landscape policies.

The interest in these areas has also been reflected during recent years by associative networks that emphasise the role of the peri-urban space, e.g. the Terres en Villes, Purple, Arco Latino Fedenatur and PeriUrban Parks. Therefore, from the first reference about peri-urban agriculture given by the OECD in 1979, this issue has become an important topic, as the Opinion of the European Economic and Social Committee on "Agriculture in peri-urban areas" (EESC 2004) also highlights.

More recently, an interesting initiative has been developed in Spain through the Red Agroterritorial, the Parc Agrari del Baix Llobregat and the Fundación Agroterritori with their proposal for a Charter on peri-urban agriculture for the conservation and management of peri-urban agricultural spaces.

Modelled by water, *vegas* and *deltas* are agricultural landscapes and also water landscapes given that water has had a decisive role in their structure and functioning from the viewpoint of their ecology, management and in their own origin and evolution as landscapes (Frolova 2007). Thus, water is present in these landscapes in two main forms: as a fluvial agent in the geomorphological origin of *vegas* and *deltas* and flowing through the irrigation systems, among which networks of historic water channels constitute one of the main features characterising *vegas* and *deltas* in the Mediterranean as a legacy from the Muslim period. These channels together with other structures such as weirs, water mills, *albercas* (open water tanks), etc. are actually examples of built elements related to water management in Al-Andalus (Trillo 2003).

This ancient origin has been put forward as an argument for the protection of some of these spaces, and the consideration of *vegas* and *deltas* as agricultural heritage has found two principal ways forward. The first one meets the idea of agricultural heritage as the set of architectural and other built elements linked to agricultural activity. The second one is more recent and enhances the agricultural landscape itself as heritage. It concerns the evolution of the term heritage itself, broadening towards other dimensions, being understood as the materialistic and also the non-materialistic legacy of the experience and effort of a community (Sabaté 2006) and emerging new concepts such as the cultural landscape coined by Carl Sauer (Sabaté 2006; Silva 2009).

Under this new approach, we perceive the "heritagisation" of territories and landscapes (Mata 2011). In recent years, agriculture is seen as a supplier of heritage (Silva 2009) and agricultural landscapes are being recognised as an important part of European heritage (Kruse 2010) where specific features of European culture are expressed (Pungetti and Kruse 2010).

Vegas and *deltas* in Andalucía gather together many qualities to be considered for their patrimonial value. The Vega de Granada (location 1 in Fig. 18.1) has been claimed as Heritage of Cultural Interest by different associations and institutions, first with the instrument named Historical Site and later (after the Andalusian Historical Patrimony Law of 2007) with the instrument of Patrimonial Zone. Nevertheless, none of the proposals have been accepted. One of the inconveniences relates to the difficulty for defining precise limits to the space that contains the cultural values worthy of protection by the cited instruments.

Multifunctionality in the context of agriculture may be defined in relation to non-trade concerns, the joint production of public goods and positive externalities (OECD 2001). The multifunctionality of landscape entails a wider socioecological system (Selman 2009) including the agricultural landscape as a whole and not merely the agricultural activity itself. From this viewpoint, *vegas* and *deltas* are especially multifunctional landscapes, since they constitute the result of a close human-space relationships throughout the centuries, generating particular agricultural landscapes.

These landscapes contain valuable elements such as those mentioned previously in this section. Even their own plot and irrigation network structure constitute a legacy from the Muslim period, when specific, sometimes subtle, irrigation techniques where developed to face periods of drought in the Mediterranean.

Vegas and *deltas* are also peri-urban landscapes in many cases. Near human settlements is where the multifunctionality of *vegas* and *deltas* may be more interesting, as they provide ecosystem, economic, cultural and social services which can be perceived and used more directly in a peri-urban context.

There are not many studies dealing with the multifunctionality of agricultural landscapes in Andalucía (Silva 2010). In Pérez (2013b), multifunctionality maps are presented for the specific case of Vega del Guadalfeo (location 2 in Fig. 18.1), showing the multiple dimensions of this *vega-delta*, including ecological, cultural and social functions. All these functions, together with the economic one, should be considered when reflecting on possible protection schemes for agricultural landscapes.

At national level, nature protection has its main regulation in Law 42/2007 for Natural Heritage and Biodiversity. Nevertheless, agricultural landscapes do not have full support in this law, which is more oriented towards protection and conservation purposes applied to natural, outstanding landscapes. This issue has encouraged the Fundación Agroterritori in Cataluña to debate the need for a specific law for the protection of agricultural spaces, which is being discussed both within and outside Cataluña.

In Andalucía, the protection schemes for nature come under Law 2/1989 (Inventory of Protected Natural Spaces). Similar to the Spanish regulation on protected areas (Law 42/2007), this regional regulation does not regard agricultural space either. Indeed, just 4.5 % of *vegas* and *deltas* are included in the Andalusian Network of Protected Areas (Pérez 2013a), but this percentage represents the protection of specific riparian ecosystems and other spaces which are in the domain of larger protected areas.

That is perfectly understandable considering that regulation on protected areas is especially focused on natural landscapes with minimum levels of anthropogenic intervention or where ecosystems are managed to help in their maintenance as natural ecosystems. But in this sense, some reflections on the relationship between agricultural landscapes and protected areas are presented by Harrop (2007) for the case of traditional agricultural landscapes and in the context of the international law and policy, showing that this type of agriculture may contribute to maintain biological diversity among other purposes. Thus, in recent years we perceive a growing recognition about the role of agricultural areas for landscape and biodiversity conservation. The Communication from the Commission to the European Parliament, the Council, the European Economic and Social Committee and the Committee of the Regions named "Options for an EU vision and target for biodiversity beyond 2010" (CE 2010) underlines the function of agricultural spaces to produce and maintain the green infrastructure and the biodiversity. The Partnership for European Environmental Research (PEER) has undertaken an assessment of ecosystem services (linked to the idea of multifunctionality) provided by agricultural and semi-natural ecosystems (Maes et al. 2011, 2012). In this context, the role of agricultural spaces is becoming more recognised, evolving from conceptions in which agriculture was just a threat for nature protection purposes towards a new acknowledgement of agricultural landscapes per se and also as sometimes crucial sites to maintain the necessary structure for protected areas. This opening may have to do with new paradigms concerning protected areas (Phillips 2003), where also innovative objectives, perceptions, management techniques, etc. may be put into place.

Finally, despite the fact that there is not a specific regulation framework, there are some examples of protection under discretionary proposals. This is the case of Cataluña, which in 1992 included some agricultural spaces as protected natural spaces in its Plan for Spaces of Natural Interest.

Depending on the planning scale, *vegas* and *deltas* in Andalucía may find different protection and management possibilities. At a subregional level, territorial plans can consider them as environmental protection zones, once they have already protected by sectorial regulation on protected areas (which has already been seen as not very common). The other option is for *vegas* and *deltas* to be considered as territorial protection zones, based on previously existing protection schemes set up by other spatial planning instruments or by the territorial plan itself.

For the first possibility, as we have previously explained, *vegas* and *deltas* hardly fit into this regulation. For the second possibility, the Regional Government has already drawn up a Special Plan for Protection of the Physical Environment in each province (the latter was finished in 2007). In some of these plans, specific *vegas* and *deltas* were protected as Singular Agricultural Landscapes. Regardless of this protection, not all territorial plans have fully incorporated the delimitation of Singular Agricultural Landscapes and have largely been ignored, and even in some cases no references to them have been found in the plans either. In other cases, Singular Agricultural Landscapes have changed their land use in the temporal lapse between the adoption of the Special Plan and the Territorial Plan.

At local level, plans should incorporate the specifications given by subregional territorial plans, adapting them to their own regulations at local scale. Local plans could then set different protection levels for *vegas* and *deltas* (if they exist in the municipality considered) classifying them as non-urbanisable land, which entails a relative protection, since this classification may be modified whenever the land is required for building purposes.

Another specific support for these landscapes comes from the agricultural parks movement. Beginning almost simultaneously in Italy (Parco Agricolo Sud Milano and Parco di Palermo) and Spain (Parc Agrari del Baix Llobregat), agricultural parks offer a planning, management and protection solution for agricultural landscapes in peri-urban areas. Since the 1990s, agricultural parks and attempts to create them have spread across Europe under different instruments depending on specific national or regional regulation frameworks (see a compilation in Simón et al. 2012). In Andalucía a proposal for an agricultural park in the Valle del Guadalhorce (Málaga) and the Special Plan for the Vega de Granada are currently being studied.

18.3 Conclusions

We have presented a brief overview concerning *vegas* and *deltas* as agricultural peri-urban landscapes and some examples of possibilities for protecting and promoting these landscapes.

The special entity *vegas* and *deltas* represent, determining the territorial structure of Andalucía, despite their limited extension, within the total area of the region has been examined.

The main arguments for their protection come from the growing recognition of the role these areas may play, performing economic, social and environmental functions (multifunctionality of *vegas* and *deltas*). As far as this concerned, the backing given by the European Landscape Convention, which implies at least a minimum framework and conceptual consensus, has been fundamental. In addition, for the specific case of *vegas* and *deltas* and given their origin and structure, these particular landscapes usually include important patrimonial elements especially linked to water management. Nevertheless, the isolated protection of elements separately from their territorial context does not offer an appropriate solution, and despite the cultural value that is being more and more recognised in agricultural landscapes as a whole, there is no a great interest in Andalucía to provide a specific protection scheme for these places. Regulations on nature protection have not offered a specific frame for these spaces either, but they are becoming more aware of their role. Vegas and deltas may offer even more arguments for their protection than other agricultural spaces, since *vegas* and *deltas* still develop traditional agriculture and may fulfil objectives approached from nature protection. Finally, spatial planning in Andalucía has made some attempts to protect *vegas* and *deltas* (e.g. through their definition as Singular Agricultural Landscapes), but it is still mainly focused on urban and building purposes, *vegas* and *deltas* remain without any specific attention. Nevertheless, it seems to be a matter of will, more than a gap in the spatial planning system, and we perceive an incipient interest in these spaces, though we need to wait for some time to evaluate the results of this interest.

References

CE (2010) Options for an EU vision and target for biodiversity beyond 2010. Communication from the Commission to the European Parliament, the council, the European Economic and Social Committee and the Committee of the Regions, Brussels

EESC (2004) Opinion of the European Economic and Social Committee on agriculture in periurban areas. NAT/204. European Economic and Social Committee, Brussels

Frolova M (2007) El estudio de los paisajes del agua en una cuenca vertiente: Propuesta metodológica. Rev Estud Reg 82:21–47

González-Bernáldez F (1992) Los paisajes del agua: terminología popular de los humedales. J.M. Reyero, Madrid

Harrop S (2007) Traditional agricultural landscapes as protected areas in international law and policy. Agric Eco Environ 121:296–307

Kruse A (2010) Agricultural landscape classification as tool for implementing the European Landscape Convention in research and planning. Results from the EUCALAND Project. In: Vv.Aa (ed) Living landscape. The European Landscape Convention in research perspective. UNISCAPE, Florence

Kruse A, Centeri C, Renes H, Roth M, Printsmann A, Palang H, Jordá LB, Velarde MD, Kruckenberg H (2010) Glossary on agricultural landscapes. Tájökológiai Lapok 8(Special Issue):99–127

Maes J et al (2011) A spatial assessment of ecosystem services in Europe: methods, case studies and policy analysis – phase 1. PEER report. Partnership for European Environmental Research, Italy

Maes J et al (2012) A spatial assessment of ecosystem services in Europe: methods, case studies and policy analysis – phase 2. PEER report. Partnership for European Environmental Research, Italy

Mata R (2011) Heritagisation of everyday landscapes and sustainable development. An experience in the Cantabrian mountains (Nansa Valley, Cantabria, Spain). Colloque International Paysages de la vie quotidienne. Regards croisés entre la recherche et l'action. Perpignan/Girona

Mata R, Fernández S (2010) Paisajes y patrimonios culturales del agua. La salvaguarda del valor patrimonial de los regadíos tradicionales. Scripta Nova. Revista Electrónica de Geografía y Ciencias Sociales, 337. Available via http://www.ub.edu/geocrit/sn/sn-337.htm. Accessed 15 Nov 2013

Meeus JHA, Wijermans MP, Vroom MJ (1990) Agricultural landscapes in Europe and their transformation. Landsc Urban Plan 18:289–352

OECD (1979) Agriculture in the planning and management of peri-urban areas. Organization for Economic Co-operation and Development, Paris

OECD (2001) Multifunctionality, towards and analytical framework. Organization for Economic Co-operation and Development, Paris

Pérez R (2013a) Evolution and planning of vegas and deltas in Andalucía: focusing on peri-urban agriculture. I Quaderni di Careggi, Uniscape 2:47–51

Pérez R (2013b) La Vega del Guadalfeo como paisaje agrario periurbano. Transformación, ecoestructura y multifuncionalidad. PhD thesis, Universidad de Granada

Phillips A (2003) Un paradigma moderno. Conservación Mundial. Boletín de la IUCN 2:6–7

Pungetti G, Kruse A (2010) European culture expressed in agricultural landscapes. Perspectives from the Eucaland Project. Palombi Editori, Rome

Regional Government of Andalucía (2012) Landscape strategy for Andalucía. Consejería de Agricultura, Pesca y Medio Ambiente, Sevilla

Sabaté J (2006) De la preservación del patrimonio a la ordenación del paisaje. In: Mata R, Tarroja A (eds) El paisaje y la gestión del territorio. Criterios paisajísticos en la ordenación del territorio y el urbanismo. Diputación de Barcelona, Barcelona, pp 329–342

Selman P (2009) Planning for landscape multifunctionality. Sustain Sci Pract Policy. Available via http://sspp.proquest.com/archives/vol5iss2/communityessay.pselman.html. Accessed 15 Nov 2013

Silva R (2009) Agricultura, paisaje y patrimonio territorial: los paisajes de la agricultura vistos como patrimonio. Boletín de la Asociación de Geógrafos Españoles 49:309–334

Silva R (2010) Multifuncionalidad agraria y territorio: algunas reflexiones y propuestas de análisis. EURE 36:5–33

Simón M, Zazo A, Morán N (2012) New spatial planning approaches to protect peri urban agrarian spaces. Ciudades, Revista del Instituto Universitario de Urbanística de Valladolid 15:151–166

Trillo C (2003) Agua y paisaje en Granada. Una herencia de al-Andalus. Diputación de Granada, Granada

Chapter 19
Linking Landscape Protection and Nature Conservation: Switzerland's Experience with Protected Mire Landscapes

Thomas Hammer and Marion Leng

Abstract Since 1987, Switzerland's Federal Inventory of Mire Landscapes of Particular Beauty and National Importance has provided an instrument for the integration of nature conservation and landscape protection. Mires and mire landscape protection are strictly regulated. However, research results show that neither the goals of mire protection nor those of mire landscape protection are being achieved. The reasons for this are manifold and, in particular, have to do with a lack of coordination between the various policy areas that shape mire environments and mire landscapes. There are several key challenges involving different political and administrative levels. At the national level, mechanisms must be devised that enable differentiated regional implementation of national sectoral policies. In the context of cantonal structure planning, regional nature conservation and landscape protection priorities should be established based on existing regional potentials vis-à-vis the natural environment and landscapes (including protected biotopes and landscapes). At the regional level (spanning multiple communes), integrated planning instruments and governance structures should be developed so that implementation of national and cantonal sectoral policies may be harmonized under the umbrella of regional and integrated development plans. These adjustments to Switzerland's institutional system are necessary to enable far-reaching integration of nature conservation and landscape protection when setting regional policy priorities. This would strengthen the protection of mire landscapes and other integrative instruments such as regional nature parks of national importance.

Keywords Landscape protection • Nature conservation • Mire landscapes • Switzerland

---

T. Hammer (✉) • M. Leng
Centre for Development and Environment CDE, University of Bern/Switzerland, Hallerstrasse 10, CH-3012 Bern, Switzerland
e-mail: thomas.hammer@cde.unibe.ch; marion.leng@cde.unibe.ch

R. Gambino, A. Peano (eds.), *Nature Policies and Landscape Policies*, Urban and Landscape Perspectives 18, DOI 10.1007/978-3-319-05410-0_19

19.1 Landscape Protection and Nature Conservation in Switzerland

Switzerland has had an official national policy on nature conservation and landscape protection since 1966 (Bürgi et al. 2008; Knoepfel et al. 2010). That year saw the adoption of the Federal Act on the Protection of Nature and Cultural Heritage, which has been revised and amended many times since. It requires the Swiss Confederation to engage itself in four areas (Bisang et al. 2008, pp. 14–15): first, the protection of indigenous animal and plant species in all their diversity; second, the protection of natural and near-natural biotopes that host indigenous species; third, the protection of landscapes of particular beauty and natural monuments; and fourth, the protection of sites of local character, heritage sites, and cultural monuments of particular importance. All are to be protected from unwanted interventions.

Depending on the area, this national legal framework also specifies explicit preservation goals (such as preservation of biotopes, landscapes, and sites of local character). Especially in the areas of landscape protection and protecting sites of local character, goals such as preserving, maintaining, and nurturing are pursued. To achieve these goals, the Confederation developed a wide range of instruments (Leimbacher 2000; Munz et al. 1996). These include the following: in species protection, red lists of threatened or already extinct species; in biotope protection, various inventories of biotopes of national importance (such as alluvial zones, amphibian spawning areas, fenlands, raised bogs and transitional mires, hay meadows, and pastures); in landscape protection, the Federal Inventory of Landscapes and Natural Monuments of National Importance (ILNM) and the Federal Inventory of Mire Landscapes of Particular Beauty and National Importance; and in heritage protection, the Federal Inventory of Swiss Heritage Sites (ISOS). Experts from the Confederation and the cantons determined the objects worthy of protection according to scientific criteria ("top-down instruments"). The Confederation and the cantons ensure that corresponding measures are implemented at the federal level, in the cantons, and in the communes, in order to achieve the specified goals.

In addition, there have been instruments developed in policy areas outside nature conservation and landscape protection that are nevertheless highly relevant to conservation and protection goals (Munz et al. 1996). These include establishment of game reserve zones, water and bird sanctuaries in regards to hunting policy, and the introduction of ecological compensation areas and networks as regards to agricultural policy. Legal foundations have been established in many other policy areas – such as environmental protection, transport policy, and regional development planning – to address concerns of nature conservation and landscape protection in an integrated, crosscutting manner.

Overall, Switzerland has a good deal of long-term national-level experience with instruments of nature conservation and landscape protection (Longatti and Dalang 2007). The majority of these instruments are viewed as having been implemented well (Leimbacher 2000). Nevertheless, the overarching goals of nature conservation and landscape protection are usually left unfulfilled, achieved only in part, or not achieved at all (Bauer et al. 2004; Leimbacher 2000; OECD 2007, pp. 91–115; Roth et al. 2010): Species lost continues unabated, many preservation-worthy biotopes deteriorate in quality, and protected landscapes are subject to creeping change due to the combined effects of numerous small unwanted interventions and inappropriate use. With respect to nature conservation and landscape protection, it is essential to figure out how the relevant instruments can be further developed in order to improve the achievement of goals. In particular, the question arises of how best to integrate the concerns of nature conservation and landscape protection, which typically point in a similar direction (Duelli 2006; Hammer et al. 2008; Hampicke 2013). It is this issue – the integration of nature conservation and landscape protection to achieve overarching goals – that we researched in the context of a European Cooperation in Science and Technology (COST) Action A27 project (Understanding Pre-industrial Structures in Rural and Mining Landscapes). The following discussion is based on the results of this project (Hammer et al. 2008, 2009, 2011; Hammer and Leng 2008, 2011; Leng and Hammer 2009).

19.2 Experiences in the Integration of Nature Conservation and Landscape Protection: The Example of Mire Landscapes of Particular Beauty and National Importance

Switzerland's Federal Inventory of Mire Landscapes of Particular Beauty and National Importance has provided an instrument for the integration of nature conservation and landscape protection since 1987. In a national referendum held that year, Swiss voters opted to protect and preserve mire biotopes *and* mire landscapes. They were declared to be "mire landscapes of particular beauty and national importance" (referred to in the following simply as mire landscapes). As a direct result, hundreds of mire biotopes – many only a few hectares in size – and 89 landscapes, totaling 87,500 ha, were placed under protection. The mires are to be preserved, in particular, on behalf of the flora and fauna they host and to serve as an archive of vegetation history (on behalf of biodiversity and research goals). By contrast, the mire landscapes – characterized by fenlands and raised bogs – are to be preserved as near-natural, extensively farmed, and attractive cultural landscapes (in other words, on behalf of aesthetic, economic, and cultural goals as well). Thus, the goals complement one another.

In the course of strict enforcement of mire protection and mire landscape protection (Seitz and Zimmermann 2008; Waldmann 1997), various measures were implemented at the national, cantonal, and communal level and, depending on the canton, at the regional level. Today, mire landscapes are by far the most stringently protected landscapes in Switzerland that are simultaneously subject to habitation and use. For this reason as well, they are prime objects of study, enabling analysis of the integration of nature conservation and landscape protection, the effects of safeguarding provisions, and success factors regarding integration of protection and use.

Various studies were conducted in the COST Action project between 2006 and 2011, examining the implementation of goals at the communal, regional, and cantonal level, the effects of implementation, the reasons why certain goals were not achieved, and the possibilities for improving achievement of goals (Hammer and Leng 2008; Hammer et al. 2011).

The results show that, despite the strict regulations and various instruments, most of the goals of protecting mires and mire landscapes have been achieved solely in part or not at all. The reasons for this are manifold. At a conceptual level, protection of the mires was given far more emphasis than preservation of mire landscapes. With regard to mire protection, land managers – typically farmers – are compensated for avoiding use and acting as stewards. By contrast, preservation of mire landscapes is generally restricted to preventing big interventions that do not conform to stated goals. Largely absent are development measures such as promoting appropriate land use, harmonizing protection, preservation, and use outside the protected mire biotopes, and introducing special measures to network the biotopes within the mire landscapes.

One key insight is that the undesirable transformation of mire landscapes particularly stems from sectoral policies outside of nature conservation and landscape protection. Agricultural policy, alpine farming policy, and forest policy are shaping the transformation of mire landscapes much more than the ostensibly superordinate policy of protecting mire landscapes. It was found that these different sectoral policies were being implemented in mire landscapes with little or no effort toward coordinating them with the goals of protecting mire landscapes. Similarly, nature conservation within the mire landscapes was not found to be much more strictly enforced than outside these landscapes, despite the declared goals. Overall, it appears that no or only very little – and thus insufficient – coordination of sectoral policies occurs under the umbrella of landscape protection.

19.3 Challenges with Regard to the Integration of Nature Conservation and Landscape Protection

The lack of regional-level coordination of sectoral policies, under the umbrella of superordinate regional goals like protection of mire landscapes, has a systemic cause. Implementation of national sectoral policies at the cantonal and regional levels is usually very constrained by nationally prescribed criteria. These criteria typically do not allow establishment of regional priorities. As a result, new instruments such as ecological networking projects (agricultural policy) have been introduced in recent years. These are tied to the existence of regional networking concepts. A similar pattern is occurring with regards to service agreements on nature conservation reached between the Confederation and the cantons. The introduction of cultural landscape projects is a current example (also belonging to agricultural policy). These strive for regionally coordinated valorization of cultural landscapes.

A key initial challenge, therefore, is that of enabling – at the national level – differentiated implementation of national sectoral policies at the cantonal and regional levels, and even promoting this where possible. Enabling national policies of agriculture, alpine farming, and forestry to be adapted to the goals of nature conservation and landscape protection at the cantonal and regional levels is especially important. In the case of mire landscape protection, this would mean aligning these policies with its goals. In this way, sectoral policies would be calibrated to the goals of protecting mire landscapes and would contribute to their achievement.

Switzerland already has instruments that could be used to promote the establishment of regional priorities and the coordination of sectoral policies. These include land use planning in the cantonal and regional structure plans as well as regional policy and spatial planning policy. These could be further strengthened and used to coordinate cantonal and sectoral policies under the umbrella of regional goals.

At the national level, specific instruments exist that are suited to establishing regional priorities of nature conservation and landscape protection and suited to coordinating sectoral policies. Besides the inventory of mire landscapes, these include the Federal Inventory of Landscapes and Monuments of National Importance Bundesinventar der Landschaften und Naturdenkmäler von nationaler Bedeutung (BLN) introduced in 1977, the Swiss Landscape Concept (SLC) introduced in 1998 (SAFEL 1998), and the Parks of National Importance introduced in 2007. These instruments could be correspondingly strengthened. For example, implementation of national and cantonal sectoral policies in parks of national importance could be calibrated to the goals of the parks themselves.

In Switzerland, the cantons are responsible for implementing national policies of nature conservation and landscape protection. Although the Confederation designates protected landscapes, species, and biotopes of national importance and establishes the overall goals, it is the cantons that bear responsibility for implementation. The cantons also pursue their own policies of nature conservation and landscape protection. These complement the national policies. In order for strengthened coordination of national and cantonal sectoral policies to occur at the regional level, corresponding instruments must be strengthened at the cantonal level.

The cantonal structure plan is ideally suited to this task. In most cantons, the cantonal structure plan outlines valued regional nature areas and landscapes and specifies protected landscapes. In principle, this would allow establishment – within a broader cantonal perspective – of regional priorities regarding nature conservation and landscape protection as well as formulation of specific regional goals. At the same time, these regional goals could be used to mold implementation of national and cantonal sectoral policies in the regions, adapting the goals where necessary.

Switzerland features spatial planning regions and the instrument of regional structure plans. Depending on the canton and the region, these structure plans are more or less comprehensive. They cover supra-communal and cantonal infrastructure projects and regional economic goals and usually encompass projects and goals in the area of nature conservation and landscape protection. In order to better coordinate implementation – at the regional level – of national and cantonal sectoral policies, according to superordinate regional goals, the instrument of regional development planning could be further expanded and refined. The regional structure plan could be elevated to a sort of managerial umbrella or regional coordination of all sectoral policies, enabling regional development planning that is comprehensive and integrated.

This would give the planning regions significantly more responsibility than they are currently granted. One possibility would be to establish regional management authorities charged with coordinating the regional-level implementation of different national and cantonal policies, in collaboration with the various national, cantonal, and communal administrative offices and other individuals concerned. Accordingly, new regional governance structures would (necessarily) be formed, and the current top-down implementation of sectoral policies would be relativized, that is, policies could be adjusted according to regional conditions. This would fundamentally strengthen the regional level.

However, this type of integrated regional development planning is only possible if national and cantonal sectoral policies may be flexibly implemented at the regional level. In principle, this requires strengthening the regional instruments within the national and cantonal policies as well as adapting the sectoral policies and regional policies as needed. The sectoral policies would not lose their fundamental importance, but it would be possible to adjust them according to specific regional conditions and regional policies.

19.4 Conclusions

Switzerland's experience with protection of mire landscapes shows that the coordination of national nature conservation and landscape protection largely depends on the degree of flexibility in implementation of sectoral policies. Switzerland faces three key challenges on three different policy/administrative levels. First, at the national level, reforms changes should be introduced to enable differentiated regional implementation of national sectoral policies. In mire landscapes and parks of national importance, this would make it possible to adapt agricultural policy, alpine farming policy, forest policy, and other relevant policies according to the goals of nature conservation and landscape protection. Second, at the cantonal level, regional priorities should be established – for instance, in the framework of the cantonal structure plan – to enable utilization of cantonal policy-based support services on behalf of regional goals. Third, corresponding integrated planning instruments and governance structures are needed at the regional level to enable coordination of sectoral policies under the umbrella of regional and integrated development plans. Only when the relevant adjustments are made will it be possible to achieve extensive integration of nature conservation and landscape protection according to regional policy priorities.

References

Bauer B et al (2004) Biodiversität in der Schweiz. Zustand, Erhaltung, Perspektiven. Haupt Verlag, Bern/Stuttgart/Wien

Bisang K, Moser TA, Zimmerman W (2008) Erfolgsfaktoren in der Naturschutzpolitik. Beispiele aus vierzig Jahren Natur- und Landschaftspolitik in der Schweiz. Rüegger Verlag, Zürich/Chur

Bürgi M, Atmanagara J, Stuber M, Egli HR (2008) Management and protection of landscape in Switzerland. In: Fairclough G, Moller PG (eds) Landscape as heritage. The management and protection of landscape in Europe. University of Berne, Berne, pp 249–268

Duelli P (2006) Biodiversität als Landschaftsqualität. In: Tanner KM, Bürgi M, Coch T (eds) Landschaftsqualität. Haupt Verlag, Bern/Stuttgart/Wien, pp 103–116

Hammer T, Leng M (2008) Moorlandschaften im Bedeutungswandel. Zur gesamtgesellschaftlichen Aufwertung naturnaher Kulturlandschaften. Allgemeine Ökologie zur Diskussion gestellt, vol 10. Schriftenreihe der Interfakultären Koordinationsstelle für Allgemeine Ökologie (IKAÖ), Bern

Hammer T, Leng M (2011) Wie lassen sich naturnahe Kulturlandschaften erhalten? Vorschläge für innovatives Handeln am Beispiel der Moorlandschaften der Schweiz. GAIA 20(4):265–271

Hammer T, Egli HR, Atmanagara J (2008) Cultural landscape in conflict between economy, ecology, and institutional steering. The example of the UNESCO biosphere Entlebuch (Switzerland). In: Bartels C, Ruiz del Arbol M, Londen Hv, Orejas A (eds) Landmarks – profiling Europe's historic landscapes. L'Harmattan, Paris, pp 99–112

Hammer T, Leng M, Raemy D (2009) Swiss Moorland landscapes and the significance of economic and technological change. Origins, threats and new patterns of use. In: Lévêque L, Ruiz del Arbol M, Pop L (eds) Heritage, images, memory of European landscapes. Deutsches Bergbau-Museum, Bochum, Paris, pp 329–339

Hammer T, Leng M, Raemy D (2011) Moorlandschaften erhalten durch Gestalten. Nutzen und Schützen naturnaher Kulturlandschaften am Beispiel der UNESCO Biosphäre Entlebuch (UBE). Allgemeine Ökologie zur Diskussion gestellt, vol 10. Schriftenreihe der Interfakultären Koordinationsstelle für Allgemeine Ökologie (IKAÖ), Bern

Hampicke U (2013) Kulturlandschaft und Naturschutz. Probleme – Konzepte – Ökonomie. Springer Spektrum, Wiesbaden

Knoepfel P, Nahrath S, Savary J, Varone F (2010) Analyse des politiques suisses de l'environnement. Editions Rüegger, Zürich/Chur

Leimbacher J (2000) Bundesinventare Schweiz. Die Bedeutung der Natur- und Landschaftsinventare des Bundes und ihre Umsetzung in der Raumplanung. Schweizerische Vereinigung für Landesplanung, Bern

Leng M, Hammer T (2009) Moorland landscapes in Switzerland – the changing significance of near-natural cultural landscapes. eco.mont 1(2):57–60

Longatti P, Dalang T (2007) The meaning of "Landscape" – an exegesis of Swiss Government texts. In: Kienast F, Wildi O, Ghosh S (eds) A changing world. Challenges for landscape research. Springer, Dordrecht, pp 35–46

Munz R, Bryner A, Siegrist D (1996) Landschaftsschutz im Bundesrecht (Schweiz). Verlag Rüegger, Chur/Zürich

OECD (2007) Umweltprüfbericht Schweiz. Bundesamt für Umwelt (BAFU), Bern

Roth U, Schwick C, Spichtig F (2010) Zustand der Landschaft in der Schweiz. Bundesamt für Umwelt (BAFU), Bern

SAFEL (1998) Swiss landscape concept. Swiss Agency for the Environment, Forests and Landscape (SAFEL), Bern

Seitz A, Zimmermann W (2008) Bundesgesetz über den Natur- und Heimatschutz NHG: Bundesgerichtliche Rechtssprechung 1997–2007. Umweltrecht in der Praxis 22(2):103–206

Waldmann B (1997) Der Schutz von Mooren und Moorlandschaften. Inhalt, Tragweite und Umsetzung des "Rothenthurmartikels". Arbeiten aus dem Juristischen Seminar der Universität Freiburg Schweiz, vol 162, Freiburg/Schweiz

Chapter 20
Putting the Park-Landscape Alliance to the Test: Protected Landscapes as a Proving Ground

Emma Salizzoni

Abstract Today, there is widespread hope that it will be possible to achieve an alliance between the policies developed inside protected natural areas and landscape policies, in that such an alliance is considered to be beneficial for the conservation of both nature and the landscape. One place that may have the right qualities to host this alliance is the Protected Landscapes (category V protected areas, according to the IUCN classification system), where natural and cultural values are closely connected. This paper uses three Protected Landscapes along the Spanish, French, and Italian coasts in its case studies to verify the methods used to implement policies for the landscape. A rather varied picture emerges, in which more and less positive signs of the "Park-landscape" alliance are identified, and these signs in turn indicate some possible paths to follow to promote this link.

Keywords Protected natural areas • Protected Landscapes • Landscape policies

20.1 Hopes for a Park-Landscape Alliance

Similar evolutionary processes took place in the concepts of nature and landscape and the policies associated with them during the closing decades of the last century. In relation to the concept of nature, conservation policies – of which protected natural areas are the main expression – have undergone a veritable theoretical and operative "revolution" (Phillips 2003, p. 13). This change of route, which marked the progress beyond an "insular" conception of protected areas, towards their opening up to the wider socio-economic and territorial context (Gambino 2005), found formal expression in the so-called new conservation paradigms launched during the fifth International Union for Conservation of Nature (IUCN) World Parks Congress held in Durban in 2003. In short, the new paradigms promote policies for the conservation of nature that are:

E. Salizzoni (✉)
Interuniversity Department of Regional and Urban Studies and Planning (DIST), Politecnico and Università di Torino, Viale Mattioli 39, 10125, Turin, Italy
e-mail: emma.salizzoni@polito.it

R. Gambino, A. Peano (eds.), *Nature Policies and Landscape Policies*, Urban and Landscape Perspectives 18, DOI 10.1007/978-3-319-05410-0_20

- *Complex*: aiming not only at conservation but also at socio-economic development
- "*Territorialised*": extended beyond the boundaries of the protected areas
- Of considerable *social value*: mainly addressed to local communities
- *Dynamic*: not limited to protecting, but extended to restoring and enhancing, and, therefore, requiring planning

This evolutionary process has found important confirmation in what has distinguished the concept of landscape in more recent years and has led to the issue, in 2000, of the European Landscape Convention (ELC), which promotes landscape policies that are:

- *Complex*, in relation to a holistic concept of landscape seen as a synthesis of different values (ecological, aesthetic, cultural, social, and economic)
- "*Territorialised*", with reference to an "extended" concept of landscape which "is an important part of the quality of life for people *everywhere*" (ELC, Preamble)
- Of considerable *social value*, in relation to a concept of the landscape as "an essential component of people's surroundings (...), a foundation of their identity" (ELC, art. 5a)
- *Dynamic*, with reference to an interpretation of the landscape as the evolutionary entity par excellence (Phillips 2005; Romani 2008)

These evolutionary processes concerning the concepts of nature and landscape and the policies associated with them are similar to one another, being characterised by a progressive convergence (Gambino 2010) which is currently at the basis of international hopes[1] for an alliance between the policies developed within protected natural areas and landscape policies. Generally, a number of reasons are given in support of this alliance: in short, the supporters of a "Park-landscape" convergence propose protected areas as valuable experimental laboratories for the landscape and the landscape as a "means" of conserving nature along the lines of the "new paradigms".[2]

Yet despite the tendencies mentioned above and the relative hopes nurtured on several sides in relation to a "Park-landscape" convergence, there are still strong divisions between the two disciplinary and operative spheres. It is the world of "nature" – namely, protected natural areas – in particular that has the most difficulty in opening up to the landscape or cultural dimension in the broader sense. Within the context of the IUCN, for instance, this "expansion", sanctioned by the aforementioned "new paradigms", has generated several adverse reactions (e.g. Locke and Dearden 2005) and, paradoxically, a climate of renewed attention to the topic of biodiversity conservation, reiterating its priority in the management aims of

[1] See, for example, the Resolution issued after the third IUCN World Conservation Congress (Bangkok 2004): "A landscape/seascape approach to conservation" (Res. 3.065).

[2] For a closer examination of the matter, see Gambino (2005), Peano (2008), and Phillips and Borrini-Feyerabend (2009).

protected areas compared to the other possible objectives (such as those of a cultural, social, and economic nature). Therefore, it cannot be automatically assumed that protected areas today – however much they may be focussed on the aims of sustainable development – implement landscape policies. Rather than an assumption, this continues to be a hope that could, however, have a greater possibility of being put into practice within those protected areas classified by the IUCN in category V (Protected Landscapes).[3] These are lived-in, working landscapes (Beresford and Phillips 2000; Brown et al. 2005), structured over time by an interaction between natural and anthropic factors, the defence of which, as indicated by the definition of the category,[4] is vital to the survival of the self-same biodiversity values.

20.2 Three Cases of Protected Landscapes: Experimental Laboratories for Landscape Policies?

The three Parks[5] in the case studies – all classified as Protected Landscapes – are analysed with the aim of verifying whether they act as experimental laboratories for landscape policies as intended by the ELC. It is no mere coincidence that they are protected areas situated in the Euro-Mediterranean region – along the coast of Spain, France, and Italy (Fig. 20.1) – a context in which the alliance between policies for the conservation of nature and for the landscape is considerably desirable in relation to the close link existing between natural and cultural values.

The picture that emerges from the analysis of the approaches to the issue of landscape and of the methods used to implement landscape policies[6] in the three Parks is rather varied.

First of all, important differences are found with respect to the role that the landscape plays in *national and/or regional legislation for the conservation of nature* and in particular in the definition of the protection category of the three protected areas that is assigned by the relevant laws (all three are classified as Regional Natural Parks).

---

[3] For an analysis of the IUCN protected area management categories, see Dudley (2008) and Dudley and Stolton in this volume.

[4] "A protected area where the interaction of people and nature over time has produced an area of distinct character with significant ecological, biological, cultural and scenic value, and where safeguarding the integrity of this interaction is vital to protecting and sustaining the area and its associated nature conservation and other values" (Dudley 2008, p. 21).

[5] *Parque Natural de la Albufera de Valencia, Comunidad Valenciana*, Spain, established in 1986; *Parc Naturel Régional de la Narbonnaise en Méditerranée, Languedoc Roussillon*, France, established in 2003; and *Parco Naturale Regionale del Conero, Marche*, Italy, established in 1987.

[6] All the information given in the text refers to state-of-the-art regulations, plans, and interventions as of May 2011, with some detailed updates as of July 2012.

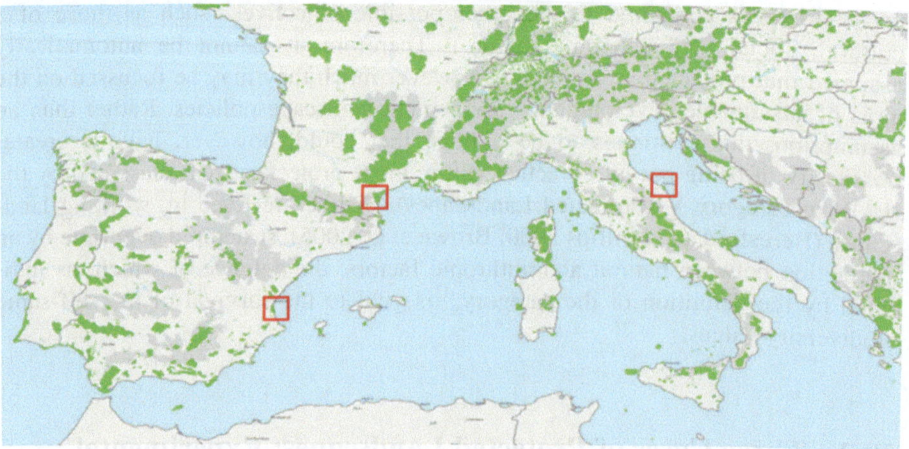

Fig. 20.1 In the context of the Euro-Mediterranean protected areas, see the three Protected Landscapes considered as case studies (Parque Natural de la Albufera de Valencia, 21,000 ha, Spain; Parc Naturel Régional de la Narbonnaise en Méditerranée, 70,000 ha, France; Parco Naturale Regionale del Conero, 6,000 ha, Italy) (Source: author's elaboration)

The French law and the Italian law, for instance, form extreme and opposite cases. The former (*Loi Paysage,* no. 93–24, 1993, adopted since 2000 in the *Code de l'Environnement*) sanctions the "Park-landscape" alliance, defining the *Parcs Naturels Regionaux* as privileged contexts for the application of landscape policies. On the contrary, the latter (Framework Law 394/1991) almost entirely ignores the landscape topic, having as its main and general aim the "conservation of natural and environmental values" and presenting only a fleeting mention of the landscape values in the definition of the Regional Natural Park category. The Spanish case takes an intermediate stance: the legislation applied in the Valencia Region (Ley 11/1994)[7] shows a relative sensitivity to the matter, acknowledging in the definition of the management aims of the *Parques Naturales* the relevance of the defence of landscape values in addition to the conservation of naturalistic ones.[8] In any case, in both the French and Spanish legislation (as mentioned, the Italian legislation is not particularly significant on the matter), the concept of landscape is rather limited, with the landscape being considered different from nature (ecological aspects) and mainly aesthetic/visibility related.

With regard to this national/regional legislative situation, the three Parks, particularly the Italian and the Spanish ones, have made some significant steps forward

[7] The legislation referred to by the management tools of the *Parque de la Albufera* (dated 2004). However, it should be pointed out that since 2007, there has been a new national law in Spain for the conservation of nature (Ley 42/2007), which, specifically implementing the principles of the European Landscape Convention, takes an important step forward in relation to local laws, which are still far removed from the themes of the ELC.

[8] Moreover, the law also envisages a protected area category specifically aimed at conserving landscape values ("*Paisajes protegidos*", art. 13).

in the definition of their policies, opening them up to the landscape dimension. The *management objectives* of the protected areas identified in the Management Plans are indicative on the matter.[9] Here, not only does landscape conservation take on a crucial role, but the concept of landscape is considered, unlike in the aforementioned legislations, and fully in line with the ELC, holistic, "extended", "social", and dynamic (see par. 1.1 in this paper). This said, the Management Plans of the three Parks are all quite recent documents – the French and Italian documents dating from 2010 and the Spanish one from 2004 – closer in time and concept to the directions taken by the ELC than to those of national/regional legislation on nature conservation.

The declarations of intent contained in the management objectives of the protected areas are fully reflected, in the French and Italian cases, in the definition of the *processes of knowledge and assessment* of the protected territory. The landscape is proposed by both Management Plans as the main filter for the interpretation of the area and assumed to be the best tool to describe ecological, socio-economic, cultural, and scenic aspects in their evolutionary (paying attention to the dynamics of the landscape and the relative criticalities), "extended" (with cognitive analyses that concern the entire Park landscape and not only those areas of exceptional value), and "social" meaning. It is this last characteristic that emerges most evidently in the two Parks, for which public meetings between the Park Authority and public and private territorial bodies have marked the definition of the cognitive-evaluative processes. The case of the Spanish Park is very different: while the Management Plan contains a reading of the area's landscape values, this is developed according to purely perceptive-visual parameters. This is in contrast with the holistic feature of landscape acknowledged in the definition of the management objectives, realigning with the restrictive interpretation provided within Spanish legislation on the matter of nature conservation.

The approach of the three Parks to the landscape, which emerges in the definition of the management objectives and in the processes of knowledge and assessment, is substantially confirmed by the Plans at the time of establishing the *strategies and measures* to be implemented. The French Park identifies a significant number of policies that explicitly propose the landscape as the main subject of the action. It should, however, be noted that, despite the holistic vision of landscape that had characterised the cognitive processes contained in the Plan, at the time of the definition of the policies, the ecological component is excluded from the concept of landscape, strategies, and measures being largely focussed on scenic and socio-economic aspects of the landscape and not on naturalistic ones. Even more consistent with the definition of the management objectives and processes of knowledge and assessment are the strategies and measures identified in the Italian Park Plan. The Park, after defining the *landscape units* that make up the area, lays them at the

[9] *Plan Rector de Uso y Gestión del Parque Natural de la Albufera* (PRUG) (2004), *Charte du Parc Naturel Régional de la Narbonnaise en Méditerrané 2010–2022* (2010), *Variante generale al Piano del Parco Naturale del Conero* (2010).

basis of the territorial-regulatory articulation of the Park, defining for each of them not only rules but also strategies and measures that are always permeated by the landscape aim. Once again, the Spanish case is very different from the French and Italian ones. The Spanish Park, in the wake of the cognitive analyses, all aimed at measuring the perceptive-visual features of the landscape, relegates the explicitly landscape-related strategies to a set of legislative measures that regulate the detailed visual impact of certain structures in the landscape (such as billboards).

The operative outcomes relating to the landscape topic, meaning the *actions* actually carried out for the landscape in the three Parks, are strictly consequential to the theoretical setting contained in the Plans. The Spanish Park, for example, in line with the lack of landscape strategies, does not develop specific actions for the landscape, there being a prevalence of sectorial actions of a biological-naturalistic kind. The French case takes an opposing stance, partly thanks to a generally high level of operative effectiveness, developing several actions for the landscape that integrate scenic and socio-economic aspects (excluding those of a naturalistic kind) and also using special methods and tools (such as the definition of landscape indicators for monitoring policies or the drawing up of guidelines for the construction of residential buildings within the landscape). In the case of the Italian Park, it is necessary to "hold judgement" on this matter, as the most recent Management Plan – the most focussed on landscape themes of all those that have succeeded each other since the Park was first created – was only approved in 2010.[10]

20.3 Emerging Evidence and Further Steps Towards a Park-Landscape Alliance

With reference to the three cases considered above, it is possible to identify more and less positive signs of the desired alliance between Parks and landscape. The following positive signs emerge:

- Although national and/or regional legislation on the theme of the conservation of nature is, in some cases, only relatively sensitive to the landscape theme, the experiences within the different case studies "go beyond", where necessary, the laws themselves. The Park Authorities have proven that they are able to see past the often obsolete regulatory frameworks, contemplating among the values of the area those related to landscape as well and considering them – at least in the definition of the management objectives – in a much more complex way than

[10] However, we can say that some tools envisaged by the Plan, such as the *Metodo di Valutazione Integrata* (Integrated Assessment Method) – a compulsory instrument for the approval of plans and projects implemented by municipalities, which jointly considers natural, historical, and perceptive aspects – should favour an effective development of landscape-oriented actions in the Park area.

that proposed by the laws. The protected areas therefore "grow" and evolve even in the presence of static legislation.

- Moreover, it should be pointed out that the main driving force behind the establishment of the protected area in all three cases was the fear of loss, especially on the part of residents, not only of a valuable environment, but also, and most importantly, of a valuable *landscape*. The protection and management of the landscape and of its cultural and identity-related values, as well as those of a scenic nature, served as the first and informal driving force for setting up the three protected areas (subsequently supported by "expert" assessments with a more specific focus on naturalistic values).

However, less positive signs also emerge. Some of the most significant are:

- The concept of landscape tends to be considered mainly a cognitive-evaluative tool. There are definitely fewer *strategies* and *measures* of a landscape nature (as a result, most of the strategies prefigured in the planning phase by the three Parks are sectorial, lacking that fundamental character that distinguishes landscape policies: that is, the integration between different thematic-operative aspects).
- Moreover, especially in the Spanish and French cases, nature, and therefore ecology, continues to be a different matter from landscape, which is interpreted in a limited way within the Parks' landscape-oriented strategies and measures, focussing only on the visual, cultural and socio-economic features but not on the ecological-naturalistic ones. This is not irrelevant because considering the landscape different from nature, failing to understand that it comprises nature – a nature conceived of as integrated with other dimensions and values – does not allow the managers of the protected areas to "see" the benefits that the landscape paradigm can bring to protect nature itself more effectively and risks relegating landscape to an accessory element, secondary to the other aims of the protected areas.
- Finally, especially in the Spanish and Italian cases, there is a big gap between the theoretical formulation, in the planning phase, of strategies and measures and their implementation. While this is true in relation to the overall set of strategies defined by the Plans, it is also, consequently, true with regard to the landscape strategies, which, where contemplated, are rarely implemented. In this way, Parks fail to act as experimental laboratories, carrying out "good practices" of conservation, management, and planning of the landscape.

Weighing the positive and negative aspects, it is possible to say that, today, the protected areas, and particularly the Protected Landscapes – at least within the limits of the Euro-Mediterranean cases analysed – are, rather than effective and existing experimental laboratories of landscape policies, *potential* laboratories. There is, in fact, no doubt that there are positive signs regarding a Park-landscape convergence, but it also clearly emerges that this is still a developing process, influenced by important conceptual and operative shortcomings. Probably, however, the convergence of nature and landscape in the management of the protected areas could be fulfilled more quickly if at least the following conditions were met:

- Contemplation of the landscape dimension within the legislation regarding the conservation of nature. Despite the efforts made by some individual Parks to go "beyond" the legislative dictates, it is obvious that legislation on the conservation of nature plays a fundamental role in directing the policies of the protected areas.
- Presence of specific skills in the matter of landscape within the Park teams. The lack of attention to the landscape theme during the definition of strategies and the difficulties in implementing specifically landscape-oriented actions are due also to the lack of landscape-related skills within the Park Authorities that manage the protected areas. This lack was particularly found in the Italian and Spanish Parks, while in the French Park, a section of the team responsible for managing the area consists of landscape experts.
- Strengthening of the role of civil society in the definition of the Park policies. Within the scope of the three case studies – and especially in the Spanish and Italian cases – residents and environmental organisations proved decisive in recognising the need for the protection of the *landscape* values of the areas. It would seem possible to assume therefore that the more the Parks open up towards society, involving it in the definition of their policies, the more they can be characterised by greater sensitivity towards landscape themes.[11]

References

Beresford M, Phillips A (2000) Protected landscapes: a conservation model for the 21st century. George Wright Forum 17(1):15–26

Brown J, Mitchell N, Beresford M (eds) (2005) The protected landscape approach. Linking nature, culture and community. IUCN, Gland

Castelnovi P (ed) (2000) Il senso del paesaggio. Ires Piemonte, Torino

Dudley N (ed) (2008) Guidelines for applying protected area management categories. IUCN, Gland

Gambino R (2005) La gestione del paesaggio nelle aree protette. In: Sargolini M (ed) Paesaggio, territorio del dialogo. Edizioni Kappa, Roma

Gambino R (ed) (2010) Parchi e paesaggi d'Europa. Un programma di ricerca territoriale. Ri-Vista 14:3–20

Locke H, Dearden P (2005) Rethinking protected areas categories and the "new paradigm". Environ Conserv 32:1–10

[11] The concept of "landscape" is closer to non-expert types of knowledge than that of "environment": while the recognition of environmental values "is subject to a gaining of a culturally complex civil and political awareness", the landscape values make more direct reference "to a system of implicit skills, that naturally form part of the common identity and opinion, which only has to be allowed to emerge" (Castelnovi 2000, p. 34).

Parc Naturel Régional de la Narbonnaise en Méditerranée (2010) Charte du Parc 2010–2022. Le projet de territoire adopté par tous. Available via www.parc-naturel-narbonnaise.fr. Accessed 20 Jan 2014

Parco Naturale Regionale del Conero (2010) Variante generale al Piano del Parco Naturale del Conero. Available via www.parcodelconero.eu. Accessed 20 Jan 2014

Parque Natural de la Albufera de Valencia (2004) Plan Rector de Uso y Gestión del Parque Natural de la Albufera (PRUG). Available via www.albufera.com. Accessed 20 Jan 2014

Peano A (2008) Aree Protette e governo del territorio. In: Gambino R, Talamo D, Thomasset F (eds) Parchi d'Europa. Verso una politica europea per le aree protette. Edizioni ETS, Pisa

Phillips A (2003) Turning ideas on their head. The new paradigm for protected areas. George Wright Forum 20(2):8–32

Phillips A (2005) Landscape as a meeting ground: category V protected landscapes/seascapes and world heritage cultural landscapes. In: Brown J, Mitchell N, Beresford M (eds) The protected landscape approach. Linking nature, culture and community. IUCN, Gland

Phillips A, Borrini-Feyerabend G (2009) Embracing diversity, equity and change in the landscape. In: Gambino R, Negrini G (eds) Parchi e Paesaggi d'Europa. Urbanistica, 139:52–57

Romani V (2008) Il paesaggio. Percorsi di studio. Franco Angeli, Milan

Chapter 21
Participatory Planning Tools for Ecotourism in Protected Areas of Morocco and Tunisia: A First Experience

Carla Danelutti, Ángeles De Andrés Caramés, Concha Olmeda, and Almudena De Velasco Menéndez

Abstract According to its programme and priorities, the IUCN Centre for Mediterranean Cooperation (IUCN-Med), with the support of the Spanish Agency for International Cooperation and Development (AECID) and in collaboration with the Ministry of Environment of Tunisia and the High Commission for Water and Forests and the Fight against Desertification of Morocco, implemented between 2011 and 2012, a project for improving decision making and capacities for planning and managing ecotourism activities in and around protected areas in two countries (IUCN Med 2008). The specific objective of the project was the elaboration of a Strategy and Action Plan for ecotourism in natural protected areas in the Maghreb and the development of guidelines for its implementation in two pilot areas: the Talassemtane National Park in Morocco and Jebel Zaghouan National Park in Tunisia. To achieve this aim, the approach of the European Charter for Sustainable Tourism was selected. This paper describes the methodology followed by the project, providing, in its final section, an analysis of the main achievements and concerns through its development, focusing on the need of improved participation in territorial planning.

Keywords Ecotourism • European charter of sustainable tourism • Participatory planning • Protected areas • Tunisia • Morocco

---

C. Danelutti (✉)
IUCN Center for Mediterranean Cooperation, C/Marie Curie, 22, 29590, Campanillas (PTA), Málaga, Spain
e-mail: carla.danelutti@iucn.org

Á. De Andrés Caramés • A. De Velasco Menéndez
ECOTONO, Equipo Consultor Turismo y Desarrollo, S.L., C/ Marcenado, 24 Bajo Dcha, Madrid 28002, Spain
e-mail: adeandres@ecotono.es; avelasco@ecotono.es

C. Olmeda
ATECMA, Isla de la Toja 2, 28400 Villalba, Spain
e-mail: concha.olmeda@atecma.es

R. Gambino, A. Peano (eds.), *Nature Policies and Landscape Policies*, Urban and Landscape Perspectives 18, DOI 10.1007/978-3-319-05410-0_21

21.1 Introduction

According to the World Tourism Organization (WTO), ecotourism and adventure tourism are growing fast as people seek diverse new experiences. This is also helping destinations to protect local communities and the environment in a responsible and sustainable manner (ITB 2012). Despite the latest political instability, North Africa has shown a strong rebound in the number of tourist arrivals in 2012 (+9 %) after a decline in 2011. Tunisia in particular (+24 %) has started to recover from the negative demand trends following the Arab Spring transition (UNWTO 2013).

It is, therefore, critical that tourism be carefully planned to ensure that such developments and activities do not compromise the natural and cultural values in the areas where tourist activities are carried out. This can be ensured through suitable and sustainable management of these areas, and emphasis also needs to be placed on the development of strong partnerships between all stakeholders, including governmental agencies (environmental and tourism administrations), civil society and the tourism sector.

21.2 What Is the European Charter for Sustainable Ecotourism?

The European Charter for Sustainable Tourism (ECTS) is an accreditation system for sustainable tourism in protected areas in Europe. It has been created to bring together both conservation and recreational objectives, to allow enjoyment of the natural, cultural and aesthetic values of an area in a sustainable way. These two objectives can be considered as universal in the creation of any protected area as well as a potential source of conflict that might arise in the area at some point.

The ECTS is not a conventional certification scheme; what the charter "guarantees" or proves is that there is a commitment to improve and progress in the sustainability of tourism. In fact, the ECTS is a participatory planning tool for sustainable tourism in protected areas, based on ten principles, which is embodied in a Strategy and Action Plan defined by the local stakeholders to be implemented in 5 years; after that, the area must renew the accreditation with a new Action Plan.

The ECTS was originally developed within a LIFE project co-funded by the European Commission (1995–1999) and includes three phases: the first phase involves the accreditation of the protected area, the second phase concerns the involvement and adhesion of tourism enterprises, and the third phase addresses the involvement and adhesion of tour operators (the latter still not enforced). The ECTS is managed by the EUROPARC Federation and it is now the largest network of protected areas accredited with a system that commits to sustainable tourism in Europe.

21.3 Why Choose the ECTS Approach?

As a general rule, the establishment of a protected area usually includes the planning of land uses which, whereas respecting the traditional uses that do not conflict with conservation objectives, can be perceived as a limitation of rights by the local population and as an obstacle to "development", if an adequate work of communication has not been implemented.

In the Maghreb region, existing protected areas were frequently established in a top-down fashion with little or no consultation to local communities (and in some cases involving the restriction or exclusion of some traditional uses that had been vital to the survival of local communities, such as grazing for herds of goats or the extraction of firewood for domestic fuel) with the result that, in many instances, communities do not understand the value, relevance and importance of these areas and all too often resent their creation (ACHPRI Work Group for Indigenous Affairs 2005).

Given the above reasons and considering in particular the need of reinforcement or even establishment of participatory processes in the region, after a revision of the main guidelines developed at international level,[1] the approach of the ECTS was selected for the implementation of this project, based on the following considerations:

- The first of the ten principles[2] of the ECTS is the cooperation between public and private actors involved in tourism in a protected area.
- From the first accreditations in 2001, the ECTS has emerged as an extremely useful tool. Some parks have renewed their system twice, therefore applying it for more than 10 years. This shows that one of its most remarkable results is achieving the active cooperation of the actors involved.
- The project schedule coincided with the social phenomenon known as the "Arab Spring", a fact that was considered as an opportunity to try to seize participatory management systems for the territory, focusing on the conservation of biodiversity and improving the economic and social conditions of the local communities in protected areas.

[1] That is, Global Sustainable Tourism Council Criteria, Guidelines on Biodiversity and Tourism Development, Guidelines of the CBD on Biological Diversity and Tourism Development and The European Ecotourism Labelling Standard.

[2] http://www.european-charter.org/become-a-charter-area/charter-principles. Accessed 17 Nov 2013.

21.4 Methodology

The methodology was adapted taking into account the specific circumstances and existing conditions in the countries and areas concerned. The project was implemented with the technical assistance of experts[3] with considerable knowledge and experience in the implementation of the ECTS, the management of natural protected areas and international cooperation projects.

The project implementation, based on the ECTS approach, used a participatory approach, which tried to involve in the most effective way all relevant stakeholders in its main steps.

As a starting point, for the selection of the working areas and to ensure the appropriation of the project, a national inter-ministerial steering committee was set up.

Later on, a local working group was created in each project area (Talassemtane National Park in Morocco and Jebel Zaghouan National Park in Tunisia) with representatives from local authorities, the managers of the National Park, tourism administrations, NGOs, private companies and associations with relevant experience and initiatives under development in the project areas.

In order to properly guide and orient the project implementation, a background analysis of the legal and institutional framework in relation to ecotourism and protected areas in each country was prepared.

Further on, a diagnosis of each of the two pilot areas was carried out to identify their main strengths and weaknesses for the implementation of ecotourism activities. This diagnosis involved field visits, meetings and interviews with the most relevant actors and target groups from the two project areas. The diagnosis document was discussed with the working group established in each project area and, based on its results, strategic lines for the definition of the Ecotourism Action Plans for these areas were defined. Through several meetings with the working groups and stakeholders and the assignment of responsibilities to some of their members, Ecotourism Action Plans to be implemented in the next 5 years were defined, including the proposed actions.

The plan was discussed and revised during the final meeting of the working group for each of the two project areas, and the results were presented in a regional workshop held in Tunisia at the end of the project, which allowed sharing experiences and discussing about the next implementation of the Action Plans. Competent authorities from the two countries and the main organisations involved in the project implementation, including members of the two working groups, took part in the workshop, where recommendations for the effective implementation of the Action Plans were elaborated, distributed to the national authorities and relevant organisations and made available to the public through the IUCN channels.

[3] The companies ATECMA and ECOTONO (Spain) provided technical support for the implementation of the project and the elaboration of the Strategic Plans of Ecotourism in the two pilot areas.

Fig. 21.1 View of the National Park of Jebel Zaghouan, Tunisia (Source: ©IUCN-MED)

This pilot project thus ended with the delivery of a Strategic Plan for the Development of Ecotourism in each of the two protected areas targeted and the creation of a local Permanent Forum for the coordination of future activities. In the ECTS process, this is the starting point for the implementation of the Action Plan for the next 5 years. In parallel a "Code of Conduct" on ecotourism for local administration, private sector and visitors was redacted and distributed to local and national stakeholders.

The most relevant adaptations made to the ECTS approach were necessary simplifications to the Diagnostic (due to the lack of accessible information or contributions) and to the contents of the Activities of the Action Plan, such as cost estimates or indicator definition, which could only be defined for some actions due to lack of sufficient data and time (Fig. 21.1).

21.5 Results

The following results concern the common achievements of the project in the region. At the end of the section, a schematic comparison of the specificities encountered in each of the countries and areas targeted is provided. This section has to be examined taking into consideration that some of the results are evaluated on a qualitative basis using the authors' knowledge and experience. A more detailed evaluation would be advisable. Common strategic project achievements include the following:

- The existing planning instruments for each National Park, their degree of implementation and the difficulties (political, legislative, technical, economic) for their effective implementation have been made public to relevant actors, as well as the applicable environmental and tourism legislation, other programming documents (territorial plans or strategies) and relevant findings from previous cooperation projects that had been hitherto inaccessible or unknown to many of the participants in the process, especially to private organisations.

- The concepts of sustainability, sustainable tourism, ecotourism and protected areas have been discussed and clarified. Participants became more aware of the necessity of healthy ecosystems.
- The protected area has increased visibility among local stakeholders and started to play a pivotal role in the implementation of socio-economic activities in the area.
- For the first time public and private initiative representatives, sometimes not aware of each other, have discussed and contrasted from an equal position and complete freedom their awareness, sensitivity and interests, about the present and future of tourism development in the territory.
- A first collaboration initiative around a Strategy and an Action Plan shared between national administrations of Environment, Forestry and Tourism, along with local administrations, has been settled.
- The participation of private entities has been promoted and activated and NGOs have even come to take the responsibility of the Permanent Forum.

Table 21.1 Main weaknesses for the development of ecotourism in PAs in the legal and institutional frameworks of Morocco and Tunisia (Sbai 2012; Ferchichi 2012)

	Morocco	Tunisia
Juridical level	Lack of a juridical text regulating ecotourism	Lack of a juridical text regulating ecotourism
	Lack of implementation of management plans for protected areas	Mismatch between the legal texts in terms of protected areas management
	Interferences regarding competence attributions	Lack of implementation of management plans for protected areas
		Absence of implementing legislation for marine protected areas law
Social level	Lack of coordination between existing initiatives on ecotourism	Absence of ecotourism culture among residents
		Existence of activities and behaviours threatening natural resources in protected areas
Technical level	Lack of sufficient funds and personnel responsible for management of National Parks	Lack of sufficient funds and personnel responsible for management of National Parks
	Lack of sufficient technical support and adequate specifications/rules for implementation of ecotourism activities	Lack of sufficient technical support and adequate specifications/rules for implementation of ecotourism activities

Table 21.2 Main features identified in the diagnosis of the two National Parks

Talassemtane National Park (Morocco)	Jebel Zaghouan National Park (Tunisia)
Creation: October 2004	Creation: March 2010
Size: 58,950 ha	Size: 2,010 ha
Director: Yes	Director: No
High level of tourism development, around the town of Chefchaouen. Existence of tourism businesses within the park	Low degree of tourism development
Large and varied tourism services offer (several resorts, guides), three associations of tourism entrepreneurs, two travel agencies	Limited tourism services offer: 1 hotel, 2 restaurants, 1 guide, 3 associations of speleology
Many cooperation projects funded by foreign agencies, in collaboration with local and national associations in the country and in the park	One project of cooperation at national level and none in the park
Many organisations working on tourism and local development but with a low level of cooperation	Limited number of associations

- The participatory approach of ECTS, with the necessary adaptation to the local conditions and circumstances, has proven to be valid, motivating and even exciting for the public and private actors in the pilot areas.

Table 21.1 presents a comparative summary description of the limitations found in the institutional and legal framework for the implementation of ecotourism activities, while Table 21.2 presents the main features found in the two pilot areas.

21.6 Conclusions

In this final section, the main challenges identified in the implementation of the project in the two natural areas are examined and where possible, some suggestions to address them are provided from the experience raised during project implementation.

Clarifying the concept of Ecotourism to clarify the objectives. Ecotourism was sometimes confused with other types of tourism, from sports without any control (caving), to the construction of a cableway in a National Park to access a reserve area. Some of these activities and projects were supported by the private sector and environmental local authorities and obviously defended in all cases, if not driven, by the tourism administration. Authorities and NGOs can have a major role in this especially as regards informing civil society and the private sector. The definition of a shared and participatory strategic plan, through numerous meetings and discussions and the establishment of the coordination Forum could help tackling these misunderstandings. The redaction and diffusion of the "Codes of conduct" helped as well in clarifying the concepts.

Social participation to be considered in national territorial planning policies. The current lack of participation is mainly due to contradictions or obstacles in the existing legislation, to the bad perception by the local population generated by the forced expropriations in the process of declaring a National Park and to the poor historic tradition in participatory processes. Nevertheless, the situation is quite different in the two countries. Whereas a clear desire of participation to the process was expressed by civil society in both pilot areas, in Tunisia – perhaps as a consequence of the democratisation of the country – this was better accepted and promoted by the authorities at all levels. In Morocco, on the other hand, though not leading to an impediment of the initiative, there has been no real support from the administrations involved to foster public participation. The creation of the Ecotourism Forum at local level is the governance tool that the ECTS approach contemplates for the coordination of the activities. Both pilot areas established their respective Forums, coordinated, for the moment, by NGOs. It would be necessary that local administrations (i.e. the Park's Direction) take the lead in the process and maintain its spirit in the future.

Focus on the conservation role of protected areas. The general attitude of the institutions responsible for the management of National Parks in both countries showed some level of resistance in promoting a real involvement of civil society in the management of decision-making processes concerning protected areas. Making the Park Direction responsible for all the participatory processes, as well as the coordination of the Ecotourism Forum, could let the park maintain its role of supervision of the activities in the area and share with other target groups potential sustainable initiatives.

Need of coordination and cooperation among national and regional authorities and environmental- and tourism-related actors. No reference to ecotourism as a sector is provided from a legal point of view in either country, and neither have the competences and responsibilities in that sense been defined, leading to problems of communication between administrations. Despite this, in Tunisia the project managed to settle a steering committee involving the Environment, Forestry and Tourism ministries at national level. In Morocco, this had been only partially possible at the local level. It is recommended to strengthen cooperation between the institutions responsible for tourism, environment and agriculture and those responsible for the management of protected areas in the execution of these ecotourism strategic plans developed. Maintaining regular meetings of the steering committees established remains very useful in this regard.

Strengthen the implementation of environmental planning inside protected areas. The management plans of the two National Parks were not really operative, only partially implemented and no carrying capacity studies had been conducted. This situation is partially the consequence of a lack of financial means and technical staff for the management of the areas as well as the limited capacity for decision making of the Park management staff with respect to central or regional administrations. At the end of the project, during the final seminar, it was agreed to proceed with the immediate implementation of the actions identified in the plan regarding strengthening capacities for planning and management activities in the two National Parks and including the appointment of the Director of Jebel Zaghouan NP and the creation of the Scientific Advisory Commission (as forecast in the order of declaration of the park), the capacity and skills of the management of the National Park Talassemtane, and the updating and implementation of the development and management plan for each park.

Project timeframe prevented integration into the national process. The feeling was that stakeholders, especially at the administrative level, viewed the project as another pilot initiative of cooperation to be added to the numerous ones already implemented, without taking advantage of the operative tool for implementation of ecotourism (the Action Plan) delivered. However, the Ministry of Environment in Tunisia applied this same method to a new project for ecotourism in natural areas funded by the World Bank. It would be important that project participants in both countries should commit to support the continuation of project activities and the implementation of results. The maintenance of steering committees and working groups established for the achievement of strategic plans should be pursued.

Though all these conclusions might seem to contradict what has been exposed in the results, the situation must be analysed taking into account the situation created by the "Arab Spring".

In fact, in the final seminar of the project, held in Tunisia in November 2012, the process was described by the participants as the first in recent history in which there had been genuine citizen participation.

Fig. 21.2 Panoramic view of Talassemtane National Park, Morocco (Source: ©IUCN-MED)

It is important to remark that cultural differences must be considered in the planning steps, and more flexible timeframes are often required. This project in particular should have included provision for at least a further 6–12 months, for the consolidation of the Permanent Ecotourism Forum and starting the implementation of the Action Plan for the next 5 years, providing advice and support during this period.

The main concern for a real impact of a participatory model into ecotourism protected areas planning strategies remains the support of national authorities and involvement of all local stakeholders.

The approach applied for obtaining a participatory strategic plan for Ecotourism in protected areas of the Maghreb was satisfactory and the method has been considered to be effective. The approach of ECTS to develop a Protected Area Ecotourism Strategy represents a successful experience that could be used and reproduced in both countries, with the necessary adaptations (Fig. 21.2).

References

African Commission on Human and Peoples' Rights and International Work Group for Indigenous Affairs (2005) Report of the African Commission's working group of experts on indigenous populations/communities. ACHPR/IWGIA, Banjul/Copenhagen

Ferchichi W (2012) Evaluation du cadre juridique et institutionnel relatif à l'écotourisme et aux aires protégées en Tunisie. IUCN-Med. Available via https://cmsdata.iucn.org/downloads/anexo_3_2_estudio_institucional_sobre_ecoturismo_y_areas_protegidas_de_tunez_2.pdf. Accessed 17 Nov 2013

ITB (2012) World travel trends report. Messe GmbH, Berlin

IUCN Med (2008) North Africa sub-regional programme 2008–2012. Available via https://cmsdata.iucn.org/downloads/na_sub_regional_programme_english_2.pdf. Accessed 17 Nov 2013

Sbai L (2012) Évaluation du cadre juridique et institutionnel relatif à l'écotourisme et aux aires Protégées au Maroc. IUCN-Med. Available via https://cmsdata.iucn.org/downloads/anexo_3_1_estudio_institucional_sobre_ecoturismo_y_areas_protegidas_de_marruecos_1.pdf. Accessed 17 Nov 2013

UNWTO, World Tourism Organization (2013) Tourism highlights. Available via http://dtxtq4w60xqpw.cloudfront.net/sites/all/files/pdf/unwto_highlights13_en_lr_0.pdf. Accessed 20 Nov 2013

Chapter 22
Tourism and Conservation in Protected Areas: An Economic Perspective

Massimiliano Coda Zabetta

Abstract In recent times, increasing interest has been focused on recreational aspects of protected areas and landscape in general. Such interest derives from both the importance of this issue itself and its connection with tourism. Furthermore, the management of protected areas has considerably evolved over the years. The turning point in Italy was the promulgation of the Framework Law on protected areas (Law no 394/1991). Thanks also to this law, which breaks with the dominant previous approach, gradually the protected areas started to be considered not as a limitation or a constraint to the development of the territory but as a source of competitive advantages. The protection of the territory and the environment through the creation of protected areas may be considered as a special case of supply of a public good, which is not ensured by the market. In fact, since individuals and enterprises are encouraged to behave as free-riders when dealing with the environment, the intervention of the public actor is necessary in order to create punitive policies and/or incentives aiming at limiting the exploitation of natural resources and safeguarding their quality. Therefore, natural parks and protected areas are an important element to be valued, but, since it is not possible to evaluate them through a price as for market goods, it is necessary to use economic instruments and evaluation methods able to quantify their value indirectly.

Keywords Tourism • Protected natural areas • Environmental economics

Goods and services used in a certain economic system by the community of consumers, as it is well known, can be distinguished according to their fruition, and the related categories of goods are distinguished by two characteristics, namely, the excludability and the rivalry of use. Private goods, which are traded on the market, shall have both these characteristics at the same time; on the contrary public goods must display the contemporary absence of both of them. Most environmental resources are public goods (the quality of air and water, biodiversity, open spaces, climate stability) even if it is often controversial to assign them the status of "pure" public goods. While the market efficiently allocates private goods, the public actor

M. Coda Zabetta (✉)
SiTI Istituto Superiore sui Sistemi Territoriali per l'Innovazione, Turin, Italy
e-mail: massimiliano.codazabetta@siti.polito.it

R. Gambino, A. Peano (eds.), *Nature Policies and Landscape Policies*,
Urban and Landscape Perspectives 18, DOI 10.1007/978-3-319-05410-0_22

must intervene to ensure, at least, a reasonable allocation to public goods (Pindyck and Rubinfeld, 2009).

The main characteristics of environmental public goods consist in the difficulty to determine a price to estimate directly the value of the good and in the relevant presence of the phenomenon of opportunism among the consumers of these goods, "free-riding" (a free-rider is a consumer who benefits from public goods without paying the corresponding price).

The creation of protected areas may be considered as a special case of supply of a public good that is not ensured by the market: the public actor can intervene to correct this bias by imposing authoritatively constraints and objectives.

Protected natural areas represent both means of defense and protection of nature and an important source of sustainable economic development, in its fullest meaning: a source of social growth and employment. The establishment of national parks and protected areas has been opposed for a long time by local resident populations because of aversion and resistance, often very tough, to some of the restrictions, such as the constraint on land use (Cannizzaro and Corinto 2011). In Italy, for a long time a conservative view in the management of natural parks has prevailed. This view made the local decisionmakers keep unaltered the aesthetic, historical and artistic aspects of natural areas in order to protect them from any activity that could put their conservation at risk. Gradually, however, the protected area started to be considered not only as a limitation, as a constraint, to the development of the territory but as a source of competitive advantages of the area, the keystone to base a new model of territorial organization, a model of protection to reconcile the need to protect the environment with the need of socioeconomic development (Marangon et al. 2004).

This was precisely the significant innovation brought by the Framework Law 394/91 on protected areas. In line with the European policy, the law provides the tools to implement the conservation of nature fostering, a significant increase in parks and reserves. In its general principles, the law gives the first legal definition of natural heritage:

> the physical, geological, geomorphological and biological formations, or groups of them, which have significant natural and environmental value.

The territories that have the values listed above are subject to a special regime of protection and management: these are the protected natural areas. In this perspective tourism plays a crucial role for the protected areas. And particularly, ecotourism bases the choice of the destination on specific environmental characteristics. This entails several problems, in particular with regard to the "natural park" tourism product. .

A first set of problems relates to the fact that (Lindberg 1991) the ecotourist is essentially "dislocal," as he tends to satisfy his desire for discovery and knowledge, always seeking new places. Furthermore, it is not always the tourists' environmental behavior that is of key importance or acquires a real awareness at the cognitive level. Often this type of tourists is not animated by a real interest in the conservation and preservation of the environment, while he is only looking for a

> visual consumption, linked to the spectacle of nature and filled with hedonistic connotations. (Martinengo and Savoia 2007)

These considerations are even more important when considering a landscape to be protected. There is indeed a strong dichotomy between landscape preservation and tourism policies in the management of Protected Landscapes. In line with the European Landscape Convention:

> the landscape has an important public interest in the cultural, ecological, environmental and social fields, and constitutes a resource favorable to economic activity and whose protection, management and planning can contribute to job creation. (Council of Europe 2000)

On one side of this dichotomy, it is possible to say that the landscape provides the "touristicity" of a destination, meaning its touristic potential, which is driven by several factors (such as natural heritage, scenery, geography, etc.), while on the other hand, tourism causes several negative externalities (congestion, pollution, creation of touristic infrastructure, etc.) so that tourism can be considered "predator" of ecosystems (Lanza and Pigliaru, 2004).

Tourism, in fact, is influenced by the environment, and it influences the environment itself. Perhaps in no other activity such as tourism, the link between the quality of environmental resources and the economic outlook is so evident: tourism demand is in fact mainly, though not exclusively, the demand of environmental values (Bimonte and Pagni 2003). Therefore, paradoxically, an excessive and uncontrolled development of tourism can lead in the long term to the destruction of those resources, which represent themselves the attractive potential of one area.

Therefore, there is a need for preserving the landscape through a management system, which has to be respectful of the nature, while on the other side an approach devoted to the sustainability in tourism is also a priority. Hence, there must be an integration of tourism within landscape policies and planning and back and forth, from landscape management to tourism policies.

In such a context, and even more at present, when the available economic resources are becoming scarcer and scarcer, the need of implementing policies that will improve the performance of a tourist destination without compromising the environmental protection has become a priority. To undertake an initiative policy, policy makers must be able to quantify with an acceptable level of confidence which benefits and which costs could be the result of this. Even in the case of the creation of parks and protected areas, it is necessary to analyze very carefully the costs and benefits of public intervention, in order to promote efficiency and effectiveness, since the amount of financial resources to be used for this purpose is always limited and their inefficient use inevitably leads to a less effective protection policy for the environment (Marangon and Tempesta 2003).

Policy makers who have the task of undertaking environmental policy decisions have at their disposal several techniques, helpful to assess positive and negative effects deriving from different interventions. The best known and most used among these is undoubtedly the cost-benefit analysis (CBA). The CBA allows to analyze, in a rational way, economic decisions considering and comparing all relevant alternatives in order to select the one to which the greatest amount of net social benefits expressed in monetary terms is associated. In the case of nonmarket goods, the size of these benefits or costs must be estimated through indirect methods. In the economic theory, the settled practice for the monetary valuation of environmental goods is based on the principle that they can be evaluated on the basis of market goods, i.e., basing on the value deriving from individual preferences of consumers (Freeman 2003). In fact, despite the intrinsic impossibility of putting a price on environmental goods, unless it just has a purely symbolic meaning without any relation with the actual value of that good, in recent years, the importance for this category of goods along with their recreational services has greatly increased.

As it has been already underlined, since there is no market price associated with environmental goods and services and since there may be a considerable uncertainty about their value, in order to make comparisons between different interventions in the field of the use of environmental and territorial resources, the assignment of a monetary value to these resources is necessary, this being based on the measure of the willingness to pay of consumers. This is a point of capital importance, in fact:

> Monetary valuation of environmental non-market goods can be more or less perfect; nevertheless, an explicit evaluation assessment realized for the considerations of policy makers and the public is always better than nothing, since in this second case actions are taken on the basis of some implicit valuation, hidden to the public opinion. (Turner et al. 2003)

Therefore, it is nowadays accepted the idea that decisions about public economics should be subjected to a requirement of efficiency, according to which an intervention can be conducted only if the deriving benefits exceed, or at least match, the corresponding costs.

It is possible to distinguish two main families of monetary valuation methodologies which can be used to evaluate landscape benefits and environmental goods using demand curves and the concept of consumer surplus: the methods based on revealed preferences and the ones based on stated preferences. Methodologies belonging to the first type analyze the behavior of individuals and can evaluate an environmental good, or a particular aspect of it, on the basis of the cost that people claim they will bear in order to enjoy it. The two main techniques, which refer to this category of methods, are the "hedonic price method" and the "travel cost method." Methods belonging to the stated preferences group, on the other hand, consider the definition of a hypothetical market in which individuals can directly express their willingness to pay for an environmental good.

The most known and used methods in this case are the "contingent valuation" method and the conjoint analysis. These techniques are able to provide useful data for cost-benefit analysis in order to estimate the value of environmental policy decisions. The quantification of costs and benefits is therefore an essential step in order to properly evaluate the work of parks' managing bodies as well as the opportunity to undertake development actions for the touristic and recreational function of the territory.

However, it is difficult to define on an abstract basis the actions, which can be implemented in order to maximize the social welfare in the field of recreational management of protected areas. The simultaneous presence of several objectives, often even very different from each other, makes essential the clear definition of the purpose of every intervention by those public bodies that carry out actions about touristic-recreational management and evaluation. Otherwise, such actions would be subject only to the immediate budget constraint and would risk a withdrawal or a reduction in the not absurd case of a decrease in the availability of funds due to exogenous causes beyond their control (e.g., public debt) rather than a change in the orientation of political strategies. In this way the investments would not generate benefits, inevitably causing a waste.

The development and the conservation of natural resources are not mutually necessary. The conservation of the environment answers, in fact, mainly to ethical reasons which could not necessarily be translated in monetary terms, while the valorization responds to the criteria of efficiency, especially on the social level, which means that it takes somehow into account also the resulting costs and benefits. In other words: while basically to the conservation follows an opportunity cost in terms of a sacrifice of private income, the valorization entails costs which will most of all weight on the society.

It is thus evident that, in such a context, it is very important and strategic to have information related to touristic demand, which gravitates on a protected area and, consequently, the social benefits associated with it, when interventions are designed. Hence, it is necessary to overcome a shortsighted perspective, linked exclusively to the analysis of the immediate spending constraints, and instead to adopt a scheduled view, anchoring policy decisions to the analysis of costs and benefits, taking into account the fact that targets which, from a social point of view, may seem perfectly plausible and desirable, thanks to a more careful and rigorous analysis, may result in conflict among them.

As we have argued, one of the priorities in these days is the sustainable development and the economic support in the protected areas. It is evident that the interrelation between tourism and a protected area plays an important role. Indeed, the fact that landscape and tourism mutually affect each other does not mean that the impact of this interrelation has to be negative. It is possible to exploit this relation in order to turn challenges into opportunities and produce positive externalities thanks to a tourism policy that contributes to the economic and social growth of territories preserving, at the same time, the cultural and natural heritage. At this point of the national experience about protected areas, with nearly 20 % of the Italian territory under protection, it is imperative to start a reflection on the management, conservation, and valorization of parks, particularly regarding the problem of finding the economic resources.

References

Bimonte S, Pagni R (2003) Protezione, fruizione e sviluppo locale: aree protette e turismo in Toscana, IRPET-Regione Toscana, Florence

Cannizzaro S, Corinto GL (2011) Turismo sostenibile nelle aree protette. Il caso del Parco Naturale dell'Etna. Rivista di Scienze del Turismo 6(3)

Council of Europe (2000) European landscape convention. Treaty series no. 176, Florence

Freeman AM III (2003) The measurements of environmental and resource values: theory and methods. RFF Press, Washington, DC

Lanza A, Pigliaru F (2004) Economia del turismo: crescita, qualità ambientale, sostenibilità, Equilibri, 8(1)

Lindberg K (1991) Policies for maximizing nature tourism's ecological and economic benefits. World Resources Institute, Washington, DC

Marangon F, Tempesta T (2003) Evoluzione gestionale e valorizzazione economica dei parchi. Working paper series in economics, Università degli Studi di Udine, Dipartimento di Scienze Economiche, Udine

Marangon F, Tempesta T, Visentin F (2004) La domanda di eco-turismo nell'Italia Nord-Orientale. Turistica, 3

Martinengo MC, Savoja L (2007) Il parco naturale: da attrazione ambientale a prodotto turistico integrato. Sociologia urbana e rurale, 81

Pindyck RS, Rubinfeld DL (2009) Microeconomics. Pearson, London

Turner KR, Pearce DW, Bateman I (2003) Environmental economics. The Johns Hopkins University Press, Baltimore

Chapter 23
Participation and Regional Governance. A Crucial Research Perspective on Protected Areas Policies in Austria and Switzerland

Norbert Weixlbaumer, Dominik Siegrist, Ingo Mose, and Thomas Hammer

Abstract Current management of large protected areas is faced with the twofold challenge of fulfilling its core mission of nature conservation and landscape protection while also responding to more complex societal expectations. The latter refer to regional development in particular and shaping of the future in general. Accordingly, research into protected areas should shift its focus towards regional shaping of the future within the framework and by means of large protected areas, if expectations are to be met. This would reflect the paradigm shift in area protection we have witnessed, transferring its emphasis increasingly towards the societal significance of protected areas on one hand and the integration of protection and development on the other. This contribution follows up on the results of a workshop organized by an international panel of researchers, managers of protected areas and representatives of protected area networks, which was held in St Pierre de Chartreuse (France, October, 2011). The following paragraphs will discuss the role of structures and processes of participation and the emergence of configurations of regional governance in the context of protected area development in view of the

Prof. Thomas Hammer's email ID was incorrect. See the "Erratum" chapter near the end of the book for his correct email ID.

N. Weixlbaumer (✉)
Department of Geography and Regional Research, University of Vienna, Universitätsstraße 7, 1010 Vienna, Austria
e-mail: norbert.weixlbaumer@univie.ac.at

D. Siegrist
Institute for Landscape and Open Space, HSR University of Applied Sciences, CH-8640 Rapperswil, Switzerland
e-mail: dominik.siegrist@hsr.ch

I. Mose
ZENARiO – Center for Sustainable Spatial Development, Applied Geography and Environmental Planning Research Group, Carl von Ossietzky University Oldenburg, P.O. Box D-26111, Oldenburg, Germany
e-mail: ingo.mose@uni-oldenburg.de

T. Hammer
Centre for Development and Environment CDE, University of Bern/Switzerland
e-mail: hammer@ikaoe.unibe.ch

R. Gambino, A. Peano (eds.), *Nature Policies and Landscape Policies*,
Urban and Landscape Perspectives 18, DOI 10.1007/978-3-319-05410-0_23

results mentioned above. These issues have gained relevancy in the current phase of protected area policies of the early twenty-first century and open up an array of important questions to be explored.

Keywords Protected area policies • Participation • Regional governance

23.1 Participation

There is ample consensus on the necessity for systematic involvement of stakeholders into the planning and management of protected areas (referred to as PA below). Generally, an involvement starting as early as possible is recommended. With the increasing role that PA are playing as instruments of regional development, demands placed on the conceptualization of approaches to planning that rely on systematic participation of stakeholders are growing, thereby minimizing the mistakes that tend to occur in following a top-down approach. These circumstances call for a primary focus on implementation-oriented research, which distinguishes between the participation of stakeholders, i.e. entities that have a vested interest in the PA, and the participation of the general public.

As can be seen from a diverse array of case studies, participatory planning is more than justified based on the different concerns of stakeholders, the goal of ensuring acceptance and the added value gained from the combined knowledge of stakeholders (Stoll-Kleemann and Welp 2008). From the perspective of both research and planning, determining the appropriate mechanism of participation is paramount in order to identify stakeholders, record their demands and expectations and motivate them to contribute on a voluntary basis to concrete projects and initiatives in the various PAs (Reutz-Hornsteiner 2002; Clark and Clarke 2011; Mehnen et al. 2013).

A model for successful participation in the preparatory stages of establishing a PA within the Austrian Alpine region can be found in the UNESCO Biosphere Reserve[1] Großes Walsertal (established in 2000) located in the province of Vorarlberg. In the following paragraphs, the structures (arrangements) and mechanisms conducive to successful participation of stakeholders and the general public will be discussed based on the model of this reserve.

[1] In Austria, biosphere reserves are called biosphere parks, which is not in line with UNESCO terminology.

23.2 Regional Governance Structures as Illustrated by Biosphere Reserve Großes Walsertal

Market development, production and distribution of an Alpine cheese labelled *Walserstolz* (Wals pride), a brand currently marketed in the BR area, may serve as an example of a participatory way of regional and PA development. A traditional cultural landscape and the quest for innovation of local stakeholders are the relevant settings for this initiative. Living with and within a biosphere reserve for the past 10 years, the local population has come to realize that the perceived crisis of their valley can be transformed into a future worth living. Large parts of the Wals population as well as experts agree that humans, nature and the economy can coexist harmoniously within the valley community and that the biosphere reserve represents an appropriate tool for creating a model landscape of international renown.

In the fourteenth century, the region now called Großes Walsertal was settled by the Wals people moving in from the West. This move caused an increase in the production of milk and cheese among other developments. By the middle of the eighteenth century, a surplus production started to develop and the local hard cheese was exported to the South. 1905 saw the founding of the first dairy cooperative of the region. This clearly demonstrates that cheese from the Wals valley has always constituted an important economic factor for the local population. Cheese helped create organizational structures and was part and parcel of local identity in the valley. However, that was a long way off from the joint labelling of Walserstolz, which came about mainly due to economic constraints emerging in the late twentieth century.

Over time, the realization that both cultural landscape and population had to be preserved in order to provide future prospects to the inhabitants of the valley took hold. The weakness of economic structures on one hand and a traditional affinity for the villages located in the valley on the other had resulted in the population moving closer together. This development created a need for stakeholders and vessels. By way of discussions and mission finding processes, the joint cheese brand and the biosphere reserve were conceived as the main carriers. These are indeed vessels that can be linked easily. Nowadays, both initiatives enjoy a deeply rooted positive self-image and a favourable external image at the same time (Coy and Weixlbaumer 2009).

The following stakeholders play a central role in the heterogeneous network underlying these initiatives, which serves the auto-regulatory processes of its territorial subsystems.

1. Local opinion leaders and politicians – also on the provincial level
2. Local inhabitants open to motivation and self-motivated
3. Dairy farmers and the newly established dairy cooperative
4. Wholesale and marketing entities as well as the dairy cooperatives themselves
5. Management of the biosphere reserve

6. Regional planning community (REGIO) consisting of the six municipalities and their long-time chairperson

All in all, these structures and stakeholders are interwoven in multiple ways and should be considered complex webs of participation. According to Fürst et al. (2006), such kinds of arrangements or structures differ from simple intergroup cooperation in their internal system of institutionalization: functionally shared labour, establishment of rules and/or contracts, and sustainability beyond the lifetime of a specific project. These complex arrangements have led to an additional place-making[2] internally while at the same time strengthening the external perspective, moving Großes Walsertal into the limelight of public attention.

These kinds of structures can be regarded as the substratum that provides fertile soil for putting into practice initiatives of regional development such as creating a cheese brand. Even greater importance should be attached to such webs of governance going beyond simple structures of cooperation if implementation of a complex project such as a Biosphere Reserve is the goal. In the end, one is hard-pressed to spell out how and to which extent the creation and functioning of individual structures that form this complex whole interact.

According to Fidlschuster (2009), three qualities that affect the functioning of participatory structures are crucial: trust (among stakeholders), generosity (towards the partner) and curiosity (regarding innovation – "curious regions"). As Mehnen et al. (2013) note, transparency, flexibility and the willingness to reach a consensus also play a major role in the engineering of successful participatory structures in PAs.

The following have been identified as factors particularly conducive to the success of regional development: regional potential, involvement of the general public (participation and cooperation), the recruitment of key figures as promoters, sufficient financial, personnel and infrastructure resources as well as professional conduct, management and marketing. These ingredients constitute the framework so to speak of successful regional development.

In this respect, the Großes Walsertal region is scoring points with its project plans for the biosphere reserve and Walserstolz. Both projects can be considered catalysts for the regional development process as it stands today. A typical example can be found in the joint project of Haus Walserstolz, which was not set up until after a second round of discussions had taken place (Coy and Weixlbaumer 2009; Weixlbaumer 2010). Built in 2009, it now serves as a demonstration cheese factory, information booth and marketing tool of the biosphere reserve, as well as architecturally progressive outpost of the valley.

If we consider the long years of settlement and development of the valley, Walserstolz cheese brand is nothing more than an intermediate stage. Yet this project clearly demonstrates that living trust, generosity and curiosity as development factors can feed a network in a sustainable way.

---

[2] The term here refers to socio-economic appropriation of space by participants, which creates a feeling of responsibility (for the region).

23.3 Multilevel Governance

In the wake of a more pronounced orientation towards stakeholders and participation, new forms of governance are emerging and have been observed in various PAs (Borrini-Feyerabend 2003; Mehnen et al. 2013). State-run institutions, private enterprise and civil society stakeholders jointly take on regulatory tasks concerning the planning and management of PAs. Yet, research on the governance of PAs has only just begun. Among the existing research, a heavy focus on theory has developed, relegating empirical studies to a position of lesser importance. Most noticeable is the interest in the dynamic PAs that can be classified as Category V of IUCN, whose increasing importance for regional development of rural areas would seem to suggest the formation of governance structures. However, research is still needed regarding stakeholders involved in governance, their interests and strategies, the operating principles of governance as well as its legitimacy within the dominant political system (Fürst et al. 2006). The latter concerns both the institutional interplay between new forms of governance and already established democratic processes of political decision-making and emergent characteristics of multilevel governance that involve local, regional, national and European levels of governance (Thompson 2005; Keulratz and Leistra 2008).

The development of nationally recognized parks in Switzerland after 2007 may serve as an example of the interplay between new forms of governance on one hand and already established processes, democratic processes of decision-making, on the other. The Swiss model is pursuing a twofold purpose in its most common category of nature parks:

1. Conservation and valorisation of nature and landscape
2. Strengthening of a sustainable economy

These objectives are implemented based on the principle of voluntary participation on the communal level embedded in a system of financial incentives and monitoring tools. In the future, the label applied to products generated in the nature park operated by the park in cooperation with private stakeholders and businesses will play an important role. As of 2013, 16 parks are open to the public, others are in the planning stages.

Similar to other countries in Western Europe, Switzerland is characterized by its federal system, which does not encourage top-down processes of PA creation and management. This is one of the main reasons why the new regional nature parks have come into being via a process that was clearly bottom-up. In contrast, the two new nature parks currently in the planning stages seem to be less amenable to bottom-up processes. In order to be able to impose the protective regulations necessary for operating a national park, a stronger top-down oriented process would be desirable. However, such an approach is meeting with resistance among the population living in the villages and regions. People do not wish to be lectured by the federal government on how to manage their landscape. Whereas nature parks represent lived multilevel governance involving municipalities, regions, cantons, the federal government and private stakeholders, national parks would require a change in political frameworks on the national level in order to guarantee their successful development. Concrete measures under consideration include higher-level financial incentives in the form of compensation for refraining from the use of lands in the core zone (Broggi 2013).

It is worth noting that the Swiss model meets with interest in those countries which enjoy a comparable federal system of government. Norway is currently implementing a model for creating landscape parks that follows a similar bottom-up process. Like Switzerland, this process is characterized by a significant component of regional development and is being supported by means of a dynamic network with funding from the central government (Haukeland 2010). This similarity in approaches has led to the establishment of close contacts between the national networks of Switzerland and Norway. A similar kind of cooperation has been developing over the past few years among the national park networks of Austria, Germany, Switzerland, Luxemburg and other countries. What is still missing is academic support for these novel processes (Wallner 2012).

23.4 Methodological Approaches to Comparative Research

For the purpose of examining the processes of participation and governance in PAs, researchers can rely on a series of methodological concepts that are well known among scholars and have been used successfully in the past. The majority of approaches have been developed in the social sciences and can easily be adapted to fit research needs in PAs. Three paradigms worth mentioning in this context are action research (Castellanet and Jordan 2002), social network analysis (Harteisen et al. 2010) and perceptual research in geography (Coy and Weixlbaumer 2009), to name just the most widely used. While a noticeable increase in studies on stakeholder participation is directing attention to this area of research, those studies have mostly been focusing on individual case studies, neglecting a necessary comparative perspective that would clearly be desirable from a European point of view (Mose 2007 and 2009). The same can be said about research on structures and processes of governance among PAs in Europe, characterized by their conspicuous heterogeneity. As Mehnen et al. (2013) point out, an advancing integration of the PA missions of conservation and development coupled with the range of different political and legal frameworks existing in Europe has resulted in the development of various types of PA governance that should not go unnoticed.

If we want to develop a comparative approach to PA research that actually reflects the reality described above, we need to develop an understanding of how the different structures of governance and participation came into being, how they are laid out in detail, how they operate and what have been the experiences of all participants concerned, stakeholders and the general public alike. From a European perspective, the potential for PAs to learn from one another in joint processes that could be stimulated by comparative research is of prime interest. At this point it must be noted, however, that comparative case studies are among the most complex research endeavours both due to their methodological challenges and the need for extensive resources (Belina and Miggelbrink 2010).

23.5 Future Prospects

As can be seen from the examples discussed above, participation of stakeholders from the private sector and civil society can lead to new structures and processes of governance that can satisfy broader societal demands on PAs while complementing more traditional forms of decision-making. New forms of governance do not replace participatory mechanisms currently in use. In fact, they facilitate the inclusion of additional stakeholders who can be made to share responsibility for the implementation of objectives defined for and by PAs.

This is to say that PA management needs to include dynamic strategies for enhancing participation and cooperation. This requires relevant management skills as well as the will to collaborate on the development of all sorts of projects with stakeholders from the private sector and civil society.

In order to support this process, the following key questions warrant our attention from the point of view of research:

- Based on the objectives formulated by the respective PAs, who are the relevant stakeholders and what expectations might they harbour? How can we integrate them into the process?
- Why do stakeholders participate or fail to participate? How can stakeholders be motivated to participate and cooperate in PA management and also with more stakeholders?
- How can we shape participatory and cooperative processes? How does the knowledge base of local stakeholders influence the design of such processes?
- How can we integrate local stakeholders into research and project development (action research, participatory research)?
- How can relevant stakeholders be involved in governance systems that have been adapted to fit objectives and local conditions?
- What are the experiences that have been gathered with different governance set-ups? Which models are most apt for different PA categories?
- What role does PA management play within the respective multilevel governance mechanisms?
- Which governance structures are more likely to be successful?

To the extent that the dialogue about governance is being conducted both from an analytical and a normative perspective, interest in Best Practices is on the rise. Against the backdrop of the paradigm shift currently taking place in the realm of PA policy, discussion and debate on successful approaches towards participatory PA management that will take into consideration new societal demands is key. Researchers can avail themselves of a series of practical tools (e.g. Focus groups, future workshops, Open Space) that in the realm of PA development have only been used in a very limited way up to now (Hammer et al. 2012).

References

Belina B, Miggelbrink J (eds) (2010) Hier so, dort anders. Raumbezogene Vergleiche in der Wissenschaft und anderswo. Raumproduktionen, vol 6. Westfälisches Dampfboot, Münster

Borrini-Feyerabend G (2003) Governance of protected areas innovations in the air. Policy Matters 12:92–101

Broggi M (2013) Nationalparks zu errichten ist schwer. Eine Herausforderung für die Schweiz. Nationalpark 1:22–25

Castellanet C, Jordan CF (2002) Participatory action research in natural resource management. A critique of the method based on five years' experience in the Transamazonica Region in Brazil. Taylor and Francis, New York

Clark JRA, Clarke R (2011) Local sustainability initiatives in English National Parks: what role for adaptive governance? Land Use Policy 28:314–324

Coy M, Weixlbaumer N (eds) (2009) Der Biosphärenpark als regionales Leitinstrument. Das Große Walsertal im Spiegel der Nutzer. Alpine space – Man Environ 8. Innsbruck University Press, Innsbruck

Fidlschuster L (2009) Gesellschaftliche Vielfalt. Eine endogene Ressource. Ausblicke – Magazin für Ländliche Entwicklung 09(1):20–23

Fürst D, Lahner M, Pollermann K (2006) Entstehung und Funktionsweise von Regional Governance bei dem Gemeinschaftsgut Natur und Landschaft. Analysen von Place-making – und Governance-Prozessen in Biosphärenreservaten in Deutschland und Großbritannien. Beiträge zur räumlichen Planung, vol 82. Institut für Umweltplanung, Hannover

Hammer T, Mose I, Scheurer T, Siegrist D, Weixlbaumer N (2012) Societal research perspectives on protected areas in Europe. Ecomont 4(1):5–12

Harteisen U, Neumeyer S, Schlagbauer S, Bizer K, Hensel S, Krüger L (2010) Grünes Band – Modellregion für Nachhaltigkeit. Universitätsverlag Göttingen, Göttingen

Haukeland PI (Red) (2010) Landskapsøkonomi [Landscape economy]. Bidrag til bærekraftig verdiskaping, landskapsbasert entreprenørskap og stedsutvikling. Med eksempler fra regionalparker i Norge og i Europa. Telemarksforsking, vol 263. Telemark Research Institute, Telemark

Keulratz J, Leistra G (eds) (2008) Legitimacy in European nature conservation policy: case studies in multi-level governance. Springer, Berlin

Mehnen N, Mose I, Strijker D (2013) Governance of protected landscapes and its implications for economic evaluation. In: Heide MH, Heijman WJM (eds) The economic value of landscapes. Routledge, London, pp 101–119

Mose I (ed) (2007) Protected areas and regional development in Europe – towards a new model for the 21st century? Ashgate, London

Mose I (ed) (2009) Wahrnehmung und Akzeptanz von Großschutzgebieten. Wahrnehmungsgeografische Studien, vol 25. BIS-Verlag, Oldenburg

Reutz-Hornsteiner B (2002) Entwicklung von unten – der Weg des Biosphärenparks Großes Walsertal, Österreich. In: Mose I, Weixlbaumer N (eds) Naturschutz – Großschutzgebiete und Regionalentwicklung. Naturschutz und Freizeitgesellschaft, vol 5. Academia Verlag, Sankt Augustin, pp 40–55

Stoll-Kleemann S, Welp M (2008) Participatory and integrated management of biosphere reserves. Lessons from case studies and a global survey. GAIA 17(S1):161–168

Thompson N (2005) Inter-institutional relations in the governance of England's national parks: a governmentality perspective. J Rural Stud 21(3):323–334

Wallner A (2012) Thematic catalogue of research topics on Swiss parks. Eco mont 4(2):37–40

Weixlbaumer N (2010) Ein Bergkäse beweist Regional Governance. RAUM 80:37–39

Chapter 24
Old and New Conservation Strategies: From Parks to Land Stewardship

Federica Barbera, Marzio Marzorati, and Antonio Nicoletti

Abstract The role of Nature Parks today is not only linked to biodiversity protection, but they are also territories where to experience a sustainable economy. A culture, which combines nature conservation and local sustainable development, led to the growth of protected territories in Italy (from 3 % to 11 % over 20 years). Now we have to make a step further: while we need to improve the quality of protected areas management, we also have to increase awareness among people that nature conservation requires our active participation. People themselves, in fact, are the most important custodians of the land and its naturalness. The Land Stewardship project has been a successful way to make owners directly involved in nature conservation and enhancement of biodiversity. Becoming a real worldwide movement, Land Stewardship is a tool to stop building extension in green spaces, thus protecting the most important resource for every country.

Keywords Parks • Biodiversity • Stewardship • Conservation • Landscape • Nature • Active citizenship • Participation

24.1 Role of Parks in Italian Conservation Policies

The incredible number of species (67,500 flora and fauna species) hosted in Italy[1] makes this country one of the richest in biodiversity among Europe, preserving about 43 % of European species, about 4 % of those on the planet. The geological and biogeographic features and its central position in the Mediterranean basin have

---

[1] Italy's biodiversity at risk. A call for action (IUCN/SSC) May 2013.

F. Barbera (✉)
Protected Areas Department, Legambiente Onlus, Via Salaria, 403, Rome 00199, Italy
e-mail: f.barbera@legambiente.it

M. Marzorati
Project Land Stewardship, Legambiente Lombardy, Milan, Italy
e-mail: marzio.marzorati@legambiente.org

A. Nicoletti
Protected Areas Department, Legambiente, Via Salaria, 403, Rome 00199, Italy
e-mail: a.nicoletti@legambiente.it

R. Gambino, A. Peano (eds.), *Nature Policies and Landscape Policies*,
Urban and Landscape Perspectives 18, DOI 10.1007/978-3-319-05410-0_24

determined the conditions to implement this richness, despite the world's biodiversity is under threat from various dangers (pollution, overexploitation, habitat loss and fragmentation, etc.), the majority of which have been caused by humans.

As defined by the Pan-European Biological and Landscape Diversity Strategy,[2] biodiversity loss and the decline of the natural environment and landscape are closely related and both caused by human pressure. For this reason, conservation must fall within policies of socioeconomic planning, as a prerequisite to recover and maintain biological diversity and landscape.

A crucial point introduced by the Strategy was the concept of "landscape diversity" as a cultural heritage of humanity, a sign of the existing relationships between man and land in different periods, hence the need to integrate the protection of the landscape with the protection of nature.

In the Italian context, the most relevant response, from an organizational and institutional point of view, to protect our delicate ecosystems, landscapes and biodiversity has been seen in Nature Parks. Their expansion, in line with the objectives of the Framework Law 394/91, over the last 20 years and over, has realized one of the most important experiences of nature conservation in Europe. The actual percentage of protected areas on the national surface is not an arrival point, as the Nagoya Protocol[3] stated that "By 2020, at least 17 per cent of terrestrial and inland water, and 10 per cent of coastal and marine areas, especially areas of particular importance for biodiversity and ecosystem services, are conserved through effectively and equitably managed, ecologically representative and well connected systems of protected areas and other effective area-based conservation measures, and integrated into the wider landscapes and seascapes."[4]

Therefore, to establish parks in order to protect threatened species and habitats remains a priority and is still considered the most effective way to stop the loss of biodiversity. How to achieve this in our country?

To date, Italy has 871 natural areas listed on the official list which involve more than 2,000 municipalities (about 1/4 of the total and mostly with a population of less

[2] The Pan-European Biological and Landscape Diversity Strategy (PEBLDS) was set up following the Rio Earth Summit and the adoption of the United Nations "Convention on Biological Diversity". It was adopted at the 3rd Ministerial Conference "An Environment for Europe" held in October 1995 in Sofia, Bulgaria. The principal aim of the Strategy is to find a consistent response to the decline of biological and landscape diversity in Europe and to ensure the sustainability of the natural environment.

[3] The Nagoya Protocol on Access to Genetic Resources and the Fair and Equitable Sharing of Benefits Arising from their Utilization (the Protocol) is a global agreement that implements the access and benefit-sharing obligations of the Convention on Biological Diversity (CBD). It was adopted in Nagoya, Japan, in October 2010, after 6 years of negotiations. In accordance with its Article 32, the Protocol was opened for signature from 2 February 2011 to 1 February 2012 at the United Nations Headquarters in New York by parties to the Convention. The Protocol has currently 92 signatures.

[4] Target 11, strategic goal C.

than 5,000 inhabitants). Protected areas extend across 3 million hectares of protected land and 2 million 800 thousand hectares of protected sea.[5]

On one side, we should improve the quality of protected areas management to revive a positive feeling from the people and revamp, both economically and politically, these areas in order to attract investment and boost local economy.

Over the last years, the National and Regional Parks have pushed many territories to compete with local innovative development policies based on environmental quality. It can be said that the Parks have strengthened and rejuvenated many territories. Thanks to their action, typical and biological production – recognized by the European Union – increased, like so many certifications and international awards. Italian protected areas have become a showcase of quality, as shown by 4 sites and 7 MAB (Man and Biosphere) UNESCO included by Parks and more than 50 products protected by the European Union and produced by hundreds of companies that operate in the Parks.

Through the Parks were made basic infrastructure (visitor centers, museums, accommodations) to revitalize villages, hamlets and many small towns. In the end, Parks ensured direct employment and fostered the activities in many strategic sectors (tourism, agriculture, farming, handicrafts, trade and services).

On the other side, we have to increase the awareness of the role that nature conservation can also exert over the current forms of protection, thanks to active participation. Protected areas, in fact, are not closed and limited by their institutional boundaries, but can be considered as territorial laboratories of a fair and sustainable local development. This model of active participation "bottom-up" has been perfectly expressed by the project of Legambiente "Land Stewardship."

24.2 Land Stewardship: What Is? For What Areas? By Whom?

Land Stewardship is a tool to limit the consumption of soil and to enhance the territory and its resources. Above all, it is an innovative practice whereby landowners and users are protagonist in the preservation of the natural, cultural and landscape values of their territory. The project is directed to people who care the territory and brings together different stakeholders: private landowners motivated by a passion to preserve and "value" their land, farmers growing food crops and institution which on behalf of their community want to act and preserve the natural landscape.

The system of land conservation through Parks cannot be the only tools at our disposal; it is necessary to involve individuals, private and public administrations in an active and responsible manner: this is the essence of Land Stewardship.

[5] These figures don't correspond to the ones cited in other chapter of this book, from different sources, particularly those gathered from CED PPN for European protected areas (see Foreword).

The project in fact stimulates processes of improvement and enhancement of the territory and of preservation of natural resources, integrating with local, regional, national network of protected areas. It allows individuals to exercise the protection of the territory and above all to establish ecological corridors and natural networks.

This project fosters the extension to private individuals of the responsibility for conservation and enhancement of the territory, a common good essential for human life.

How to become Land Stewards? Through Land Stewardship, we want to create a network of landowners and public institutions willing to preserve permeable land. A priority for Italy is to limit soil consumption or for buildings and infrastructures. Every day hectares of agricultural and natural land are lost: as a consequence, there is an impoverishment of nature and biodiversity. Our objective is to stimulate an increased number of Land Stewards in order to create a capillary network connecting the protected areas. Each participant will accept the responsibility of precise engagements for a specified time period by signing a Land Steward contract. Land Stewardships will be identified with a logo and by detailed commitments. Each subscriber will be part of a common responsibility.

During the project a database will be created and training programs will be developed in order to enhance the cultural value of Land Stewardship. We will designate Land Steward promoters who, after a proper training, will be the reference point in a specific area for the various initiatives leading to the fostering of knowledge and the diffusion of the Stewardship practice. Each promoter will have an identity card. A Land Steward, if so desired, could become a promoter in his or her area.

In this way, Land Stewardship will increase through its own capability of spreading; it enables the establishment of a network of experiences: each area will have the capability of managing, caring, maintaining, farming and preserving nature. Land Stewardship is not limited to the conservation of soil; also it relates human experiences connected to the land and its care as evidence of the interrelationship between the territory and the process of human development.

We propose to turn private landowners protagonists of land preservation and to create a network of local authorities that on behalf of their community preserve land and natural resources. Further, these initiatives match a desire more and more diffuse: the desire of an improved scenery, the respect and appreciation of the place where we live and in which we recognize ourselves.

24.3 The Experience in Lombardy

Land Stewardship deals with the ratio between permeable and impermeable soil. To simplify, one may call this process overbuilding of the territory. The use of the soil for residential, industrial buildings and infrastructures (roads, railways, bridges, etc.) causes a definitive impermeable ground and generates an irreversible process which prevents future recovery and consequently the possibility to generate new

Fig. 24.1 Colli Briantei Park, Municipality of Usmate Velate (Monza Brianza, Italy) (Source: Luca Fantoni)

environmental value. Construction has been driven by the economic resources rather than the real needs of the collectivity.

An agglomerate of buildings, industrial sites, urban areas, extensive parking lots and dismissed areas are a complex network of covered ground, which definitively loses its environmental potential (Fig. 24.1).

Through an "observatory" elaborated with the Italian National Institute of Urban Planning (INU) and the Polytechnic University of Milan, Legambiente has conceived a land consumption indicator. We have established this scientific approach since there is not sufficient awareness on the concerning data. The "observatory" has become a valid instrument to measure the loss of permeable soil. As a result every day the loss of agricultural and natural land in Lombardy has been calculated to be about 13 ha.

Furthermore, in Lombardy region, Legambiente has promoted a campaign to collect signatures to submit to the Regional Parliament a proposal for a draft law. With this law we ask for the implementation of a preventive compulsory compensation for the land being used in construction with an equivalent size of land to be preserved forever as an agricultural/natural site. This proposal is expected to reduce the use of permeable land for urban and infrastructures, regulating the land market, increasing the cost of permeable land and encouraging the restructuring and recovery of urban spaces.

References

CBD (2010) Nagoya protocol on access to genetic resources and the fair and equitable sharing of benefits arising from their utilization to the convention on biological diversity. United Nations, Nagoya. Available via http://www.cbd.int/abs/text/default.shtml. Accessed 19 Jan 2014

Council of Europe (1995) Pan-European Biological and Landscape Diversity Strategy (PEBLDS). 3rd Ministerial conference An environment for Europe, Sofia

IUCN (2013) Italy's biodiversity at risk. A call for action. The European Red List project. Available via https://cmsdata.iucn.org/downloads/italy_s_biodiversity_at_risk_fact_sheet_may_2013.pdf. Accessed 19 Jan 2014

Chapter 25
Between Nature and Landscape: The Role of Community Towards an Active Conservation in Protected Areas

Rita Salvatore

Abstract The application of governance practices in the Italian Gran Sasso and Laga Mountains National Park (Abruzzo region) represents the case study on the base of which this paper explores some issues about conservation and enhancement of landscape. The analysis of the pros and cons expressed by local stakeholders with regard to some infrastructure initiatives the Park managed on its territory highlights the need to combine these types of actions with processes of more inclusive regional development. A central role can be played by the activation of a participative relation between local populations and the parks for the management of territory. Whereas the former are to be meant as important custodians of local knowledge, for their position in between nature and landscape, the latter may represent "bridge institutions", potentially prone to mediate between the global instances (related to the need to assert sustainability in the ongoing development paths worldwide) and territorial specific social needs.

Keywords Community-Led Local Development • Active conservation • Landscape policies • Protected areas

25.1 Introduction

In 2003, the Italian Gran Sasso and Laga Mountains National Park (GSLMNP) started a new governance management process by dividing the wide territory of the park (covering about 150,000 ha in 3 regions, with 44 municipalities) into 11 different cultural and environmental tourist districts on the base of homogeneity in cultural, historical and natural identity. The main objective of this kind of management has been to operate some coordinated interventions in enhancing the tourist opportunities, in harmony with local economic and social situations. This was to be obtained basically through a material infrastructure (visitors' centres, well-

R. Salvatore (✉)
Faculty of Science of Communication, University of Teramo, Viale Crucioli 122, 64100 Teramo, Italy
e-mail: rsalvatore@unite.it; ritasalvatore00@gmail.com

R. Gambino, A. Peano (eds.), *Nature Policies and Landscape Policies*,
Urban and Landscape Perspectives 18, DOI 10.1007/978-3-319-05410-0_25

equipped tracks, information offices, park's guesthouses, eco-museums) aiming to improve tourist offer and competitiveness.

The first project the Park carried on in one of these districts ("The Main Road District"[1]) became the case study of a sociological research (Salvatore 2008). By subscribing partnerships with the other public bodies of the region, the Park got extra ministerial funds and realised infrastructures with a total amount of 9,000,000€.

This case study represents an occasion to reflect about some issues related to the conservation and the enhancement of landscape according to a possible integration of policies and management. The analysis of the pros and cons expressed by the local stakeholders with regard to the way the Park managed this project highlights the need to combine this type of intervention with processes of Community-Led Local Development to be put on a larger scale, where the local dimension joins regional/national/international networks involved both in nature and landscape policies.

A central role can be played by populations as custodians of a local knowledge (Geertz 1973) able to foster more active and reflexive interventions in ecosystems conservation, also in accordance with the inner sense of places (Carter 2001). In their position in between nature, landscape and territory, they can be considered as the connecting link among more integrated policies, as far as their activation might help in balancing the needs expressed by the vulnerable areas in terms of financial improvement with the simultaneous protection of biodiversity at risk and cultural heritage. In turn, protected areas could play the role of "bridge institutions", potentially prone to mediate between the global instances (related to the need to assert sustainability in the ongoing development paths worldwide) and territorial specific social needs, which – especially in mountain regions – are often expressed in the aspiration to renovate rural economies.

25.2 The Pros and Cons of a Park's Infrastructure Project

The case study conducted in the District of the Main Road in GSLMNP soon appeared to be quite exemplary and representative in order to observe empirically what could have happened in a marginalised area after the intervention of a park. The territory under observation in the past decades (and particularly in the mid-1980s) was a victim of a heavy emigration flow after the opening of a new highway that moved all the traffic away from the main road passing through the

[1] This District includes 7 municipalities on a 39,600 ha wide area, whose 74.7 % is protected. Its altitude goes from a maximum of 1,420 m a.s.l. to a minimum of 210 m a.s.l. According to statistical data referring to year 2010, it has 15,485 inhabitants with a very high old-age index, that in the worst case (Pietracamela) reaches 594, (232 points higher than the regional one). With the exception of Pietracamela – which is a quite popular ski-mountain resort with a rate of tourist function equal to 203 – in the rest of the area, also the tourist accommodation capacity is very low, with the same rate going from 2 to a maximum of 11.

mountain area, by deviating it in the direction of important urban centres. That eventually led to a complete abandon of the old routes from travellers, visitors, passers-by, traders and commuters. The result was a sudden and quick decline, in terms of economic and sociocultural turndown. This grim reality represented the reason why the park decided to actuate its district infrastructures starting from this one.

How local actors would have responded to such a re-transformation and re-invention of their territory under the conservation and enhancement of landscape promoted by the Park? What sort of attitude would have they developed after the start-up of these infrastructure actions? Which were the persons' reasons for either agreeing or disagreeing with the Park's choices? In order to answer to all these questions, some key "informants" (19 in total, including bars and restaurant keepers, local administrators and "pro–loco associations" members) have been involved in the research using different data collection techniques such as in-depth interviews, ethnographic and visual notes and focus groups.

The whole amount of collected data has shaped a complex and heterogeneous picture which can be systematised nor easily or in a unique way. All the people included in the target range cannot be perfectly divided into a population of Park's objectors and a population of its supporters. This result is to be attributed to the complete freedom that was guaranteed to the informers in expressing their positions. The possibility of defining a specific indicator – aiming to place them on the one front rather than on the other one – was never taken into account during the research. Thus, the same person could have expressed both a position of agreement and disagreement, depending on the semantic field he/she was treating in his/her discourse.

The materials were elaborated and reorganised into a chart (Table 25.1) where the first column is reserved to the semantic sphere of the argument, the second one to the dimensions related to those spheres, and the third one to the possible declinations of those dimensions. Here is a brief description of the semantic fields, which were defined out of the argumentation context and their contents.

1. Economic field: under this section the arguments concerning economic effects and outcomes related to the Park's actions on the territory have been grouped
2. Policies and management: in this category all those arguments which concern the management actions and the modalities the Park has undertaken both in terms of relation with residents and in terms of organisation and political choices have been placed. It also includes people's response to the infrastructural interventions realised in the District.
3. Socio-economic field: this category has to do with those arguments in which the Park is represented as an actor able to make either better or worse social and cultural local communities' conditions.

By reading through the data and analysing informants' attitudes, it is possible to discover that only 6 people out of 19 express a completely disagreeing position against the Park's actions, whilst the remaining ones (13) show both agreeing and disagreeing positions, according to the different issues they deal with.

Table 25.1 Agreeing attitudes and arguments

Semantic fields	Dimensions	Topics
Economic (7 occurrences)	Benefits	In economic activities (4)[a]
		In tourism (2)
	Development of competition	Development of competition (1)
Policies and management (13 occurrences)	Local resources enhancement	Through infrastructural interventions (6)
		Through marketing (1)
		Through the constitution of districts (1)
	Information implementation	About call for projects and funding (1)
		About chances related to development (2)
	Participation	Participation in decision making (2)
Sociocultural (13 occurrences)	Social improvement	In trust level (3)
		In the quality of life (2)
	Enhancement	Of environment and landscape (5)
		Of local culture (3)

[a]The number in *brackets* refers to the single topic's occurrences

Notwithstanding, most of the arguments (77 out of 110) show a disagreeing position (Table 25.2).

Some interesting cues for reflection come from the detailed analysis of the distribution arguments occupy throughout the semantic fields. The most critical issues seem to be represented by the way the Park managed its policy choices (see Policies and management field). Within 41 occurrences (expressed by 9 people), it is assumed that the Park has failed in its communication for different reasons. If for some of the informants the image of the Park has not been promoted sufficiently in big urban areas, for others the works of what has been sponsored (tracks, traditions and infrastructures) have not been completed yet and tourists often get disappointed. According to their statements, the inefficiency of such marketing campaigns would be demonstrated by both the missing increase of the touristic flow and by some tourists' discontent. "The Park divulged too much information, but most of it was wrong. This is disinformation" some of them say. In many cases, that would have caused a boomerang effect, eventually leading to tourists' disappointment and operators' unease. Moreover, for some other interviewees, the Park has wrongly spent its financial resources. Most of the projects it realised (i.e. the restoration of some old buildings to make new guest houses, as well as picnic areas) do not actually meet local needs. At the same time very pressing problems, like those related to viability and to accessibility, still have to be sorted out.

Table 25.2 Disagreeing attitudes and arguments

Semantic fields	Dimensions	Topics
Economic (15 occurrences)	Limitations and prohibitions	In hobbies and sport (2)[a]
		In economic activities (5)
		In private initiatives (3)
	Labour market	Inefficiency in labour market (2)
	Market competitiveness	Unfair competition (3)
Policies and management (41 occurrences)	Choices inadequateness/ inefficiency	In marketing (9)
		In resources management (5)
		In infrastructural interventions (5)
		In territory management (7)
	Lack of involvement	In political choices (5)
		In activities organisation (3)
		In marketing (2)
	Information gap	Lack of information campaigns (4)
		Short availability (1)
Sociocultural (21 occurrences)	Ecocentrism	Too strict environmentalism (4)
	Enhancement deficit	Towards local tradition (4)
	Short interests sharing	Short interests sharing (8)
	Disadvantage position	In demography (1)
		In education (1)
		In communication (1)
		In socio-economy (2)

[a]The number in *brackets* refers to the single topic's occurrences

Ten occurrences have reported about a lack of involvement, both in political choices (most of which are not focused on residents' needs) and in the organisation of the activities: "the Park organises its own promotional activities without valorising the already existing ones". This situation has had some effects also on the sociocultural level; it has missed to implement trust, leading to a general "cold and offish atmosphere" and to people's dissent (in informants' own words). According to their positions, the Park's actions would have worsened that fragmentation process which in the past was first caused by emigration. In fact, its nature preservation constraints – when meant too strictly – could obstruct the persisting of some important local rural traditions like those related to wood works.

Most of the bar and restaurant owners also underlined their disagreement respect to the Park's actions, in terms of economic effects. According to them, these conservation prescriptions would further reduce communities' possibilities of development. Firstly, the prohibition to realise some touristic infrastructures would obstruct their business growth. Secondly, the imposition of high quality standard in the restoration of the buildings has made costs too expensive and unaffordable, so constraining private initiatives. Thirdly, some of these limits would also discourage visitors and tourists who are interested in hobbies and sport activities such as fruit picking in the woods, hunting and fishing. Within the economic field, some protest arguments have been expressed from those who think that the building of Park's tourist accommodation could generate an unfair competition.

Even if according to the general view, most interviewees have not approved the Park's actions, albeit (except 6 persons out of 19), they also express some consent arguments. Despite the negative criticism they manifest towards the way this National Park has been managing governance and its relations with local stakeholders, some interviewees still believe the conservation cause is the right one. Quite representative to this end is an expression stated by one of the pro–loco members: "I am not against Parks. I think Parks are a good thing, but only if they are not managed in this way". Speaking about the future, informants think the Park can represent:

- A good opportunity for a social improvement and change, in terms of improvement of the quality of life and cultural growth
- A tool for the enhancement of landscape, both on a natural level and on a cultural one

Within the panorama of agreement attitudes, we also find some appreciation arguments for the infrastructural interventions the Park undertook on the margins of the main road, with special concern to the restoration of some old and crumbling "road workers' houses". According to the informants' statements, thanks to these valorisation actions, the landscape has gained a new and less desolated aspect, which could invite travellers and passers-by to stop.

Looking at the economic field, despite their worries related to limits and prohibitions, most of the bar and restaurant keepers (five out of seven) think about their future in the park in a hopeful and optimistic way. They do believe the development of this protected area will represent both a launch window for the promotion of their business and a pull factor able to appeal eco-touristic demand.

Ten years have passed since the first launch of the project, but the material infrastructure has not been followed by an actual development in cultural-environmental tourism yet. Despite the several calls the Park issued both on a local and on a national level, neither the picnic areas nor the guest houses or the information office management has been entrusted to some company able to run them throughout the year. They have been available only for few weeks during the summer time and still remain underused.

25.3 Towards Active Conservation

The case study's main highlights seem to underline the necessity for protected areas to face a still far-reaching change of paradigm. As the above-cited occurrences partly reveal, now parks have to deal with a multitude of tasks, many of which go beyond pure environmental protection (Hammer et al. 2013). As a matter of fact, compared to the old generation parks, contemporary PAs have been gaining a "full value" (Harmon and Putney 2003) according to which environmental issues can no more be set apart from economical, societal and cultural arising new demands.

Also because of this perspective, some Italian scholars (Calafati 2004; Cassola 2005) have suggested a shift in the meaning of protection and conservation. Whereas the former has to do with an old approach aiming to preserve nature, especially from the human activity intrusion, the latter implies instead an active involvement from local societies in taking care of biodiversity. Accordingly, the landscape Parks inherit is but the product of the diverse patterns of land use that resident populations have adapted to environmental conditions, particularly in agriculture and forestry. "Past human activity has markedly contributed to the diversity which we today consider worth protecting", Leng et al. (2013) assert. Thus, the space dimension shows an interdependent link with the socio-relational one.

In their position between nature, landscape and territory local communities can be considered as custodians of a local knowledge (Geertz 1973) potentially able to make more flexible the processes of ecosystems conservation. Notwithstanding, the simple fact that some people live on the same space does not necessarily mean that they chase the same interests or that they share the same goals for the development of their territory. An appropriate path of cooperation and interpretation should be started aiming at stimulating social responsibility, and at disclosing that inner sense of place (Tilden 1957; Carter 2001) whose enhancement and communication might help in balancing the needs expressed by rural areas in terms of financial improvement with the simultaneous protection of their natural and cultural heritage.

All this clearly requires a wider integration of policies, able to combine nature protection with regional development and eventually leading to the desirable "territorialisation of nature conservation", that is, with a more articulated idea of sustainability which integrates the cultural and procedural principles of equity (Haughton 1999; Salvatore and Maretti 2012) into a cohesion policy.

An important turning point in this direction could be represented by the share of the approach European Commission (2012) is adopting for the 2014–2020 programming period, which assigns a central role to Community-Led Local Development (CLLD) as a single method to be used within Common Strategic Framework Funds. By taking into important consideration local needs and potential, this method is carried out through integrated and multi-sectorial area-based local development strategies, which can contribute to achieve a more inclusive growth (Wallner and Wiesmann 2009). The hypothesis is that it could increase the effectiveness of environmental policies especially in those vulnerable areas (like most of the protected ones are) where there is a need to respond to territorial and local challenges requiring some structural change, in demography as well as in economics and cultural aspects.

In this panorama, national parks are invested with a very complex mission which is both an ambitious and a binding one because aims to re-establish a new right balance between natural capital and sociocultural one, towards a more innovative idea of "countryside capital" (Garrod et al. 2006) and in the direction of the approach the European Landscape Convention (along with the proposal of an International Landscape Convention) have conceived.

These goals need the sharing of responsibilities as well as the commitment from all actors possibly involved, citizens and economic stakeholders on the one side and institutions on the other. It is an arena where parks ought to act as "bridge institutions", as "societies in the middle" (Bonomi 2002), that is, as intermediate subjects able to explicit a linkage function between local populations and national/international institutions. Moreover, their territorial sovereignty would allow them to turn into practice important strategies in order to strengthen social relations and to invite societies in investing in social capital. At the same time, they can encourage interpretation of local cultural values and heritage whilst keeping on respecting the general objectives of conservation defined by national and supranational nature, and landscape policies.

The application of this new approach is what is meant here as "active conservation", that is, a strategy which (through parks' participative management) implies an acquisition of consciousness and in turn a deeper commitment and responsibility from local communities/institutions towards those collective actions aiming at recognising, interpreting, consolidating and communicating the sense of territorial ownership and the local identity, within landscape and environment policies.

References

Bonomi A (2002) La comunità maledetta. Viaggio nella coscienza di luogo. Edizioni di Comunità, Turin

Calafati AG (2004) Conservazione e sviluppo nei parchi italiani: un'agenda di ricerca. Riv Geogr Ital 1:2–22

Carter R (2001) A sense of place. An interpretive planning handbook. Scottish Interpretation Network. Bridge House, Iverness

Cassola P (2005) Turismo sostenibile e aree naturali protette. Concetti, strumenti e azioni. ETS, Pisa

European Commission (2012) Community-led local development. Cohesion Policy 2014–2020. Available via http://ec.europa.eu/regional_policy/sources/docgener/informat/2014/community_en.pdf. Accessed 25 Apr 2013

Garrod B, Wornell R, Youell R (2006) Re-conceptualising rural resources as countryside capital: the case of rural tourism. J Rural Stud 22:117–128

Geertz C (1973) The interpretation of cultures. Selected essays. Basic Books, New York

Hammer I, Mose I, Scheurer T, Siegrist D, Weixlbaumer N (2013) Societal research perspectives on protected areas in Europe. Ecomont, J Protected Mt Areas Res 4(1):5–12

Harmon D, Putney AD (eds) (2003) The full value of parks. From economics to the intangible. Rowman & Littlefield, Oxford

Haughton G (1999) Environmental justice and sustainable city. J Plan Educ 18:233–243

Leng M, Hammer T, Scharrer B (2013) Opportunities and limits of the eco-system services framework in capturing the cultural dimension of sustainable regional and landscape development in protected areas. Conference volume of 5th symposium for research in protected areas, Mittersill, pp 453–456

Salvatore R (2008) Sviluppo locale e aree protette. Il Piccolo Libro, Teramo

Salvatore R, Maretti M (2012) The link between sustainable tourism and local social development. A sociological reassessment. Sociologica 2:1–21

Tilden F (1957) Interpreting our heritage: principles and practices for visitor services in parks, museums and historic places. University of North Carolina Press, Chapel Hill

Wallner A, Wiesmann U (2009) Critical issues in managing protected areas by multi-stakeholder participation. Ecomont, J Protected·Mt Areas Res 1(1):45–50

Chapter 26
The Contractual Communities' Contribution to Cultural and Natural Resource Management

Grazia Brunetta

Abstract Defined as voluntary territory-based organisational forms (i.e. tied to a specific tract of land) by which members join on the basis of a contract unanimously underwritten, and in light of the benefits it will guarantee them, the contractual communities can foresee actions to revitalisation and enhance the landscape. This type of civic associationism comprises various types of 'voluntary social forms' capable of realising and managing collective territorial resources, placing the rediscovery of ethical, individual and collective values by the members of a community at the centre of their action. This contribution proposes some recent experimentation, at national and international level, of forms of voluntary association of citizenship, with a view to recognising the contribution for the triggering of new actions of active conservation of the landscape. The perspective of research is that of highlighting the elements of innovation within a new model of governance inspired by the principle of horizontal subsidiarity. In conclusion, themes and matters to develop in order to innovate landscape planning will be proposed.

Keywords Landscape planning • Contractual communities • Social self-organising • Citizenship • Subsidiarity

> The very notion of the commons implies a resource is owned, managed, and used by the community. A commons embodies social relations based on interdependence and cooperation. There are clear rules and principles; there are systems of decision-making. Decisions are made jointly by the members of the community. Vandana Shiva (2005)

G. Brunetta (✉)
Interuniversity Department of Regional and Urban Studies and Planning (DIST), Politecnico and Università di Torino, Viale Mattioli 39, 10125, Turin, Italy
e-mail: grazia.brunetta@polito.it

R. Gambino, A. Peano (eds.), *Nature Policies and Landscape Policies*, Urban and Landscape Perspectives 18, DOI 10.1007/978-3-319-05410-0_26

26.1 Landscape and Social Self-Organisation: A Premise

Several decades have passed since the first international treatises on biodiversity and the landscape promoted a radical change in perspective in the approaches to landscape planning.[1] Partly anticipating the logic of the European Landscape Convention (ELC, 2000), in the 1990s, the International Union for the Conservation of Nature qualified the category of Protected Landscapes as *areas of active conservation*, proposing the launch of dynamic conservation policies for the creation of *new landscapes*. In actual fact, the concept of 'dynamic conservation', on which there is now widespread consent, is still struggling to move on from the state of theoretic enunciate to that of widespread practices effectively implemented within territorial governance policies. Among other things, in Italy today, this aim is made even more necessary by the lack of effectiveness of policies and projects to transform the territory and the landscape (just think of the vulnerability of the hydrogeological situation) and also by the global economic crisis which refers to the need to rethink the growth model, starting with actions to enhance the landscape, focused on the sustainability of development.

The landscape recognised by the ELC as an 'essential aspect of the life of the populations, identity, local diversity and economic resource (...), basis of the wellbeing of the populations', should be subject, in its 'entirety' to policies for protections, planning and management, in order to promote the social-economic development of the territories[2] (ELC, articles 1f, 2, 5a, 6). At the focus of this notion of landscape, seen as an expression of the cultural identity of the populations, are the social values of belonging to natural and anthropic factors of each community, as a key resource for planning requalification and development actions. In line with the rationale of the Convention, the strategies for the landscape should stem from new approaches and working methods, capable of accompanying the actions for the protection on ecological and environmental resources with those of the enhancement of the culture of a territory, thus contributing to the creation of *new landscapes* and models of shared fruition by the community.

The point of view here is meant to reflect on the innovation potential of an emerging and rapidly growing phenomenon, related to a model of self-organisation of social action aimed at landscape planning and management. The recent emergence, in the national and international sphere, of forms of voluntary association which experiment new methods of planning, organisation and management of collective natural resources encourages us to ask whether and in which conditions

[1] See the Pan-European strategy for biological diversity and the landscape (1995) and the European strategy for biological diversity (1998) which highlighted the need to develop a vast knowledge of landscapes, including the natural and cultural aspects, for their sustainable management, integrating the aim of the conservation of biological diversity with that of landscape diversity.

[2] '[...] the Convention applies to the whole territory and regards natural, rural, urban and peri-urban spaces (...) concerning landscapes that are considered exceptional, whether they are the landscapes of everyday life or run-down landscapes' (Art. 2).

these processes can favour the reacquisition of values of social cohesion, opening up to a fertile process of institutional change. The theoretic viewpoint, which is assumed here, recognises the centrality of self-organisation mechanisms[3] and their role, insofar as they are agents for meaningful and effective changes in the scope of landscape policies. In no way, however, it is assumed that everything that is voluntary is good regardless. It is simply considered that the same possibility to experiment different options is crucial to every society, in that it is a source for the institutional innovation. The view here is meant to overturn the consolidated conventional viewpoint, in order to highlight new local landscape care actions within a new model of governance inspired by the principle of *horizontal* subsidiarity.

26.2 The Contractual Communities as a Collective Action to Enhance the Landscape

The contractual communities have been defined in a previous research (Brunetta and Moroni 2012, p. IX) as

> The voluntary territory-based organisational forms (i.e. tied to a specific tract of land) by which members join on the basis of a contract unanimously underwritten, and in light of the benefits it will guarantee them.

The contract – the agreement – establishes a set of commitments and rights for the members of the contractual community. They are a variety and a number of *territory-based social forms* (linked to a specific portion of territory) of a private, non-profit nature, characterised by two elements: the *self-organising* and the *contract*. I would particularly like to investigate the first of the two elements, in order to shed light on the innovation potential for the landscape project.

We are talking about an organisational phenomenon to be placed historically within a sphere of extensive and heterogeneous experimentations regarding the different forms of voluntary associationism, which are beginning to re-emerge all over the world. On this point, we ought to mention that the rediscovery of forms of 'civic activism', which has been increasing since several years, is characterised by people who 'do on their own', in compliance with organisational models that are separate from the traditional paradigms of political participation and also from the consolidated models of corporative associationism, exercising independent and original powers that concern the care and enhancement of common assets and collective resources.[4] The contractual communities are heading in this same

[3] The background of these considerations regards the neo-institutionalism approach, which interprets the symbiotic and dialectic relationship between institutions and organisations as the origin of the institutional transformations (North 1990).

[4] On the relationship between civic activism and horizontal subsidiarity, see Arena (2006) and Moro and Vannini (2008).

direction insofar as they are forms of social self-organisation characterised by two aspects: (i) the complete self-organising of the actions, not encouraged by the public institutions, and (ii) the recognition of mechanisms of *reciprocity* and *solidarity*, present in varying degrees of importance at the basis of the collective action. In other words, the interaction between the individuals who decide to undertake these types of voluntary action can be aimed at achieving goals of mutual support among the members of community, but it can also promote actions of solidarity towards others. In this sense, reciprocity and solidarity are not necessarily present exclusively, but can interact in a variety of ways, allowing a plurality of combinations that are found in the three models[5] of contractual community defined (Brunetta and Moroni 2008).

This is, therefore, a particular form of autonomous social initiative, which falls within the broader phenomenon of *voluntary social formations*, i.e. a family of social initiatives that comprises a plurality of voluntary associations and organisations (Brunetta 2011, p. 104). For this very reason, the varied mix of the two mechanisms of *reciprocity* and *solidarity* is interesting for reflecting on the way the enhancement and care of natural and cultural resources could be innovated. Considering these criteria, as they emerge from the process of social self-organisation, we can recognise two groups of territory-based actions at the extremes of this umbrella of *voluntary social formations*.

The first comprises a combination of organisational forms aimed mainly at the management of collective environmental assets, recognised by society as having a particular symbolic and cultural value and for which the legal ownership of the land and usage rights to it are separate or can be determined by civic traditions and habits. This is the case, for example, of certain types of common land (Ostrom 1990) but also of the American experiments of the *community land trusts*.[6] On this matter, there has been significant experience of enhancement of environmental assets and collective urban resources (green areas and ecological corridors) created thanks to the voluntary enterprises of *community gardens*, which are growing in different American urban areas. In all these territorial actions, the mechanism of *solidarity*, i.e. the benefit for the others, prevails.

The second group comprises various types of residential associations and organisations with regard to the planning of settlements and the independent supply of collective services to the members of the community who adhere to the action. This groups comprises various community forms of private ownership which range, for example, from residential cooperative to emerging forms of aggregation to design areas and services for free time (community gardens, urban green infrastructures),

[5] Basically, there are three types of contractual communities (Brunetta and Moroni 2012) that can be termed 'leasehold contractual communities', 'freehold contractual communities' and 'common hold contractual communities'. An example of this third type is certain forms of collective private property of natural resources (certain so-called commons in Ostrom's 1990 usage of this term). Examples of collective private properties can be found in many countries, for instance, in Italy: to date, collective property in Italy concerns a total of land of around three million hectares.

[6] See Bray in this volume.

mainly characterised by mechanisms of *reciprocity* and *mutual support*. Here the advantages of the enterprise are aimed mainly at the members of the community who put the enterprise into practice, but – as shown by certain forms of 'leasehold contractual communities' – the enhancement enterprise can also be launched by a businessman with evident advantages not only for the direct users of the goods created but for the entire community (as in the case of the S. Stefano di Sessanio, looked at in the next paragraph).

26.3 A Plurality of Enterprises: Experiences in Mountain Contexts

The territorial organisation model of contractual communities does not seem univocal and defined. Its settlement characteristics cannot therefore be re-proposed indifferently with respect to the local context. It shows a plurality of forms of settlement and a variety of possible territorial dimensions that range from the rural village to the metropolitan greenbelt, community gardens and peri-urban gardens, reaching the extension of authentic linear infrastructures, as shown by the ecological corridors of New York and Seattle and the green roofs of Chicago.

In order to highlight, from the plurality of experiences in progress, the innovative potential in the management of natural resources, here I am going to look at two forms of contractual community[7] in Italian mountain contexts: the *collective properties of Cortina d'Ampezzo* in the Natural Park of the Dolomites (Belluno) and the *S. Stefano di Sessanio* – 'Sextantio Hotel Diffused' in the Park of the Gran Sasso (L'Aquila). In relation to the general theory of contractual communities, the two experiences refer respectively to the *co-ownership community* (collective properties of Cortina d'Ampezzo) and the *leasehold community* (S. Stefano di Sessanio). What makes the contractual community experiences significant in mountain contexts is the nature of the natural assets and collective services managed. The resources are inevitably linked to the management of common assets (e.g. ground, water, woods, fauna) and the relative uses (e.g. for timber, grass, rural) for local production. They are flanked also by the management of resources linked to the autochthonous cultural identity characteristic of certain territories, which have remained more or less intact in time.

In the case of the collective properties of Cortina d'Ampezzo, the ground resource and also the relative resources for herding, forestry and rural uses have been protected and preserved from potential risks of exploitation. The management method pursued with the 'Rules'[8] of the contractual community, based on specific conditions of use, has guaranteed the reproducibility of environmental resources over time. The example in this sense is the introduction of mechanisms to

---

[7] For further information of the two enterprises, see Baglione et al. (2011) and Minora (2011).

[8] In Italy the collective property is the so-called *Regole*.

compensate for deforestation, i.e. the obligation to plant minimum quantities of new shrubs in place of every shrub cut down. The combination of resources present has also been enhanced via the promotion of small tourist, reception and recreational activities aimed at promoting the knowledge of the territory and funding defence and conservation of collective properties, in keeping with and to strengthen the management strategies introduced by the Natural Park of the Dolomites (set up in 1990).

In the case of the hotel in the fortified medieval village of S. Stefano di Sessanio, the local resources, made up of architectural and cultural heritage, have been preserved from the contingent risk of abandonment, typical of many mountain areas that are hard to reach. The construction of a contractual community, active since 2009, has determined the recovery of the material dimension of the village, which is in a state of abandon and decay, and a rediscovery of its cultural heritage, made up of local productive traditions. The enhancement of similar resources, in this case too, engages the promotion of a new vocation for tourism, unknown up until now, based principally on the actions to enhance and protect cultural and natural resources. Examples are the recovery of traditional crafts and autochthonous crops, both at risk of extinction.

The two cases show a different promoter of the self-organisation enterprise for the management of existing resources. In the first case, the local people, in their capacity as member of the contractual community of co-owners, determined, adapted and stabilised the system of use, conservation and enhancement of the natural resources, proposing enterprises to promote the territory with a view to local development.

In the second case, the promoter of the enterprise is someone from outside the local context. In this experience, the synergy with the local stakeholders, particularly the public stakeholders in territorial governance, made it possible to link up the enterprise with the context and implement requalification projects extended to all areas of the rural village.

In conclusion, the two contractual community models highlight the capacity to elaborate innovative mechanisms of management, collectively shared and linked to the needs of the local communities. They are not previously defined models (i.e. re-proposed according to a set of centralised rules and therefore indifferent to the demands of the context) but forms of local governance, which stem from the desires of the members of the community and adapt to the evolution of the demands for use and management of the common assets planned.

In contractual communities, resources are protected and enhanced so that their stable quality in time is guaranteed, to be valid also for future generations, and the risks of an excessive and exclusive consumption are avoided. In particular, the system of shared rules on one hand and reciprocal supervision by the members of the community on the other guarantee a balanced management of resources and, at the same time, prevent the risk of behaviours outside the system of rules established.[9]

[9] To further this aspect, see Ostrom (1990).

26.4 Innovations for the Landscape Project: Emerging Themes

There are three dimensions of innovation through which a community project aimed at promoting landscape projects is configured. As we will see, it is around the binomial of individual/community that an active landscape management project that is the outcome of a social process of sharing values, resources and objectives gains strength.

First dimension: identity/belonging/ownership. Contractual communities redefine the primary meaning of the term 'community',[10] usually identified with the experiences of recognising ourselves in a determined place and in sharing the same interests with others, conveying value to it. This consolidated first meaning is, in our case, flanked by a second, less common and obvious one, which unites the experience of belonging to the dimension of ownership of the site in which the community experience to be proposed and created is launched (Block 2008, MacCallum, 1970). The property gains ethic value before it gains economic value, becoming a fundamental aspect for the planning and birth of a 'new' community and a new landscape. In this sense, the 'belonging' of the individuals is strengthened, generating networks of support and reciprocity among all members. There emerges a strength that the value of co-ownership (partial or total) takes on in favour of a habit in terms of interpersonal relationships, shared rules and stable decisions for the care of the collective assets.

Second dimension: voluntariness/autonomy/responsibility. Here reference is made to the autonomy of choice of those who propose single community options (and not only to the 'freedom of choice') because freedom without independence – which first of all means cultural skills, expertise and experience – cannot become a resource for society on its own.[11] Contractual communities highlight not only the value of free individual choice but also the capacity to experiment, which comes from the autonomy of those who take part in the enterprise. In this perspective, the capacity to experiment can play a crucial role in re-establishing a stronger integration between the choices made by individuals and the relative restrictions to the systems of choice. We can state that a higher level of autonomy of individuals corresponds to a higher level of responsibility. In this rationale, the notion of responsibility takes on a community value, insofar as it is the outcome of the capacity of action of individuals to measure themselves against the combination of rules and institutional values from which the single proposals for the management of local cultural resources. The notion of responsibility takes on a significant meaning to formulate proposals for the management of landscape resources in response to new local demand for fruition and maintenance.

---

[10] For an overview of the meanings of the term 'community', see Bruhn (2005).

[11] See Goodman (1972) with regard to the concept of independence, presented in the meanings of 'competence' as a necessary condition for the bottom-up construction of community projects, extending the areas of freedom.

Third dimension: collectivisation of natural and cultural resources. Contractual communities represent an experiment of collectivisation of private spaces. In these forms of self-organisation, reciprocity and solidarity being present in varying degrees of importance place the initiatives undertaken in the rationale of horizontal subsidiarity. This concept takes on a civic significance here, as widespread responsibilities in civil society achieved through direct actions in the general interest. Contractual communities achieve this civic ideal, triggering substantial improvements for the entire community and proving to be an effective form for the management of a system of natural assets through the creation of community awareness of the local landscape values. From this perspective, the two experiences shown here promote enterprises of local resource management, capable of enhancing the vocation of the territory, increasing the degree of responsibility and independence of the local people to the contexts they belong to. As Elinor Ostrom (1990) reminds us, we have been traditionally led to believe that 'individuals who use collective resources are considered capable of maximising in the short term, but not of long-term reflection in relation to common strategies to improve common results' (Ostrom 1990, p. 312). The experiences of contractual communities show, on the other hand, that, by virtue of the *voluntariness* and of the *contract* that characterises them, these forms are able to foresee long-term strategies for the management of local natural and cultural resources, representing the driving force behind local institutional learning processes. This aspect is particularly interesting to open up the way for action to promote landscape protection, management and planning, hoped for by the international treatises.

References

Arena G (2006) Cittadini attivi. Laterza, Rome

Baglione V, Caldarice O, Chiodelli F (2011) L'albergo diffuso di S. Stefano di Sessanio. In: Brunetta G, Moroni S (eds) La città intraprendente. Carocci, Rome, pp 53–60

Block P (2008) Community. The structure of belonging. Berret-Koehler, San Francisco

Bruhn JG (2005) The sociology of community connections. Springer, New York

Brunetta G (2011) Dimensioni sociali e organizzative di una teoria generale delle comunità contrattuali. In: Brunetta G, Moroni S (eds) La città intraprendente. Carocci, Rome, pp 103–110

Brunetta G, Moroni S (2008) Libertà e istituzioni nella città volontaria. Bruno Mondadori, Milan

Brunetta G, Moroni S (2012) Contractual communities in the self-organising city: freedom, creativity, subsidiarity. Springer, Dordrecht

Goodman P (1972) Just an old fashioned love song, New York. Italian version: Goodman P (2000) Amo la libertà ma preferisco l'autonomia, Libertaria

MacCallum SH (1970) The art of community. Institute for Humane Studies, Menlo Park

Minora F (2011) Le proprietà collettive di cento e Cortina d'Ampezzo. In: Brunetta G, Moroni S (eds) La città intraprendente. Carocci, Rome, pp 61–80

Moro G, Vannini I (2008) La società civile tra eredità e sfide. Rapporto sull'Italia del Civil Society Index. Rubbettino, Catanzaro
North DC (1990) Institution, institutional change and economic performance. Cambridge University Press, Cambridge, UK
Ostrom E (1990) Governing the commons. Cambridge University Press, Cambridge, UK
Shiva V (2005) Le nuove guerre della globalizzazione. UTET, Turin

Chapter 27
The Concept of Limits in Landscape Planning and Design

Francesca Mazzino

Abstract The paper examines the question of whether 'the sense of limits' can be consistent with the primary task of landscape planning. During the twentieth century and the first part of the twenty-first century, a growing awareness of the complexity of landscape has contributed to strengthening the 'sense of limits'. Globalisation, the new media and technology reduce perception of this 'sense of limits' as a sense of an awareness of the need for self-control in human interventions on the landscape. The relationships between natural phenomena and the dynamics of these huge transformations are currently a theme of great interest in various studies and theories in economics, the natural and social sciences and planning. Furthermore, these research fields show a convergence between attempts to identify the causes of the economic crisis and related predictions regarding the changes which will be necessary in post-industrial society. The development of scientific theories and social movements such as 'sustainable retreat', 'deep ecology', 'degrowth' and 'biosphere education' is based on knowledge about the effects of irresponsive behaviour and the consequent destructive effects on the environment and landscape. It is argued that it is not realistic to imagine such a radical change in human behaviour, which would enable global problems to be solved in a cooperative fashion. Nonetheless, it is possible to draw a number of connections between these theoretical frameworks and the principles of the European Landscape Convention (ELC). A number of recent projects are presented: the Prinzessinnengarten (Berlin), the Cheonggyecheon stream (Seoul) and Lausitz post-mining landscapes (Germany).

Keywords Global changes • Landscape planning • Ecological design

Landscape design is in constant evolution, a sort of 'open field' requiring an 'open mind' in which the social and natural sciences make an essential contribution to responding to contemporary society's increasing demands for a better quality of life. Its evolution is based on growing attention to the human dimension and to the

F. Mazzino (✉)
Department of Sciences for Architecture, Polytechnic School, University of Genoa, Stradone S. Agostino, 37, 16123, Genoa, Italy
e-mail: mazzino@arch.unige.it

R. Gambino, A. Peano (eds.), *Nature Policies and Landscape Policies*,
Urban and Landscape Perspectives 18, DOI 10.1007/978-3-319-05410-0_27

principles of sustainable development. Ecological and behavioural aspects are also key in landscape design because of the need to solve multifaceted problems in response to the complexity of modern society. During the twentieth century and the first part of the twenty-first century, a growing awareness of the complexity of landscape has contributed to strengthening the 'sense of limits'.

Globalisation, the new media and technology reduce this perception of a sense of limits, in the sense of an awareness of the need for self-control in human interventions on the landscape. The growth of standardised, large-scale production, the development of technology in every field of human activity and opportunities to 'fritter away' natural resources in colonised countries have all led to an absolute condition faith in unlimited growth.

The relationships between natural phenomena and the dynamics of these huge transformations are currently a theme of great interest in various studies and theories, the natural and social sciences and planning. Furthermore, these research fields show a convergence between attempts to identify the causes of the economic crisis and related predictions regarding the changes, which will be necessary in the post-industrial society.

Some espouse a radicalism based on a reversal of an economic system predicated on increasing consumption, which they see as a necessary condition for avoiding large-scale crisis and which can be compared to nineteenth-century theories and practice of Utopian socialism.

Scientific theories and social movements such as 'sustainable retreat' (Lovelock 2007), 'deep ecology' (Naess 2008) and 'degrowth' are simultaneously met with approval for their innovative approach or criticism by those who consider them impossible to put into practice.

Particularly, the French economist Latouche criticises the concept of sustainable development on the basis of the ambiguity of the term sustainability; he argues that the radical alternative of the 'degrowth' society is inevitable, rejecting the irrational pursuit of 'growth for growth's sake'. He asserts that degrowth is different from the negative growth, offering a way out of growth based on the continuing destruction of ecosystems and local cultures in the name of development. He proposes that it is not necessary to not produce but to produce better.

The need to underline the 'sense of limits' in order to achieve sustainable co-evolution between man and nature starts from the recognition of one's own 'home' landscape (Küster 2009).

Rifkin, a classical economist, also advocates 'biosphere education' (Rifkin 2010) based on awareness of the effects of irresponsible behaviour and its consequent destructive effects. He suggests a new form of development, consisting in a cooperative mode of production, which is derived from a sense of empathy towards the biosphere. An interesting aspect of Rifkin's assumption is his analysis of the most significant philosophical and scientific statements of the nineteenth century and in particular the scientific development of ecology, which is increasingly connected with landscape planning and design.

The rejection of economic growth based on consumption of natural resources is a subject of debate among anti-globalisation activists who are convinced that growth is unsustainable and harmful from the ecological, economic and social perspective. For the scientist and economist Shiva, leader of popular movements in India (Shiva 2012), economic growth threatens our planet's survival as a result of fossil fuel dependency and the control exercised by the international corporations over food production and water commons. In this sense, Karr has developed the idea of 'bio-regionalism' (Karr 2005), which affirms the urgent need for sustainability in the post-industrial societies through civic associations and networks of people in opposition to the homogenising tendency of corporate globalisation.

Some argue that it is not realistic to imagine such a radical change in human behaviour to solve global problems in cooperative way. The criticisms are that growing global population generates the climate disaster and that there are not sufficient data to assert that the pre-modern techniques used by small rural communities are an appropriate solution to the global crisis.

Nonetheless, it is possible to draw a number of connections between these theoretical frameworks and the principles of the European Landscape Convention (2000). The theme of landscape as a common good and overriding factor in the production of goods and services is a significant topic of both international research and the official European documents. The implementation of the European Landscape Convention underlines the fact that it is awareness of everyday life which can bring about an increased sense of responsibility for the place in which everyone lives. In particular the *publication Landscape and Sustainable Development: Challenges of the European Landscape Convention* (2006) admits that equality of access to natural and cultural resources for exploitation purposes poses problems of sustainability. The document points out that over the previous century, the population of Europe enjoyed rising standards of living, but that these improvements were not equally distributed. Landscape is damaged by excessive attempts to raise productivity.

The European Landscape Convention asserts that the central issue is that in every social context, the stakeholders concerned express opposing points of view, their own interests for many social, economic, cultural reasons; they are influenced by subjective considerations, which create situations of conflict Luginbühl and Terrasson (2013). The formulation of landscape quality objectives entails taking on responsibilities and making convergent decisions with regard to various aspects of landscape policies and planning. This ambitious goal can be supported by experts on different aspects of landscape, in particular those working in the landscape architecture fields.

A stimulating question is whether the principles of degrowth can be congruent with the primary task of landscape architecture oriented to create a closer relationship between communities and the landscape, which they inhabit.

The first example of a the sense of limits in landscape planning and design can be found in the plans and projects of Frederick Law Olmsted, Sr., who was concerned with wildlife preservation and with planning and design parks and green systems, anticipating modern ecological principles. He recognised that America's economic development and its massive use of natural resources could not be unlimited. The social aspects of design were one the priorities of his vast work; for example, he was

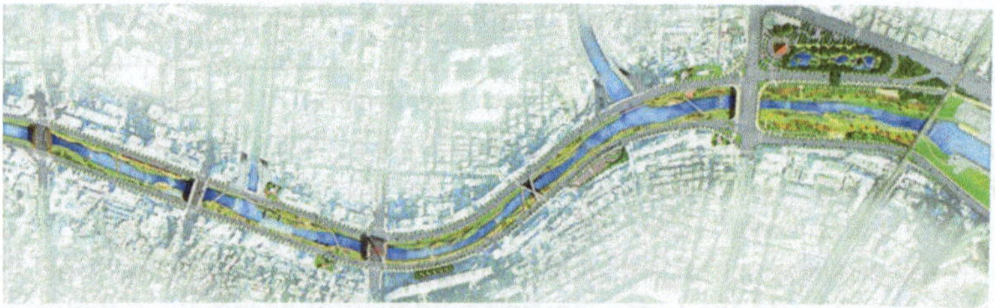

Fig. 27.1 The plan for the restoration of Cheonggyecheon stream (2003). In Seoul, the Cheonggyecheon is a tributary of the Jungnangcheon which flows into the Han River. After the Korean War people who had immigrated to Seoul lived along the stream in ramshackle houses. The stream was deteriorated by waste and pollution. Starting in 1958 the stream was covered over by an elevated highway completed in the 1970s. The project to remove the highway and restore the stream commenced in 2003. The Cheonggyecheon restoration is considered an important step towards reintroducing nature into the city, supporting ecological urban design and redeveloping the city's cultural heritage with the restoration of two historic bridges. The plan includes the construction of a network of pedestrian paths to connect historic places (the Cheonggyecheon Culture Belt)

critical of the effects of slavery and the poor quality of life of the working class. He was convinced that American democracy could be strengthened by sustaining solidarity and equality through the provision of public services for urban rehabilitation and environmental control.

The concept of the 'Land ethic' in the work of forester and philosopher Leopold (1949) can be seen in continuity with the idealism of the Conservation Movement. He compared individuals and groups cooperating to build more advanced economic and political systems to symbioses in nature. An 'ecological conscience' means the inclusion of water, soil, vegetation and animals in the 'land ethic' and pursuit of harmony between man and nature. This ecological conscience entails an awareness of individual responsibility for the health of the landscape and the capacity for self-renewal. The ecological restoration of the landscape carried out by Leopold with the plantation of pine trees and the renewal of prairies along the Wisconsin River should be compared with Donadieu's concept of nature as a product of social construction (Donadieu 2002).

Design with Nature by Ian McHarg (1969) has its roots in the nineteenth-century landscape architecture philosophy concerning ecological planning and design and had a major influence on the development of landscape planning and design. He extended the field of landscape architecture to include brownfields restoration, river corridor and coastal planning and environmental impact assessment. McHarg is in radical opposition to urban and industrial modern exploitation (Fig. 27.1).

The diffusion of such an ecological conscience increasingly informs landscape design and restricts the egocentrism and anthropocentrism of designers.

The sense of limits in landscape design is a function of two main factors: time and space. A project has to deal with a large number of natural elements, which in many cases are impossible to anticipate, unlike in architecture or engineering projects, and this fact reinforces the prudent assessment of the transformation. While other

Fig. 27.2 Re-naturalised banks on the Cheonggyecheon (2003). The project aimed to construct an environment with natural habitats and clean water for the fish, birds and insects that have populated the stream excavation. The stream contributes to the reduction of temperature in the nearby areas by 3.6 °C. The decrease in traffic volumes was made possible by the implementation of public transport and the demolition of the two heavily used roads. The project has been criticised by a number of Korean environmental organisations for its high costs and from the ecological and historical point of view

projects are 'finished' on completion of the work, landscape projects are completed over a much longer timescale and require appropriate management to 'mature'. In addition, the use of living material demands a special attitude to understand the dynamic of the natural elements. The use of plants and their interaction with other living elements such as soil, water and the atmosphere requires a sensitive and patient approach to the existing and future ecosystems. The most advanced dimension of the landscape design is related to services provided by ecosystems (Rodríguez et al. 2006); each intervention has rapid or slow consequences, at distant locations on a large scale or locally at a small scale (Fig. 27.2).

Another aspect of the 'sense of limits' is that landscape design does not necessarily consist in major transformations or large-scale projects with vast budgets (Internationale Bauaustellung Fürst-Pückler-Land 2010). Small-scale, self-organised actions in abandoned industrial and urban which recycle materials, seeds or rainwater may receive support from local administrations who, as a result of the economic crisis, come face to face with a 'sense of limits' as they are unable to invest in new large-scale projects.

From its beginnings as a 'guerrilla gardening' project, the Prinzessinnengarten, located in Moritzplatz near the Berlin Wall, in the Kreuzberg district, a wasteland which had been abandoned for more than 50 years, has been transformed into a

Fig. 27.3 The Welzow open cast mine. The media contribute to circulation of ideas, projects and plans, which succeed in relationship to the more wide diffusion. Landscape design based on low-cost interventions and social solidarity for the conservation and the regeneration of resources requires the adoption of a critical selection of contents and design models in research and education processes

non-profit urban farm, one of the first urban agriculture projects undertaken in Berlin (Fig. 27.3).

Since 2009 this biological urban farm is open to anyone wishing to tend its plants and, at same time, is a meeting place for people to exchange experiences and to reinforce a sense of community. Volunteers and residents have removed rubbish and collected recycled containers for growing vegetables in raised compost beds without pesticides or artificial fertilisers. The anonymous plaza on the top of the underground station has been turned into a new kind of green space extending over 6,000 m^2. The motivation for the project's participants is the goal of experimenting with a new form of local economy: those who work on the farm can buy the produce (vegetables) at the lowest prices and vegetables grown are consumed locally with no transportation costs, thus improving the microclimate and enhancing biodiversity with the cultivation of various varieties of plants.

This project, like others around the world, shows that according to public perception, the future challenge of landscape design is to assure a healthy environment.

Ecological design has been described (Beck 2013) as a way of limiting landscape interventions if they are not geared towards promoting the ecological functions of the site. The aspects analysed by Beck are biogeography, plant-animal interactions, material recycling and soil ecology, the application of landscape ecology, designing and managing ecosystems, building plant communities, preventing the spread of invasive species, integrating animals (birds and bees), dealing with climate change to improve biodiversity and ecosystem services.

A pioneer of ecological landscape management was the microbiologist Fukuoka (1978) who, 50 years ago, introduced natural farming. In this case the exploration of 'the sense of limits' has deep roots in Taoism, and, specifically, in the principle of nonaction or non-doing (*Wu wei*), which does not equate to laziness. On the contrary, nonaction means eliminating the arrogance of ego. Zen philosophy states that Heaven and Earth share the same roots, and that humans and the myriad things are one. Fukuoka clarifies that the agriculture of nonaction consists in cultivating as much as possible in a simple way. In the citrus orchards, rice fields and vegetable gardens, he did not plough or use machines. He did little weeding and used compost to improve the soil. This system required less labour than any other and his production was comparable with the most productive Japanese farms, which use the techniques of industrial agriculture. Fukuoka's alternative methods of cultivation respect local species and the seasonal cycle. He used various techniques, such as scattering whole straw on the soil against birds and weeds, and to conserve the soil structure, he did not use pesticides as they exterminate natural predators, and cultivated vegetables like wild plants.

The sense of limits in landscape design and planning means that environmental and cultural damages are hard to repair due to the costs, time necessary for restoration and results; this consideration should be crucial in planning landscape transformations.

References

Beck T (2013) Principles of ecological landscape design. Island Press, Washington, DC

Council of Europe (2006) Landscape and sustainable development: challenges of the European landscape convention. Council of Europe Publishing, Strasbourg

Donadieu P (2002) La Société Paysagiste. Acte Sud/ENSP, Arles

Fukuoka M (1978) The one-straw revolution: an introduction to natural farming. Rodale Press, Emmaus

Internationale Bauaustellung Fürst-Pückler-Land (2010) 2000–2010 New landscape Lusatia. Catalogue. Hrsg.ed. Iovis Verlag GmBH, Berlin

Karr M (2005) Bioregionalism and civil society: democratic challenges to corporate globalism. UBC Press, Vancouver

Küster H (2009) Schöne Aussichten. Kleine Geschichte der Lanschaft. Verlag C.H. Beck oHG, München

Leopold A (1949) A sand county Almanac. Oxford University Press, New York

Lovelock J (2007) The revenge of Gaia: Earth's climate crisis & the fate of humanity. Penguin, London

Luginbühl Y, Terrasson D (eds) (2013) Paysage et développement durable. Edition Quae, Versailles

McHarg I (1969) Design with nature. American Museum of natural history. Natural History Press, New York

Naess A (2008) Life's philosophy. Reason and feeling in a deeper World. University of Georgia Press, Athens

Rifkin J (2010) The age of empathy. Penguin, New York
Rodríguez JP, Beard TD Jr, Bennett EM, Cumming GS, Cork S, Agard J, Dobson P et al (2006) Trade-offs across space, time, and ecosystem services. Ecol Soc 11(1):28
Shiva V (2012) Making peace with the earth. Beyond resource, land and food wars, 2nd edn. Pluto Press, London

Chapter 28
Landscape and Ecosystem Approach to Biodiversity Conservation

Franco Ferroni, Monica Foglia, and Giulio Cioffi

Abstract In the evolution of the models for the establishment and management of protected natural areas, we assist to the transition from "sanctuaries" of nature, where men are considered exclusively a threat or an interference to eliminate, to tools for conservation of biological diversity in situ through the involvement of local communities. This vision of protected natural areas has been formalised in the International Convention on Biological Diversity and its Action Plans. To implement the CBD in 2000, a new method has been developed that considers the human communities as part of the ecosystems and of their governance assets, called "ecosystem approach". The ecosystem approach recognises that human activity affects ecosystems interacting with their structure and composition, resulting in an irreversible loss of ecosystem functionality once some boundaries are crossed. At the same time the ecosystem approach recognises particular importance to the role of local communities and traditional knowledge in the definition of strategies and programmes for the conservation of biodiversity. The ecosystem approach requires the definition of clear targets of biodiversity conservation and the identification of the most appropriate community to deal with its management. The introduction of territory's perception by local communities in defining the European Landscape Convention is echoed in the approach of the CBD ecosystem, particularly for Italy, where the high biodiversity is matched by a high cultural diversity of local traditions and customs. The application of the ecosystem approach to the landscape scale requires a coherent planning approach that includes a proper biodiversity' conservation plan. These targets must be identified not only for the conservation of

F. Ferroni (✉)
Biodiversity, Protected Areas and Agricultural Policies, WWF Italy, Via Po, 25/c, Rome 00198, Italy
e-mail: f.ferroni@wwf.it

M. Foglia
INEA Research Grant at Agricultural Society La Quercia della Memoria, CREDIA WWF, C. da Vallato, 62016 San Ginesio, MC, Italy
e-mail: foglia.monica@gmail.com

G. Cioffi
CREDIA WWF, Via U. Betti, 3, 62025 Fiuminata, MC, Italy
e-mail: zephiro_oct@virgilio.it

R. Gambino, A. Peano (eds.), *Nature Policies and Landscape Policies*,
Urban and Landscape Perspectives 18, DOI 10.1007/978-3-319-05410-0_28

each species or habitat but rather for the maintenance of the structure and functionality of the ecosystems.

Keywords Biodiversity • Ecosystem approach • Participation • Community • Local and global

28.1 From the "Sanctuaries" of Nature to the "Ecosystem Approach" to Nature Conservation

The International Convention on Biological Diversity (CBD), signed in Rio de Janeiro on the 5th of June 1992 and ratified in Italy by the law n. 124 of the 14th of February 1994, represents today the key document for those who, to any extent or degree, deal with the conservation of nature. The goals of the Convention are the conservation of biological diversity, the long-lasting use of their components and the fair and equitable sharing of the benefits arising from the use of genetic resources. The article 6 of the Convention refers explicitly to the necessity of integrating the conservation and the long-lasting use of biological diversity in the relevant sectorial and cross-sectorial planning and recalls for every Contracting Party to develop strategies, planning or national programmes for the conservation of biodiversity.

Article 8 of the Convention deals with the issue of in situ conservation, acknowledging the key role played by protected areas, since the second half of the nineteenth century with the first paradigm of nature conservation. If at the beginning protected areas were considered as a sort of "sanctuaries of nature" where the presence of the human being was seen exclusively as a threat or as an intrusion to remove, they evolved later as the places for the conservation of biological diversity in situ also through the active participation of local communities.

Taking into account the evolution of models that guided the creation and the management of protected areas, it is useful to underline that this evolution was part of a bigger consideration about the relationship between development of human activities and structure and functions of ecosystems. This gave birth to a scientific and cultural debate, which finally resulted in the adoption of the term "Anthropocene", widely promoted by the Nobel Prize-winning atmospheric chemist Paul Crutzen, to indicate the actual geological period in which man and his activities are considered the main cause of territorial, structural and climate changes Crutzen (2005). The ability of our species to modify the environment in which they live and adapt it to their needs and aspirations extended as far as to influence, in an unpredictable way, those same processes which are at the basis of the evolution of life on earth. Evolution is an incessant process and natural systems are continuously changing: man, with his activities, has caused such swift and substantial changes at different spatial and temporal scales and has become an essential variable in the natural processes which determine the structure and functionality of the ecosystems over time.

The CBD is the first international agreement concerning all aspects of biological diversity, acknowledging this direct human responsibility and extending the results of that analysis and consideration following the creation and management of protected areas to all natural and social systems.

For the implementation of the CBD, during the Fifth Meeting of the Conference of the Parties (COP5 of Nairobi, Kenya) in 2000, a methodology called "ecosystem approach" was set out, considering human community as an integral component of ecosystems and of the mechanisms that regulate them.

The ecosystem approach, as defined in the COP5 report (UNEP/CBD/COP/5/23, 103–109), is a strategy for the integrated management of land, water and living resources that promotes conservation and sustainable use in a fair and equitable way.

The ecosystem approach of the CBD acknowledges that human activity, interacting with ecosystems, influences their structure and composition and causes beyond limit an irreversible loss of the ecosystem functionality. At the same time, the ecosystem approach recognises a particularly significant role to local communities in the definition of strategies and programmes for the conservation of biodiversity, as stated by the article 8 of the CBD, letter j, according to which each Contracting Party shall, as far as possible and as appropriate:

> Subject to its national legislation, respect, preserve and maintain knowledge, innovations and practices of indigenous and local communities embodying traditional lifestyles relevant for the conservation and sustainable use of biological diversity and promote their wider application with the approval and involvement of the holders of such knowledge, innovations and practices and encourage the equitable sharing of the benefits arising from the utilization of such knowledge, innovations and practices.

As a consequence of this, one of the main characteristics of the ecosystem approach is the direct and substantial involvement of local stakeholders in the land management, seen as an integrated process from both a social and environmental point of view (land, water, atmosphere, living resources). This principle applies both inside and outside a protected area (the CBD defines a "protected area" as a geographically defined area which is designated or regulated and managed to achieve specific conservation objectives).

The ecosystem approach is a working methodology that can be employed at different spatial and temporal scales for the management of an ecoregion, of a protected area (from the National Park to the small Regional Reserve), or for a single-specific project of habitat and species preservation.

It is more adequate for medium- or long-term objectives, but it is also suitable for the management of projects that pursue short-term results (certainly more functional from a political and social point of view). The general purpose is anyway the achievement of the three objectives of the CBD. For these reasons, another characteristic of the ecosystem approach is to be a methodology that does not follow a fixed and guaranteed implementation; this one must be set on a case-by-case basis according to the instrument (protected area, plan, programme or project) and adapted to the context (environmental and socio-economical).

The implementation of the ecosystem approach must in any case meet the following 12 principles, complementary and interrelated, that constitute the axiom that provides its theoretical framework[1]:

1. The objectives of management of land, water and living resources are a matter of societal choice.
2. Management should be decentralised to the lowest appropriate level.
3. Ecosystem managers should consider the effects (actual or potential) of their activities on adjacent and other ecosystems.
4. Recognising potential gains from management, there is usually a need to understand and manage the ecosystem in an economic context.
5. Conservation of ecosystem structure and functioning, in order to maintain ecosystem services, should be a priority target of the ecosystem approach.
6. Ecosystems must be managed within the limits of their functioning.
7. The ecosystem approach should be undertaken at the appropriate spatial and temporal scales.
8. Recognising the varying temporal scales and lag effects that characterise ecosystem processes, objectives for ecosystem management should be set for the long term.
9. Management must recognise that change is inevitable.
10. The ecosystem approach should seek the appropriate balance between, and integration of, conservation and use of biological diversity.
11. The ecosystem approach should consider all forms of relevant information, including scientific and indigenous and local knowledge, innovations and practices.
12. The ecosystem approach should involve all relevant sectors of society and scientific disciplines.

Cross-reading the 12 principles, it becomes evident that the ecosystem approach is based on the application of appropriate scientific methodologies, focused on levels of biological organisation which involve essentially structure, processes, functions and interactions between organisms and their environment. Furthermore, the ecosystem approach recognises that human beings, with their cultural diversity, are an integral component of many ecosystems. The involvement of many and various social and economic actors, however, demanding, is necessary to the definition of a strategy, programme or project that may be effective for the conservation of biodiversity.

This participatory approach represents one of the innovations of CBD, which considers the integration of the necessities of biodiversity conservation in the social

[1] For a more detailed description of the principles and further information on the operational guidelines for the application of the ecosystem approach, you can visit the official site of the CBD: http://www.cbd.int/ecosystem/default.shtml.

and economic national policies. The ecosystem approach does not preclude other approaches for the management and conservation of biodiversity, such as the creation and management of protected areas, conservation programmes for the preservation of single species and other approaches that are adopted according to the national policies and laws; it can rather integrate these various approaches and methodologies to deal with complex situations.

28.2 Biodiversity and Landscape Protection

The ability of our species to change the environment, in which they live and adapt it to their needs and aspirations, also produced "landscapes". It is more convenient and appropriate in this case to refer to a plurality of landscapes that can have different meanings depending on whether its perceptive, cultural, territorial or ecological aspects are taken into consideration. The most recent developments in the regulatory framework include the European Landscape Convention (2000) that legally recognises landscape as an essential component of the context of people's lives and promotes its protection, its management and planning. The relationship between biodiversity and landscape is expressly mentioned in the preamble of the ELC with reference to the legal texts already existing at international level in the field of protection and management of the natural and cultural heritage, i.e. in particular the Convention on the Conservation of European Wildlife and Natural Habitats (Bern, Sept. 19 1979), and the same CBD (Teofili and Clarino 2008). The introduction of the perception of the land by local communities in the ELC definition of landscape is reflected in the CBD ecosystem approach that considers human beings, with their cultural diversity, an integral component of many ecosystems. This relationship is even more evident with the application of the 12 principles of the ecosystem approach into territorial contexts with a high population density and age-old human presence, in Europe. This is even truer for Italy, which has both a very high biodiversity levels (Italy is the European country with the highest number of endemic species and variety of ecosystems) and a very high cultural diversity.

A large debate on the relationships between biodiversity, protected areas and landscape has been ongoing for some years and makes it clear that a substantial revision of the approaches to urban planning is needed (from protected areas to landscape planning), in an effort to transcend the distinction between protected and unprotected areas, following an approach that considers a wide area, aimed at strengthening the laws for the protection of biodiversity and the other nonliving resources of the territory (Gambino 2011). Likewise, landscape planning must integrate in an appropriate way the biodiversity conservation objectives, identified not only with the protection of each species or habitat but rather with the maintenance of ecosystem structure and functionality (Battisti and Romano 2007). This proposal has been already submitted by the Italian WWF: in 2005, during the conference "*Conservazione Ecoregionale, Reti Ecologiche e Governo del*

Territorio" (Abbadia di Fiastra, Italy), a calling was set out for the inclusion of the biodiversity conservation objectives within the landscape quality objectives for single areas (landscape "units").

Despite the crisis of territorial and urban planning in Italy, and in particular of the landscape planning, the debate on the integration of biodiversity in the landscape and management planning is still lively and open (Vv.Aa 2006). Nevertheless, this debate still does not include the ecosystem approach defined for the CBD, and it is still marginal the consequent issue of the involvement and responsibility of local communities in ecosystems management. Thanks to the Italian WWF, the first experiences of urban planning and participatory planning have started in 1994 as part of a cultural collaboration agreement with the INU (National Institute for Urban Planning). During the third edition of the INU National Contest (2004) also, experiences related to ecological networks in the city and their integrations into wider territorial contexts (landscape units) and areas (ecoregions) have been taken into consideration. Participatory projects for the involvement of local communities in the strategies for sustainable development have also arisen, thanks to the methodology of Local Agenda 21s. From 2001 onwards, reflections and practices realised by WWF and other numerous partners in the context of the "Ecoregional Conservation" have contributed substantially to the definition of some aspects of the National Strategy for Biodiversity (SNB) adopted by the State-Regions Conference on October 7, 2010. The ecosystem approach and the issue of the participation of local communities in the conservation and sustainable management of biodiversity have therefore been treated (not without some difficulty) in the working area n.14 "Education, information, communication and participation" of the SNB (Italian Ministry of Environment, Land and Sea Protection, 2009a).

28.3 Role and Involvement of Local Communities

The SNB priorities in this area are the realisation and promotion of the "preservation of the cultural heritage of local communities and the participatory management of natural resources". In Italy, politicians and administrators of local institutions have often exploited, and so it happens today, the necessity for a wider local communities involvement in biodiversity conservation and management. The objective is to claim primarily a greater decision-making power over the government administration of lands for a profit-oriented exploitation of natural resources, influenced by private interests, particularly in the management of protected areas. The failure of many of the experiences of Local Agenda 21s, the controversy about the management of parks and the recent perspective of a reform of the Law 394/91 highlight the damages produced by this instrumental approach over the role of local communities. This cultural and political context has often made it complicated and affected any positive and innovative experiences for a wider participation and responsibility of local communities in the biodiversity and landscape management.

However, there have been significant experiences of participatory planning at a national scale, and, specifically, it is of particular interest the experimentation carried on by the CREDIA WWF[2] since 2008 within the Monti Sibillini National Park for the implementation of the ecosystem approach, with the creation of an eco-museum (Ferroni and Romano 2010).

The activities undertaken wanted to underline in particular the relationships, at different spatial and temporal scales, between human communities and the use of natural resources and the consequent effects on the structures and functions of ecosystems of a specific landscape unit identified by the National Park Plan.

The project, still ongoing, has confirmed that the ecosystem approach needs clear objectives of biodiversity conservation to be defined and the identification of the community which is, for various reasons, responsible of its management. It is not of lesser importance that the community understands the "value" of ecosystems and is aware of the relationships between ecosystem services and the well-being of the same community (Italian Ministry of Environment, Land and Sea Protection, 2009b). For this reason, it is necessary that the right interpretation of the attribute "local" is to be understood not specifically as "of a particular place", but rather as "in a more or less direct and more or less conscious relationship with a particular place": a relationship that has influenced the ecosystems evolution. In the Apennine Mountains of Central Italy, until almost all of the eighteenth century, when the satisfaction of primary needs depended strictly upon the availability of natural resources, the local community could be identified with the resident community. The economic-social system of the Appennino Umbro-Marchigiano was based on the *villae*, small villages situated on mountains slopes, centred almost exclusively on agroforestry-pastoral activities. The landscape was thus deeply marked by human activities, where woods, copses in particular, were productive areas exploited mainly for breeding than for harvesting timber and where secondary grasslands provided fodder.

The social and economic changes that have characterised the eighteenth and nineteenth century profoundly transformed the relationships between local communities and their territory (Gobbi 2003). From the 1950s of the twentieth century, "villae" depopulated and traditional agroforestry-pastoral activities were almost completely abandoned. Ecosystems, because of this reduced pressure of human activities, underwent substantial changes, swiftly transforming also the Apennines landscape (Foglia et al. 2007; Catorci 2007). These commonalities survived to depopulation and abandonment of the mountain agricultural are form of collective ownership that have been managing more or less large portions of land so as to supply the inhabitants of a municipality or of a village (*villa*) with the necessary resources (pasture and wood), and still do it today, preserving that relationship with the land that has guaranteed a sustainable management of resources based on rotation and periodic exploitation of soil, teaching respect for what is only

[2] A Centre for Environmental Education, Documentation and Interpretation, managed in cooperation with the Agricultural Society La Quercia della Memoria.

temporarily own. In the twentieth century, tourism becomes the most important activity for the economy of these areas. This land is not only "lived" by the local resident community but also rather by a "global community". The population of the National Apennine Mountains Park (about 70,000 ha) has halved in the last 50 years, from about 60,000 to 25,000 people, but in reverse the National Park is visited every year by 60,000 residential tourist and occasional visitors. In this new context, the stakeholder analysis interacting with the ecosystems, necessary to the implementation of the ecosystem approach, becomes more complex, because the responsible community cannot only be identified with the local resident community. It becomes necessary to consider everyone who benefits from the ecosystem services provided by the protected area. Relationships between man and the natural environment are no more mere productive activities that provide value to the traditional use of resources but also social and recreational that provide value to the "non-use" of natural resources. Landscape and ecosystem are now an occasion for new "life experiences" generating individual and collective well-being for people.

These new relationships of the "global community" with ecosystems influence their evolution as much as the direct use related to agroforestry-pastoral activities, and they do not preclude pressures and threats that can jeopardise their structure and functionality. Therefore, it is useful for the definition of biodiversity conservation objectives the analysis of ecosystems services provided by single landscape units and their relationships with the "global community" that benefit from these services. By analysing ecosystem services, according to the MEA (Millennium Ecosystem Assessment), the transformation of the communities-ecosystem-landscape relationship can be interpreted considering the type of ecosystem services variation that determine the community well-being, the ecosystem pressures and threats and the increasing awareness of the people insiders and value of the ecosystem services that they benefit from.

References

Battisti C, Romano B (2007) Frammentazione e connettività. Dall'analisi ecologica alla pianificazione ambientale. Città Studi Edizioni, Turin

Catorci A (2007) Lineamenti storico-economici del paesaggio pastorale maceratese. In: Catorci A, Gatti R (eds) Le praterie montane dell'Appennino maceratese, vol 42, Braun-Blanquetia. Dipartimento di Botanica ed Ecologia, Università di Camerino, Camerino, pp 39–45

Council of Europe (1979) Convention on the conservation of European wildlife and natural habitats, 19.IX.1979, Bern

Council of Europe (2000) European landscape convention. Treaty series no. 176, Florence

Crutzen P (2005) Benvenuti nell'Antropocene. L'uomo ha cambiato il clima, la Terra entra in una nuova era. Mondadori, Milan

Ferroni F, Romano B (eds) (2010) Biodiversità, consumo di suolo e reti ecologiche. La conservazione della natura nel governo del territorio. WWF Italia, Miur. Cogecstre Ed., Pescara

Foglia M, Sparvoli D, Catorci A (2007) Analisi multi-temporale dell'uso del suolo della dorsale appenninica marchigiana nel XIX e XX secolo. In: Catorci A, Gatti R (eds) Le praterie montane dell'Appennino maceratese, vol 42, Braun-Blanquetia. Dipartimento di Botanica ed Ecologia, Università di Camerino, Camerino, pp 39–45

Gambino R (2011) La pianificazione dei parchi in prospettiva internazionale. In: Sargolini M (ed) Piani per i parchi. Edizioni ETS, Pisa

Gobbi O (2003) La Villa un microcosmo ecosistemico. In: Vv.Aa. (eds) I Sibillini oltre il mito. Aspetti socio-economici di una regione appenninica nei secoli XV–XVIII. Amandola, Staff Ed., pp 165–185

Italian Ministry of Environment, Land and Sea Protection (2009a) Studio e analisi delle forme di coesistenza e criticità tra sviluppo economico-sociale e conservazione della natura. Il ruolo del processi partecipati. In: Proceedings of technical meeting Verso la Strategia Nazionale per la Biodiversità, Rome

Italian Ministry of Environment, Land and Sea Protection (2009b) Definizione del metodo per la classificazione e quantificazione dei servizi ecosistemici in Italia. In: Proceedings of the technical meeting Verso la Strategia Nazionale per la Biodiversità, Rome

Teofili C, Clarino R (2008) Riconquistare il Paesaggio la Convenzione Europea del Paesaggio e la Conservazione della biodiversità in Italia. WWF ONG/Onlus/MIUR, Rome

Vv.Aa (2006) Conservazione Ecoregionale, Reti Ecologiche e Governo del Territorio. In: Proceedings of the national conference on the Abbadia di Fiastra natural reserve. WWF Italia, UPI, Provincia di Macerata. Tolentino, 9–10 Jun 2005

Chapter 29
Biodiversity and Landscape Policies: Towards an Integration? A European Overview

Bianca Maria Seardo

Abstract Although outcomes of complex co-evolutionary interactions, landscape and biodiversity are often addressed by separated policies, thus undergoing the risk of being scarcely coordinated or even conflicting. The contribution discusses such an issue from a European point of view taking as a case study the EU Member States committed in the implementation both of the Convention of Biodiversity (1992) and the European Landscape Convention (2000): as a matter of fact, although almost all of the EU Member States have ratified both the CBD and the ELC, it is not obvious at all that respective sectoral policies addressed to biodiversity adhere to the new conception of landscape promoted by the ELC. The screening of 50 National Biodiversity Strategies and Action Plans (NBSAPs) finally puts in light that (a) the coherence to the multifunctional conception of landscape appears at different degrees in EU Members States, oscillating between strictly ecological approaches to more multidimensional ones, and (b) landscape is often used as a means to enlarge biodiversity conservation efforts beyond administrative borders.

Keywords Landscape policies • Biodiversity policies • Landscape multifunctionality

29.1 Converging Paradigms

Biological diversity and landscape are the product of complex dynamic co-evolutionary interactions (Borrini-Feyerabend et al. 2008; Jones-Walters 2008). Biodiversity and landscape are linked by increasingly convergent approaches: on the one side, theoretical advances about the concept of landscape turn from an "insular" to a "holistic" approach, based on the assumption that each landscape element gains its significance, importance or existence not only from its

B.M. Seardo (✉)
Interuniversity Department of Regional and Urban Studies and Planning (DIST), Politecnico and Università di Torino, Viale Mattioli 39, 10125, Turin, Italy
e-mail: bianca.seardo@polito.it

R. Gambino, A. Peano (eds.), *Nature Policies and Landscape Policies*,
Urban and Landscape Perspectives 18, DOI 10.1007/978-3-319-05410-0_29

intrinsic properties but also in accordance to its relationships with the context (Antrop 2006).

On the other hand, nature conservation paradigms have abandoned the aprioristic opposition to any form of development, "yielding more and more to the idea that conservation constitutes the face of innovation for emerging contemporary society" (Gambino 1997).

Moreover, the issue of ensuring protection by managing territorial dynamics puts emphasis on the role of landscape planning and management, as required by the European Landscape Convention (ELC, article 5.b).

Whereas, on the one hand, it can be easy to converge on abstract principles, on the other hand divergences may arise when it comes to put them into practice.

As a matter of fact, when considered as matters of policies, regulations and planning, landscape and biodiversity are likely to experience conflicting situations, for example, when they are subjects of distinct policies: while the European Union is competent in environmental policies, the Council of Europe (2000) promotes international conventions on "culture, heritage *and nature*", such as the European Landscape Convention.

The EU Member States have ratified the CBD and the ELC in a not uniform way, thus making not obvious the compliance of biodiversity policies to the new landscape conception supported by the ELC. Consequences on planning and management and interactions on the national regulative frameworks are probably not deeply investigated and differ among member states: in Italy, for example, "territory", "landscape", "environment" and "protected areas" are regulated by different laws and plans, while landscape regional plans prevail on the ones concerning protected areas.

29.2 Putting Ideas into Practice: Do Biodiversity Policies Take into Account Holistic Landscape Conception?

With the ratification of the CBD, EU Member States are committed to develop National Biodiversity Strategies and Action Plans (NBSAPs), while the ratification of the ELC does not foresee new specific national policies but the integration of its principles within existing regulatory frameworks and planning processes.

The landscape conception promoted by the ELC is based on the key concepts of landscape multidimensionality and holism. Multidimensionality suggests that landscape is the result of the action and interaction of natural and/or human factors (ELC, art.1.a) shaping its historic, natural and aesthetic visible features. Thus, landscape analysis has to be holistic and calls for interdisciplinarity (Brandt et al. 2000; Hassan et al. 2005, among the others); holism indicates that each landscape element gains its significance, importance or existence not only from its intrinsic properties but also in accordance to its relationships with the context (Antrop 2006). The attention is extended from few outstanding elements to their broader contexts, thus including "natural, rural, urban and peri-urban areas (...) land, inland water and marine areas (...) landscapes that might be considered outstanding as well as everyday or degraded landscapes" (ELC, art.2).

Starting from the assumption that policy integration is a desirable prerequisite for effectiveness, this contribution tries to answer the question: do National Biodiversity Strategies and Action Plans (NBSAPs) take into account the addresses of the European Landscape Convention?

The research[1] focused on the EU Member States committed in the implementation of the CBD and the ELC and has concerned the examination of the NBAPs. At the time of closing the research (April 2012), the state of the art was the following:

- 23 EU Member States out of 28 had ratified the European Landscape Convention.
- All EU Member States had signed the Convention on Biological Diversity (168 countries all around the world had signed it), but the development of National Biodiversity Strategies and Action Plans (NBSAPs) is at different stages of progress.

Since then some changes have occurred: Croatia has joined European Union on July 1st, 2013 (having the ELC ratified in 2003) and Switzerland ratified the ELC at the beginning of the same year.

29.3 Searching for Landscape Multidimensionality Within Biodiversity Policies

Landscape ecologists link the concept of landscape to the idea of a specific "spatial scale" through which investigating phenomena and dynamics related to ecosystems: Forman (1995) defines "landscape" as a "mosaic formed by a group of ecosystems that are repeated in space with similar shape, in a mileage range, with identifiable boundaries (...). Specific level of biological organization of life".

At the landscape scale, particular structures and processes – not detectable at other levels of analysis – emerge: ecotones, connectivity between ecosystems, porosity of the landscape matrix and metastability strategies (Ingegnoli 1999).

The conception of landscape as a specific "spatial scale" is frequent within the National Biodiversity Strategies and Action Plans (NBSAPs) across the EU Member States, in some cases representing the only approach to landscape: Bulgaria, Poland, the Netherlands and Slovakia identify main areas of national ecological

[1] This contribution is based on the results of the author's Ph.D. thesis: Seardo B. M., *Paesaggio e Biodiversità: orizzonti convergenti? Concetti, politiche ed esperienze di pianificazione* ("Landscape and biodiversity: converging perspectives? Concepts, policies and planning experiences") (tutor: Prof. Attilia Peano, Arch. Claudia Cassatella), Dottorato in Pianificazione Territoriale e Sviluppo Locale XXIV ciclo, Politecnico di Torino, 2012 (not published).

importance which, once connected to the system of existing protected areas, will be the stepping stones or the buffer areas of the national ecological network.

Another group of countries recognizes explicitly the interdependence of biodiversity conservation with the human-affected territories, interpreting the landscape as a complex matrix where both ecological and cultural values call for safeguard.

Austria, for example, focuses on the synergies among genetic biodiversity, landscape quality and economic development: the preservation of some local breeds of domestic animals is thus encouraged since the breeding activity has indirect positive effects on the preservation of the typical features of rural landscapes and also on tourism.

In the Czech Republic, the reintroduction of landscape natural elements has to be assessed also on the basis of their effects on scenic beauty, while the NBS foresees the coordination with the national programme for the rehabilitation of former quarry and industrial sites, to be carried out on the basis of the restoration ecology principles.

The UK NBS gives specific relevance to the historical role of mankind in the diversification of plant and animal species, through the traditional practices of agricultural land use. The main issue is to encourage farms multifunctionality (rural landscape cover almost the 80 % of the national area) and to apply environmentally suitable land management techniques with special regard to continuity with the historic landscape characteristics.

In Estonia, one of the issues for biodiversity policies is landscape fragmentation due to the rapid reprivatization. Thus, the NBS is used to develop a unified landscape policy aimed both at preserving natural habitats (e.g. by establishing protection zones around water courses and lakes) and maintaining the richness and diversity of the national landscape. Such objectives are pursued linking conservation to agricultural policies which can help to restore and create woodlands, wetlands, dunes, riverbeds, strips, dry stone walls etc.

Since 2009, Finland has at its disposal a new category of protected areas: National Urban Parks are aimed to safeguard special situations of coexistence between natural and cultural values and to "protect and maintain the beauty of natural or cultural landscapes, biodiversity, historic features or other social and recreational values associated to the urban environment". In Germany, the *Eingriffsregelungen*, regulating the ecological compensation of new building interventions, include not only environmental but also scenic measures for the "preservation of the aesthetic character of the landscape".

The Italian NBS focuses on the role of biodiversity within the government of territory, stating as priorities: to adopt a national framework law with a strong environmental mould; to shape a regulatory system in which the development of ecological networks is no longer left to the initiative of local administrators (albeit till now very active in proposing and integrating ecological networks into local spatial plans, as shown in Padoa Schioppa et al. 2010); and to identify an ecological network at the landscape scale to be assumed as the basis for the development of cities, planning and design.

29.4 Which Landscapes for Biodiversity?

The ELC states:

> (. . .) this Convention applies to the entire territory of the Parties and covers natural, rural, urban and peri-urban areas. It includes land, inland water and marine areas. It concerns landscapes that might be considered outstanding as well as everyday or degraded landscapes (art. 2).

Simultaneously, the CBD requires to extend the conservation efforts from protected areas to a wide range of ecosystems: agricultural areas, arid and sub-humid areas, forests, mountains, inland waters, marine and coastal environments and islands. Although the Parties are required to specify objectives and measures in regard to these environments, the cross-cutting analysis of the European National Biodiversity Strategies highlights a more complex picture of environments and landscapes, as many countries show the need to include landscapes which are characteristic to their specific national or bioregional context, encompassing geosites, wilderness, arctic landscapes, cities, productive sites and infrastructures and open landscapes. Moreover, "unique" landscapes are recognized and given proper addresses, such as the Black Sea, the Baltic Sea and the Azores.

Portugal focuses on the importance of its heritage of geodiversity landscapes safeguarded as Regional Natural Monuments of Geological Interest.

France, Germany, Latvia, Lithuania and the United Kingdom dedicate a specific part of their NBS to the urban landscape. On the one hand, Latvia focuses on the monitoring of invasive species within urban ecosystems creating inventories of genetic resources stored in urban parks and botanical gardens, seeing these places as "oasis of biodiversity" (a dualistic view of the urban environment prevails, considering only open spaces). On the other hand, Germany puts greater attention to open spaces within the urban landscapes which need to be preserved from soil consumption with a more "systemic" approach (incentivizing densification in specific areas). At the federal level, a national programme aimed to map those areas not yet fragmented by major traffic arteries is being carried out, in order to establish priority interventions for the safeguard and the restoring of ecological corridors.

Moreover, France is the only country to devote a specific action plan to the urban environment (*Plan d'action urbanisme*) focused on the reform of the planning law and of the financial instruments supporting a sustainable spatial development. The construction of Ecoquartiers and EcoCités and the elaboration of a specific action plan for the ecosystems' multifunctionality in urban areas are other specific issues included in the NBS.

Finally, some NBS reflect a scarce assumption of the landscape approach. In the first version of its National Strategy, Ireland never employed the term "landscape", although having ratified the European Landscape Convention; the biodiversity conservation policy had a very sectoral approach, for example, giving a central role to wildlife suitability of the rural landscape rather than to other aspects. Nevertheless, the current Irish NBS addresses in a more explicit way the connections with landscape issues and policies, for example, recommending the use of the national vegetation classification for the development of landscape character assessments.

29.5 Framing More Landscape-Oriented Biodiversity Policies

The influence of the new landscape conception – promoted by the European Landscape Convention – on biodiversity strategies varies depending on the country, with a general oscillation between two main approaches consisting in: (a) focusing on sectoral efforts addressed to biodiversity conservation or (b) favouring the development of multifunctional landscapes combining biodiversity conservation and the cultural demand of recreation and tourism in natural areas, beauty of everyday life contexts, maintenance of historical features and also economic issues.

In some countries, despite the ratification of the ELC, biodiversity policies are mainly shaped just on the basis of a landscape conception deriving from landscape ecology: while the implementation of scientific instruments (e.g. gap analysis, ecological network model etc.), in the development of place-based policies, has to be considered a positive fact, landscape is basically intended as a type of "geographical scale", instead of the "result of the action and interaction of natural and/or human factors" (ELC, art.1.a). The risk is a scarce consideration of the wider set of (cultural) values attached to biodiversity.

Another group of states pursues a more multidimensional conception of landscape. On the basis of the comparative analysis of the European NBSAPs, an overview of basic principles joining landscape-oriented biodiversity policies can be proposed as follows:

- Enlarging the attention from ecosystems to landscapes and from functions to values: the European Landscape Convention defines "landscape [as] an area as perceived by people" (art.1.a). As biodiversity is a considerable component of people's everyday landscapes, therefore its effective management requires also to include people's perception. As a matter of fact, besides its intrinsic value, biodiversity supports a number of ecosystem functions and services (basic material for good life, health etc.), but the consideration of people's perception requires biodiversity policies to be framed by taking into account also "values" attached to biodiversity (especially intangible ones: cultural, spiritual etc.).
- Addressing policies also to "human-affected" landscapes: biodiversity policies should be addressed not only to the set of ecosystems categories identified by the CBD but also to human-affected landscapes, basing on the evidence of the interdependence between biodiversity and human existence. Thus, cities and urban systems should be a central issue of biodiversity policies, as also highlighted by the recent set up of the "Cities and Biodiversity Outlook project", developed in the context of the application of CBD.

- Highlighting the connections between biodiversity and human (traditional) land-use/management practices, also in a co-evolutionary and historical perspective. This would be useful in identifying preferred processes and proper actions to help maintaining and increasing biodiversity.
- Early detection of the potential trade-offs (synergies and conflicts) between biodiversity policies and landscape policies: e.g. cultural ecosystem services, such as scenic beauty, could be adversely affected by actions aimed to biodiversity conservation. The aim would be a better coordination among sectoral policies and an increase in their global effectiveness.
- Preserving, restoring and creating new sustainable landscapes. A special attention to degraded contexts is needed, focusing not only on their reclamation but also to their active management in favour of new social uses. Both the Convention on Biological Diversity and the European Landscape Convention refer respectively to degraded "ecosystems" (CBD, art. 8) and to "degraded landscapes" (ELC, art. 2). Biodiversity policies willing to be more landscape-oriented should then address also to former productive territories, abandoned industrial sites and so on. As an example, former mining and quarry sites, besides being restored in a functional manner to biodiversity, should also be designed on the basis of a global landscape approach, thus figuring out people's accessibility, enhancing (not wiping out) the identity value attributed by populations and imagining new social and educational uses.

The proposed principles could be the basis for the development of more landscape-oriented biodiversity policies and as such they could be applied in different contexts, during the updating of the National Biodiversity Strategies and Action Plans (NBSAPs) as well as in the development of local biodiversity policies.

At the same time, a useful input should be addressed to the Council of Europe in its efforts to spread the application of the European Landscape Convention: the ELC should be promoted as a support to reinforce environmental conservation efforts, rather than conceived as a risk of watering them down.

Strategies and measures should be developed in order to implement the landscape-oriented approach across nature and biodiversity policies. In doing so, not to be forgotten are the very different starting situations characterizing each national context with respect to culture, policies and institutional frameworks concerning nature conservation and landscape management (Voghera 2006, 2011; Wascher 2000). Some states have already developed an integrated approach between biodiversity and landscape or among nature conservation policies and spatial development policies (consider, e.g. the English Highway Agency having adopted its own Biodiversity Action Plan – HABAP), whereas other states still present a less integrated approach.

Another big difference among EU Member States to be taken into account concerns the funding for biodiversity conservation: Wadron et al. (2013) rank 3 EU Member States among the top 40 underfunded countries in the world (Finland (3rd) but with spending accelerating rapidly, France (36th) taking into account only the mainland territories and Austria (40th) considered to have an agri-environmental-centred conservation policy, making European conservation spendig generally complex). All of these aspects are not a matter of this contribution, but should be deeply investigated in order to better interpret the presented results as well as to strengthen policies integration on a more place-based approach.

References

Antrop M (2006) From holistic landscape synthesis to transdisciplinary landscape management. In: Tress B, Tres G, Fry G, Opdam P (eds) From landscape research to landscape planning. Springer, Dordrecht, pp 27–50

Borrini-Feyerabend G, Gambino R, Phillips A (2008) Landscape dynamic mosaic. Embracing diversity, equity and change. WS, IUCN 4th World Conservation Congress, Barcelona

Brandt J, Tress B, Tress G (2000) Multifunctional landscapes: interdisciplinary approaches to landscape research and management. In: Proceedings, multifunctional landscapes, Centre for Landscape Research, Roskilde, 18–21 Oct 2000

Councile of Europe (2000) European Landscape Convention. Treaty series no. 176, Florence

Forman RTT (1995) Land mosaics. Cambridge University Press, New York

Gambino R (1997) Conservare Innovare. Paesaggio, Ambiente, Territorio. UTET, Roma

Hassan R, Scholes R, Ash N (2005) Ecosystems and human well-being: current state and trends. Findings of the condition and trends working group, the millennium ecosystem assessment series, vol 1. Island Press, Washington/Covelo/London

Ingegnoli V (1999) Ecologia del paesaggio. In: Massa R, Ingnegnoli V (eds) Biodiversità, estinzione e conservazione: fondamenti di conservazione biologica. UTET, Torino, pp 189–210

Jones-Walters L (2008) Biodiversity in multifunctional landscapes. J Nat Conserv 16:117–119

Padoa Schioppa E, Digiovinazzo P, Gentile Ficetola F, Bottoni L (2010) Oltre le reti ecologiche: il paradigma del corridoio diffuso nel paesaggio periurbano. Valutazione Ambientale 16:7–11

Voghera A (2006) Culture europee di sostenibilità. Storia e innovazioni nella pianificazione. Gangemi, Rome

Voghera A (2011) After the European Landscape Convention: policies, plans and evaluation. Alinea, Florence

Waldron A, Mooers AO, Miller DC, Nibbelink N, Redding D, Kuhn TS, Roberts JT, Gittleman JL (2013) Targeting global conservation funding to limit immediate biodiversity declines. PNAS 110(29):12144–12148. Published ahead of print July 1, 2013

Wascher DM (2000) The face of Europe – policy perspectives for European landscapes, European centre for Nature Conservation (ECNC technical report series), Tilburg

Chapter 30
From P-Arks to P-Hubs

Moving from the Idea of the Protected Areas as Noah's Arks to the Concept of Parks as Places of Meeting and Sharing (Hubs) of Wider and More Efficient Frameworks

Paolo Pigliacelli and Corrado Teofili

Abstract The experience of the Italian national parks shows what we might call, using an oxymoron, "the opportunity of limits," i.e., the potential economic and social benefits resulting from the application of a natural resources management aimed at their conservation. The opportunities created by the presence of a protected area that get across in the sustainable management of tourism and high-quality agricultural products promotion have yet to be identified and defined from different aspects (e.g., from the construction industry to services, from transport to crafts). The future of parks management and role in a wider context cannot be meant as going back to the "park = Noah's ark" that just saves animal species regardless of the context, and without the participation of the stakeholders, the future will be having a sort of "park hub" or a hub with many links that radiate throughout the territory and the stakeholders, a kind of management model of sustainable development that can support nature needs, companies, residents, and visitors. In order to run the "park hub," a methodology for evaluating the effectiveness of management as reaffirmed in Jeju is necessary. It is important also to define a clear distinction between the park role as guarantor of the scientific guidelines for the conservation of nature and its role of conveying suggestions and good practices for sustainability.

Keywords Parks management • Sustainable development • National protected areas framework • Green economy

P. Pigliacelli (✉) • C. Teofili
Federparchi-Europarc Italia (Italian Federation of Parks and Natural Reserves), Via Nazionale 230 00184 Roma, Italy
e-mail: paolo.pigliacelli@parks.it; corrado.teofili@parks.it

R. Gambino, A. Peano (eds.), *Nature Policies and Landscape Policies*, Urban and Landscape Perspectives 18, DOI 10.1007/978-3-319-05410-0_30

30.1 Knots or Nodes

The term "node"[1] refers both to the positive concept of vital fulcrum and the negative and crucial point of a problem. Considering these different points of view, the role of protected areas can be challenging; in fact parks could play the role of facilitator of virtuous policies as well as be perceived as an obstacle to the area. There is often a fine line between the two perceptions and it could be independent of the quality of the protected area managers.

A decisive factor lies in the position that the law assigns to the parks, a role that with time, even after the adoption of the Italian Framework Law of 1991, has had continuous integrations with a goal closer to the local authorities' interests instead of toward the reinforcement of the management bodies as was at the very beginning of the parks' establishment.

The management of species and habitats, as is well known, ignores the administrative limits and allowed the managers to raise considerable experiences in coordinating themselves with other subjects in the same jurisdictional area. This actually precious wealth of experiences is even enhanced by the constant practice of collaboration with a land shaped by history and ancient cultures. Biodiversity preserved by the Italian parks is strictly intertwined with the elements that characterize the land itself, and Italian protected areas work as a system but, at the same time, enhance the features of the place.

One example, among many, is the success of scientific, technical, and communication activities that allowed the return of the Apennine chamois on the entire ridge of the Apennines. A system design, not depleting the Abruzzo National Park, has instead enhanced the role of this park along with that of the other protected areas that have contributed to the brilliant result: Majella National Park, Gran Sasso Laga National Park, Sirente-Velino Regional Park, and Monti Sibillini National Park.

For the Italian parks, working in a logical framework means knowledge sharing for a common goal in the interest of all of them; in fact, if the chamois had remained in some valley of the Park of Abruzzo, its existence would have been at risk due to any trivial epidemic disease. Now, the chamois is back actually spread on all the Apennines. Moreover, perhaps, it is not a coincidence that almost 20 years ago Legambiente and Federparchi conceived, along with some administrative Regions, an even more advanced model of a logic system: the project APE "Apennines Park of Europe" (CED PPN 2003), a real "Green Infrastructure" that, starting from land conservation, is intended to create an opportunity for local development, based on Apennines features.

Therefore, working in a logical framework is not only cost saving, but, especially for issues like land management and sustainable development, it means

[1] Please note that in Italian the translation for the word "node" (*nodo*) has also the same meaning of "knot."

increasing the effectiveness and efficiency of the actions carried out with the participation and involvement of all the local stakeholders.

The recent summit of Rio +20 has already declared the "failure" of governance based on sectorial environmental policies; in essence it has been established that it cannot achieve significant results in environmental governance without the engagement, with an active role, of the institutions and actors involved.

Paradoxically, the current reorganization of the Italian public apparatus could seriously undermine the essence and the innovation, instead of learning from the experiences of the parks. Recently, in some regions and even at the national level, it is slowly growing a new model of protected areas where the management bodies, in charge for representing and sharing policies, are heavily scaled down and their representatives, including presidents, are merely reduced to volunteers.

On the contrary, in order to exploit the potential of protected areas designed as "nodes" in a positive sense (hence, as the *fulcrum* of the flow of services at several levels), it is necessary to promote the actual added value that, at local level, can be ensured only by the body responsible for the protection of natural values.

The misunderstanding done by many management bodies of protected areas lies in the perception of assigning the parks an improper role of agents of local cultural and social development neglecting, at same time, their institutional mission of nature protection; unfortunately, this misperception is quite widespread and in some cases even supported by the parks themselves.

This is an approach that not only does not take adequately in account the results of the international recent debate about the role and management of protected areas but also implies a risky depletion of the peculiar functions of the managing bodies that would become useless. Also considering the presence of other institutional bodies could perform more efficiently the same functions of promotion at local scale.

Conversely, a stronger role assigned to the protected areas as guarantors of biodiversity management and sustainable anthropogenic activities would be, as a result, also the strengthening of the parks as important points of reference for the empowerment of communities in the sustainable management of natural resources, in the enhancement of traditional knowledge, and in the definition and implementation of strategies for biodiversity conservation.

One of the issues that should be taken in account, considering the Italian current situation about the parks definition and management, is the complex arrangement and classification of the Italian parks that are established under different laws and geographical scale. Nowadays in Italy, there are dozens of different categories of PA, from the national parks to the regional parks to the urban protected areas referred to IUCN definition. Further, we must consider the Natura 2000 Sites of European interest.

And even though the IUCN has developed six PA categories for a globally standardized classification that define different levels of protection, depending on the aims and the degree of protection, these are barely applicable to the European context. The relatively recent revision of the IUCN classification system poses a challenge to the categorization of PAs that were established before the IUCN classification.

Considering the European historical development of the protected areas' national and European systems, very often the park designation by the autonomous local communities according to different conservation laws is a crucial point in explaining patterns of local current regulations. The overall weak correspondence of management practices to IUCN categories at Italian level also lead to believe that management is influenced by factors other than international standards. Among other factors, the lack of a procedure for IUCN category assignment increases the variation in management planning among different regions and by different managers.

From this point of view, the efficiency and results of park management should be assessed relying on different criteria than the IUCN standards; one of the keys to interpret the management process could be, for instance, the manager's ability to involve people into the management procedures.

In fact, there are already a number of good practices in Italy that have attempted an active involvement and participation of local communities in defining strategies of development and conservation of land, with some good practices but also with some failures.

The validity of the presence of a park that "does its job" (conserving nature) is also confirmed by the many forms of "green" employment opportunities, related to the direct or indirect management of environmental services for the conservation of biodiversity or other services to support the community.

There are yet companies or cooperatives professionally engaged in projects for the conservation, management, and monitoring of biodiversity, but the relatively new sector yet to be developed for protected areas is the so-called green economy. According to some studies, it could produce more jobs at a lower cost compared with those expected by the current Cohesion Policy and the Common Agricultural Policy (CAP) of the European Union.

Hence, protected areas could play, more effectively than any other institutional or private entities, a brilliant role of driving or experimental model for the realization of sustainable infrastructure and environmental programs in agriculture, transport, and building rehabilitation that could reach almost 30,000 employees at national level.

In order to promote this virtuous process, along with ensuring the scientific approach of the management, it needs a process of sharing and training for companies and citizens aimed to maintain the environmental sustainability of the local resources, an activity that could be promoted by the "community of the park" if planned and organized as representation of local trade associations, professional associations, and environmental and cultural associations, as well as local authorities as it is currently.

30.2 New Skills for New Contexts

Unfortunately, the process of accreditation of protected areas within the social, political, and economical contexts is still too slow, along with the definition of the relative synergies derived from collaboration with the protected areas.

The official list of Italian Protected Areas (EUAP, MATTM Italian Ministry for Environmente and Territory and Sea 2013) contains of 871 Italian protected areas with 16 different subcategories, to which other 52 categories should be added according to the regional regulations. Even the most commonly used keywords used in the regulations on protected areas indicate a complex articulation of the purposes and expertise of protected areas:

> (. . .) organize, participate, develop, replace, improve, make up, conserve, restore, sustain, develop, propose and monitor etc.

These are activities that substantially affect the objectives of a protected area and, therefore, its necessary skills that, with a considerable effort, can be synthesized and summarized as follows:

- Providing a broad and effective protection and enhancement of the environment (expertise skills)
- Organizing actions, defining resources, and measuring the results (planning skills)
- Promoting and updating the model of management (management skills)
- Encouraging participation (relationship-wise skills)

It is clear that the various skills should complement each other in a consistent and recognizable way. That is not a trivial point considering that, less than 50 years ago, during a conference on protected areas the following words have been used:

> the elementary need for a national park must be a unique district, protected by armed guards, within which the complex of plant and animal life can be maintained and evolve spontaneously, preserved by any human action.

During the more recent years, this approach has changed and finally accelerated by the Framework Law on Protected Areas (1991) and subsequent amendments; however, there was not a parallel process of updating the skills and training needs of the staff of the management bodies.

Nowadays, all the elements are available in order to define a more precise analysis of the tools, services, and facilities necessary for the establishment of a management organization that can contribute on improving the efficiency and effectiveness of the system of the Italian parks.

The ultimate goal is hence to create a close correlation between the variables that define the management of a protected area, which change over time and, if not properly rebalanced, could lead to the collapse of the entire system. In order to encourage the conditions to enable the improvement of the parks toward new challenges, it is crucial to ensure a whole consistency between the aims of a protected area (as defined by the law), the natural capital to be managed (as defined by monitoring), and economic resources available (defined by the political and economic contexts).

We should imagine a mechanism, as flexible as possible, that works according to the variables mentioned above, a model that matches the efficiency standards imposed by the needs of the natural capital and, consequently, provides the policy makers useful elements in order to regulate the other two important variables: the legislative and economic issues with the aim to ensure the achievement of the identified objectives.

The challenge is therefore having national parks with new skills and consistent professional and financial resources; those parks could become a management model to be proposed in other contexts as well.

30.3 The Protected Areas in the Country's Service

In order to put into effect the abovementioned model, it is necessary to keep in mind the context in which managers of protected natural areas operate, more and more influenced by the different fields of public management and other external circumstances. In some cases, these are local-scale pressures and hence it does not affect the institutional purposes of the protected area; in other cases, interest groups and influential stakeholder determine the management of the park itself by imposing a set of priorities that are not perfectly consistent with the objectives and the mission of a protected area. However, the growing awareness of the strategic importance of the conservation of natural resources has allowed to establish the "crucial role" of protected areas as "one of the fundamental tools and strategies essential for the conservation of biodiversity and ecological processes of the planet." The complexity of the issues and situations that make up the daily task of a national park, as well as the emergence of a work that is often underestimated, highlights the opportunity to translate the institutional purposes of conservation in a more effective and tangible contribution to new scenarios that affect the decisions of governments, businesses, and citizens.

It is well known that the growing demand for energy and natural resources for food production processes has been placed at the center of the limits of the human growth. Therefore, the environmental sustainability of the solutions proposed in order to address the economic and social processes toward sustainability can be summarized in two categories: mandatory laws and economic incentives. These two pillars are the basis of the "green economy": a system of production that takes into consideration the environmental impact and the potential environmental damage caused by the entire cycle of processing the materials from their extraction, transport, and transformation into energy and products up to the possible environmental risk produced by their definitive elimination or disposal (UNEP 2008).

The experience of the Italian national parks shows what we might call an oxymoron "opportunity of limits," i.e., the potential economic and social benefits resulting from the application of a natural resources management aimed at their conservation. The opportunities created by the presence of a protected area that get across in the sustainable management of tourism and high-quality agricultural products promotion have yet to be identified and defined from different aspects (e.g., from the construction industry to services, from transport to crafts).

In the world, more than 200,000 protected areas have been established and they are in 240 different countries, most in the UN member countries or those participating in the London Olympics Games. With these numbers, it must be admitted that the tool "park" has reached a level comparable to other basic life services, such as health, education, and justice, and as far as those services. Nowadays, parks must deal with the needs of the growing population groups with the inevitable integration of tasks and objectives that show a complex view with hundreds of different types of protected areas.

The future cannot be going back to the "park = Noah's ark" mindset that just saves animal species regardless of the context, and without the participation of the stakeholders, the future will be having a sort of "park hub" or a hub with many links that radiate throughout the territory and the stakeholders, a kind of management model of sustainable development that can support nature needs, companies, residents, and visitors.

Promoting and enhancing the political system frameworks, updating the strategies to the new scenarios of environmental, economic, and social issues, could represent a significant opportunity both at national level, with the launch of a green economy based on the areas, and at the international level according to "nature +: nature-based solutions for a new era of conservation, sustainability, and social and economic development," one of the approaches discussed at the 2012 IUCN World Conservation Congress (IUCN 2012) organized in Jeju (South Korea). The Congress of Jeju marks a substantial shift of conservation policies and assigns a crucial role to the protected areas as keystones (purposeful nodes) and promoters of actions for the protection of natural resources, which identifies the parks as real "hub" of strategies developed on several levels.

The measures identified in Jeju "parks hub" can be summarized in the following points:

1. Increasing the effectiveness of participation and knowledge for environmental strategies.
2. Intensifying the efforts to increase knowledge on species, habitats, and ecosystems in order to provide decision makers the tools for effective management.
3. Raising awareness about the conservation of biodiversity cannot be an obstacle but an opportunity to achieve broader social objectives: "protecting the environment is everyone's business and humankind is dependent on nature."
4. Valuing ecosystem services also as a way for a recognition of the role of protected areas.
5. Strengthening the three dimensions of sustainability: economic growth inclusive and equitable, development and social integration, and integrated and sustainable management of natural resources and ecosystems.

6. Increasing the involvement of the private and public sectors for the transfer of "green" technologies and to share knowledge, experience, and skills in order to integrate biodiversity and ecosystem values in production and consumption supply chains.

In order to run the "park hub," a methodology for evaluating the effectiveness of management as reaffirmed in Jeju is necessary. It is important also to define a clear distinction between the park role as guarantor of the scientific guidelines for the conservation of nature and its role of conveying suggestions and good practices for sustainability.

References

CED PPN (2003) APE. Apennines Park of Europe: creating environmental infrastructures and enhancing the Apennines' role in Europe. Alinea Editrice, Florence

IUCN (2012) World Conservation Congress. Jeju Island, 6–15 Sept 2012

MATTM Italian Ministry for Environment and Territory and Sea (2013) Elenco Ufficiale delle Aree Protette (English version: Official list of Italian Protected Areas). Available via http://www.minambiente.it/pagina/elenco-ufficiale-delle-aree-naturali-protette-0. Accessed 20 Jan 2014

Repubblica Italiana (1991) Framework law on protected areas (Italian version: Legge Quadro sulle Aree Protette). Law no. 394/1991

United Nation Environment Programme, UNEP (2008) Global green new deal – environmentally-focused investment historic opportunity for 21st century prosperity and job generation. London/Nairobi

Chapter 31
The experience of the European Landscape Observatory of Arco Latino

Domenico Nicoletti

Abstract The European Landscape Convention, in defining the "Landscape policy," aims, among other things, to organize European cooperation in this field. Arco Latino is an area of political cooperation between level II administrations of the Western Mediterranean in which joint positioning in defense of common interests are articulated. The Association territory covers 320,000 km^2 incorporating 43.5 million inhabitants. The various members of Arco Latino include 8,000 municipal bodies. Within the ambit of the Land and Sustainable Development Commission in 2005 in Barcelona, the province of Salerno proposed a process for cross-border cooperation for the Euro-Mediterranean landscape via the creation of a European Landscape Observatory (OEP) Arco Latino already recognized during the Congress of Local and Regional Authorities for the Implementation of the European Landscape Convention in Strasbourg on 27 May 2004: Document CG (11) 12. In February 2007, in Vietri sul Mare (Italy), the GT "Natural Areas" Arco Latino, subscribed an "Agreement for Landscape" between the promoters of the Observatory to organize European cooperation in this field and promote the integration of nature and landscape policies. This start-up has led to the implementation of numerous programs, such as the project by ENPI CBC Med IB/1.3/350.

Keywords Landscape policies • Cooperation • Observatory • Euro-Mediterranean cooperation

The European Landscape Convention defines "Landscape policy" as the formulation, by competent public authorities, of general principles, strategies, and guidelines which permit the taking of specific measures aimed at the protection, management and planning of landscapes, and the organization of European cooperation in this field.

D. Nicoletti (✉)
European Landscape Observatory of Arco Latino, Certosa di San Lorenzo, 84034 Padula, Italy
e-mail: domeniconicoletti@libero.it

R. Gambino, A. Peano (eds.), *Nature Policies and Landscape Policies*,
Urban and Landscape Perspectives 18, DOI 10.1007/978-3-319-05410-0_31

Arco Latino (Arco Latino 2007a, b)[1] is an area of political cooperation between level II administrations of the Western Mediterranean in which joint positioning in defense of common interests is articulated. Its association and programmatic potential allows for direct dialogue with European institutions in order to incorporate the Mediterranean dimension in the local policies.

The local administrations that adhere to Arco Latino become part of a platform able to increase the capacity of their members and to internationalize their activities and strategies, combining their efforts towards a previously defined common goal. Arco Latino is a territorial organization, which is responsive to the present and directed towards the future, which will allow for the definition of strategies and joint initiatives of communication, lobbying, and the pilot actions. At present, Arco Latino is composed of 71 active members (42 Italian, 18 Spanish, and 12 French). The Association territory covers 320,000 km^2 incorporating 43.5 million inhabitants and 8,000 municipal bodies.

Within the ambit of the Land and Sustainable Development Commission in 2005 in Barcelona, the province of Salerno proposed a process for cross-border cooperation for the Euro-Mediterranean landscape via the creation of a European Observatory of Landscape (OEP Arco Latino), already recognized during the Congress of Local and Regional Authorities for the Implementation of the European Landscape Convention in Strasbourg on 27 May 2004.[2] In February 2007, in Vietri sul Mare (Italy), the GT "Natural Areas" Arco Latino subscribed to an "Agreement for Landscape" between the promoters of the Observatory, which identifies the location at the Certosa di San Lorenzo in Padula, in the province of Salerno, Italy, as the central offices. The Territory and Sustainable Development Commission for Arco Latino met on 12 June 2008 in Viterbo as a result of the determination of the province of Salerno for a project of Observatory; the implementation of this process has allowed to initiate and stimulate programming activities in accordance with Recommendation CM/Rec (2008)3 of the Council of Europe's Committee of Ministers, on the guidelines for the implementation of the European Landscape Convention (adopted on 6 February 2008), which describes the mission of observatories, centers, or institutes:

- Describe the condition of landscapes at a given time.
- Exchange information on policies and experience concerning protection, management and planning, public participation, and implementation at different levels.
- Use and, if necessary, compile historical documents on landscapes which could be useful for knowing how the landscapes concerned have developed (archives, text, photographs, etc.).
- Draw up quantitative and qualitative indicators to assess the effectiveness of landscape policies.

[1] http://www.arcolatino.org

[2] Document CG (11) (12), the draft resolution submitted by L. Becker (Hungary, L, GILD).

- Furnish data leading to an understanding of trends and to forecasts or forward-looking scenarios.

Exchanges of information and experience between states, regions, and territorial communities, which already take place, should be based on exemplarity but should always consider the political, social, ecological, and cultural context of the original landscape.

The choice of the composition of observatories is a matter for the administrative bodies concerned but should allow for collaboration between scientists, professionals and technicians, and public authorities.

Territorial cohesion, together with the principles of local autonomy and subsidiarity, is a crucial point in order to give local governments a new role. The landscape, as stated in the European Landscape Convention (CoE 2000), in addition to being an instrument for European cooperation,

> has an important public role in the cultural, ecological, environmental and social fields, and constitutes a resource favorable to economic activity and whose protection, management and planning can contribute to job creation.

For Arco Latino, the implementation of the European Landscape Convention is a fundamental challenge for strengthening territorial cohesion in the context of the Euro-Mediterranean and in particular promotes (Nicoletti 2003):

- More conscious and stronger cooperation activities for the development of local cultures which is an essential part of the cultural and natural heritage of Europe, thus contributing to the welfare and satisfaction of humans and the consolidation of the European identity
- Recognition that the landscape is an important element for the quality of life of people everywhere: both in urban areas and in the countryside, in degraded areas and in those of high quality, and in areas considered exceptional and in those which are more familiar
- Recognition that the quality and diversity of European landscapes constitutes a common resource with a need for cooperation for their protection, management, and planning

The work carried out within the framework of the strategic orientation "Territory and Sustainable Development" follows this direction and is strengthened by actions and initiatives of awareness and analysis of quality indicators such as those of economy, population and the environment, infrastructure, and regional imbalances between rural and urban and peri-urban areas with a thorough investigation of natural protected areas. All of these tools and alliances made available by Arco Latino have allowed to build a scenario of the dynamics of a large area, which has been useful to the decision-making and for adapting policies, bearing in mind the reality of the territories concerned, as well as their connections with other territories.

These include the "Study of the economic benefits induced by the Protected Natural Areas of Arco Latino," which presents a set of indicators applicable in each of the 200 protected areas surveyed. The Observatory, in cooperation with other Arco Latino institutions, has, over the years, activated a number of operational and technical initiatives and actions to assess the environmental and sociocultural impacts of these natural protected areas on the territories of Arco Latino. These include the "Study of the economic benefits induced by PAs of Arco Latino." In this experience, three categories of processes are estimated quantitatively: firstly, the direct economic activities located there; secondly, the processes that generate the ecological services of natural systems; and the third group of indicators has the task of quantifying the benefits for the well-being of citizens, based on an evaluation of the willingness to pay. The project identified indicators for two parks in Catalonia – that of Montseny and that of Garraf – and has also produced an assessment manual to facilitate its application by the institutions managing the protected areas of Arco Latino. In 2009, the European Landscape Observatory decided to invest in the European Masters in Landscape at the Interdepartmental Center for Environmental Sciences (CISA) at the University of Salerno, in cooperation with the universities of Lyon, Strasbourg, and Granada, actualizing both the principles of the European Landscape Convention and the challenge of Arco Latino for the creation of new jobs and employment opportunities for the young graduates facing the government and care of the land. Through a complex and coherent framework of competencies (ecological, landscape, spatial planning, geomorphological, perceptual, philosophical, sociological, and anthropological), the Master has raised the issue of updating the contents of the commitment of the institutions and offering a wealth of practical steps of European experience to be implemented in the territories of Arco Latino, "Guidelines for the implementation of the ELC" approved by the Council of Europe. In the course of running the ENPI project, "Local Agenda 21 in Territorial Planning in Energy and Waste Management" (Ref Number: IB/1.3/350) has been functional in seeking regular exchanges of information and experience between member states and non-EU states based, in tune with the political, social, ecological, and cultural landscape.

For Arco Latino, the European Landscape Observatory is a link within the themes of landscape, it provides networks as the ideal platform to combine efforts and promote strategic projects at a local level which are difficult to achieve individually. It provides access to in-depth discussions, which can generate a more immediate impact in the local environment. It forms a critical mass capable of intervening in the formulation of European policies, it increases visibility among similar institutions and the authorities of the higher levels of government, and it optimizes the resources and technical capabilities and policies of each of the organizations involved.

The Observatory of Arco Latino, within the Strategic Plan of Arco Latino 2015, "A Structured and Innovative Mediterranean", investigates the issues of biodiversity and Mediterranean resources conservation in a highly polluted context, the sustainable management of water resources, the sustainable use of ecosystems, and protection of natural areas and landscape, bringing added value to the technical profiles and local politicians who are part of the association and through collaboration with other organizations that share these principles and want to create job opportunities. In fact, since the beginning, Arco Latino has endeavored to consolidate its profile in this sense, investing more and more in projects based on networking and strengthening of members and initiatives.

References

Amoroso B, Gallina A (2000) The impact of free-trade area on EU social cohesion and economic co-operation in the Wider Europe, first FEMISE conference. Forum Euro-Mediterranéen des Instituts Economiques, Padova

Andria A (2007) The European Landscape Convention. Diari dall'Europa, Gruppo ADLE, Edizioni Segno Associati

Arco Latino (2007a) L'Arco Latino: an action plan from a territorial strategy. Report 2005–2006, Nuoro

Arco Latino (2007b) Better understand our specific Mediterranean. Building a quality policy at the service of our country starting from his cards to play and its specificity Mediterranean: a Latin Arc quality. Report 2005–2006, Nuoro

Commission of the European Communities (2003) European neighborhood policy strategy paper, communication from the commission. Brussels

Council of Europe (2000) European landscape convention. Treaty series no 176, Florence

Council of the European Union (2003) Wider Europe – new neighbourhood. Brussels

Giordano A (2004) Better understand our specific Mediterranean. Building a quality policy at the service of our country starting from his cards to play and its specificity Mediterranean: a Latin Arc quality, Napoli

La Camera F (2005) Sviluppo Sostenibile. Origini, Teoria e Pratica. Editori Riuniti

Matvejevic P (2007) Il Processo di Barcellona: speranze e disillusioni nel Mediterraneo. Arco Latino, Argra Trading

Nicoletti D (ed) (2003a) Il governo delle aree naturali protette: esperienze a confronto. ESI, Napoli

Nicoletti D (2003b) Convenzione Europea del Paesaggio: dall'identità allo sviluppo sostenibile. Den, Napoli

Nicoletti D (2008a) Arco Latino e le Politiche Euro-Mediterranee per lo Sviluppo Sostenibile: il Caso del Parco Nazionale del Cilento e Vallo di Diano in Politiche del Paesaggio. Il Denaro Libri

Nicoletti D (2008b) Politiche del Paesaggio. Il Denaro Libri

Ortega M (2003) A new EU policy on the Mediterranean? In: Batt J et al (eds) Partners and neighbours: a CFSP for a Wider Europe, vol 64, ISS Chaillot papers. European Union Institute for Security Studies, Parigi, Paris, pp 86–101

Petri P (2004) Trade strategies for the Southern Mediterranean, OECD technical papers, no. 127. Organization for Economic Cooperation and Development

Philippart E (2003) The Euro-Mediterranean partnership: unique features. First results and future challenges, CEPS working paper no. 10

Prodi R (2002) A proximity policy as key to stability, sixth ECSA World conference. Brussels

Provincia di Salerno (2008) "Premessa" di Progetto OEP, Documento ufficiale presentato in occasione dell'incontro di Arco Latino. 12 June, Viterbo

Sacco G (2008) Méditerranée. Commentaire, Paris 31(131):113–118

Saitta A (2007) Il ruolo dei governi locali nelle reti internazionali. Arco Latino, Argra Trading

Vezineh A (2007) La costruzione della rete. Arco Latino, Argra Trading

Chapter 32
Crosscutting Issues in Treating the Fragmentation of Ecosystems and Landscapes

Gabriele Paolinelli

Abstract Since the 1990s, many issues have emerged concerning fragmentation in nature and landscape studies. The dominance of the ecological standpoint in dealing with these problems emphasizes the importance of conserving ecosystems and biodiversity. Encompassing ecosystems and landscape fragmentation involves identifying several types of sizable systems. A landscape concept comprehensive of natural components, processes and dynamics, and the crosscutting integration of landscapes in the policies that affect them are the primary conditions for a "bridge" between nature-oriented policies and culture-oriented policies. Sustainable relationships among communities and their habitats can be developed through a comprehensive landscape-based planning.

Keywords Ecosystem fragmentation • Landscape fragmentation • Comprehensive planning • Land policies • Integration

32.1 Fragmentation: From Nature to Landscape and Back

This paper will not undertake to review the considerable literature on fragmentation. Some references will be made in discussing the concept of fragmentation with a view to elaborating integrated policies and implementation for nature protection and landscape care.

Since the 1990s, many issues have emerged in nature and landscape studies in regard to fragmentation (Saunders et al. 1991) as "a primarily anthropogenically-driven phenomenon" (Young and Jarvis 2001); the discussion of its meaning/s is growing (e.g., Cook and van Lier 1994; Forman 1995; Dramstad et al. 1996; Collinge 1996; Fahrig 1997; Romano 2000; Haila 2002; Hidding and Teunissen 2002; Gulinck and Wagendorp 2002; Taylor 2002; Paolinelli 2003, 2012; Battisti and Romano 2007).

G. Paolinelli (✉)
Department of Architecture, University of Florence, Via P.A. Micheli, 2, 50121 Florence, Italy
e-mail: gabriele.paolinelli@unifi.it

R. Gambino, A. Peano (eds.), *Nature Policies and Landscape Policies*,
Urban and Landscape Perspectives 18, DOI 10.1007/978-3-319-05410-0_32

Fragmentation has been defined as "the breaking up of a habitat, ecosystem or land-use type into smaller parcels" (Forman 1995) and is considered a major cause of the decrease in biodiversity (Wilcox and Murphy 1985). Landscape ecology recognizes that fragmentation may also be due to natural factors, but it

> (. . .) has become an international land policy issue because of the widespread alteration of land mosaics by human activities. (Dramstad et al. 1996)

Thus, fragmentation is also present in space planning, in relation to the dynamics of natural and rural mosaics, urban patterns, and farmland and farming patterns (Romano 2000; Carsjens and Van Lier 2002; Paolinelli 2003; Battisti and Romano 2007). Pat D. Taylor (2002) has highlighted the tendencies and risks of misuse of the term "fragmentation" in landscape planning and design.

Today's reality depends on much broader and denser networks of material, energy, and informational relationships than in the past. The issues of access to resources are complex, controversial, and transcalar, in some instances acquiring global importance. The last two decades have seen the development of digital information networks in addition to the older networks for the mobility of people and goods. The new information and communication technologies facilitate living in low-density settlements far from the workplace, thus augmenting landscape fragmentation due to urban sprawl (Hidding and Teunissen 2002).

Sustainable development cannot realistically be realized as long as economic development and the conservation of natural and cultural legacies are in opposition to one another in territorial processes. Not only can differences between nature and landscape policies be managed by alliances, but they can also have greater socio-cultural influence in dealing with economic pressures.

However, a true balance in the development of integrated policies concerning nature and landscape must take into account a diversity of values: public versus private, collective versus individual, insiders' versus outsiders', industrial versus residential, recreational versus residential, etc. Thus, processes such as urban sprawl, infrastructural growth, and agricultural homogenization may or may not be considered technically as factors of landscape fragmentation. Then, the same processes may or may not be rejected socially because of their recognized negative effects. Cultural, technical, and scientific considerations often do not harmonize, and by the same token, at the social level, there may be a multiplicity of cultural perceptions. These may lead to conflicting goals and pressures. For instance, from a private point of view, landscape fragmentation due to urban sprawl may not be considered an important enough motive to warrant its prevention or limitation, if the prevalent value is to live in suburban settlements with low-density houses and gardens (Hidding and Teunissen 2002). The same holds true for biodiversity, since the effects of landscape mosaics fragmentation make themselves felt not only on spatial and temporal scales but also by types of organisms (Olffa and Ritchieb 2002).

Landscapes are complex systems, whose evolution may to a limited extent be predicted. The different perceptions of their changes are notable variables of the relative indeterminateness of the problems of managing them. Living organisms and the different ecological systems they constitute react with different capacities for adaptation and resilience to landscape dynamics and their more or less present and influential chaotic connotations. Humans exert peculiar capacities for planned action by projecting a contrast between chaos and plan (Antrop 1998) toward the goal of plan-into-chaos, with an increasing awareness of the inevitable uncertainty of their development (Brancucci and Gibelli 2006).

32.2 A Comprehensive Vision for Dealing with Fragmentation

According to Gulinck and Wagendorp (2002), fragmentation is neither a new concept nor simply an indicator of the status and dynamics of ecosystems.

> Fragmentation of the ecosystem produced by total urbanization of the territory seemed to find comparison in the chipping away of the social fabric and relationship systems, in the extreme diversification of interests, stakes, and social demand. (Gambino 2013)

The concept of landscape fragmentation has been interpreted as an extension of the basic ecological acceptance of the phenomenology to the range of principal landscape features (Paolinelli 2003). The following ten topics identify a concept of landscape fragmentation (LF), proposed here also as a tool for delineating, organizing, and encouraging possible alliances between nature and landscape policies.

1. Landscape fragmentation has historical roots, but only since the last century have they become widespread and emphatic, generating pathological effects over broad areas of the Earth. The identification and treatment of the multiple dysfunctions, which go into LF processes, have become increasingly urgent and manifest over the past five decades. Recent evidence has brought to light no proof that these phenomena have significantly decelerated. On the contrary, economic and social transformations have simply altered the causes, which, also by reason of their prevalence within the principal historical types, tend to aggravate these phenomena.
2. Landscape fragmentation can be caused by natural factors and processes as well as anthropic ones, but priority must be given to identifying and treating the latter, for these generate dysfunctional consequences which are often irreversible or only partially reversible, given that nature is less equipped to neutralize them through its normal ecological capacities. In addition, these effects also damage landscape features of purely cultural interest and as such call for appropriate safeguards and conservation.
3. Landscape fragmentation cannot be attributed exclusively to housing and infrastructural development, since it can also be caused by hydraulic, agricultural, and forestry installations, as well as by the socioeconomic exploitation of the earth's resources and land values and by the consequent modes of management and commercialization.

4. Landscape fragmentation is both a phenomenon and a concept and as such is also an indicator. It possesses a broad potential for interpreting reality. At the same time, LF is an intrinsically partial phenomenological point of view with respect to the complex of demands expressed by the processes of planning policies and actions that affect nature and landscape.
5. Landscape fragmentation is not a new concept to science and technology, but the circulation of theories, applications, and observations, which refer to it, is recent and demands empirical confirmation and theoretical-practical elaboration.
6. Landscape fragmentation is a systemic concept, which fosters the development of complex analytical and diagnostic fact-finding of landscapes. The dominance of ecological sensitivity to the problems, in regard to which LF studies have been undertaken, indicates the importance of the theme for the conservation of ecosystems and biodiversity, but this does not mean that ecological factors have exclusive pertinence in the phenomenon, whose many effects can be observed in landscapes.
7. Landscape fragmentation is understood as a critical concept, an indicator of dysfunctional transformations of landscape systems, distinct from the positive concept of landscape heterogeneity, which is sometimes used as a synonym of it.
8. Landscape fragmentation is a dependent concept and consequently is such also an indicator or rather constitutes a family of dependent indicators. The position and study of the problems can express qualitative and quantitative objectivity in relation to the criteria and parameters adopted and rendered explicit as points of reference. The estimate of criticality of the phenomena changes depending on the paradigms adopted at the scientific and technical level and/or on the perceptions expressed at the social level by the subjects involved and requires clear explication and argumentation.
9. The systemic concept of landscapes resulting from spatial-temporal relationships between environmental, economic and social factors, and processes makes it possible to formulate indicative LF properties on the basis of major complementary thematic groups, safeguarding them from the limits of sectorial analytical simplifications. In this context, LF already comprises a family of synthetic indicators at the level of the main type of landscape features, whether naturalistic and ecological, historical and archeological, scenic and panoramic; a more in-depth diagnostic synthesis of the information, which they provide, is not advisable for its intrinsic risks of inutility in planning and design applications.
10. The inappropriateness of a fact-finding synthesis of the information provided by the aforesaid main families of thematic LF indications suggests the need for their combined interpretation in the planning and design processes, which must lead the thematic formats of specific landscape systems back to a holistic view of landscapes, whereby they are considered as unified, inseparable elements, unless and only for instrumental reasons and purposes.

Since "the subject of land transformation and fragmentation is significant to all human issues that involve land" (Forman 1995), it may be significant for relationships between nature and landscape and thus also for the integration of their policies.

The systemic configurations of landscapes affect their different qualities (Paolinelli 2012) and render possible the elaboration of fact-finding and project-planning interpretations referable to the network model (Fabbri 2010; Gambino 2013). All landscapes are characterized by structures, functions, and changes and display spatial mosaics that can be traced to the systemic model's "patch-corridor-matrix" (Forman 1995). When there are interferences capable of disturbing or altering landscape systems, the fragmentation of their components and/or relationships can compromise the emergent properties because of the effects, which the reduction into parts has on the whole.

Identifying LF factors and landscape receptors sensitive to them does not require specialized analytical bases distinct from the typical ones of landscape studies for plans and projects. The techniques of overlay of analytical spatial databases and specialized metrics are useful for landscape diagnoses (e.g., Davidson 1998; Bissonette et al. 1998; Jaeger 2000; Cumming and Vernier 2002; Romano and Paolinelli 2007; Kima and Ellisb 2009; Lina et al. 2009; Poelmans and van Rompaey 2009; Yeha and Huangb 2009; Nga et al. 2011). Their combined utilization makes it possible to undertake the study of landscape fragmentation on the basis of the different spatial-temporal scales of the phenomenology.

In view of the technical application, it helps to clarify the reasons why studying a range of specific phenomena under the same explicitly critical profile can be useful in identifying priority aims of plans and projects. First of all, ecosystems alone have little chances of evolving toward improved stages without profound deterioration where and when anthropic factors have considerably compromised or weakened the landscapes' natural structures and functions. Besides, treating purely cultural phenomena needs human strategies and actions. In all such cases, the prevention of hazards, the containment of possible effects of plans and projects, and the treatment of the states of landscape fragmentation (Paolinelli 2003, 2012) are in urgent need of policies and actions. There are, ultimately, clear direct and indirect priorities for developing active policies concerning landscape criticalities. The former are related to compensation and support objectives aimed at improving the resilience and adaptation capacities of landscapes made dystrophic by alterations. The latter are related to protection and conservation objectives of excellent or ordinary positive qualities of efficient landscape matrices and systems. Protection must not exclude innovation, but, to the same extent, transformation must not exclude tradition.

Landscape planning and design must not fail to take into account consequentiality, completeness, and reiteration of processes as essential conditions for land policies efficacy (McHarg 1969; Steiner 1994; Palazzo and Steiner 2011). Landscape structures, functions, and changes must be identified as the informational bases of transparent, documented processes. The limits and possibilities of landscape transformation must be stressed as the terms of reference of processes whose goal is the sustainability of territorial models and, conjointly, the enduring use of resources. Policies and actions must be defined within processes of public participation, which enables them to take root in the sociocultural humus and favors coherent actuation, as well as by processes of multi-criteria technical evaluation, which foster an environmental, social, and economic balance essential for their sustainability.

32.3 Starting from Primary Needs and Opportunities for Alliance

Nature and culture generate landscapes and their evolution. No landscape is free of the influences of natural factors and processes, even the most "artificial." By the same token, no landscape is free of the influences of cultural factors and processes, even the most "natural." Thus, nature and landscapes are intimately related. Landscape studies can improve our knowledge of this reality, whose relationships need to be understood, not invented. Rather, there may at times be no alliance at all among nature and landscape policies, or worse, there may be conflicts. Resolving this contradiction requires constructive ideas and strategies, but problems occur at the scientific, technical, and sociocultural levels. For science, difficulties lie in the relationships between ongoing specializations and the need for integrated knowledge and ideas (Carsjens and van Lier 2002).

Concepts of integration rather than segregation must be introduced into policies and actions regarding nature and landscape care. A landscape concept comprehensive of natural components, processes, and dynamics and the crosscutting integration of landscapes in the policies that affect them (European Council 2000) are the primary conditions through which they can become a "bridge" favoring, by virtue of their intrinsic needs, integration instead of separation in territorial policies. In this way, landscapes can become an instrument of mediation between nature and culture (Osti 2013), rather than being improperly confined to cultural policies or limited to a scale of ecology.

The extension from ecosystem to landscape fragmentation involves identifying several kinds of sizable systems. This can foster a perception of the true environmental, social, and economic relevance of landscape alterations. Sustainable relationships between human communities and between these and their habitats may be designed and developed through comprehensive, landscape-based plans and projects. By means of their definition and implementation, human communities can perceive natural and cultural heritage sites as integral living components of their sociocultural identities and as fundamental resources of their socioeconomic potentials.

References

Antrop M (1998) Landscape change: plan or chaos? Landsc Urban Plan 41:155–161
Battisti C, Romano B (2007) Fragmentation and connectivity. From ecological analysis to environmental planning. CittàStudi, Novara
Bissonette JA, David JL, Hargis CD (1998) The behavior of landscape metrics commonly used in the study of habitat fragmentation. Landsc Ecol 13:167–186
Brancucci G, Gibelli G (eds) (2006) Planning uncertainty. In: Proceedings of the 7th national congress SIEP-IALE, Genoa, 30 Sept–1 Oct 2004
Carsjens GJ, van Lier HN (2002) Fragmentation and land-use planning. An introduction. Landsc Urban Plan 58(2–4):79–82
Collinge SK (1996) Ecological consequences of habitat fragmentation: implications for landscape architecture and planning. Landsc Urban Plan 36:59–77
Cook EA, van Lier HN (eds) (1994) Landscape planning and ecological networks. Elsevier, Amsterdam
Council of Europe (2000) European Landscape Convention. Treaty series no 176. Florence
Cumming S, Vernier P (2002) Statistical models of landscape pattern metrics, with applications to regional scale dynamic forest simulations. Landsc Ecol 17:433–444
Davidson C (1998) Issues in measuring landscape fragmentation. Wildl Soc Bull 26(1):32–37
Dramstad WE, Forman RTT, Olson JD (1996) Landscape ecology principles in landscape architecture and land use planning. Island Press/Harvard University Graduate school of Design/ American society of Landscape architects, Washington
Fabbri P (2010) Paesaggio e reti. Ecologia della funzione e della percezione. Franco Angeli, Milano
Fahrig L (1997) Relative effects of habitat loss and fragmentation on population extinction. J Wildl Manage 61(3):603–610
Forman RTT (1995) Land mosaics. The ecology of landscapes and regions. Cambridge University Press, Cambridge
Gambino R (2013) Foreword. In: Sargolini M (ed) Urban landscapes. Environmental networks and quality of life. Springer, Milan
Gulinck H, Wagendorp T (2002) References for fragmentation analysis of the rural matrix in cultural landscapes. Landsc Urban Plan 58(2–4):137–146
Haila YA (2002) Conceptual genealogy of fragmentation research: from island biogeography to landscape ecology. Ecol Appl 12(2):321–334
Hidding MC, Teunissen ATJ (2002) Beyond fragmentation: new concepts for urban–rural development. Landsc Urban Plan 58(2–4):297–308
Jaeger JAG (2000) Landscape division, splitting index, and effective mesh size: new measures of landscape fragmentation. Landsc Ecol 15:115–130
Kima J, Ellisb CD (2009) Determining the effects of local development regulations on landscape structure: comparison of the woodlands and North Houston, TX. Landsc Urban Plan 92(3–4): 293–303
Lina YP, Verburgb PH, Changc CR, Chena, Min-Hua Chena HY (2009) Developing and comparing optimal and empirical land-use models for the development of an urbanized watershed forest in Taiwan. Landsc Urban Plan 92(3–4):242–254
McHarg IL (1969) Design with nature. NHP, New York
Nga CN, Xiea YJ, Yua XJ (2011) Measuring the spatio-temporal variation of habitat isolation due to rapid urbanization: a case study of the Shenzhen River cross-boundary catchment, China. Landsc Urban Plan 103(1):44–54
Olffa H, Ritchieb ME (2002) Fragmented nature: consequences for biodiversity. Landsc Urban Plan 58(2–4):83–92
Osti G (2013) Afterword. In: Sargolini M (ed) Urban landscapes. Environmental networks and quality of life. Springer, Milan

Palazzo D, Steiner F (2011) Urban ecological design. A process for regenerative places. Island Press, Washington

Paolinelli G (2003) Peri-urban landscape fragmentation. Project criteria for requalifying the Florentine plain. Firenze University Press, Florence

Paolinelli G (2012) Landscape fragmentation: an interpretive project key. In: Falqui E et al (eds) Walking through landscapes. ETS, Pisa

Poelmans L, van Rompaey A (2009) Detecting and modelling spatial patterns of urban sprawl in highly fragmented areas: a case study in the Flanders–Brussels region. Landsc Urban Plan 93(1):10–19

Romano B (2000) Environmental continuity. Planning for the ecological re-organization of territory. Andromeda, Colledara

Romano B, Paolinelli G (2007) The interference of human settlements in eco-systemic structures. Models for the ecological network of the Veneto Region. Gangemi, Rome

Saunders DA, Hobbs RJ, Margules CR (1991) Biological consequences of ecosystem fragmentation: a review. Conserv Biol 5:18–32

Steiner F (1994) Costruire il paesaggio. McGraw Hill, Milan

Taylor PD (2002) Fragmentation and cultural landscapes: tightening the relationship between human beings and the environments. Landsc Urban Plan 58(2–4):83–92

Wilcox BA, Murphy DD (1985) Conservation strategy: the effects of fragmentation on extinction. Am Nat 125:879–887

Yeha CT, Huangb SL (2009) Investigating spatiotemporal patterns of landscape diversity in response to urbanization. Landsc Urban Plan 93(3–4):151–162

Young CH, Jarvis PJ (2001) Measuring urban habitat fragmentation: an example from the Black Country, UK. Landsc Ecol 16:643–658

Chapter 33
Multi-scalar and Inter-sectorial Strategies for Environment and Landscape

Paolo Castelnovi

Abstract Aspects of governance are very important in territorial policies; the role of best practices is growing and their success is only sustainable if others share them. Paying attention to the perception of environmental and landscape issues is essential in order to implement policies that can be accepted by the environmental and landscape sensibilities of inhabitants and can become a physiological part of their normal behaviour. It is therefore important to examine both the physical extent of a territory that adequately suits the sense of "cultural landscape ownership" in order to organise the choice of management policies at a level that will be comprehensible and that will involve local sensibilities, as well as the criteria that are generally used to condense the various expertise and encourage general and specific value judgements on the conditions of each territory and each landscape.

Keywords Best environmental practices • Best landscape practices • Scalar relationships • Inter-sectorial integration

33.1 The Required Level of Sensibility

Even though territories are changing at an unprecedented pace, processes reorganising collective culture following those transformations remain painfully slow. We metaphorically call the results of these processes 'sensibility'. For example, we can assert that environmental sensibility has risen higher up the league table of Western European values from the late 1980s until today. The increase in sensibility towards landscape is less widespread, though awareness of traditional natural and rural landscapes is reaching significant levels.

Nevertheless, we are dealing with collective cultural mechanisms that are based on intrinsic aspects of quality of life: the combined loss of widespread conditions of easy access to the natural environment and the loss of the perception of the traditional landscape as one's own identity defining habitat have been essential in heightening sensibility. Indeed, we only realise the importance of aspects of quality of life when we lose them, as is the case of household chores. We suddenly notice

---

P. Castelnovi (✉)
ex-Politecnico di Torino, Via della Basilica, 3, Turin, Italy
e-mail: p.castelnovi@libero.it

R. Gambino, A. Peano (eds.), *Nature Policies and Landscape Policies*,
Urban and Landscape Perspectives 18, DOI 10.1007/978-3-319-05410-0_33

the value of clean air and water or the sense of wellbeing that we gain from the places we love when they disappear from our horizon, when we need to invest time and money to seek them out, as they are no longer part of our daily surroundings. Like textbook marginalists, we are only capable of measuring the value of something based on a cost: as long as an object has no cost, it has no value.

On the other hand, the theoretical debate and the interest of academics in these issues have accelerated the perception of the intrinsic mechanisms involved among an elite. Some intellectuals realised – two or three decades before the indisputable evidence emerged – that processes of change affecting territories are damaging environmental quality and render the landscape uniform, thus destroying a particular place's identity. As ever, the elite sought to deviate the course of such historic events: a one-sided and heroic battle that nevertheless has led to increasing awareness in the past few years. In some cases, decision-makers were successfully influenced using strategies or even legislation that predated popular belief. This led to attempts that saw varying degrees of success, attempts at conserving special features of excellence or elements that hoped to stem the more negative consequences of the mechanisms of change, with or without popular consensus.

However, although these are positive actions, they only affect particular aspects directly and do not influence the general situation. Generally speaking, the established models of behaviour – and consequently widespread sensibilities – prevail, evolving much more slowly than physical changes. Widespread sensibilities directly influence the day-to-day management of a territory that is the result of an extensive network involving hundreds of thousands of public and private operators. Therefore, where an awareness of environmental and landscape values is lacking – values based on forward-thinking public strategies and laws – the tendency to erode the spirit of regulations comes automatically, when they clash too much with current opinions.

Of course, the effects of public government activities over time (programmes, regulations, schemes) not only affect physical transformation processes; they affect collective sensibilities as well. However, the effects are more long lasting and intrinsic when – alongside the apparatus of procedural rules – we activate a process of cultural assimilation, of sharing benchmark environmental and landscape values and of activating best practices that are imitated and spread through conviction and not through obedience to formal obligations.

For over a century in European countries, particularly in Anglo-Saxon nations, territorial management – and subsequently environment and landscape management – has been the result of a systematic partnership between government and governance, between rules and the creation of a generally agreed behaviour adopted by most of the local population. In Italy, the aspect of governance was never cultivated as a necessary complement to that of government; on the contrary, the fact that it was indifferent if not actually in conflict with the existing body of regulations raised no concern.

In such conditions, collective sensibilities do not evolve in one particular direction; instead, they develop contradictory aspects, as they have no reference to shared and established values, aspects that cannot be reversed, as usually occurs with slow evolutionary processes that arise through daily practices.

Evidence of this crisis becomes clear with the proliferation of protests against more far-reaching transformation programmes. Starting with local protests (the NIMBY effect), the debate takes on enormous ideological significance, which tends to entrench dialectically opposite positions, each side mixing scientific fact with hypothetical theories. The players operate in a field of opinions where nothing is accepted as a certainty any longer, as it cannot be verified by fact.

Accumulated sensibilities lose their power to unite a community: as a matter of fact, a decision is no longer necessary. The 'political' stance taken (i.e. a position that concerns the common goods, the *polis*) no longer refers to sensibilities built up through practice, but rather to ideal or ideological choices that the various 'external' watchwords put forward. It is a vicious circle that is created when 'internal' reference sources lose their authority. A general distrust is generated that is difficult to counteract, given that we can no longer count on the tools that are fundamental to all kinds of social communion: widespread sensibilities, generally accepted common sense. In an ideological cultural climate that has no roots in common practice, arguments concerning common assets divide rather than unite.

In short, wherever they occur, territorial government policies are largely entrusted to governance aspects, and their success is only sustainable if such policies are generally accepted. In such a scenario, the issue of developing environmental and landscape sensibility cannot be limited to a particular niche, but is put forward as a fundamental tool for operating in a way that focuses on lasting results.

33.2 Conditions for Environmental and Landscape Best Practices

The terms 'raising awareness' and 'participation' have been present in every European programme announcement for decades, and yet the operational conditions needed to optimise environmental and landscape sensibilities have not been studied in depth and are still limited to the virtuous sphere of best management practices.

However, it takes an enormous effort to fine-tune management practices, a process that each time is undertaken in different conditions that are tied to specific human or territorial resources and are therefore difficult to get adopted to any extent; only rarely do they activate positive processes of emulation. Interesting environmental or landscape practices rarely, if ever, automatically spread to other areas, given that a few, simple operational criteria that can easily be reproduced are hardly ever enough.

The success and sustainability of a territory's good management depend on the complex interaction of various different factors: inter-institutional organisation, widespread entrepreneurial spirit, attachment to the local area, a traditional sense of local identity and awareness of the common assets to be promoted.

On the other hand, we can't imagine that when we implement the good management of a common asset, it will magically result in collective participation and agreement, founded on awareness and sensibility. Quite the contrary, experience has shown that the process is interactive: lasting good management cannot happen without prior sensibility and participation and that an analysis of the conditions for developing environmental and landscape sensibilities can help enormously when implementing successful schemes in such a complex operation.

For the past two generations now, the need to consider the objective and historical conditions of a territory has made headway in the technical training of Italian urban planners: we have studied natural and cultural resources, mechanisms of change and the situations encountered in hinterlands. This is a technical expertise that Italy developed before other European countries did and to a greater extent and which has had positive results in terms of the quality of planning technique in many sectors, for example, in the difficulties encountered when dealing with old town centres where we have certainly fine-tuned operational models that have been imitated in many other countries.

However, we have hardly ever studied the subjective conditions that would have served as models for managing territories after planning: neither those of institutional, cultural and economic players nor declared or implied operational and project development capabilities, nor the sense of landscape identity.

To a large extent, this failure in Italy derives from a long-standing negligence in territorial government as regards managerial aspects, partnerships unregulated by legislation and operational best practices. So it is clear that as far as this aspect is concerned, there have been no particularly notable innovative programmes. In Italy, planning capabilities have never tackled those organisational, persuasive, coordination or verification aspects that have been required in all industrial or administrative management strategies for the past century.[1]

Therefore, during the phase where environmental and landscape aspects emerge in all their intrinsic importance, we have successfully and usefully applied the fact-finding techniques for objects and their systems to them, techniques perfected in urban planning. However, we have not studied the subjective conditions that generate collective sensibilities concerning environmental issues nor the psychological and cultural processes through which the common perception of a landscape unites, rather than divides, a territory's local population and users.

Considering the difficulties encountered by the first generation of environmental and landscape plans and programmes, we can outline a general framework listing the primary subjective conditions that should be examined. We need to implement policies that can hope to be accepted by the environmental and landscape sensibilities of local people and that can therefore become an integral part of the normal behaviour of those who produce and use the area on a daily basis.

[1] See Fayol (1916).

First and foremost, what emerges is the need to study the physical extent of a territory that adequately suits the sense of 'cultural ownership' of landscapes associated with it, so as to be able to choose the management policies in view of this extent, on a scale that is comprehensible to local sensibilities and involves them.

While on the subject of scalar relationships, in order to facilitate the exchange of views and participation in decision-making, we need to study the criteria that are most commonly used (and are generally considered to be common sense) in order to condense and unify the different sectorial fields of expertise that accumulate in investigations on a territory's objective aspects. Indeed, the interdisciplinary complexity of accumulated expertise makes specific and all-encompassing value judgements on the conditions of each territory and landscape difficult, if we do not fine-tune holistic simplifying techniques that can be applied to the sea of information and specific evaluations. When best practices for condensing facts and information are missing, it is difficult to discuss the pros and cons of complex choices and we end up being dominated by general preferences formed a priori on certain kinds of information rather than others (e.g. environmental aspects more than cultural aspects, functional aspects rather than those related to identity, etc.).

33.3 The Right Level of Multi-scalarity

The question of what physical extent is appropriate when discussing a territory is worthy of Bertoldo, the astute peasant: as in the short story, it requires an ability to meet apparently opposite requirements at the same time.

We need to take into account the physical extent of a territory affected by current changes (which often alter vast expanses of land, particularly in metropolitan areas), while the reference to the landscape identity of each community (which is often of a modest size and never more than a couple of boroughs) must be clear.

The increasing extent of networks of relationships (even those that seem fragmentary compared to the physical extent of a territory) and transport speed and mobility are all important, but visual relationships are also essential, those featuring the immediate proximity of different elements that make the perceived landscape the perfect place for serendipity, that pleasant surprise that is such an important part of the charm of travel and visiting new locations.

Furthermore, we need to take into account the administrative borders that define institutional powers and therefore the operational headquarters managing plans and projects, as well as the lack of precise borders when dealing with a generally accepted sense of "cultural ownership" associated with the landscape or with environmental relationships (take, e.g. those aspects linked to rivers or mountains).

In such a complex framework, only those who possess a precise hierarchy of benchmark values can have the right tools to make decisions, but this does not stop conflict from arising a priori, often aggravated not so much by the merits of the issues themselves as by the contrast between points of view, starting with different distances from the places involved and therefore from different levels of scale.

Clashes over the project for the high-speed rail link in the Susa Valley (Piedmont, Italy) are a perfect example of the problems that arise when those managing complex projects neglect to consider the difference in scale of points of view: in this instance, it is the difference between those who are thinking in terms of a European transport network and those who are thinking in terms of the cultural ownership of a local territory.

The failure to consider the complexities of the physical size of a territory involved encourages a sense of common purpose where one adopts one way of thinking and resists others, losing any possibility of deciding 'rationally', i.e. by considering the systematic coexistence of effects with different extents with each process of change. Even those operating from within one particular sector, for example, the field of environmental or landscape values, can come into conflict on different levels of scale. Schemes that help guarantee the general continuity of environmental networks can clash with the local need to protect specific indigenous features, and strategies promoting a tourist area can be counterproductive for the landscape value of specific areas prized for their very isolation and tranquillity.

In order to overcome the intrinsic difficulties posed by the subjective stances of users and decision-makers, we need to get used to being informed and to deciding things on different scales at the same time: the right choices on territorial issues are those that derive from the practice of inter-scalarity.

In order to become more familiar with this variable scale of knowledge and judgement, it is important to set up government action, right from the first territorial investigations and environmental and landscape interpretations, on the right basis.

On the one hand, the combination of multidisciplinary fields of expertise in a framework of holistic interpretative simplification must always be evident; on the other hand, the scale of each system of information must always be clear, highlighting in each case the effects on levels above and below it.

For example, in a regional landscape plan, the reference scale of regulatory maps must be quite detailed (1:50,000 or 1:100,000). The effect of regulations, applied in a formal way based on maps at that scale, is inevitably contradictory: the large size of print ends up imposing limits in areas where they are not necessary, and on the other hand, we neglect action in specific areas that actually require it. The problem arises even where specific approaches have been defined for smaller territorial areas, taking into account the historical and geographical differences of the various areas that make up a region, but entirely neglect to verify local awareness and the generally shared sense of landscape that should support each situation.

This contradiction can only be overcome if other criteria and interpretational background information are available, allowing us to correct large-scale general indications, verifying situations on a smaller scale. However, this method necessarily refers us back to 'peripheral' modes of governance, the application of plans over time and management scales. This method not only consists of explicit descriptions and the application of detailed objective rules, but implies a comparison with the generally accepted sense of places and their use, an examination of interpretational subjectivity and its widespread adoption which, for better or worse, will lead to different interpretations and evaluations with each case. When operating at this level of detail, we need to carry out a careful evaluation of the generally accepted sense of landscape and local reactions to change in order to manage plans, because the widespread cultural attitude regarding these aspects is essential in order to implement the strategies of more general policies in the best possible way in such situations.

Piedmont's landscape plan has attempted a first step in this direction, identifying the landscape units (550 out of over 1,100 boroughs in the region) notable for the level of change and the importance of their landscape resources, in order to encourage the development of consistent attitudes and appropriate forums for debate, on a local scale, that matched the generally accepted sense of landscape. The landscape units should therefore be the best points of reference to make the effects of the issues that lead to intrinsic transformations of a territory comprehensible and therefore to be able to discuss them technically and politically, both in terms of ordinary policies to be managed on a local level as well as much more large-scale projects connected to regional or even international policies.[2]

A second example, which remains unexplored, is the planning generation potential of nature conservation areas, which in the turn of this century allowed Italy to conduct experiments that were not possible in the planning programmes of other institutional bodies. Parks are still the only institution where, by statute, administration is conducted by a supra-borough management body, which the law supplements with a 'park community'. This body of involved players, set up to manage complex problems, has practically imposed the experimentation of multi-scalar projects and investigations. In such organisations, the novelty of issues concerning landscape quality and ecological networks – which have padded out environmental policies that were formerly too protectionist – has encouraged unusual levels of interaction both in the immediate areas near parks (hence, the vast surrounding area) and within the populations involved (hence, a local point of view). It has been a scientific and methodological challenge – in some cases, even a political challenge – that has not yet been sufficiently examined in terms of its useful effects for general planning in Italy, founded for the first time on the marked involvement of local people and a constant debate regarding inter-scalar issues.

A third example worth pondering is the Landscape Award Programme: entirely honorary awards, set up by the European Landscape Convention and ever since 2009 on a biennial basis. The award's requirements – as set up by the Council of Europe – focus on raising awareness, participation and good management. In the space of three editions, over 300 organisations of public interest have applied for the award, presenting examples of well-managed programmes at very different levels of scale; organisations that have only restored one asset that qualifies its environment and organisations with UNESCO World Heritage Site or conservation area management plans or with local development programmes or strategic urban

[2] See Vv.Aa (2007).

areas with a strong landscape element have all participated. In all cases, the awards have generated an unprecedented level of enthusiasm that has not even been seen when large sums have been awarded in the past, and with an authentically 'political' expression of interest: to establish positive relationships between the experience gained by 'institutional capillaries' (such as, e.g. small mountain boroughs) and the centre, when it recognises the value of local practices and relaunches them right up to European levels as examples which should be imitated. In short, the requirements for a real exercise of positive, unimposed inter-scalar relationships take shape.

References

Fayol H (1916) Administration industrielle et générale; prévoyance, organisation, commandement, coordination, controle (in French). Dunod et Pinat, Paris

Vv.Aa (2007) Per il Piano paesaggistico regionale. Quaderni del Governo del Territorio. Regione Piemonte, Turin

Chapter 34
Urban Landscapes and Nature in Planning and Spatial Strategies

Massimo Sargolini

Abstract Monitoring the effects that programmes for sustainability, or plans for the landscape and large parks, have had on the design of the city, on both city-wide and large-area scales, should be discouraging. Principles and new paradigms raised by the environmental question that lack the necessary practical and operational implications regarding design and management, and which are postponed to another time, have little impact on the relatively rapid changes caused by diffuse urbanization and become simply a refrain of good intentions. Elsewhere, this great responsibility regarding the landscape is deliberately and specifically entrusted to urban planning by the European Landscape Convention. The true revolution introduced by this directive is to invite the landscape to square with the matters of the territory and the city in all of its many facets. At the same time, "protected areas", which directly or indirectly touch more than a third of Italian territory, could become (together with the environmental infrastructure network and the system of residual and decommissioned areas) new spatial anchors in urban and territorial reorganization, provided that these elements become components of the ecosystem of the city and not just cosmetic dressing. The pervasiveness of these themes should cause those interested in territorial government to reflect, going so far as to consider parks and landscape planning as the foundation of urban planning in this special historical moment, particularly if we are able to manage to detach it from the sectoral vision to which it is so attached.

Keywords Urban landscapes • Quality of life • Urban landscapes • Decision support systems

34.1 Protected Areas and Urban Planning

The separation between protected areas and territorial context (highlighted in Italy by a framework of park laws that are not extensive enough to manage such deep interactions) has developed parallel to a separation between the landscape and the

M. Sargolini (✉)
School of Architecture and Design, University of Camerino, Piazza Cavour 19/f, Camerino 62032, Italy
e-mail: massimo.sargolini@unicam.it

R. Gambino, A. Peano (eds.), *Nature Policies and Landscape Policies*,
Urban and Landscape Perspectives 18, DOI 10.1007/978-3-319-05410-0_34

territory (accentuated in Italy by the Cultural Heritage and Landscape Code, which still keeps separated conservation from development). In both cases, the fear of having anything to do with "diabolic" actions of the plan has prevailed. All of this is the result of a deplorable myopia in the world of conservation. Only recently have people realized that merely conserving resources is not the best choice to preserve the goods themselves. Everyone now seems to agree that without a plan, there can be no conservation.

In Italy, however, the older fringes of urban planning still do not seem to understand that it is necessary to introduce a proactive rapport in matters of protected areas and the landscape in general; they consider these matters to be something outside of urban planning. It seems like they are not able to understand that proactive intervention on parks and ecological networks can contribute to regenerating degraded urban landscapes, increasingly at the mercy of atopy, and can therefore aid in managing the confused and knotted city through planning.

The profound relationships between the urban landscape and the quality of life on the global scale were analysed at the Venice Biennial of Architecture directed by Richard Burdett in 2006. At the 10th International Architecture Exhibition, qualitative and quantitative assessments of the environmental situation in major metropolis around the world were clearly presented, along with specific relationships regarding the health of the planet and living conditions of the inhabitants. The direct relationship between environmental health, the natural spaces available and the quality of life was evident.

Internal debates at the INU (National Urban Planning Institute), the SIU[1] (Italian Urban Planning Society), or on international levels[2] reach encouraging levels when considerations are made regarding the quality and sustainability of the city. In such a sense, when we speak about the quality of the city, we should refer not only to the built city but also to open spaces and large natural parks. PRIN research from 2006[3] studied the quality of the Adriatic city, the operational unit UNICAM assumes as a central theme when giving meaning to the free areas; particular attention was made to decommissioned industrial areas in the Adriatic region, which constitute a large part of areas available for strategic interventions in the requalification of the coastal

[1] On the National level, the theme of quality of life and the sustainable city is confronted through the following INU conferences: 2007, "Pianificazione energetica e politiche del clima nel nuovo piano"; 2010, Effetti del consumo di suolo sul governo del territorio; 2011, Città senza petrolio; i seguenti congressi INU: XXIII del 2000, "Il progresso della città contemporanea: domanda sociale, politiche piani"; XXVII del 2011, "La città oltre la crisi: risorse, welfare, governo; le seguenti conferenze SIU: VII del 2003, "Il progetto di territorio e paesaggio"; XII del 2009, "Il progetto dell'urbanistica per il paesaggio"; XIII del 2010, "Città e crisi globale: clima, sviluppo e convivenza"; XV del 2012, "L'Urbanistica che cambia. Rischi e valori".

[2] On the European level, I mention only a report by the EU from 2005, "Ensuring quality of life in Europe's cities and towns", which speaks specifically about the quality of life in urban areas and cities.

[3] "Opere pubbliche e città adriatica. La qualità del progetto nelle interazioni tra costa e sistemi vallivi marchigiani: infrastrutture ed aree dismesse", coordinated by Barbieri P., University of Chieti, Pescara.

landscape. "Settlement dispersion" in the urbanized strip from Venice to Pescara has generated a state of "porousness" in which full and empty, finite and open and used and decommissioned live together in overlaid and conflicting functional systems. These spaces are virtual voids that constitute an extraordinary resource, in the same way as parks and protected areas, in which it is possible to construct a system of new collective values. They are now focal points in the urban sprawl, possible factors in urban morphogenesis and unique areas of reserve and building reorganization that have in fact reduced to a minimum those spaces defined as "distance vacuums" in the contemporary urban sprawl complex. The new urban project should know how to start from an innovative planning interpretation of these "vacuums", rediscovering the disciplinary mandate through the convergence of multiple skills. This means recognizing in the city's open spaces a key (or important) role as a reference in the design of the city and territory, as Frederick Law Olmsted, Sr. prophesized 150 years ago when he set about cutting out a large "green hole" in the heart of Manhattan's urban fabric, removing 778 acres of land from building speculation. At that time, some considered Olmsted to be crazy for this "loss of value". Today, could we even begin to imagine that there is a single New Yorker who would renounce Central Park and the quality of life it brings to the inhabitants of this chaotic city?

But are we really so different from the society in the USA, where parks have become a reference for national identity? Paul Bray (in this book) reminds us that after September 11, Americans felt lost and spontaneously found themselves seeking out the large national parks. But these are other sensibilities, coming from other roots. We should, however, ask ourselves if we as urban planners have some responsibility in not being able to inculcate parks in the conscience of a people or more simply in the territory.

The breadth of the question regarding the relationship between the city and nature, along with its problems, has been used and understood up to now as an argument that is shared but comes from on high. It harkens back to a complexity outside of urban planning that is not very pertinent to contingent local administrative and productive problems, and it lacks any impact on operational, localization, actuation and detailed choices. Cultural and disciplinary advancement in matters of "territorial government", recorded in Italy and Europe in recent years, has directed attention at environmental armour intended as a constitutional element in new settlement forms, both on the urban and landscape scales. Finally, we discover that it is not possible to work on natural environments only with defensive or compensative actions. On the contrary, we see that nature (and open spaces in general) should assume the dimension of an effective cognitive paradigm regarding urban and territorial forms, even in economic operations, with planning effects on the multiple territorial scales.

34.2 Parks and the Quality of the Urban Landscape

The landscape is probably the right approach to establish fertile contacts between parks and cities. Various authors in this volume have remarked on this idea. But why is it so difficult to start from the landscape?

On the one hand, management of the landscape requires recognizing and developing the energies expressed by nature, which deal with animal and vegetation resources; on the other hand, it requires a careful look at the "soul of places", that extraordinary heritage of urban and rural culture that constantly interfaces with nature and produces that landscape described by the European Landscape Convention (2000) as the combination of natural and human activities. However, the course of landscape interpretation has occurred from different disciplinary fronts, highlighting a large difference between two different schools of thought:

1. The first, arising from the art of gardens, which records historical attention on open spaces (gardens and parks). It has opened the way to landscape architecture but does not have the necessary background to manage a systemic complex system that should be at the heart of territorial transformations and therefore landscape planning.
2. The second one, connected to landscape ecology. It is developed as system ecology, but lacks contact with the territorial question and, in particular, with collective feeling, intuition, and the unconscious with subjective interpretation.

New urban planning should start precisely from these two ideas in order to lead them to maturity in the territorial approach, taking advantage of the drive to improve the quality of life, which is common to both visions. A new interpretive and planning framework emerges, in which aesthetic and ecological goals come together. This framework forms the basis for a new strategy in spatial planning that has landscape assessment, and therefore the quality of life, at its core.

The difficulties and efforts made towards synthetic evaluation of the landscape are evident.[4] European directives regarding environmental and landscape evaluation have produced confusing interpretations for at least two reasons:

1. On different levels of government, the environment is linked to EU directives and the landscape is linked to the Council of Europe treaties. Disorganization continues within these two centres that produce, respectively: (i) methods of implementation and (ii) policy orientations.
2. In evaluation systems used up to now in Europe, the landscape component is understood solely from the aesthetic and perceptual point of view and is even described as a component internal to the "environmental" domain. This has occurred even in light of the fact that the first types of quality evaluation were produced exclusively for the environmental aspect and that the Council of

---

[4] See, in particular: Cassatella and Peano (2011), Sargolini (2012).

Europe has tried in the last decade to stimulate assessment regarding the landscape.

It is evident that it can be rather difficult to give a quantitative value to the landscape in all its complexity, whilst it is possible to identify and measure one of its many aspects (perceptual, ecological, historical, cultural, etc.). Because of this, in the experiences examined, the interpretations that stand out are aimed at reducing the complexity of a single aspect or they are holistic visions entrusted to the overall judgement of experts that, however, make use of a subjective evaluation method. With the ELC, a new policy plan for the landscape begins that is aimed at (i) recognizing its overall meaning, (ii) gathering its relevance for the entire territory, (iii) favouring integration in territorial planning and design and (iv) involving the community and populations interested in identifying the quality objectives to pursue. Recommendations made by the Committee of Ministers of the Council of Europe on 06/02/2008 (Guidelines for the Activation of the ELC. All. 1, point 10) foresee the opportunity to create landscape observatories that either are specific or form part of a wider observation system that, in addition to describing the state of landscapes at a specific time, should tune quantitative and qualitative indicators to evaluate the effectiveness of landscape policies and produce data useful to define tendencies and possible future scenarios. The Catalonia Observatory, one of the first of such experiences regarding this, along with some still non-institutionalized Italian experiences in the regions of Abruzzo, Calabria, Umbria and Sardinia, conceived the observatories as meeting places in which expert knowledge encounters diffuse, common knowledge, calling together scientists, technicians, administrators and representatives of civil society. This means considering not only social processes through which communities are manifest but also the specific values that the subjects and interested populations attribute to the landscape. It is clear that this implies a substantial difference with respect to the deterministic approach, which looks at the environmental question.

An important step forward can be made in synthetic assessment of the landscape if it is linked to the goal of improving the quality of life. In the interdisciplinary and interuniversity[5] research "QLand QLife", using the Adriatic city as a case study, advancement has been made in this direction, looking at the relationship between landscape and the quality of life and keeping the great theme of sustainability as a basis.

It is clear that this begins with an idea of a polysemous and complex landscape that highlights the deep interactions between humans and the territory and includes, on the one hand, the objectivity of the environmental paradigm and, on the other, the subjectivity of the perceptions, consciousness, emotions and expectations of the "interested populations". The case study was used to investigate thematic areas of

[5] The working group for UNICAM research "QLand QLife", coordinated by Massimo Sargolini, is composed of several researchers from Italian and foreign Universities that work in the field of Ecological Sciences, Landscape Architecture, Geology, Biological Sciences, Economy, Sociology, and Geography.

landscape interpretation that influence the quality of life; comparison methods between the different analyses were then tested to determine synthetic evaluation indices.

The landscape, therefore, is also studied in relation to the urban ecosystem in order to reach the goal of improving the quality of life. Indeed, this is the large index the result of landscape politics is evaluated with respect to. It is this approach to the landscape that new urban planning should always take.

34.3 A Decision Support System for Increasing the Quality of Life

We therefore study the city, highlighting the role of nature and its resources in improving landscapes and the quality of life. In this view, it is clear that climatic, geological and biological factors, which define the urban ecosystem, are structural references that anyone intending to give a form to the city should confront. Interaction with these factors during design becomes a complex operation, essential for the urban project and everything to be tested.

The relationship between the form of the city, urban sustainability and the quality of life is not new. For some decades, this has been the object of an intense debate at both the academic and political levels; however, the search for a shared theory that is applicable and usable by planners is still underway. The European Union, with its Green Paper on the Urban Environment (1990) and later with the European Charter II (2008), has indicated the dense, compact city as the best solution for reaching energetic efficiency and urban quality; it is also the most sustainable from the economic point of view, in that it requires less resources to function, guaranteeing more efficient access to services. However, such a claim has not been automatically accepted or shared.

Even if density shares a deep relationship with urban morphology, this alone is not enough to guarantee great quality and efficiency in the contemporary city; research aimed at supporting higher density, in contrast to greater settlement dispersion, still shows partial results that do not take into account systemic visions and which obviously recall different concepts and measures in different countries and in different cultures. From the ecological point of view, the scientific debate in the last decade has expanded from strictly environmental questions to the quality of life linked to the nature in cities. Some arguments strictly linked to conserving biodiversity seem to support the compact city (Jim and Chen 2003); others argue for the capacity of green areas to counteract the heat island, decreasing the temperature and allowing for significant energy savings necessary to the operation of air conditioning systems (Santamouris 2001). The presence of nature in cities is, finally, explored in the project Living Cities in Birmingham, analysing the relationships—not only human ones—that characterize cities and looking with particular attention at activating the biological chain between different animal species in the city.

According to R. Burdett and D. Sudjic (2008), from the transportation and energy flow systems point of view, the compact city maximizes the efficiency of mechanized urban transport systems. In fact, a high construction density allows for "circular metabolism" to be activated, which the urban ecologist Herbert Girardet indicates as a strategy for reducing the environmental impact of urban areas. On the other hand, it seems to be precisely territorial dispersion that produces the maximum values of capture and use of alternative clean sources, obviously facilitated by the greater availability of open surfaces and the reduced presence of obstacles.

On the social level, the importance recognized in the urban density/proximity binomial is sustained by Reale (2008), who discusses the consideration that the urban quality of a city is also surely linked to the proximity of people and activities and the relationship between the city and the functions it provides.

One last important study in the strictly urban-planning field is "The cost of sprawl", which was led by the US Government between 1970 and 1990. This analysis did not look for the "best density", but rather the "best density for each city" (Roaf 2010). The SUME (Sustainable Urban Metabolism for Europe) project, financed by the EU, was designed along these lines, but, however, does not bridge the deep divide between scientific knowledge and other types of skills held by "city users" (citizens, professionals, public officials). On the contrary, the contribution by S. Bertuglia, L. Staricco, F. Rota (2004) seems to place more attention on problems of daily life in the city, also opening up a comparison between these and political and social actors.

The research "QLand QLife" by the University of Camerino, centred on several case studies in the Adriatic city, begins with this current state of the field in order to try and produce disciplinary advances related to:

- The interpretation and assessment of urban complexity, by researching new "transdisciplinary" indicators for the quality of urban life that, going beyond the vertical, mono-disciplinary approach, start from the synthetic landscape vision and develop mutual correlations and adaptability to the context of reference.
- The definition of a new integrated, multi-objective conceptual model that is able to manage new parameters for the urban quality of life, to be applied to different ideal profiles of the existing and planned Adriatic city, going beyond the binomial "compact city/diffuse city" vision.
- The construction of integrated awareness of the city which can direct policies and strategies at different levels of governance, in order to increase the quality of the landscape and, with it, the quality of life of the city's inhabitants.

The research therefore concentrates on constructing a dynamic, integrated decision support system that uses parametric optimization and is able to gather the complex interconnections between different ideal urban configurations and possible drivers of the quality of life. The abstraction necessary to conceptually define the system places different "attributes" of the urban form (density, complexity, centrality, compactness, porousness, etc.) next to exemplary "formal models" (ideals) of the European city inferred from the European Commission's GMES "Urban

Atlas" (made explicit through the construction of an "urban to rural" transept). The model will take into consideration the connections, the mutual dynamic relationships and adaptability amongst the different "attributes" of the urban form with each of the ideals examined, and it will be able to guide public administrations in identifying the most appropriate strategic scenarios to respond to the needs and expectations of the territory, which are, at the same time, the most sustainable with respect to efficiency and energy savings within aesthetic, environmental, social and economic goals.

References

Burdett R, Sudjic D (2008) The endless city. Phaidon Press, London

Bertuglia CS, Rota FS, Staricco L (2004) Pianificazione strategica e sostenibilità urbana. Concettualizzazioni e sperimentazioni in Italia. Franco Angeli, Milan, pp 1–127. ISBN 9788846453402

Cassatella C, Peano A (2011) Landscape indicators. Assessing and monitoring landscape quality. Springer, Dordrecht

Jim CY, Chen SS (2003) Comprehensive greenspace planning based on landscape ecology principles in compact Nanjing city, China. Landsc Urban Plan 65:95–116

Reale L (2008) Densità Città Residenza. Tecniche di densificazione e strategie anti-sprawl. Gangemi editore, Rome

Roaf S (2010) Transforming markets in the built environment: adapting to climate change, Architectural science review series. Routledge, London. ISBN 13: 978-1849710886

Santamouris M (2001) The role of green spaces. In: Santamouris M (ed) Energy and climate in the built environment. James & James, London, pp 97–109

Sargolini M (2012) Urban landscape. Environmental networks and quality of life. Springer, Dordrecht

Chapter 35
Integrated Planning for Landscape Protection and Biodiversity Conservation

Alessandro Tosini

Abstract The definition of landscape adopted in the European Landscape Convention implies the pursuing of awareness raising and public involvement as a primary instrument for planning and policy implementation. As in the Italian tradition, national institutions prove slow in applying such principles and coordinating them with traditional urban planning. The main problem consists in understanding the positive role of landscape, as a dynamic synthesis of cultural and ecological features. In many cases, policy-makers and professional actors still consider the theme of landscape conservation and biodiversity protection as limitative entities. Acts and policies related to Protected Areas worldwide can represent a precious background of experiences for the implementation of an operative procedure of territorial management, which will consider landscape and biodiversity as relevant as economic features. Other effective suggestions come from the outcome document of Rio +20, the latest international convention on sustainable development. Integrating landscape and biodiversity in current national laws may prove inadequate. A successful application of the most recent tools of planning based on holistic approaches and including public-participated processes will be achieved only through a radical reflection about traditional policy-making, which is still linked to the division of the matter in obsolete compartments.

Keywords Holistic strategies • Landscape and biodiversity management • Sustainable development • Integrated planning

35.1 Introduction

The purpose of this paper is to highlight some peculiar features of the actual lawmaking and institutional background, compared to the state of the art determined by international conventions and the academic debate, related to landscape protection and biodiversity conservation. Moreover, it will put forward some considerations regarding the enhancement of integrated policies.

A. Tosini (✉)
Interuniversity Department of Regional and Urban Studies and Planning (DIST), Politecnico di Torino, Via Levata 100, 15122 Spinetta Marengo, AL, Italy
e-mail: arch.atosini@libero.it; dr.arch.alessandro.tosini@gmail.com

R. Gambino, A. Peano (eds.), *Nature Policies and Landscape Policies*,
Urban and Landscape Perspectives 18, DOI 10.1007/978-3-319-05410-0_35

The Italian situation is an interesting case study, as it still shows substantial flaws vis-à-vis the general requisites established by the Council of Europe. The European Landscape Convention (ELC) states the need to widen the meaning of *landscape*, underlining the role of *awareness raising* (ELC, art. 6.A) and the active involvement of the public – meaning both private subjects and local institutions – as fundamental for the success of integrated environmental policies at all scales. Similar topics have been recently faced in international talks and conventions on sustainable development, such as Rio +20, which emphasized the importance of considering biodiversity and cultural heritage-related themes as necessary elements of any advanced form of policy-making.

The focus will be both on the main role of awareness raising and public participation as decisive tools for present and future planning and on the operative definition of landscape as intended by the ELC, which still meets many difficulties to be accepted and applied, both in European and Italian initiatives.

35.2 The Operative Definitions of Landscape

The definition of landscape worldwide is quite difficult and composite, involving a number of possible meanings according to different cultures, regarding different approaches to sensorial experience (Bruns 2013). The ELC highlights the importance of landscapes due to their contribution to individual and social well-being, their role in Europe's heritage and their significance as the environment of towns and countryside (Ward Thompson and Sarlöv Herlin 2004).

The Convention suggests the necessity to overcome other traditional attitudes towards landscape, widening the meaning of the term and linking it to the active role of the European population: it is considered as "an area, as perceived by people, whose character is the result of the action and interaction of natural and/or human factors" (ELC, art. 1.a). The definitions adopted in various European strategies lack coordination with the ELC, since the word "landscape" may be referred to:

- The connective tissue of natural/paranatural habitats, which counters the fragmentation of the landscape itself and the threatening of biodiversity (*Benefits beyond boundaries*, IUCN 2003)
- The sociocultural dimension of landscape, mostly related to policies for rural development and cohesion (e.g. Carmona-Torres et al. 2011) which, according to UNESCO strategies (*Vienna Memorandum* 2005; *Management guidelines* 1998), are broadening the categories of excellent "cultural landscapes", including strategies to attract tourists, inhabitants and investments (Voghera 2011)

The coordination of these definitions appears to be vital even for the construction and implementation of the operative idea of landscape proposed in the text of the ELC. The task could be achieved through a joint effort by the EU members: yet such coordination seems quite complex. On this behalf, the internal situation of any

European countries presents political, institutional and cultural problems. Overcoming the existing contradictions is sometimes hindered by the actual condition of national legislations. In these regard, the Italian case study shows some of the major difficulties, as the operative definition of *landscape* in the Italian Cultural Heritage and Landscape Code, Leg. Decree no. 42/2004, shifts between the ELC text and other traditional definitions, taken from dated national laws. The current version of the text does not take into account the role played by the public in the process of codification of what a landscape is (Priore 2009), as it says:

> (. . .) landscape (. . .) a homogeneous part of territory, whose features descend from nature, from human history and from their mutual interactions (Cultural Heritage and Landscape Code, art. 131, c. 1)

while the same article used to link the concept of landscape to the identity of the people who lived on the related territory, following the ELC (art. 1.a). Conservation and enhancement are then limited to the values of those parts of territory, which can be considered as "perceivable manifestations of identity" (Cultural Heritage and Landscape Code, art. 131, c. 2). This passage is quite tricky, as the statement about the cultural aspect of the landscape implies in many experiences the attribution of an exceptional cultural value. Meanwhile, the traditional conception of landscape and the related demand for beauty has not been mitigated through the widening and updating of its meaning. On the contrary, it has been partially substituted by a renewed interest for the environment (Savio and Paludi 2005), yet without achieving coordination among the various institutional fields involved.

Some of the problems inherent in the Italian case can be found in other contexts, where attempts are underway to coordinate national laws with the ELC. Much must be done, not only in Italy but also across the whole European Union, in order to receive a positive feedback from institutions and social actors. First of all, the traditional conservative conception of landscape (strongly linked to touristic offer) must give way to the new conception of a *multilayered public infrastructure* (Bunge 2011), characterizing its planning as a composite and positive phenomenon based on new holistic aesthetics. The entity *landscape* itself encompasses both eco-centric and cultural dimensions; the adoption of cross-cultural approaches, as elaborated in landscape scholarship, is fundamental for the formulation of effective and up-to-date strategies (Stokols 2011). As shown infra, then, any operative definition of landscape should be conceived as a feature of an integrated policy-making aimed at sustainable development. The participation and awareness of the public are the crux of the matter.

35.3 Awareness Raising and Public Involvement: A Neglected Task?

The ELC insists on awareness raising and public involvement as decisive features of any planning strategy. Article 6 of the Convention underlines the importance of allocating resources to the information and involvement of citizens, multi-disciplinary training of technical staff and local/regional/national authorities and development of related school and university courses (*ECL explanatory Report*, 52–53). The achievement of such objectives is essential, yet it encounters many difficulties. Social and political actors should pursue these tasks for many reasons. First of all, public awareness is fundamental for the definition of bottom-up landscape concepts, which could become the rationale of effective policies (Groening 2007). Any planning operation should be characterized by the constant dialogue among the proposing subjects and the public, so that bottom-up proposals can be harmonized with top-down decisions. Following Daniel Burnham:

> when the majority of the people of any town come to think that convenience and its consequent beauty are essential, they will have them, for a democracy has full power over men, land and goods, and can always make its laws fit its purpose. (Burnham 1910)

Traditional frameworks of European national institutions are not generally ready for the implementation of such principles. The Italian case proves useful to underline particularly critical features. In Italy planning processes are still characterized by top-down approaches, by traditional conception of the regulation processes (mainly based on rigid set of rules) and by the small places left to local communities and stakeholders. The effects are aggravated by the confusion deriving from the coexistence of different plans (local urban plans, regional and territorial plans, landscape plans, PAs plans and others).

In Italy, design activities have to face an eminently limitative attitude by public institutions. Top-down planning is often conceived as the elaboration of multi-layered landscape and zoning plans, which set standards and limits according to composite criteria of territorial management. The involvement of citizens is limited to the possibility of presenting written remarks during the process of formulation of the plans, while awareness-raising activities are almost ignored, especially at a local/regional scale. The institutions play a rigid role in the process of landscape transformation: after the formulation of the plan, they mainly have to express acceptance or denial of the proposals put forward by private actors. Meanwhile, update periods for the plans themselves are often too long, and the related procedures too cumbersome. It should be added that both officers and professional actors are seldom trained to consider landscape or biodiversity-related themes as active tools of urban design: the various fields of intervention still appear disconnected (Isman 2004). These limits are particularly evident when it comes to infrastructure designing and building. In such cases, economic interests tend to overshadow all other aspects, although many European experiences speak in favour of the possibility of integrating infrastructure building, landscape planning and biodiversity conservation (Shannon and Smets 2010, 2011). The problem is inherent in the system: the Italian legislation conceives most governance tools and rights as an emanation of the central state. Any attempt to promote bottom-up policies is hindered by the substantial aim of the law: syncretism finds stringent limits in the current separation of the various aspects of territorial management. Meanwhile, the institutional relations among the central government and local/regional authorities are continuously being debated (Bonaudo 2005). Designing political measures implies that outstanding cultural values are identified and highlighted, in such a way as suggested, for example, by the codification of UNESCO sites. This tendency risks to clash with the intent to assure adequate "knowledge, conservation, planning and management" of the "whole territory", as stated by the *Codice dei Beni Culturali e del Paesaggio* (Cultural Heritage and Landscape Code, art. 135, c. 1). Italian landscape planning, as conceived in the same article 135 of the *Codice,* cannot be considered as an effective implementation of the ELC, since it is often linked to obsolete notions of traditional urban zoning. Although the coordination of different specializations is achieved through various forms of multidisciplinary actions, the state of the art of the academic and political debate would still recommend a greater effort towards a transdisciplinary model (Doble and King 2011; Stokols 2006; Linehan and Gross 1998), allowing the elaboration of holistic approaches, which could help preserve and enhance the various aspects of a territory effectively.

Nevertheless, interesting suggestions for the development and enhancement of holistic policies may derive from experiences related to the management of outstanding areas and buildings. In these contexts, authorities have to consider an advanced degree of interaction between biodiversity conservation and landscape protection as a standard approach, since these features represent a priority. Protected Areas, such as transnational, national and regional parks, represent excellent case studies. In particular, interesting hints come from the Anglo-Saxon cultural area. The US *National Park Service Organic Act* of 1916 held the conservation of "the scenery and the natural and historic objects and the wildlife therein" as vital to the enjoyment and well-being of present and future generations of citizens. Similar principles are stated in the UK *National Parks and Access to the Countryside Act of 1949* (II, 5 and 11a), which considers the creation of opportunities for the understanding and enjoyment of "natural beauty, wildlife and cultural heritage" as relevant as their conservation and enhancement. The statement also includes the social and economic well-being of local communities as a primary task of national parks. These previous experiences represent good precedents for the formulation of the ELC text, thus explaining quite well why England has been one of the first countries to promote an institutional framework for the enhancement and implementation of measures related to public involvement and awareness raising (Butler and Berglund 2014) after the ratification of the Convention. Recent formulations linked to Protected Areas at a regional and local scale have specified the main terms of public involvement. A particularly clear manifesto about the relevance of public involvement could be found in 2011 on the Web portal of the natural parks of the Australian state of Victoria:

> The purpose of Park Victoria is to conserve, protect and enhance natural and cultural values, provide quality experiences, services and information to customers, provide excellence and innovation in park management, and contribute to the environmental, social and economic well-being of Victorians. (Parks Victoria 2009)

The experience related to the institution of the Park as a protected area could prove very precious for what regards the conception of holistic landscape policies: the latest approaches cannot do without considering Protected Areas as contexts, in which planning related to biodiversity and landscape preservation have to coexist in a syncretic way. In similar contexts, we see different attempts to enhance the application of bottom-up models of information and intervention, like the Public Participation Geographic Information Systems, or PPGIS (Brown and Weber 2011), and to overcome traditional political obstacles, such as the difficult dialogue among institutions, which can be verified, for instance, in the processes of constitution of transnational parks.

35.4 Concluding Remarks: Integrating Landscape and Biodiversity

The international debate at a strategic scale is trying to define a set of criteria for the definition of truly holistic approaches, which may involve conservation and protection of global cultural, natural and biological features, organized in "landscapes" and "ecosystems". On this behalf, a very important point was marked by the Rio +20 convention on sustainable development. The outcome document (*Future We Want* (United Nations 2012)) recognizes that "ecosystems… their livelihoods, their economic, social and physical well-being, and their cultural heritage" are all deeply connected in the life of human beings (art. B.30), while the "conservation, as appropriate, of the natural and cultural heritage of human settlements" is considered as a primary task (art. V.A.134). The document also underlines the relevance of "natural and cultural diversity" as active contributors to sustainable development (art. B.41). The section devoted to "sustainable tourism" (art. V.A.130) states the need to "conserve and protect the environment, respect wildlife, flora, biodiversity, ecosystems and cultural diversity, and improve the welfare and livelihoods of local communities". In the end, art. V.A.197 affirms the "intrinsic… ecological, genetic, social, economic, scientific, educational, cultural, recreational and aesthetic values of biological diversity", thus aiming at an ideal definitive overcoming of the barriers between biodiversity and cultural heritage as separate institutional fields.

It is clear that the document favours econometric-like definitions and avoids any open mention of complex terms, such as "landscape", due to their different meaning in the various cultures worldwide and to the resulting ambiguity (Bruns 2013). The document pleads the adoption of a holistic attitude towards policy-making and planning, as recently claimed also by international associations of designers (Moore and Marques 2013). Such principles will work only when considering the public as a protagonist of the multilayered process of territorial management and transformation at all scales. Many tools are available, considering international conventions and political orientations and the results of the academic research; interesting examples can be found among coordination experiences of bottom-up and top-down policy-making in the USA (Doble and King 2011; Gray 2007; Forrester 1999; Healey 1997; Hester 1989), where aware researches on this themes have been conducted since the 1960s (Arnstein 1969). Various experiences can be found in the Italian context as well (Cassatella et al. 2010); yet, due to the discontinuous support given by national institution, the implementation of up-to-date principles is still perceived as a sort of extraordinary cost. Policy-makers reason in an easier way about trying to integrate landscape policies in the existing tools of governance, as requested by the guidelines for the implementation of the ELC (*Recommendation CM/Rec* (2008)*3*, I.1.D-E-F); landscape should be used as a fundamental peculiarity of the policy itself. Compromises are sometimes too difficult and many actors could look at the ELC as a sort of limitation. Since landscape is an

> essential component of people's surroundings, an expression of the diversity of their shared cultural and natural heritage, and a foundation of their identity (ELC, art. 5.a)

landscape planning has to be faced together with other aspects of territorial planning, namely, those regarding the traditionally "less outstanding" parts of the land. Available instruments have to be re-thought according to a new holistic attitude, which shall look at landscape and ecology as winning moves for an innovative policy-making.

References

Arnstein S (1969) A ladder of citizen participation. J Am Inst Plann 35(4):216–224

Benefits beyond boundaries (2003) 5th World Park congress. International union for the conservation of nature, Durban. Available via http://data.iucn.org/dbtw-wpd/edocs/2005-007.pdf. Accessed 14 Jul 2013

Bonaudo M (2005) La Normativa sulla tutela del paesaggio. La Rivista dell'Urbanistica 4:10–11

Brown G, Weber D (2011) Public participation GIS: a new method for national park planning. Landsc Urban Plan 102(1):1–15

Bruns D (2013) The multicultural meanings of landscape. Topos 82:16–20

Bunge J (2011) New aesthetics for modern landscapes. Topos 74:94–97

Burnham D (1910) City of the future under a democratic government. Transaction of the town planning conference. R.I.B.A, London

Butler A, Berglund U (2014) Landscape character assessment as an approach to understanding public interests within the European landscape convention. Landsc Res 39(3)219–236

Carmona-Torres C, Parra-López C, Groot JCJ, Rossing WAH (2011) Collective action for multiscale environmental management: achieving landscape policy objectives through cooperation of local resource managers. Landsc Urban Plan 103(1):24–33

Cassatella C, Devecchi M, Gambino R, Larcher F (eds) (2010) Landscape education and research in Piedmont: for the implementation of the European landscape convention. Politecnico di Torino, Turin

Doble C, King M (2011) Plural planning at multiple scales. From local communities to Statewide change. Landsc J 30(1):72–85

European Landscape Convention (2000) Council of Europe, ETS n° 176. Florence, 20 Oct 2000. Available via http://conventions.coe.int/Treaty/en/Treaties/Html/176.htm. Accessed 20 May 2013

Forrester J (1999) The deliberative practitioner: encouraging participatory planning processes. MIT Press, Cambridge

Gray B (2007) The process of partnership construction: anticipating obstacles and enhancing the likelihood of successful partnerships for sustainable development. In: Glasbergen P, Biermann F, Mol A (eds) Partnerships, governances, and sustainable development. Elger, Northampton, pp 29–48

Groening G (2007) The "Landscape must become the law" – or should it? Landsc Res 32(5): 595–612

Healey P (1997) Collaborative planning: shaping places in fragmented societies. University of British Columbia Press, Vancouver

Hester R (1989) Community design today: from the inside out. Landsc J 8(2):128–137

Isman F (2004) Less Tuscany, more Berlusconi. Topos 47:54–60

Linehan JR, Gross M (1998) Back to the future, back to basics: the social ecology of landscapes and the future of landscape planning. Landsc Urban Plan 42:207–223

Management guidelines for World cultural heritage sites (1998) International Center for the Study of Restoration and Preservation of Cultural Property, Rome

Moore K, Marques B (2013) Breaking down silos and compartments. Topos 82:21–23

Parks Victoria (2009) About us: vision, purpose, values. Available via http://www.parkweb.vic.gov.au/1process_content.cfm?section=17&page=18. Accessed 1 Mar 2011

Priore R (2009) No people, no landscape. La convenzione europea del paesaggio: luci e ombre nel processo di attuazione in Italia. Angeli, Milan

Recommendation CM/Rec (2008) 3 of the Committee of Ministers on the guidelines for the implementation of the European landscape convention. Adopted by the Committee of Ministers on 6 Feb 2008 at the 1017th meeting of the Ministers' Deputies. Available via https://wcd.coe.int/ViewDoc.jsp?Ref=CM/Rec(2008)3&Language=lanEnglish&Ver=original&Site=CM&BackColorInternet=C3C3C3&BackColorIntranet=EDB021&BackColorLogged=F5D383. Accessed 17 May 2013

Savio A, Paludi G (2005) Il Concetto di paesaggio. La Rivista dell'Urbanistica 4:12–13

Shannon K, Smets M (2010) The landscape of contemporary infrastructure. Nai Publisher, Rotterdam

Shannon K, Smets M (2011) Towards integrating infrastructure and landscape. Topos 74:64–71

Stokols D (2006) Towards a science of transdisciplinary research. Am J Community Psychol 38(1–2):63–77

Stokols D (2011) Transdisciplinary action research in landscape architecture and planning. Prospects and challenges. Landsc J 30(1):1–5

United Nations (2012) The future we want. Outcome document of the United Nations conference on sustainable development in Rio de Janeiro from 20 to 22 Jun 2012. Available via http://sustainabledevelopment.un.org/futurewewant.html. Accessed 25 Jun 2013

Vienna Memorandum (2005) Managing the historic urban landscape. World Heritage Convention, Durban

Voghera A (2011) Dopo la convenzione europea del paesaggio. Politiche, piani e valutazione. After the European landscape convention. Policies, plans and evaluation. Alinea, Florence

Ward Thompson C, Sarlöv Herlin I (2004) Die Europäische Landschaft convention. The European landscape convention. Topos 47:44–53

Chapter 36
An Assessment of the Role of Protected Landscapes in Conserving Biodiversity in Europe

Nigel Dudley and Sue Stolton

Abstract Protected landscapes (IUCN category V) make up over half the area of protected areas in Europe and are thus a critical part of Europe's conservation strategy, but critics have raised serious challenges about their usefulness to conservation. We present information on existing research into their conservation effectiveness. This is used, along with additional case studies from Spain, Germany and Croatia, to provide an initial assessment of biodiversity conservation within category V. Our research suggests that protected landscapes can be effective tools for conservation, but that this is not invariably the case and depends to a large extent on whether they are well planned and effectively managed. This management approach will work better for some species and ecosystems than for others and is not suitable for all conservation tasks. The contribution reviews the available evidence, makes some recommendations about what is needed to increase the effectiveness of conservation within protected landscapes and outlines areas requiring further research.

Keywords Protected area • Category V • Biodiversity • Spain • Croatia • Germany

36.1 Introduction

In 2008, IUCN redefined a protected area as "A clearly defined geographical space, recognised, dedicated and managed, through legal or other effective means, to *achieve the long-term conservation of nature* with associated ecosystem services

N. Dudley (✉)
School of Geography, Planning and Environmental Management, University of Queensland, Brisbane, QLD, Australia

IUCN World Commission on Protected Areas and Equilibrium Research, Bristol, UK
e-mail: nigel@equilibriumresearch.com

S. Stolton
IUCN World Commission on Protected Areas and Equilibrium Research, Bristol, UK
e-mail: sue@equilibriumresearch.com

R. Gambino, A. Peano (eds.), *Nature Policies and Landscape Policies*,
Urban and Landscape Perspectives 18, DOI 10.1007/978-3-319-05410-0_36

and cultural values" (Dudley 2008, our emphasis). This new definition addressed a long-standing debate about the primacy of nature conservation within protected areas. It made clear that in IUCN's view, *within protected areas* nature should take precedence over other important benefits such as ecosystem services and cultural values. An associated principle made this even more explicit: "For IUCN, only those areas where the main objective is conserving nature can be considered protected areas; this can include many areas with other goals as well, at the same level, but in the case of conflict, nature conservation will be the priority" (Dudley 2008).

This agreement made little difference to areas already managed as dedicated nature reserves. But it was hugely significant for the "less strict" management approaches, including particularly protected landscapes and seascapes (IUCN management category V). Many of these were originally designed for recreation or landscape values and their conservation role emerged much later (Phillips 2002). Managers were faced with the task of integrating nature conservation within areas that had originally been agreed with stakeholders for different reasons. The challenges that these changes created have seldom been fully explored.

Protected landscapes have increased relatively fast compared with other IUCN protected area categories, driven by two competing pressures: to increase national protected area estates in line with the Convention on Biological Diversity's *Programme of Work on Protected Areas* and to give local stakeholders a greater say in how such lands are designated and used. Category V, with its settled human communities and emphasis on traditional management, is socially and politically more acceptable than stricter forms of protection. It represents over half the area of national parks and nature reserves in Europe (Gambino et al. 2008). But does it work as a tool for conservation?

36.2 Management Options in Protected Landscapes and Seascapes

IUCN defines a protected landscape as "A protected area where the *interaction of people and nature over time* has produced an area of distinct character with significant ecological, biological, cultural and scenic value: and where *safeguarding the integrity of this interaction* is vital to protecting and sustaining the area and its associated nature conservation and other values" (Dudley 2008, our emphasis). The concept is predicated on the conviction that some long-managed habitats contain biodiversity that has become reliant on the particular forms of management: the European Mediterranean is a classic example. Proponents argue that biodiversity is richer than it would be without the presence of human management (Atauri and Lucio 2001) and that reducing or changing management (e.g. by setting up a "strict" nature reserve) would reduce biodiversity (González Bernáldez 1992).

However, the problem with this approach, both philosophically and practically, is that it is difficult to halt all aspects of development. Attempts to maintain traditions indefinitely risk creating a "museum landscape". In practice the reality is often messier. Around our home in the Snowdonia National Park in Wales, a large category V area of mountains and woodlands, upland farming has changed dramatically over the last 30 years. Shepherds on foot have been replaced by farmers on quad bikes; hundreds of new drivable tracks have been developed in the uplands, paid for by the European Union; and the density of sheep fluctuates dramatically depending on subsidy payments. Today, tourism earns far more money than agriculture. Nature conservation values have existed alongside these other changes, sometimes suffering in consequence of the changes and sometimes being supported and regaining ground. Most other protected landscapes can recount similar stories.

Today, a new protected landscapes model is also emerging: areas that are deliberately planned and managed to combine the needs of human populations and other species. Here the emphasis is less on maintaining traditions and more on developing "landscape approaches" to conservation and sustainable development. Such approaches have gained high visibility and support, although to date they remain rather theoretical.

Not everyone subscribes to the protected landscapes vision. Many conservation planners tacitly ignore category V and VI reserves in their analyses. Harvey Locke and Phil Dearden wrote a damning critique of the protected landscape concept (2005), kick-starting a long debate about effectiveness, which continues today. Governments have a lot riding on the success of protected landscapes, in terms of their conservation strategies, but little hard evidence of success or failure.

We made a start by carrying out a literature survey of biodiversity conservation in protected landscapes and commissioning a series of examples from around the world (Dudley and Stolton 2012).

36.3 What Do We Know About Protected Landscapes and Biodiversity Conservation?

Interest in management effectiveness of protected areas is growing generally. Unfortunately, category V has been under-represented in this research and results are still inconclusive. A study across 49 protected areas in 22 countries found category to be insignificant in predicting amount of land clearing (Nagendra 2008). Research from the World Bank found that multiple use-protected areas were more effective than strict protected areas at protecting forests from fire (Nelson and Chomitz 2011). Conversely, research using the Management Effectiveness Tracking Tool found category V and VI to be the least effective protected area approaches, although the category V sample was very small (Dudley 2007). A study in Catalonia, Spain (Mallarach 2008), found protected landscapes provided habitat even for rare species like the bear and that their large size made them more effective than small, strictly protected areas. The Royal Society for the Protection of Birds found quantifiable benefits for wild species in UK category V national parks (Robins 2008). Research in the Lombardy plain in Italy found that bird diversity was significantly higher in protected landscapes than in areas outside protection (Canova 2006). But these are fragmentary examples and the gaps in our understanding are alarming. In the remainder of this chapter, we focus on three case studies from protected landscapes in Europe and draw some preliminary conclusions about effectiveness and biodiversity.

36.4 Three Case Studies from Europe: Croatia, Germany and Spain

The following case studies look at a range of protected landscapes in Europe where staff or researchers have made attempts to monitor effectiveness in delivering biodiversity conservation.

Croatia: Lonjsko Polje National Park is one of the last seminatural floodplains in Europe, including 200,000 ha of lowland riparian forest (Gugíc et al. 2012). The Central Sava River basin was never subjected to engineering solutions to flooding through use of dykes and levees, in part because the Habsburg Empire saw it as a defence against the Ottoman Empire (the old border runs through the park). Settlements and vegetable growing areas are on high ground that does not flood, while the lowland is used mainly for pasture. Unique varieties of livestock have been bred that are capable of surviving the harsh conditions of the floodplain, including seven indigenous breeds of horse, pig, cattle and geese and a variety of hound. Some farmers even use floating pig sties during the wet periods, which can occur throughout the year. Protected area managers work with local farming communities to maintain traditional management systems that also provide habitat for wildlife. At the same time, alternative livelihood strategies are appearing in the region, linked to the presence of white storks (*Ciconia ciconia*) nesting on roofs in one of the villages and various tourism and ecotourism opportunities.

Many individual research projects attest to the success of the approach for biodiversity conservation. BirdLife International recognised the site as an Important Bird Area, and the area retains bird species rare elsewhere in Europe including the corncrake (*Crex crex*) (Dumbović 2003), black stork (*C. nigra*) and spoonbill (*Platalea leucorodia*) (Schneider-Jacoby 2002). Other studies confirm the existence of rich populations of a range of groups including flowering plants, fish (Mrakovčić et al. 2002) and ground beetles (Brigić et al. 2003). Bird populations are monitored annually by park staff to assess the success of conservation management strategies. Here the classic category V model is being employed, through encouraging the continuation of long-established management practices that incidentally support wildlife.

Germany: Lüneburger Heath Nature Park protects 5,600 ha of heathland area within the larger Lüneburger Heath habitat. The main conservation efforts are directed in particular towards recreation of suitable habitat for the black grouse (*Tetrao tetrix*), which has declined throughout Western Europe due to a combination of habitat decline and abundance of predators. Only a few birds remained on the heath at the turn of the century, with a low of 20 birds recorded in 1998 (Stiftung Naturschutzpark Lüneburger Heide 2009). Management has two main strategies: predator control programmes to reduce the number of foxes and wild boars and control of emergent trees to retain 500–600 ha of heath. In particular emergent woodland is controlled partly through grazing by sheep and goats, mowing and occasional use of fire (Wormanns 2010) and partly by mechanical removal; in the latter case the timber is used for chips to power a cogeneration plant that supplies district heating to nearby homes.

Since active management was introduced, grouse numbers have increased, in Lüneburger Heath as a whole from 78 birds in 2007 to around 220 birds in 2010 out of a total German population estimated at around 2,000. Within the nature protection area of the Lüneburger Heath Nature Park itself, estimates in 2011 put the population at 66 birds (Porzelt and Liesen 2012). The heath supports other rare birds, including the nightjar (*Caprimulgus europaeus*) and woodlark (*Lullula arborea*) (Liesen 2008). Here management is much more active than in Croatia, with conservation staff planning and making interventions to create more suitable conditions for particular target species.

Spain: Somiedo Natural Park (Alba 2012) was established in 1988 and covers 30,000 ha of mountainous territory, made up of five valleys and steep slopes in between. Around 40 % is woodland, dominated by oak (*Quercus petraea*) and beech (*Fagus sylvatica*) and 45 % is grassland pasture. The area traditionally has a rich biodiversity (Marquínez et al. 1986), including the brown bear (*Ursus arctos arctos*), the Cantabrian capercaillie (*Tetrao urogallus cantabricus*) and a regional endemic member of the gentian family, *Centaurium somedanum*. The region also has a long-established human population, with about 1,400 people in 41 settlements (IAE 2011), although population is declining. The natural park was set up after long consultation and negotiations as a way of bringing fresh investment and arresting population decline. Tourism has since had an enormous impact, with jobs in service industries doubling since the establishment of the park and disposable income increasing significantly. Farming is still the main employer, although the service industries provide almost three quarters of the income.

Some 60 social and conservation indicators have been identified for monitoring (Álvarez 2006). Most trends are positive, with the exception of health of freshwater fish populations, presence of some invasive plant species and continuing decline of capercaillie, which is currently not recorded as breeding from within the natural park. The brown bear population is increasing (Fernández-Gil et al. 2010); the bear was formerly regarded as an enemy but is now viewed with greater pride as one of the park's assets. Here we have a hybrid approach, with elements of traditional management remaining but a fresh emphasis on ecotourism.

36.5 Conclusions

It is clear that the category V approach can work in terms of biodiversity conservation, but not totally, not always and not unless a wide range of stakeholders take conservation seriously (Dudley and Stolton 2012). There is still a great deal to be learned. We need a concerted effort to look at the conservation role of category V protected areas within Europe. The IUCN World Commission on Protected Areas already has a joint task force with the IUCN Species Survival Commission, examining long-term biodiversity trends in protected areas, and a special focus on category V is overdue. Research suggests that much of the data needed to do this already exist (Nolte et al. 2010).

More perhaps than in most protected areas, conservation in protected landscapes does not result automatically from designation but requires further work, usually including further negotiations with stakeholders. Trade-offs between human priorities and biodiversity needs are often not clear-cut, particularly when there is the potential for human wildlife conflict. In some cases, active management interventions are needed as well as more traditional land management, particularly when critical species or habitats have declined. Careful monitoring is essential to check if contemporary land management practices support or damage biodiversity. This need for a continually evolving, adapting approach will become greater under conditions of climate change.

Social attitudes are also critically important. Sometimes the main role of "protection" seems to be as a vehicle to raise awareness within resident communities about the value of biodiversity. An influx of newcomers, including people with a direct stake in the success of conservation such as those involved in tourism, can help to foster a new relationship with nature. Visitors influence residents who learn to see their natural environment from other perspectives, which is not to suggest that local communities are unaware of these values. One lesson learned from the study is that motivations for establishment of a category V reserve are seldom narrowly utilitarian. Pride of place and a desire to maintain cultural values feature alongside deep-rooted respect for natural values within many communities. Building on and developing from these core values remains at the heart of the protected landscapes approach.

References

Alba J (2012) Spain: Somiedo Natural Park "Bringing everything to life". In: Dudley N, Stolton S (eds) Protected landscapes and wild biodiversity. Values of protected landscapes and seascapes, vol 3. IUCN/GIZ/BMZ, Gland

Álvarez MA (2006) Elaboración de un sistema de indicadores de desarrollo sostenible para la Reserva de la Biosfera de Somiedo. INDUROT Consejería de Medio Ambiente

Atauri JA, De Lucio JV (2001) The role of landscape heterogeneity in species richness distribution of birds, amphibians, reptiles and lepidopterans in Mediterranean landscapes. Landsc Ecol 16: 147–159
Brigić A, Vujčić-Karlo S, Stančić Z (2003) Carabidae fauna in different terrestrial habitats of Krapje Đol. Lonjsko Polje Nat Park Bull 5:1–2
Canova L (2006) Protected areas and landscape conservation in the Lombardy plain (northern Italy): an appraisal. Landsc Urban Plan 74:102–109
Dudley N (2007) Tracking progress in managing protected areas around the world: an analysis of two applications of the management effectiveness tracking tool developed by WWF and the World Bank. WWF, Gland
Dudley N (ed) (2008) Guidelines for applying protected area management categories. IUCN, Gland
Dudley N, Stolton S (eds) (2012) Protected landscapes and wild biodiversity. Values of protected landscapes and seascapes, vol 3. IUCN/GIZ/BMZ, Gland
Dumbović V (2003) The corncrake (Crex crex), a globally endangered bird species continues to nest successfully in Lonjsko Polje Nature Park. Lonjsko Polje Nat Park Bull 5:27–35
Fernández-Gil A, Ordiz A, Naves J (2010) Are Cantabrian brown bears recovering? URSUS 21(1): 121–124
Gambino R, Talamo D, Thomasset F (eds) (2008) Parchi d'Europa. Verso una politica europea per le aree protette. ETS Edizioni, Pisa
González Bernáldez F (1992) Ecological consequences of the abandonment of traditional land use systems in central Spain. Options Méditerr 15:23–29
Gugíc G, Zupan D, Zupan I (2012) Croatia: the floodplain ecosystem of the Central Sava River Basin. In: Dudley N, Stolton S (eds) Protected landscapes and wild biodiversity. Values of protected landscapes and seascapes, vol 3. IUCN/GIZ/BMZ, Gland
IAE Instituto Asturiano de Estadística – SADEI (2011) Datos básicos de Asturias 2011. IAE
Liesen J (2008) Wärme für Wilsede schützt Birkhuhn und Ziegenmelker. Land in Form – Magazin für Ländliche Räume 3:32
Locke H, Deardon P (2005) Rethinking protected area categories and the new paradigm. Environ Conserv 32(1):1–10
Mallarach JM (ed) (2008) Protected landscapes and cultural and spiritual values, vol 2, Values of protected landscapes and seascapes. IUCN/GTZ/Obra Caixa Social Catalunya, Heidelberg
Marquínez J et al (1986) Estudio ambiental del Concejo de Somiedo. Universidad de Oviedo, Oviedo
Mrakovčić M, Mustafić P, Ćaleta M et al (2002) Characteristic of fish communities in Lonjsko Polje Nature Park. Lonjsko Polje Nat Park Bull 4(1–2):9–56
Nagendra H (2008) Do parks work? Impact of protected areas on land cover clearing. Ambio 37(5):330–337
Nelson A, Chomitz KM (2011) Effectiveness of strict vs. multiple use protected areas in reducing tropical forest fires: a global analysis using matching methods. PLoS ONE 6(8):e22722
Nolte C, Leverington F, Kettner A et al (2010) Protected area management effectiveness in Europe: a review of application, methods and results. BfN Skripten 271a, Bundesamt für Naturschutz, Bonn
Phillips A (2002) Management guidelines for IUCN category V protected areas: protected landscapes/seascapes. Best practice protected area guidelines series no. 9. IUCN, Gland/Cambridge
Porzelt M, Liesen J (2012) Germany: managing biodiversity in nature parks. In: Dudley N, Stolton S (eds) Protected landscapes and wild biodiversity. Values of protected landscapes and seascapes, vol 3. IUCN/GIZ/BMZ, Gland
Robins M (2008) Protected landscapes: sleeping giants of English biodiversity. Ecos 29(1):74–86
Schneider-Jacoby M (2002) Croatia, home of the last Central European Spoonbill population in alluvial wetlands. In: Report of the 68th EUROSITE workshop 19–22 Apr 2002, Texel, Wetland Management for Spoonbills and Associated Waterbirds, vol 68, pp 17–21

Stiftung Naturschutzpark Lüneburger Heide (ed) (2009) Artenschutzprojekt zum Schutz des Birkhuhns im Naturschutzgebiet Lüneburger Heide. Stiftung Naturschutzpark Lüneburger Heide, Bispingen

Wormanns S (2010) Projekt zum Schutz des Birkhuhns im Naturschutzgebiet Lüneburger Heide. Stiftung Naturschutzpark Lüneburger Heide, Bispingen

Chapter 37
Lessons Learned from U.S. Experience with Regional Landscape Governance: Implications for Conservation and Protected Areas

Daniel Laven, Nora J. Mitchell, Jennifer Jewiss, and Brenda Barrett

Abstract It is generally acknowledged that protected areas do not encompass the scale necessary for effective conservation of socio-ecological systems. Consequently, there have been repeated calls for a "new paradigm" for conservation that transitions from "islands" to "networks." By extending conservation to reflect wider landscape perspectives, this approach integrates community development and economic and quality of life interests, thereby forging productive relationships between protected areas and their regional context. This broadened agenda involves many more landowners, organizations, and levels of government and requires coordination, partnerships, and new forms of governance. Drawing from nearly a decade of research, this contribution examines US experience with this new paradigm for conservation and models of network governance. The findings from this research program indicate that three key dimensions are fundamental to governance: engaging a diversity of stakeholders and building consensus, creating and sustaining ongoing networks of partners, and developing a central hub for the network. This central coordinating and facilitating function appears to be an essential governance element as it is the activity of these networks of private and

D. Laven (✉)
Department of Tourism Studies and Geography/European Tourism Research Institute, Mid Sweden University, Kunskapensväg 1, Ostersund 83125, Sweden
e-mail: daniel.laven@miun.se

N.J. Mitchell
Rubenstein School of Environment and Natural Resources, University of Vermont, Burlington, VT, USA
e-mail: norajmitchell@gmail.com

J. Jewiss
Department of Leadership and Developmental Sciences, University of Vermont, Burlington, VT, USA
e-mail: jennifer.jewiss@uvm.edu

B. Barrett
Living Landscape Observer, Harrisburg, PA, USA
e-mail: brendabarrett88@gmail.com

R. Gambino, A. Peano (eds.), *Nature Policies and Landscape Policies*,
Urban and Landscape Perspectives 18, DOI 10.1007/978-3-319-05410-0_37

public partners that deliver accomplishments. This contribution suggests that despite their challenges, networked-based models can strengthen social capital at regional levels, thereby increasing capacity for innovation, adaptation, and resiliency.

Keywords Governance • Regional landscape conservation • Evaluation research case studies • Networks

37.1 Introduction

It is generally acknowledged that protected areas do not encompass the scale necessary for effective conservation of socio-ecological systems (Brown et al. 2005; Thompson et al. 2011), particularly in the face of rapid social and economic shifts as well as climate change (Gunderson and Holling 2002; Rands et al. 2010). Consequently, there have been repeated calls for a "new paradigm" for conservation that transitions from "islands" to "networks" to better reflect regional landscape dynamics (Rössler and Mitchell 2005; Mose 2007). By extending conservation to the wider landscape, this approach helps forge productive relationships between protected areas and their regional landscape context.

This shift requires integration of conservation goals with community development, economic vitality, and quality of life interests. This broadened agenda involves many more landowners, organizations, and levels of government and requires coordination, partnerships, and new forms of governance (Dudley 2008; Kothari et al. 2013). Existing governmental structures, based on political jurisdictions, often do not coincide with landscape systems or match the dynamics associated with new threats or opportunities.

Many countries now have experience in integrating regional perspectives into management of natural and cultural landscapes. The IUCN has, for many years, recognized protected landscapes and seascapes as one management category (Brown et al. 2005; Dudley 2008). Cultural landscapes with traditional land practices often extend over large areas (Mitchell et al. 2009). Similarly, the European Landscape Convention recognizes the value of all landscapes and calls for close collaboration between the public, local, regional, and national authorities along with private organizations to achieve sustainable development at the landscape scale (Jones and Stenseke 2011).

In the USA, there has been increasing interest in regional landscape strategies (Regional Plan Association 2012; McKinney and Johnson 2013). One example is the National Heritage Area (NHA) model, which has grown substantially over the last two decades; today, 49 areas have been established (US National Park Service (USNPS) NHA n.d.). This chapter examines the US experience with this new paradigm for conservation and, in particular, new models of governance associated with NHAs. Our findings and conclusions are drawn from over a decade of research on NHAs and the related Chesapeake Bay Gateways and Watertrails Network (Chesapeake program).

37.2 US National Heritage Areas and the Chesapeake Bay Gateways and Watertrails Network

NHAs are congressionally designated places where natural and cultural resources form a cohesive landscape that tells significant stories reflecting the diverse heritage of the US. NHAs integrate natural and cultural resource conservation goals with economic and community development objectives across multiple sites within a defined region (Barrett and Mitchell 2003; USNPS NHA n.d.). Even with national designation, the areas retain their existing landownership, which can include parks and protected areas.

Because NHAs often cross multiple jurisdictions, their governance structures differ from the traditional national park approach. NHA enabling legislation creates a management entity that generally includes representatives from local, state, and federal levels of government, including the USNPS, nonprofit organizations, and resident, business, and other stakeholder groups. Members of the NHA management entity work together to identify and conserve important heritage resources, improve the local economy, create recreational opportunities for residents and visitors, and guide the region's planning for the future. To be successful, the NHA management entity functions as a facilitator since it creates and sustains networks of private and public partners. It is the activity of these networks that delivers NHA accomplishments.

The Chesapeake Bay watershed covers 64,000 square miles and extends into six states – Delaware, Maryland, New York, Pennsylvania, Virginia, West Virginia – and the District of Columbia. The Chesapeake program was established in 1998 to coordinate interpretation and stewardship throughout the region. The Chesapeake program is a network of over 150 gateways, including museums, historic sites, parks, wildlife refuges, and water trails. Like NHAs, the Chesapeake program relies on partnerships for its governance and management.

NHAs and the Chesapeake program represent similar versions of the "new paradigm" for conservation. They are lived-in landscapes at regional scales and the strategy integrates a range of management objectives (natural, cultural, social, and economic) and employs partnership-based governance.

37.3 Evaluation Research Conducted at NHAs and the Chesapeake Program

Between 2003 and 2009, the USNPS Conservation Study Institute worked with the University of Vermont and the QLF/Atlantic Center for the Environment to conduct evaluation research that included case studies at three NHAs and one similar case study of the Chesapeake program. Each case study included a qualitative component; 30 key stakeholders at each site participated in in-depth interviews.

The qualitative data for each site were analyzed and reported individually (Tuxill et al. 2005; Copping et al. 2006; Jewiss 2007; Tuxill et al. 2008). In addition, the analyses were designed to be comparable across sites and cross-case analyses of the NHA findings identified many common elements (Laven 2006; Laven et al. 2010a, b). As a result, a theoretical framework for NHAs was developed (Jewiss et al. 2008; Laven et al. 2013).

Similar governance patterns were observed across the NHA and Chesapeake study sites. The following section presents key governance findings illustrated with qualitative data from the NHA and Chesapeake studies. This discussion broadens the scope of previously published work on this topic by delving deeper into governance, which is a critical element for this "new paradigm" of conservation.

37.4 Key Findings from a Governance Perspective

Findings from the Chesapeake and NHA studies suggest that the following governance dimensions are essential for regional landscape initiatives: (i) engaging a diversity of stakeholders and building consensus, (ii) moving from partnerships to networks, and (iii) developing a central hub and a platform for network governance.

Engaging Diverse Stakeholders and Building Consensus. Numerous study participants indicated that engaging a diversity of stakeholders and building consensus on a wide range of issues is critical to effective governance. Laven et al. (2010b) report that doing so is essential for "development and pursuit of a common, regional mission" (p. 204). These authors further note that NHAs have "helped find points of consensus while facilitating dialogue" between an array of interest groups and actors (p. 204), such as the business community, nature conservation, historic preservation interests, and recreation groups. Study participants also repeatedly emphasized the importance of working across institutional, political, and sectoral boundaries. For many respondents, their perception of "good governance" was closely linked to crossing boundaries, overcoming obstacles, and finding synergies. One NHA partner characterized this dimension of governance as unintended yet very positive:

> The unintended consequence [of the NHA approach] is to take the strong liberal environmental groups and to integrate them more thoroughly with the business community.... I never expected to see [that]. (Laven 2006, p. 59)

Similarly, partners in the Chesapeake program described the ability to unite a broad spectrum of organizations in promoting stewardship of the vast Chesapeake Bay watershed. The following viewpoint was expressed by many others from around the region: "This is the first time I've seen any organization try to bring together all the disciplines across the Bay to create a real force for stewardship" (Jewiss 2007, p. 27).

Moving from Partnership to Networks. A related pattern was the shift in thinking about collaboration from a partnership to a network perspective.

Partnerships are typically established for a defined period of time to achieve a specific purpose or objective. Networks, by contrast, are often thematically driven and offer members the opportunity to engage in strategic issues that exceed the capacity of any single actor (or any single partnership between two actors) within a region to address. Indeed, many study participants underscored that NHAs can "transcend" a host of boundaries because their networks are thematic rather than political or interest specific (Laven 2006, p. 70). The following NHA partner clearly reflected this network perspective:

> There used to be all these individual groups running around on their own, and they were not really connected. The [NHA] has since given them the ability to connect to each other, and I think there have been tremendous cooperative efforts as result.... I think the [NHA] has really become the focal point for all of these efforts over the years. (Laven 2006, p. 55)

This shift towards a network perspective has strong implications for governance because of the very different ways networks behave compared to project-based partnerships.

Developing Network Governance and the Central Hub. In terms of networks, the most important governance implication is the central hub concept. In the NHA studies, interviewees identified over 30 distinct roles that the NHA organization plays within their respective networks, while also serving as the actor that maintains network function (Tuxill et al. 2005, 2008; Copping et al. 2006; Laven 2006; Laven et al. 2010a, 2013). One study participant described how the NHA helped his organization find its place within a complex and constantly evolving regional conservation strategy (Laven 2006, p. 76). The following interviewee used an umbrella analogy to describe how NHAs interact with different partners while maintaining overall network function:

> Having the heritage area to pull everybody together lets us think about these things.... It brings value to the community because I'm just one partner in this whole area, and it gives all of us a better umbrella to work under instead of just being an individual entity among many others. (Laven 2006, p. 105)

Moreover, almost all of the study participants indicated that no other organization within their region could replace this "central hub" function played by the NHA organization (Laven et al. 2010a). Our analyses suggest that one reason for this response is that network governance models – including the central hub function – can be effective at blending top-down and bottom-up approaches, while also ensuring that decision making activities are conducted in open and inclusive ways. A member of the governing body (the working group) of the Chesapeake Bay Gateways Network echoed this notion by describing how the USNPS brought together diverse stakeholders as a network:

> As the program did come into being ... it became clear that not only was our input sought, but also our experience in dealing with the museum and historic site community. We could have a real impact on how the program developed.... (Jewiss 2007, pp. 29–30)

These findings indicate that effective network governance (i) enables the sharing of risks and rewards of project development (Laven et al. 2013), (ii) facilitates communication flow between previously unconnected actors (Laven et al. 2010a), and (iii) works to ensure that network activities reflect shared, regional interests, rather than the interests of any specific set of actors (Laven et al. 2010b).

The findings reported in this section parallel some of the key themes in the broader literature on network governance. For example, research suggests that regional actor networks rely on (1) reciprocity between partners, (2) relationships between actors, (3) organizational learning throughout the network, and (4) the ability to be creative and flexible (Innes and Booher 2003; Bogason and Musso 2006). Other research has suggested that interorganizational networks tend to be more effective in the presence of a central coordinating entity (Mandell 1984; Provan and Milward 2001). Consequently, a central coordinating function appears to be an essential governance element for NHAs and the Chesapeake program and has important implications for a "new paradigm" for conservation and the future of protected area management.

37.5 Implications for Conservation and Protected Area Governance

Recent Additional Evaluations Reinforce Research Findings on Governance. These findings on network-based governance have been reinforced by recent evaluations of nine NHAs (Alliance of National Heritage Areas 2013). These studies generally conclude that NHAs have extensive networks of partners and that the NHA coordinating organization plays a significant role as a central hub. Overall, the evaluations demonstrate that the NHAs are effective in conserving and interpreting cultural and natural resources on a regional scale. The evaluations also document their success in leveraging public financial investments and supporting recreational and economic development and overall community vitality.

Implications for the Future. The findings of the evaluation research on NHAs and the Chesapeake program indicate that three key dimensions are fundamental to governance: engaging a diversity of stakeholders and building consensus, creating and sustaining an ongoing network of partners, and developing a central hub for the network. The combination of these characteristics creates an ongoing participatory process that builds a shared vision and related broad-based strategies, which can be integrated across public and private sectors to meet conservation and other community goals. Consequently, natural and cultural conservation becomes part of a larger and more relevant strategy for the region and its communities. This governance approach also places protected areas within a regional landscape that results in mutual benefits to the protected area and the surrounding territory. This participatory model of governance complements existing governmental structures and helps navigate tensions between local, regional, and national interests in ways that empower collective action. This shift to networked governance is, however, not without its challenges. This form of governance requires new perspectives on leadership. It requires communities and protected area leaders within the region to recognize the importance of governance through a networked approach and embrace their roles within the network. This shift can be challenging for protected area managers as regional landscape approaches represent a substantial shift in conservation thought and practice. Developing leadership for a network hub is another key challenge, as this role is generally unprecedented, unrecognized, and therefore unfamiliar to many protected area managers. Encouragingly, the evaluation research provides evidence that some NHA leaders have strong partnership skills and some NHA entities have performed the role of the network hub with success. Building on the experience to date, there is clearly much to be learned and shared regarding leadership within networked governance for regional landscapes. Learning from experience with networked models of governance at a regional scale can support innovation in other settings, advancing conservation and community vitality. Through inclusive ongoing engagement in a network, communities can build the social capital and capacity at the regional level that will strengthen their ability to cooperate and take coordinated action individually and collectively. As regions face new and unprecedented challenges, this form of network-based governance can provide capacity for innovation, adaptation, and resiliency.

References

Alliance of National Heritage Areas (2013) National Park Service evaluations find national heritage areas are achieving their purpose and accomplishing their goals. Available via http://www.nationalheritageareas.us/pdf/Final-NHA.pdf. Accessed 20 Nov 2013

Barrett B, Mitchell N (eds) (2003) Stewardship of heritage areas. Thematic issue of The George Wright Forum 20(2):1–88

Bogason P, Musso JA (2006) The democratic prospects of network governance. Am Rev Public Adm 36:3–18

Brown J, Mitchell N, Beresford M (eds) (2005) The protected landscape approach: linking nature, culture and community. IUCN, Gland

Copping S, Huffman P, Laven D, Mitchell N, Tuxill J (2006) Connecting stories, landscapes, and people: Exploring the Delaware & Lehigh National Heritage Corridor partnership. USNPS Conservation Study Institute, Woodstock

Dudley N (ed) (2008) Guidelines for applying protected areas management categories. IUCN, Gland

Gunderson LH, Holling CS (eds) (2002) Panarchy: understanding transformations in human and natural systems. Island Press, Washington, DC

Innes JE, Booher DE (2003) Collaborative policymaking: governance through dialogue. In: Hajer M, Wagenaar H (eds) Deliberative policy analysis: understanding governance in the network society. Cambridge University Press, Cambridge, UK, pp 33–59

Jewiss J (2007) A qualitative evaluation of the partnership system of the Chesapeake Bay Gateways and Watertrails Network. Report submitted to the Chesapeake Bay Gateways and Watertrails Network, Annapolis

Jewiss J, Laven D, Mitchell N, Huffman P (2008) Development of a National Heritage Area evaluation strategy: report on phase I. USNPS Conservation Study Institute, Woodstock, VT in cooperation with the USNPS National Heritage Areas Office and the Alliance of National Heritage Areas

Jones M, Stenseke M (eds) (2011) The European landscape conventions: challenges of participation. Springer, Dordrecht

Kothari A, Camill P, Brown J (2013) Conservation as if people also mattered: policy and practice of community-based conservation. Conserv Soc 11(1):1–15

Laven D (2006) Evaluating national heritage areas: theory, methods, and application. Doctoral dissertation, University of Vermont, Burlington

Laven D, Jewiss J, Mitchell N (2013) Toward landscape-scale stewardship and development: a theoretical framework of United States national heritage areas. Soc Nat Resour 26(7):762–777

Laven DN, Krymkowski DH, Ventriss CL, Manning RE, Mitchell NJ (2010a) From partnerships to networks: new approaches for evaluating US National Heritage Area effectiveness. Eval Rev 34:271–298

Laven D, Ventriss C, Manning R, Mitchell N (2010b) Evaluating US national heritage areas: theory, methods, and application. Environ Manag 46:195–212

Mandell MP (1984) Application of social network analysis to the implementation of complex project. Hum Relat 37:659–679

McKinney M, Johnson S (2013) Large landscape conservation in the rocky Mountain West: an inventory and status report. University of Montana, Missoula

Mitchell N, Rössler M, Tricaud PM (2009) World heritage cultural landscapes: a handbook for conservation and management (World heritage papers 26). UNESCO World Heritage Centre, Paris

Mose I (ed) (2007) Protected areas and regional development in Europe: towards a new model for the 21st century. Ashgate Publishing, Aldershot, pp 3–20

Provan KG, Milward HB (2001) Do networks really work: a framework for evaluating public-sector organizational networks. Publ Admin Rev 61:414–423

Rands MR, Adams WM, Bennun L, Butchart SH, Clements A, Coomes D, Vira B et al (2010) Biodiversity conservation: challenges beyond 2010. Science 32(5997):1298–1303

Regional Plan Association (2012) Landscapes: improving conservation practice in the Northeast Megaregion. Available via http://www.rpa.org/library/pdf/RPA-Northeast-Landscapes.pdf. Accessed 5 Nov 2013

Rössler M, Mitchell N (2005) Landscapes linkages without boundaries? In: World Heritage reports 16. UNESCO World Heritage Centre, Paris, pp 23–26

Thompson JD, Mathevet R, Delanoë O, Gil-Fourrier C, Bonnin M, Cheylan M (2011) Ecological solidarity as a conceptual tool for rethinking ecological and social interdependence in conservation policy for protected areas and their surrounding landscape. C R Biol 334(5):412–419

Tuxill J, Huffman P, Laven D, Mitchell N (2008) Shared legacies in Cane River National Heritage Area: linking people, traditions, and culture. USNPS Conservation Study Institute, Woodstock

Tuxill J, Mitchell N, Huffman P, Laven D, Copping S, Gifford G (2005) Reflecting on the past, looking to the future: sustainability study report. USNPS Conservation Study Institute, Woodstock

USNPS National Heritage Areas (n.d.) Available via http://www.cr.nps.gov/heritageareas/htm. Accessed 5 Nov 2013

Chapter 38
Park, Perception and the Web

Visual Preferences on *Digital Native* Urban Parks as a Tool for Investigating Urban Landscape Perception in Europe

Caterina Franchini and Elena Greco

Abstract The contribution identifies new digital sources for assessing and monitoring visual preferences of landscape, in order to strengthen the tools of specific active protection and planning regulations. The subject of the research is the European urban parks of the twenty-first century (that we called digital native) because of their belonging to the digital era and for the crucial role they play in nature and landscape policies. Urban parks represent a bridge between nature and culture, a place of encounter between human projects and the environment, especially since the nineteenth century, when they are conceived to regenerate cities, to improve urban landscapes and biodiversity. An effective photo-based method of investigation is proposed by this research. It deals with users' visual preferences related to scenic values by focusing on three categories of features: context, structure and functional system. A combination of free and directed sorting procedures based on visual communication analysis was applied to pictures taken and freely published on the web by park users. Quantitative and categorical data derived from the multiple sorting methods resulted from the research. The data allow to actualize sustainable development policies based on shared values. Preference ratings, showing that uniqueness and identity values are the most shared ones, while the values related to nature are less perceived by users, underline the importance of strengthening the natural values perception in future environmental policies.

Keywords Digital views • Landscape perceptions • Urban parks • Visual preferences • Scenic values

C. Franchini (✉) • E. Greco
Interuniversity Department of Regional and Urban Studies and Planning (DIST), Politecnico and Università di Torino, Viale Mattioli 39, 10125, Turin, Italy
e-mail: caterina.franchini@polito.it; elena.greco@polito.it

R. Gambino, A. Peano (eds.), *Nature Policies and Landscape Policies*, Urban and Landscape Perspectives 18, DOI 10.1007/978-3-319-05410-0_38

38.1 The Research: Subject and Purposes

This paper presents the results of a pilot research project on urban landscape visual perception. The research proposes an investigation through the analysis of visual preferences related to the most recent urban parks that we call *digital native,* to indicate their belonging to the digital era.

According to Simmel (1968), modern social life implies an increasing significance of purely visual impressions. From the 1980s, photo-based studies have been established in the field of landscape (Shuttleworth 1980; Scott and Canter 1997). In the digital age, due to the rise of digital photography, visual impressions are pervasively fixed by the photo shot.

This study uses the digital pictures which are published on the popular and largely used software *Google Earth*[1] and on the website *Panoramio*[2] as the primary source of analysis.

The web offers new possibilities to collect a huge quantity of pictures directly taken and published by the users of urban spaces. These pictures are a medium to express user preferences about a given space. Moreover, each picture shows a particular feature of interest that can be sorted by its specific visual landscape indicator, through visual communication analysis.

The first step of the investigation focused on contemporary urban parks since they are located in confined areas where urban regeneration has taken place. However, the same method can be adopted to broader research projects related to landscape. The purpose of the analysis is to assess the visual perception as a supporting tool for urban policies and more specifically, for sustainable development, nature and landscape conservation. If it is accepted that the perception of the image of the city is an instrument for urban space interpretation and design (Cullen 1961; Marchigiani 2002), the resulting data can be useful to develop specific strategies for enhancement, renovation and quality improvement of urban landscapes and the natural environment.

Urban parks are one relevant component of the *urban landscape* and greatly contribute to the perception of the overall urban and environmental quality. We consider the *urban park* as a particular kind of garden situated in an urban area. According to Bernard Tschumi, the concept of the park as an open space ended in the 1980s and the so-called *cultural park* was born (Barzilay et al. 1984). In fact, until the 1980s, the concept of public space was strictly functional, i.e. squares for events, enclosures for sports, aromatic gardens.

Following the economic changes in urban areas, citizens asked for cities to be regenerated, with gardens, greens and squares, which could reflect the collective imagination. Urban parks changed and some include the multifunctional uses, erasing clear boundaries and indications of exclusive use (Capezzuto 2004). This

[1] Google Earth. Google Inc. 2013. Accessed May 2013.

[2] Panoramio. Photos of the World: http://www.panoramio.com. Accessed May 2013

is the reason why, since the 1980s, for the new planned parks, the term *garden* has been replaced by the term *urban park*.

The contemporary urban park thus becomes the *agora* of a multi-ethnic society, a place of tolerance and multicultural active space. It ethically reacquires the *genius loci* (Venturi Ferriolo 2002), to become a centre for urban regeneration and sustainable development. Beyond the multiple classifications of urban parks (thematic park, industrial park, memorial park), the constant feature is its relationship with landscape and nature. If *garden*, which is purely contemplative, can be considered a perceptual threshold of an area of *loisir* with predominantly aesthetic fruition (Assunto 1973; D'Angelo 2001), *urban park* is an interactive space. Here, the user becomes the protagonist, collaborating with the life of the park. Thus it is crucial to investigate the user perception focusing on how, and to what extent, a detail of a park strikes the visitor. At times a park will be remembered just for a detail: its shape, its colour, its location and the feeling that transmits through its artificial and natural components.

The concept of landscape includes physical reality as well as a representation, which exists in the imagination. Thus, the landscape cannot be simply observed; it must be experienced (Rodwell 2010). In order for a landscape to exist, there must be a subject that perceives it. This assumption derives from the definition of landscape provided by the European Landscape Convention (Council of Europe 2000). When this area is an urban area, we can use the term *urban landscape*. The fundamental question, which arises, is how to assess the urban landscape perception. Section 1.b of the ELC provides part of the answer: "taking into account the particular values assigned to [landscape] by the interested parties and the population concerned".

The Guidelines for the Implementation of European Landscape Convention, Recommendation CM/Rec(2008)3 on the application of the ELC, help to investigate the subject:

> The sensory (visual, auditory, olfactory, taste) and emotional perception which a population has of its environment and recognition of the latter's diversity and special historical and cultural features are essential for this respect and safeguarding of the identity of the population itself and for the enrichment of the individual and of society as a whole. (Council of Europe 2008)

As previously stated, this study proposes a method of analysis based on the visual sense, which is just one component of sensory perception. Nevertheless, this component is involved in new instruments of communication, belonging to the web, in which subjectivity emerges, even unconsciously (Ware 2013). The study on the visual preferences can be useful even in a conservation perspective of historic cities, as European cities are (Roth 2006).

Even the most recent European urban parks are an integral part of the *historic urban landscape* (Ippolito 2006), as defined by the Recommendation on the Historic Urban Landscape (World Heritage Centre 2011). The same Recommendation proposes a kind of approach that considers the perception of local communities.

The assessment of urban landscape perception is fundamental to formulate sustainable development policies addressing to nature and culture (Smardon 1987), as recommended by both the international documents European Landscape Convention and Historic Urban Landscape.

38.2 Method of Analysis

Visual preferences expressed on digital pictures are based on specific components and features of the landscape scene and on subjective or objective factors of perception. Visual preferences of urban parks were sorted using visual indicators selected among those classified by Cassatella (2011). The indicators are grouped in three categories, *context*, *structure* and *functional system,* defined as follows.

Context includes pictures taken within the park and focus on the surrounding area. The constructed surroundings can include streets, avenues, corners, monumental buildings and bridges. The natural surroundings can display hills, mountains, seas, rivers and lakes. The perceptive relationship between the park and the city highlights the public significance of various landscape values. The indicators of landscape values, which were selected for this category, are panoramic views, vantage points and atmospheric views.

- *Panoramic or scenic views* express a holistic opinion on the landscape surrounding the park showing natural or anthropic features. They convey values like naturalness, historicity or modernity and beauty related to the landscape surrounding the park.
- *Indoor vantage points* on natural or anthropic features underline a more selective perception on the relationship between the park and the outside landscape. Landmarks – as tall buildings or monumental historic buildings – situated outside the park and well visible from it contribute to the perception of the city identity values.
- *Atmospheric views*, which comprise *panoramic views* and *vantage points*, show the landscape outside the park under a particular weather condition or under an unusual light (at dusk, during a lightning storm, under the snow, at night). For this indicator, the landscape value is mainly aesthetic.

Structure category classifies the pictures framing the park shapes and colours. The landscape indicators, which were identified for the structure are aesthetic views, atmospheric views and geometric views. These indicators disclose subjective aesthetic preferences related to identified portions of the park, or natural and artificial elements as, for instance, trees, flowers, path, walls and flowerbeds. For this category, the park landscape values are natural beauty, naturalness and artificiality.

- In particular, *geometric views* are those focusing on park design. They establish a direct relationship between visual elements of the park in terms of geometry, colours, textures, lights and shadows – of paths, flowerbeds, lighting, street furniture, fountains, gangways and footbridges.

Functional system is related to the park fruition both active and contemplative (artistic and memorial). The assumed indicators are fruition views and imageability. The first is associated to the value of use and to the social value, while the second is correlated to the values of uniqueness and identity.

- *Fruition views* focus on mobility, accessibility, signage and functional infrastructures – such as sport facilities, playgrounds – and in general depict people in the park.
- Pictures related to *imageability* show distinctive elements of the park such as work of arts, monuments, remains of previous uses of the area and landmarks inside the park. Works of art can also be of spontaneous nature, for instance, graffiti, murals or tags. The historic remains can be intentional or unintentional such as train rails, tunnels, reservoirs, mills, and smokestacks.

The selection of the case studies was carried out following several parameters. First of all, it was decided to choose the *digital native* urban parks built in the twenty-first century. In fact, these parks were built at the time of the spread of web software such as *Google Earth*, chosen for our analysis.

The source of this first selection was the website of *Landezine – Society for Promotion of Landscape Architecture*[3] in which, under the parks section, many of contemporary urban parks are published.

Once the European urban parks were selected, the sample of this study included only those parks for which a large number of pictures shared by park users were available on *Google Earth* and *Panoramio*. In all there are 25 case studies (see references)[4]. Each picture has been classified in its category and in the related indicator, previously described.

Finally, the percentage of pictures for each indicator was calculated for every park. Thus, it was possible to make a statistic investigation about user visual preferences of urban *digital native* parks. The results of this analysis are summarized in tables (Tables 38.1 and 38.2).

---

[3] Landezine – Society for Promotion of Landscape Architecture: http://www.landezine.com. Accessed April 2013.

[4] The 25 selected parks are:

- *Campa de Los Ingleses Park*, Bilbao, ES (2012); *Mangfallpark Rosenheim*, Bavaria, DE (2010); *Development Bank Of The Meurthe RaonL'Eetape*, Vosges, FR (2012); *Schöneberger Südgelände Park*, Berlin, DE (2009); *Riverside Origami – Millennium*, Budapest, HU (2011); *Plaza Del Milenio*, Valladolid, ES (2001); *Garden of Giants*, Lille, FR (2007); *Park am Gleisdreieck*, Berlin, DE (2011); *Sa Riera Park*, Palma de Mallorca, Illes Balears, ES (2007); *Madrid RIO*, Madrid, ES (2011); *St Andrew Square*, Edinburgh, Scotland, UK (2009); *Montjuïc Heights*, Barcelona, ES (2011); *Wernigerode Horticultural Show 2006*, Wernigerode, Harz, Sachsen-Anhalt, DE (2006); *Park Camillo Tarello*, Brescia, IT (2007); *Fontsanta Park*, Barcelona, ES (2002); *Anchor Park Västre Hamnen*, Malmö, SE (2001); *Dania Park*, Malmö, SE (2000); *Galindez Slope*, Bilbao, ES (2007); *Water Mirror*, Bordeaux, FR (2009); *Platz der Einheit*, Potsdam, DE (2001); *Parque da Devesa*, Castelo Branco, PT (2007); *Park Diagonal Mar*, Barcelona, ES (2002); *Park La Pineda*, Tarragona, ES (2008); *Parc Central de NouBarris*, Barcelona, ES (2004); *MFO Park*, Zürich, CH (2006).

Table 38.1 Summary of the quantitative results for categories

	Context		Structure		Functional system	
Total N. photos	N. photos	%	N. photos	%	N. photos	%
1038	311	30.0	351	33.8	397	38.2

Regarding the categories, the output is significant (Table 38.1). The percentage difference between the three categories is less than 10 %, proving the substantial balance among the user preferences.

The most significant category is the *functional system* (38.2 %), meaning that the value of use, the social value, the uniqueness and the identity values are the most favoured with park users.

Among the indicators, *imageability* certainly results as the most important (24.2 %). It is followed, among them almost in parity, by the *geometric* (14.8 %), the *panoramic* (14.6 %), and the *fruition* (14.1 %) indicators.

It is interesting to note that each indicator belongs to a different category. This output proves that the landscape values related to each of these indicators assume the same importance in user visual preferences.

The lowest indicator is the *atmospheric* (4.7 %), concerning the *context*. Its percentage of pictures is very far from that of *vantage point* (10.6 %) and of *panoramic views* (14.6 %). This outcome shows how user preferences in regard to the *context* are linked to landscape values such as historicity or modernity and beauty. The naturalness itself is perceived in terms of beauty. A river or a lake increases the charm of a panorama or of a vantage point, while other aspects of the naturalness of the park, such as the biodiversity, are not so relevant.

Comparing the two *atmospheric* indicators, we note that the atmospheric changes in the landscape surrounding the park are less considered than the interior ones. Even if the *atmospheric* indicator is the lowest in the *structure* category, its percentage (9.2 %) of pictures approaches to the percentage of the *aesthetic* indicator (9.8 %).

Regarding the structure of the parks, the *geometric views* (14.8 %) decidedly attract the visual preferences of the users. This fact reveals that landscape values of natural beauty and naturalness are less preferred than the artificiality. It would be interesting to investigate the reason why natural values are less preferred or perceived than the cultural ones.

38.3 Conclusions

As shown in the table (Table 38.2), the most characteristic elements of the visual preferences about urban parks are those concerning what we defined as *imageability*.

Even when parks are very attractive from a design point of view, what really captures the interests of the users are the distinctive elements. These elements can

Table 38.2 Summary of the quantitative results for indicators

Context			Structure			Functional system		
Indicators	N. photos	%	Indicators	N. photos	%	Indicators	N. photos	%
Panoramic views	152	14.6	Aesthetic views	102	9.8	Fruition views	146	14.1
Vantage points	110	10.6	Geometric views	154	14.8	Imageability	251	24.2
Atmospheric views	49	4.7	Atmospheric views	95	9.2			

be quite modest in their dimensions – such as works of art and monuments – or, alternatively, they can be very imposing and act as landmarks for the park, such as high-rise and contemporary architectures (i.e. Fontsanta Park in Barcelona, by Xavier Vendrell and Manuel Ruisánchez Capelastegui).

This data indicates how important the aspect of identity in urban landscape is. This identity can be given by elements, which are not necessarily linked to the geographic or cultural background of the city where the park is located. What really matters is the uniqueness of the park itself. At the same time, *imageability* is a very important indicator also in those parks in which identity is given by a historical sign, which becomes the distinctive element of the park (i.e. Schöneberger Südgelände Park in Berlin, by the artist group Odious).

Investigations on the visual preferences prove to be helpful in determining the perception of the landscape by the general population and can give suggestions for future projects on urban conservation and regeneration, for instance, in the frame of biodiversity and natural conservation in the city. Furthermore, if applied to natural parks, the method may consider other indicators related to biodiversity, such as biological species, natural landmarks, panoramic routes, etc.

As a result of this study, which demonstrates the most perceived aspects of a place, and the most shared values, urban policies can be oriented towards a sustainable development, interpreted as a synergetic link between healthiness and beauty of the urban environment (Calace 2011).

Future research with larger numbers of pictures taken from photographic search engines (*Google images*, *Bing image*, etc.) and websites (*Picasa*, *Flickr*, *Pbase*, etc.) might be necessary to verify the preliminary conclusions of the present study. The method can be developed considering more specific visual indicators and it can be integrated with other research instruments belonging to disciplines such as sociology, anthropology and digital communication, for future interdisciplinary research.

References

Assunto R (1973) Il paesaggio e l'estetica. Arte. Critica e Filosofia. Giannini, Naples

Barzilay M, Hayward C, Lombard Valentino L (1984), L'invention du Parc, Parc de la Villette Paris. International competition 1982–1983, Graphite – E.P.P.V., Paris

Calace F (2011) La sostenibilità nelle pratiche di progettazione Urbana. Urbanistica Informazioni 233–234:7–32

Capezzuto R (2004) Forum 2004 Barcelona. On Està en Roig? Dove è Roig? Where is Roig? Domus 871:40–49

Cassatella C (2011) Assessing visual and social perception of landscape. In: Cassatella C, Peano A (eds) Landscape indicators. assessing and monitoring landscape quality. Springer, Dordrecht, pp 105–139

Council of Europe (2000) European landscape convention. Treaty series no 176, Florence

Council of Europe (2008) Guidelines for the implementation of the European Landscape Convention, Recommendation CM/Rec (2008)3

Cullen G (1961) Townscape. The Architectural Press, London

D'Angelo P (2001) Estetica della natura. Bellezza naturale, paesaggi, arte ambientale. Laterza, Roma-Bari
Ippolito AM (2006) Il parco urbano contemporaneo. Notomia e riflessioni. Alinea Editrice, Florence
Marchigiani E (2002) I molteplici paesaggi della percezione. Cullen, Townscape, 1961. In: Di Biagi P (ed) I classici dell'urbanistica moderna. Donzelli editore, Roma, pp 163–192
Rodwell D (2010) Historic urban landscapes: concept and management. Managing Historic Cities, World Heritage papers 27:99–104
Roth M (2006) Validating the use of Internet survey techniques in visual landscape assessment: an empirical study from Germany. Landsc Urban Plan 78:179–192
Scott MJ, Canter DV (1997) Picture or place? A multiple sorting study of landscape. J Environ Psychol 17:263–281
Shuttleworth S (1980) The use of photographs as an environment presentation medium in landscape studies. J Environ Manag 11:61–76
Simmel G (1968) Soziologie: Untersuchungen über die Formen der Vergesellschaftung. Duncker & Humblot, Berlin
Smardon RC (1987) Perception and aesthetics of the urban environment: review of the role of vegetation. Landsc Urban Plan 15:85–106
UNESCO WHC (2011) Recommendation on historic urban landscape. Paris
Venturi Ferriolo M (2002) Etiche del paesaggio. Il progetto del mondo umano. Editori Riuniti, Roma
Ware C (2013) Information visualization: perception for design. Elsevier, Philadelphia, pp 1–30

Chapter 39
Landscape Scenic Values: Protection and Management from a Spatial-Planning Perspective

Claudia Cassatella

Abstract The aesthetic dimension distinguishes the concept of "landscape" from other concepts such as "environment" and "territory" and is a recurrent justification for conserving both natural and cultural landscapes. However, scenic beauty remains particularly difficult to define and protect under specific regulations. Moreover, the topic is seldom dealt with in spatial planning literature and practice. This lack of a systematic approach severely limits the capacity of public administrations to protect and enhance scenic resources. The paper highlights the need for further research into technical tools and suggests a number of perspectives which call for international collaboration. As a worked example of how planning can deal with scenic resources, an Italian case is illustrated, the Piedmont Region's "Guidelines for the analysis, protection and enhancement of the landscape scenic characters". The guidelines focus on the protection of visual relationships, which connect designated heritage assets and outstanding features with their settings and the area as a whole. The implementation measures within the planning regulatory system are discussed, as well as the role of different actors, planning levels and phases.

Keywords Scenic landscape • Natural beauty • Landscape planning • Landscape protection • Italian landscape heritage

39.1 Protecting Nature for Its Scenic Beauty, Protecting Landscape for Its Natural Values: Common Roots, Different Prospects

The protection of nature and landscape was born as one. The world's first National Park was established in the USA in 1872 to protect not merely natural areas but also areas which represented the values of national identity. In 1994, the *International*

C. Cassatella (✉)
Interuniversity Department of Regional and Urban Studies and Planning (DIST), Politecnico and Università di Torino, Viale Mattioli 39, 10125 Turin, Italy
e-mail: claudia.cassatella@polito.it

R. Gambino, A. Peano (eds.), *Nature Policies and Landscape Policies*, Urban and Landscape Perspectives 18, DOI 10.1007/978-3-319-05410-0_39

Union for Conservation of Nature (IUCN) introduced the designation "Protected Landscapes" to classify areas with "significant ecological, biological, cultural and scenic value", and the use of this category is increasing. In many western countries, the categories of protected assets reveal the coexistence of scientific and aesthetic instances. This derives from the legislative framework at the beginning of the twentieth century, for example, Areas of Outstanding Natural Beauty in England, Natural Monuments in Germany, Remarkable Viewpoints in France and National Landmarks and Wild and Scenic Rivers in the USA. In Italy, along with the birth of the first National Park in 1922, a law was established "For the protection of natural beauty and buildings of particular historical interest" (Kingdom of Italy, Law no. 778/1922). The subsequent "Regulations on the protection of natural beauty" (Kingdom of Italy, Law no. 1947/1939) introduced the expression *quadri della natura* [literally "pictures of nature"] and stated that

> [The following] are subject to this law because of their high degree of public interest: 1) sites that have substantial character of natural beauty or geological singularity; (...) 4) panoramic beauties seen as pictures of nature as well as those viewpoints accessible to the public, from which the sight of those beauties can be enjoyed.

For more than a century, what has to be protected has been stated but "how to protect" it is not clear. Indeed, scenic beauty remains difficult to define and to protect by means of specific measures. Moreover, it is now clear that nature-oriented actions can have controversial effects on scenic landscape and vice versa.[1] Dealing with such controversy requires us to move beyond old paradigms: for example, Visual Impact Assessment methods generally assume that the most natural landscape is the most scenic (Daniel 2001; Cassatella 2011; Churchward et al. 2013).

The attention that is currently being paid to ecosystem services, which include cultural services, such as spiritual and aesthetic experiences, seems to encourage the consideration of the scenic values of landscape in environmental policies. Nevertheless, a clear distinction of concepts is needed in order to avoid ambiguities and eliminate the potential risks of actions which claim to be "multifunctional". Indeed, landscape can simultaneously provide multiple benefits, such as biodiversity conservation and public enjoyment, simultaneously. However, this is not always the case: "multifunctionality" is an option, a possible goal and not an intrinsic characteristic of landscape actions. As a consequence, in order to better understand and manage the interferences and synergies between nature-oriented and landscape-oriented policies, specific tools for identifying, assessing, planning

[1] For example, planting new forests for ecological reasons can mean compromising the openness of the landscape. Similarly, cutting down trees in a park in order to create a panoramic view and keeping a notable and appreciated "natural scene" can be considered environmental damage. Respecting the natural processes of vegetation change can mean eradicating traces of traditional agriculture: what priority should be given to the natural and cultural values recognised by people? (See the approach of the Scenic Vista Management Plan for Yosemite National Park in California and other examples in Cassatella 2012a).

and managing the scenic character of landscape must be developed, both in theory and in practice (Cassatella and Seardo 2014).

39.2 Scenic Landscape in Spatial Planning and Management Tools

The literature concerning scenic landscape is rich in paradigms and in methods of analysis and assessment (Daniel 2001; Ode et al. 2008; Cassatella 2011; Nijhuis et al. 2011; Churchward et al. 2013). Although there has been a recent surge in interest in the general area of perception studies by social and environmental sciences, the literature that connects these areas of study with planning issues is generally weak. The majority of studies concern the visual impact of interventions in sensitive areas. Moreover, due to the need for field surveys and interviews, perception studies are often carried out at the local scale, while landscape planning concerns a variety of scales.

While assessment methods can be easily found in international literature, planning measures and techniques regarding scenic quality are seldom considered. Although the value of scenic beauty has been explicitly referred to in European and US law since the start of the twentieth century, spatial planning measures to protect and manage landscape heritage of scenic value have never been systematised, and more scientific evidence of their application within the regulatory system and implementation measures is needed. This lack of a systematic approach severely limits the capacity of public administrations to protect landscape resources.

The variety of institutional and legislative frameworks hinders a systematic study of planning and regulation systems from an international perspective. To carry out an international overview of these issues using significant local cases for in-depth examination and discussion, a research perspective is urgently needed, requiring international cooperation. In fact, a structured overview of protection categories and legislation, assessment methods (including participatory approaches), sectoral plans and other forms of regulation at a local level that concentrate on scenic-perceptive issues would be relevant to landscape, park, regional and local planning and for the assessment and establishment of control procedures regarding both natural and cultural heritage. Specifically, in the European context, an international overview of regulatory systems could contribute to the harmonisation of landscape policies (thus implementing the European Landscape Convention), provide an effective consideration of cultural services in the EU agri-environmental schemes and help to manage transnational protected areas.

Despite the lack of general theories and methods, a number of examples demonstrate that scenery can be subject to regulation, planning and management. The following list identifies a number of fields and topics that may be useful for further studies:

- Scenery management in the context of nature park planning. In particular, the manuals and experiences of the US National Park Service and Bureau of Land Management
- Cultural heritage protection. For example, recent guidelines issued by English Heritage deal with the visual relationships between historic assets and their surroundings (English Heritage 2011). Management Plans of World Heritage Sites, which implement the recent UNESCO *Recommendation on the Historic Urban Landscape* (2011)
- Townscape appraisal and town development regulations. Many western cities have adopted sophisticated rules for protecting their skyline or for creating impressive new ones (Cassatella 2012b). Geographic information systems techniques for controlling the visual effects of urban development, notably the rise of high buildings (Nijhuis et al. 2011)
- Visual Impact Assessment methods, in the context of Environmental and Landscape Impact (a wide and updated review in Churchward et al. 2013)
- Protection and enhancement of Scenic routes; guidelines for interventions on existing routes or for designing of new transport lines
- Design codes and standard requirements, in particular for designated areas and assets
- Landscape plans

The overview set out above shows that scenic features can be a subject of consideration in many kinds of spatial planning instruments, dealing with different landscape characters (natural, rural and urban), scales (regional, local, site-specific) and normative and design approaches (guidelines, strategic plans, statutory plans, regulations).

The following section illustrates a worked example of a set of rules for the preservation of scenic assets, related to an Italian regional landscape plan. Inspired by an international review of practices, it focuses on those visual characteristics which can be subject to planning control in the Italian context. Natural and cultural resources are integrated within the same scenic perspective.

39.3 A Case Study in Italy: Guidelines for the Analysis, Protection and Enhancement of the Landscape Scenic Characters

In Italy, landscape beauty deemed a strategic asset tied to national identity and to economic development. Protection of landscape by national laws dates back to the beginning of the twentieth century (see Sect. 39.1 above). However, for a long time the protection acts, which imply public control over development, were not accompanied by land-use plans or specific requirements. A recent law (The Cultural Heritage and Landscape Code 2004) states that protective designation acts must associate the description and appraisal of landscape assets with regulative measures and requirements (i.e. limitations on its use and transformation). The consequent process of "ruling" (which also applies to the thousands already designated landscapes) is still underway, in connection with the formation of new statutory regional landscape plans, which cover the entire territory of each region.[2] This process entails collaboration between Regional Authorities in charge of landscape planning and the Regional Departments of the Ministry for Cultural Heritage and Activities (MiBAC), in charge of landscape protection.

Italian landscape plans usually pay limited attention to scenic features, instead merely considering designated scenic roads and views and restricting building activity to narrow buffer zones. Sometimes, aesthetic and cultural values are explored in the analyses, but they are not subject to planning measures.

The Piedmont Region Landscape Plan (Regione Piemonte 2009), drawn up with the technical support of the Politecnico di Torino,[3] introduced a wider-than-usual set of categories of scenic features, namely, viewing places (belvedere, panoramic routes and others), scenic features (such as landmarks, skylines and profiles and others), visual relations, areas with specific characters and visual detriments. These categories are applied to natural, rural or urban landscape, with appropriate specifications.[4] The scenic features have been mapped at the regional scale (1/100,000) and dealt with systematically by the plan's regulatory system. A crucial aspect is that the same measures can and must be applied to any similar scenic feature in the landscape, whether it is a designated asset or not. In this way, the scenic character of a landscape can be planned and managed as a whole, beyond the boundaries of conservation areas (Fig. 39.1).

Due to their scale, the regional planning measures in question are mostly guidelines and directives which requires more detailed indications provided by local statutory plans. In order to help their implementation at the local scale, the MiBAC Regional Department of Piedmont commissioned the Politecnico di Torino to conduct a further study on landscape scenic assets, which outlined the *Guidelines for the analysis, protection and enhancement of the landscape scenic characters* (G-SCL) (Cassatella 2014). These address, in particular, the protection of designated landscapes (which require prescriptive regulations), the application of regional directives in local planning and visual assessment in occasion of procedures regarding interventions on conservation areas and assets. Nevertheless, as will be explained below, they are also intended to be used for spatial planning purposes in relation to the ordinary landscape.

The G-SCL provides a glossary, criteria for identification and representation (including GIS-based analysis of viewshed and visual sensitivity), planning

[2] Before 2004, landscape plans were compulsory for conservation areas only.

[3] Politecnico di Torino, Studies for the Regional Landscape Plan, research programme commissioned by Regione Piemonte, Scientific Director Prof. R. Gambino, 2006–2008. The author participated in the research group which investigated the perceptive values associated with landscape. A brief summary of the study approach can be found in Cassatella and Gambino (2011).

[4] For example, "landmarks" comprise the subcategories "natural" and "man-made" landmarks.

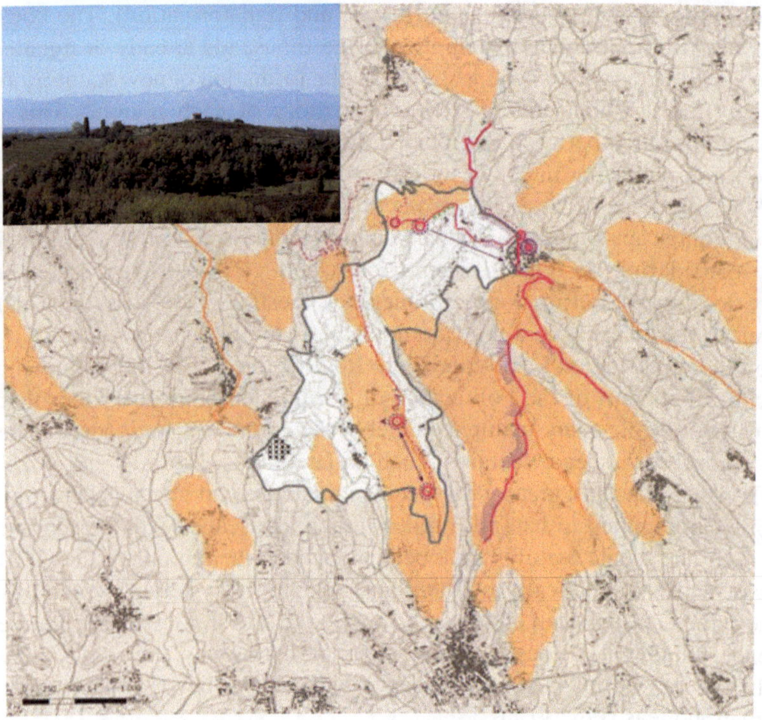

Fig. 39.1 Scenic landscape characterisation of the Abbey of Santa Maria di Vezzolano and surroundings. A National Designation Act concerns the area in the view, but not the scenic routes from which it can be seen and experienced. The map represents the main visual features and the intangible visual relationships between observation points and landmarks, thus extending the focus from the area under protection to the surrounding landscape (Source: Cassatella 2014)

measure proposals and impact assessment criteria, using worked examples and illustrations of Piedmont landscapes. In so doing, the G-SCL supplies both regional and local authorities (and, of course, professionals) with a shared language and method.

The approach is pragmatic. As the history of landscape planning in Italy shows that applicable regulations are few in number and recurrent, drawing up a list of potential planning measures has helped define the field of focus. Therefore, when it is appropriate to distinguish between planning measures, the scenery categories (in the glossary) are also distinguished[5] (Table 39.1).

Regional authorities may use the suggested standard measures in the process of defining the requirements for each designated landscape asset, while municipalities

[5] For example, "Isolated landmark" is a subcategory of landmarks, which requires a ban on building in the nearest area, while a generic landmark merely requires restrictions on building height and envelope.

Table 39.1 Guidelines for the analysis, protection and enhancement of the landscape scenic characters (Cassatella 2014). General assessment criteria and planning requirements, related to scenic character and features

Category	Subcategories	Assessing criteria and/or planning requirements
Viewing places	Viewpoint; equipped viewpoint; indoor vantage points (providing public access)	Accessibility, maintenance, protection of the view cone (ban on building, maximum building envelope, tree-cutting, etc.)
	Panoramic route	Protection of the viewshed (as above)
	Paths and minor routes	Maintenance of material and sensorial characters, improving public access
	Axial line	Protection of the view cone, continuity, design standards for lateral screens (alignments, height requirements, etc.)
Scenic features	Landmarks (at regional/local scale; built/natural; isolated)	Protection of the view cones which consent the appreciation of a landmark from viewing places; protection of prominence effect; avoiding competition (volumetric proportions, height, backdrop effects, etc.) in the zone of visual influence
	Natural profile; skyline	Integrity
	Other elements (tree lines and hedgerows; built/natural screens; areas with specific characters, such as textures)	Conservation, design codes and/or standards on materials, colours, etc.
Visual relations, views and panoramas	Inter-visibility; focal view; axial line; viewing corridor/opening; enclosure	Building control (envelope, maximum height, screening effects), vegetation-cutting
	Viewshed [of a viewing place] (foreground; middle ground; second ground; background)	Visual Impact Assessment (avoiding obstruction and intrusion; proportions of building, texture effect, lightness, etc.)
	Zone of visual influence [of a landmark]	(see Landmarks)
Visual detriments and degraded areas	Visual detriments (punctual, linear, areal; high-distance visibility)	Elimination, remodelling, mitigation
	Alteration (intrusion; obstruction; disorder; de-connotation)	Requalification, mitigation

may use the same ones in their overall local plans, dealing with features which are protected or not, if necessary in a “softer” manner, e.g. as design guidelines.

The main challenge is to involve local authorities in a process of awareness-raising with regard to landscape scenic values, leading to a specific focus in their statutory plans, which cover their entire area, instead of merely relying on prescriptive restrictions imposed by national and regional authorities. Local communities in Italy are frequently aware that landscape beauty enhances their quality of life and can be an economic resource. However, they lack an awareness of the technical measures that need to be taken in order to protect and enhance their scenery. The G-SCL aims to provide a set of tools for this purpose.

Special attention is paid to guiding the identification of visual relationships between viewing places and observed scenic features (such as landmarks or skylines), in order to avoid a fragmented panorama of “points” and “lines” and to underline the syntactic sequences of the landscape. The visual relations connect outstanding features with their surroundings (the ordinary landscape) and go beyond the administrative boundaries of conservation areas (Figs. 39.1 and 39.2). For example, it is argued that the visual cones towards a protected landmark should be protected too, as well as the viewshed of a protected vantage point (although this is not explicitly stated in protective acts). However, in the context of Italian legislation, such form of indirect protection is a difficult task. The mapping of such intangible relations is a significant initial step towards effectively taking them into consideration. Maps of scenic character and visual sensitivity covering the overall territory contribute to planning decisions (e.g. regarding the location of potential visual detractors, or land-use transformations) (Fig. 39.2). Finally, they contribute to the assessment of development proposals by establishing assessment points which are not dependent on the individual case study, but on the overall scenery. For example, if a new high building is proposed, it is possible to know whether it would be visible from selected existing viewpoints and whether it affects actual landmarks. Moreover, the impact of visual detractors can be evaluated in the planning process, when the debate may concern their location, and not just their design.

The technical process is fundamental in order to gather and provide decision-makers with correct information. Items produced by this process, such as cartographic representations, are designed to facilitate the debate between stakeholders. The G-SCL method is expert based, but public opinion may be taken into consideration during the assessment process. It is the responsibility of the authorities to involve local people in a public debate, in order to establish the significance of, or (preferably) identify, a scenic resource.

Fig. 39.2 Santa Maria di Vezzolano Abbey, map of visual sensitivity. The darkest areas are the most visible ones from the sum of the selected viewing places (Source: Cassatella 2014)

39.4 Discussion and Research Prospects

The case study of the Piedmont Guidelines shows how a systematic consideration of scenic features might be introduced to spatial planning, in a trans-scale process linking regional and local planning, outstanding and ordinary places, and natural and cultural assets. The G-SCL proposes basic categories, which correspond to specific requirements and, thus, can be integrated with spatial planning tools, such as landscape plans, urban development plans and park plans, and also criteria for landscape assessment procedures.

With regard to policies and plans for natural protected areas, the importance of considering scenic resources is supported by the history of nature conservation (Sect. 39.1), by theories on environmental services and by concrete experiences of scenery management. Integrating scenery into the toolbox may help to:

- Expand the consideration of and draw public attention to the interaction between man and nature, thus increasing awareness of and attention to natural environments
- Emphasise the visual relations between protected areas and their surroundings, thus helping to identify and manage buffer zones

- Enrich the identification of the multiple values of an area, e.g. cultural and aesthetic ones (which mean amenity, recreation, spirituality, memory and so on), with a specific focus on tangible features which support their perception by local population, visitors and stakeholders
- In so doing, reveal the potential conflicts between uses, functions and values deriving from different perceptions of the same places and elements

The process of scenery analysis and assessment is a technical contribution to public debate and decision-making, as it enhances the information and its transparency. It aids understanding and management of potential conflicts regarding landscape resources (as well as synergies between them), thus fostering an alliance between landscape and environmental policies.

Once the objectives of such policies regarding scenic quality have been identified, their effective implementation in the regulatory planning system remains a challenge (Sect. 39.2; Cassatella and Gambino 2011). In conclusion, future research should concentrate on the clear gap between knowledge and action by shifting the focus from scenic-perceptive landscape assessment methods to implementation methods within spatial planning regulation systems.

References

Cassatella C (2011) Assessing visual and social perceptions of landscape. In: Cassatella C, Peano A (eds) Landscape indicators. Assessing and monitoring landscape quality. Springer, Dordrecht

Cassatella C (2012a) The spectacle of nature. In: Cassatella C, Bagliani F (eds) Enjoy the landscape. Celid, Torino, pp 50–57

Cassatella C (2012b) Townscapes. In: Cassatella C, Bagliani F (eds) Enjoy the landscape. Celid, Torino, pp 92–105

Cassatella C (2014) Linee guida per l'analisi, la tutela e la valorizzazione degli aspetti scenico-percettivi del paesaggio, MiBACT Direzione Regionale del Piemonte, Regione Piemonte, Politecnico di Torino-DIST [Guidelines for the analysis, protection and enhancement of the landscape scenic characters, Research report, Italian Ministry for Cultural Heritage and Activities and Tourism MiBACT – Piedmont Regional Department, Piedmont Region, Politecnico di Torino – DIST] Available via http://www.regione.piemonte.it/territorio/dwd/paesaggio/seminario15mag/paesaggioScenografico.pdf. Accessed 30 June 2014

Cassatella C, Gambino R (2011) Landscape perception and regulation: opposition or integration? In: Living landscape – the European Landscape Convention in research perspective, vol I papers. Uniscape, Firenze/Bandecchi & Vivaldi, Pisa

Cassatella C, Seardo BM (2014) In search for multifunctionality. The contribution of scenic landscape assessment. In: Rega C (ed) Landscape planning and rural development. Key issues and options towards an integration. Springer, Dordrecht

Churchward C et al (2013) Evaluation of methodologies for visual impact assessments. US TRB's National Cooperative Highway Research Program (NCHRP) report 741, Washington

Daniel TC (2001) Whiter scenic beauty? Visual landscape quality assessment in the 21st century. Lands Urban Plan 54:267–281

English Heritage (2011) Seeing the history in the view. A method for assessing heritage significance within views. London

Italian Republic, Cultural Heritage and Landscape Code, Legislative Decree n. 42, 2004 (amended in 2006 and 2008)

Nijhuis S, van Lammeren R, van der Hoeven F (eds) (2011) Exploring the visual landscape. Advances in physiognomic landscape research in the Netherlands. IOS Press, Amsterdam

Ode A, Tveit M, Fry G (2008) Capturing landscape visual character using indicators: touching base with landscape aesthetic theory. Landsc Res 33(1):89–117

Regione Piemonte, Piano Paesaggistico Regionale, adottato con DGR n. 53–11975 del 4 agosto 2009 [Piedmont Region, Regional Landscape Plan, adopted by Decree number 53–11975, 4 August 2009]

UNESCO (2011) Recommendation on the historic urban landscape. Available via http://portal.unesco.org/en/ev.php-URL_ID=48857&URL_DO=DO_TOPIC&URL_SECTION=201.html. Accessed 31 Jan 2014

Chapter 40
European Cultural Routes: A Tool for Landscape Enhancement

Silvia Beltramo

Abstract This contribution deals with the topic of the European Cultural Routes as a tool for landscape enhancement – a tool that seeks to build a commonly shared, identity-based perception of local areas and communities – through an analysis of the recent European policies addressing the issue of cultural heritage. This is an extremely broad topic that has seen a significant growth in the past few decades among European and national institutions seeking to finalise and implement decisions aimed at defining and promoting Cultural Routes. Through an analysis of the evolution of the concept of "Cultural Routes" in the Council of Europe, it emerges that this tool can act as a driving force for developing a new approach to promote the cultural and natural landscape conservation. These routes are part of the EU assets linked to local traditions and identities, history and culture. In such a systematic vision, our heritage – both physical and intangible – must be understood as a key element in building these routes at a European level, as well as the local identity of the areas they cross.

Keywords Cultural routes • Landscapes • Cultural heritage • European cultural policies

The analysis of Cultural Routes as a tool for landscape development and enhancement is one of the topics of most recent interest for the scientific community, in line with the evolution of concepts of landscape and cultural heritage as discussed at a national and international level within organisations such as ICOMOS, UNESCO-WHC and the Council of Europe (ICOMOS 2008; Fisher 2007; Penette 1997; Ghersi 2007, Itinerarios Culturales Europeos 2007, Zhiu 2005; Lombardi and Trisciuoglio 2013; Beltramo 2013). It is possible to find a complex historiography tied to the social, historical and economic aspects of Cultural Routes (Carta 1999; Nappi 1998; Scazzosi 2001; Baldacci 2006; Mautone and Ronza 2010; Peano 2011).

In recent years, there has been a growing – though inconsistent – interest in Cultural Routes in Italy as well, as shown by the creation of the *Consulta Nazionale*

S. Beltramo (✉)
Interuniversity Department of Regional and Urban Studies and Planning (DIST), Politecnico and Università di Torino, Viale Mattioli 39, 10125, Turin, Italy
e-mail: silvia.beltramo@polito.it

R. Gambino, A. Peano (eds.), *Nature Policies and Landscape Policies*, Urban and Landscape Perspectives 18, DOI 10.1007/978-3-319-05410-0_40

degli Itinerari Culturali (the national committee on Cultural Routes, now defunct), large-scale investment (both in financial terms and in terms of public policies) and the proliferation of new cultural trails, not all of which have been recognised by the Council of Europe, that are found throughout the country (Mariotti 2012).

40.1 Historical Landscape and Cultural Heritage: The Role of European Cultural Routes

The topic of European Cultural Routes as a tool for landscape enhancement – which work by building a common identity-based perception of local areas and communities – stems from a view of cultural assets that expands to include the many different expressions of a community. This wider vision of cultural heritage leads us to complex diachronic readings that allow to understand the origins and development of European cultures. Within this approach, Cultural Routes created from an analysis of the relationship between the land (in its physical form) and physical and intangible assets encapsulate the cultural context and identity of the places they cross (Tosco 2009).

European institutions seem particularly interested in the need to tackle the difficulty of defining a common cultural identity. The search for this common identity in cultural and landscape heritage can be rewarded by these routes, which provide an opportunity to identify a common cultural element through what a traveller experiences in places, areas and landscapes without physical borders and cultural barriers (Valentino 2003; Volpiano 2011). Cultural Routes are a rich atlas of historical records, traces marked by the variety of landscapes that the European continent offers. The landscape – the common origin and keeper of deeply felt meaning for the communities involved, derived from the structure of places, the layering of traces that mankind, over time, has left behind when adapting and surviving – takes on historical value: this is collective heritage and it contributes to creating an individual's sense of belonging to society. The importance of the landscape and its perception as expressed in 2000 by the European Landscape Convention (ELC) indicates the link between Cultural Routes, understood as a combination of itineraries and landscapes we pass through. Themes such as the interpretation of the landscape, democratic participation in its protection and the population's re-appropriation of local areas – which lie at the heart of the Convention – are interwoven with the meaning and the potential of Council of Europe Cultural Routes.[1] These routes, which are by their very nature transnational – at times characterised by uninterrupted trails while in other cases involving single localities or areas and which are tied to the history of the agricultural, urban,

[1] There are currently 24 routes recognized by the Council of Europe; the complete list can be consulted at www.coe.int/t/dg4/cultureheritage/culture/Routes/default_en.asp Accessed 20 June 2013.

natural, ordinary or extraordinary landscape crossed in each case – appear to be places where the application of the ELC can take place effectively (Beltramo 2011a; Voghera 2011).

40.2 Cultural Routes in European Planning

There are several complex interpretations of Cultural Routes depending on the ultimate aims of the organisation promoting them.[2] The Council of Europe sees Cultural Routes as a tool designed to demonstrate – through cross-border trails that touch upon various themes – that the cultural heritage of European countries is in actual fact a commonly shared heritage.[3] In the Council of Europe, the debate concerning Cultural Routes has a long history: a working group set up in the 1960s presented a report that highlighted the presence of important places of cultural value that could help enhance their local areas. The guiding concept of this working group – which particularly focused on European cultural cooperation and development – was the awareness of a common heritage that could be rediscovered through travel, the exchange of cultures and ideas, inter-faith dialogue, the protection of minorities and the landscape[4] (Berti 2012).

Thus, Cultural Routes became the main focus of a specifically designed programme launched in 1987 – the Council of Europe's Cultural Routes Programme – whose first approved itinerary was the Santiago de Compostela Pilgrim Route, one of the Middle Ages' main *peregrinationes maiores*. Two subsequent resolutions – no. 4 in 1998 and no.12 in 2007, currently in force – defined the criteria for recognising Cultural Routes of the Council of Europe[5] (European Institute of Cultural Routes 2011).

The first resolution adopted by the Council of Europe's Committee of Ministers on Cultural Routes (4/1998) identifies the characteristics that Cultural Routes should possess as regards cultural and educational exchanges among young people, cultural tourism and sustainable cultural development through the drafting of European cooperative projects and research tied to promoting heritage, culture and the arts.

The increasing interest in Cultural Routes led to a new European Resolution in 2007 (CoE 2007/12), which specifically defined the criteria that must be met by new European Cultural Route projects. Their inherent themes should cover European

[2] Other international institutions have special programmes for cultural routes. For example, in the early 1990s, UNESCO launched a programme of cultural routes which gave rise to projects such as the Slave Route Project, the Silk Roads Project and The Ksour Route (http://whc.unesco.org/archive/routes94.htm. Accessed 20 June 2013).

[3] www.coe.int. Accessed 20 June 2013.

[4] Parliamentary Assembly Recommendation 987 of 28 January 1984 and Mandate of the Council of Europe's Committee of Ministers.

[5] CoE 12/2007 https://wcd.coe.int/ViewDoc.jsp?id=1194679. Accessed 20 June 2013.

memory, history and heritage; they should refer to at least one of the themes for reflection proposed by the programme – peoples, migrations and the spread of major European currents of civilisation – and contribute to the interpretation of modern-day Europe. Routes should therefore focus on promoting physical and intangible assets (Zabbini 2012; Martorell Carreno 2003) with a historical approach, highlighting similar elements in different European countries; however, they should also promote and carry out their work in accordance with European charters on landscape protection and the increased restoration of monuments as dictated by the Council of Europe, UNESCO and ICOMOS. The Resolution of 2007 considers an approach that promotes and implements the ELC as one of the key elements: the landscape is a palimpsest crisscrossed by Cultural Routes which, thanks to their various different themes, provide just as many ways of interpreting it.

Thus, the landscape becomes a fundamental cultural feature with profound meaning for the population that inhabits it and is associated with the structure of places, the layers of traces that mankind has left behind over time there. The landscape takes on historical meaning, the value of a collective asset, a common cultural legacy for those who come into contact with it. The landscape contributes to creating an individual's sense of belonging to a community, taking on values that pertain to intangible spheres, such as traditional cultural practices passed on from generation to generation, which have shaped and created today's landscapes and are part of a population's identity (Salvarani 2005). We can therefore assert that every landscape is the result of an overlapping of interactive processes involving the land, lifestyles and specific types of settlements, developed and practised by the local population: the understanding of this complex system must come about through diachronic analyses that take into consideration, among other things, the symbolic and cultural values assumed in different eras by places.

In December 2010, the Council of Europe's Committee of Ministers adopted the Enlarged Partial Agreement (EPA) in order to encourage dialogue and cooperation between the countries particularly interested in developing Cultural Routes.[6] The agreement's aims focus on a strengthening of the potential of *Cultural Routes* for the sustainable development of local areas and incentivising social cohesion and cross-border cultural cooperation (Khovanova-Rubicondo 2011). Moreover, it also went on to define Cultural Route of the Council of Europe certification, awarded to those Cultural Routes that meet the criteria implied and defined in the Council of Europe's Resolution CM/Res(2010)52.[7]

[6] Resolution CM/Res (2010)53, Enlarged Partial Agreement (EPA).

[7] Council of Europe (2010), Resolution CM/Res (2010)52 on the rules for awarding "Cultural Route of the Council of Europe" certification, Strasbourg.

40.3 Prospects and Project Development for European Cultural Routes

According to George Steiner (2006), the landscape is one of the four pillars of European cultural identity, based on the concrete involvement of European citizens. Common roots are found in the various different types of landscape that have become the destinations of travel itineraries. Cultural Routes help turn these concepts, tied to European landscape identity, into tangible reality, with the added advantage of overcoming the limitations posed by borders and exclusive attitudes such as nationalistic ideas of identity. This theme is highlighted by the ELC, where in its Preamble it states:

> The landscape contributes to the formation of local cultures and [. . .] is a basic component of the European natural and cultural heritage, contributing to human well-being and consolidation of the European identity.

The participatory policy achieved through the active involvement of citizens, as welcomed by the ELC, is a fundamental step towards the re-appropriation of the rural or urban environment by the inhabitants who live there, work there and spend their leisure time there. By involving citizens in political and strategic decisions, two important aims are achieved: to raise awareness of, and promote, the numerous local identities involved and make inhabitants responsible for their environment.

Hence, Cultural Routes become a tool for understanding, interacting with and preserving the landscape, allowing us to create a continuity in how the landscape is understood which goes beyond political and administrative borders and allowing us to associate and highlight fragments of historical landscape, offering an understanding of the route's theme and the history of the area. Thanks to Cultural Routes, we can redefine and place fragments of landscape in context, fragments that have lost their meaning due to policies and mechanisms that failed to read the deep-seated signs left on the local environment, as well as implement processes that create new shared landscapes and cross-border cooperation policies. One of the priorities worth highlighting is the creation of an observatory studying changes and mechanisms affecting the European landscapes involved (Beltramo 2011b). In setting up these routes that cross several different national borders and follow broad themes of all kinds, the landscape encapsulates the history of places and the ways with which mankind, over time, has related to it while at the same time making them understandable to observers, whether they be "insiders" or "outsiders", local inhabitants or visitors.

The landscape is the unifying element of the complex system of "routes", a hypertext with several layers of meaning, defined by a complex network of relations between the elements involved. In setting up landscape-based routes, we come across the problem of the scale of analyses, which depends on the range and types of routes planned as they arise. It is necessary to start with a hierarchical understanding of the landscape, considering changes of scale, spatial continuity, the complementary nature of the information we may collect, the landscape's innate evolving dynamics and an understanding of the different levels of traces of this constant evolution.

The landscape-based project of Cultural Routes must cover different levels of analysis on different scales with the following phases: establishing the kind of route so as to understand the potential and problems posed by its layout; determining the extent of the route; defining the scale of the work to be done, with the understanding that the route will not be limited to a road or an isolated monument or part of a region, but rather that it will be identified with the network of physical and cultural relationships that exist between elements of the route and the landscape of which they are a part; implementing the local framework's participatory structure; identifying and interpreting the relationship between the landscape and elements of the route which take on particular meanings depending on the characteristics of the route (type and theme) and the landscape (geographical layout, climate, perception, historical and cultural features); carrying out specific analyses in order to evaluate the route's strengths and weaknesses; selecting one or more solutions that can address the problems identified; and identifying organisations involved in the planning phase and appropriate forms of funding (Berti 2012).

During the project phase – given that we are dealing with plans for a territorial network – the landscape constitutes the main unifying element in this complex system of "routes". If the landscape is considered a hypertext, we can extract several layers of meaning and grasp a complex network of relationships between different elements.

The fundamental prerequisite for a route that encapsulates all the various, complex features of assets, both physical and intangible, connected with the route's way and/or theme will have to take into account some important operations, including: implementing projects through landscape-based aims, developing processes and methods for launching projects on a local level, exchanging expertise and good governance practices along the routes, and supporting multidisciplinary research projects and programmes that take into account the involvement of the local population, focusing on residents' appreciation of diversity and their perception of the landscape.

40.4 Conclusions

Given their fundamental relationship with the landscape, Cultural Routes must be considered instruments that can foster Europeans to take care of their environment, preserving the characteristics that reflect a certain kind of culture shaped by centuries of history. Through this combination of elements, Europeans will also gain self-awareness in all their differences.

The Cultural Route project's potential consists in understanding the elements that interact with the landscape with the changing social, economic and cultural environments they cross. A step in this direction could be considered a way of counteracting fragmentation and the increasing level of land use, achieved by seeking a degree of continuity in the networks of relationships that develop between routes and landscapes (Berti 2013).

Cultural Routes can unleash a process whereby the residents of – and visitors to – a local area see the identity of such areas restored. By travelling, people can rediscover and learn to recognise the tangible facets of a common history. It is essential that we harness elements tied to memory as found in the landscape, the real potential that a route – depending on the theme and the type – possesses for creating new, shared landscapes.

References

Baldacci V (2006) Gli itinerari culturali Progettazione e comunicazione. Guaraldi Universitaria, Rimini

Beltramo S (2011a) Itinerari artistici e percorsi storici in Piemonte: la scoperta del paesaggio storico nel turismo culturale. In: Levi Sacerdotti S, Mauro S, Gasca E (eds) Visitor management. Turismo, Territorio, Innovazione. Celid, Torino, pp 77–80

Beltramo S (2011b) Percorsi culturali in Valle di Susa: ricerca storica e mappe interattive degli itinerari. In Id., pp 86–95

Beltramo S (2013) Itinerari culturali e reti di conoscenza: identità e valorizzazione del patrimonio culturale. Il caso studio del Piemonte. AlmaTourism 08:12–15

Berti E (2012) Itinerari culturali del Consiglio d'Europa tra ricerca di identità e progetto di paesaggio. Firenze University Press, Florence

Berti E (2013) Cultural routes of the Council of Europe: new paradigms for the territorial project and landscape. Almatourism 7:1–10

Carta M (1999) L'armatura culturale del territorio. Il patrimonio culturale come matrice di identità e di strumento di sviluppo. Franco Angeli, Milan

European Institute of Cultural Routes (2011) History of the cultural routes, report presented to the Council of Europe, 5

Fisher R (2007) A cultural dimension to the EU's external policies from policy statements to practice and potential. Boekman studies, Amsterdam

Ghersi A (ed) (2007) Politiche europee per il paesaggio: proposte operative. Gangemi, Roma

ICOMOS (2008) The charter on cultural routes, Québec. http://www.international.icomos.org/charters/culturalroutes_e.pdf

Itinerarios culturales europeos (2007) 6 volumes, Grupo Editorial Planeta-Council of Europe Publishing

Khovanova-Rubicondo K (ed) (2011) Impact of European cultural routes on SMEs' innovation and competitiveness, European Commission's Competitiveness and Innovation Framework programme (CIP). Available via http://ec.europa.eu/cip/index_en.htm. Accessed 20 June 2013

Lombardi P, Trisciuoglio M (2013) Itinerari, networking e liste di eccellenza. In: Barosio M, Trisciuoglio M (eds) I paesaggi culturali. Egea, pp 237–252

Mariotti A (2012) Sistemi Locali, Reti e Competitività Internazionale: dai Beni agli Itinerari Culturali. AlmaTourism 5:81–95

Martorell Carreno A (2003) Cultural routes: tangible and intangible dimensions of cultural heritage. In: 14th ICOMOS general assembly and international symposium: place, memory, meaning: preserving intangible values in monuments and sites, 27–31 Oct 2003, Victoria Falls. Available via http://openarchive.icomos.org/473/. Accessed 20 June 2013

Mautone M, Ronza M (eds) (2010) Patrimonio culturale e paesaggio. Gangemi editore, Roma

Nappi MR (ed) (1998) Il paesaggio culturale nelle strategie europee. Electa, Napoli

Peano A (ed) (2011) Fare paesaggio: dalla pianificazione di area vasta all'operatività locale. Alinea, Florence

Penette MT (1997) L'Europe en Bref. Les itinéraires Culturels. Actes Sud, Arles

Salvarani R (2005) Storia locale e valorizzazione del territorio: dalla ricerca ai progetti. Vita e pensiero, Milan

Scazzosi L (2001) Politiche e culture del paesaggio. Gangemi editore, Roma

Steiner G (2006) Una certa idea d'Europa. Garzanti, Milan

Tosco C (2009) Il paesaggio storico. Le fonti e i metodi di ricerca tra medioevo ed età moderna. Laterza, Roma-Bari

Valentino PA (2003) Le trame del territorio. Sperling e Kupfer, Milan

Voghera A (2011) Dopo la Convenzione Europea del Paesaggio. Politiche, piani e valutazione. Alinea, Florence

Volpiano M (ed) (2011) Territorio storico e Paesaggio. Conservazione Progetto Gestione. L'Artistica Editrice, Savigliano

Zabbini E (2012) Cultural routes and intangible heritage. AlmaTourism 5:59–80

Zhiu L (2005) Cultural routes and the network construction of the world cultural heritage conservation, section IV: cultural routes: the challenges of linear settings for monuments and sites. In: Proceedings of the scientific symposium: monuments and sites in their setting – conserving cultural heritage in changing townscapes and landscapes. ICOMOS

Chapter 41
Economic Valuation of Landscape at Risk: A Critical Review

Marina Bravi and Emanuela Gasca

Abstract Identifying which aspects should be considered at risk is very important to preserve the environment and landscape. Although the literature on this issue is currently centered on climate changes, the preservation of landscape appears to be closely connected to property rights and local policies. A wide range of applications focused on landscape evaluation has provided a methodological framework, but the inclusion of the economic point of view appears to be more recent. At this regard, the contribution considers the economic valuation of landscape with the aim of highlighting the role played by risk and uncertainty in individual choices. It reviews the limits of expected utility theory and reclassifies according to the method, over a period of about 20 years, a certain number of experimental studies in relation to their valuation goals. Finally, it shows the various sources of risk that affect landscape in order to encourage future applications in this field.

Keywords Economic valuation • Landscape evaluation • Risk and uncertainty valuation • Risk factors

41.1 Introduction[1]

Human pressures and technological progresses can lead to significant changes in the environment and landscapes over time. Their preservation is therefore utopian, even if it does form a central part of public policy. As a result, the risk of not being

[1] The authors have shared the design of the entire work but it should be noted that, while E. Gasca has selected the case studies and has elaborated the Table 41.1, M. Bravi has written the final version of the chapter.

M. Bravi (✉)
Interuniversity Department of Regional and Urban Studies and Planning (DIST), Politecnico and Università di Torino, Viale Mattioli 39, 10125, Turin, Italy
e-mail: marina.bravi@polito.it

E. Gasca
Interuniversity Department of Regional and Urban Studies and Planning (DIST), Politecnico and Università di Torino and SiTI Istituto Superiore sui Sistemi Territoriali per l'Innovazione, Viale Mattioli 39, 10125, Turin, Italy
e-mail: emanuela.gasca@polito.it

R. Gambino, A. Peano (eds.), *Nature Policies and Landscape Policies*, Urban and Landscape Perspectives 18, DOI 10.1007/978-3-319-05410-0_41

able, in the long term, to keep the environment in which we live intact, or of not being able to preserve the unique characteristics of a landscape, is very high.

"Environment" and "landscape" represent broad categories, and it is first of all important to identify which aspects are considered to be *at risk*. As a result of the catastrophic events and irreversible consequences to which climate change gives rise, the literature on the issue of environmental risk is currently centered on it. Economists describe generically human-induced climate change as an externality and the climate as a public good. However, climate change is a global phenomenon with persistent, long-term effects. In addition, there are a number of uncertainties that prevent a precise quantification of the economic impacts, and there is a serious risk of irreversible changes with non-marginal economic effects (Stern 2007). Actually, as a result of uncertainty not only over future costs but also over the behavior of producers and consumers and the degree of policy flexibility worldwide, the full range of cost estimates is wide (Barker et al. 2006).

By contrast, landscape appears to be more closely connected to property rights and local policies, although it is also subject to the effects of environmental changes on a large scale. It is therefore necessary to make a distinction between the risks associated with global changes and those which pertain to land-use-related decisions. However, considering the heterogeneity of individual attitudes and preferences, particular relevance should be accorded to the subjective perception of risk in relation to this change.

Moreover, the term landscape clearly entails a focus on the visual properties of the environment, including natural, human, and biological elements, which could be identified visually. As a result, non-visual characteristics, such as cultural/historical values, wildlife species, wilderness value, and economic opportunities for recreational activities, are not included. The scenario of possible change is even more complex to describe because perceived quality is often defined as including a wide range of environmental/ecological, sociocultural, and psychological factors.

A long period of landscape assessment has provided a range of different methods for attempting to rate landscape quality, some of which were targeted at setting out landscape policy priorities. This research did not however obtain a full consensus. Price (2012) outlines the uncertainty that exists around whether or not landscape evaluations should be subjective or objective. In any cases, the possibility of implementing a robust, complete economic approach has been limited by the difficulties involved in comparing landscape quality scores and the incomplete basis of data. In fact, an economic value would allow for a comparison of impacts across different landscapes in time and space. It therefore seems necessary to first examine what is meant by an economic valuation[2] of landscape.

[2] The broader context of ecosystem services can be analyzed from physiological, psychological, and economic perspectives. However, "evaluation" and "valuation" cannot be treated as synonyms. The first constitutes the process of scoring, or rating, the landscape quality, while the second assigns a monetary value to landscape or its attributes (Antrop et al. 2012).

41.2 Economic Valuation of Landscape

The inclusion of economic thinking in landscape evaluation appears to have been a relatively recent phenomenon (Schaeffer 2008), even though it should be recognized as a resource in land-use decisions. Landscape planners have employed various economic tools to provide decision makers with information relating to the land and its features. Actually, an economic valuation can provide useful information in relation to the relative values of environmental and recreational services. For example, agriculture is a multifunctional activity, which produces food, but also sustains rural landscapes, protects biodiversity, and generates employment in collateral sectors such as ecotourism.

Van der Heide and Heijman (2012) have argued that the basic approach to an economic valuation of landscape is anthropocentric; in other words, it emphasizes the utilitarian aspect and the benefits obtained for men. Nevertheless, in European countries, landscapes are characterized by the historical coevolution of social systems and ecosystems. This leads not only to a situation of greater complexity but also to increased benefits for society.

The definition of landscape by the European Landscape Convention (2000) underlines the necessity of understanding how people perceive and value landscape as a component of land-use policies. In this context, studies focusing on public preferences have followed two methodological approaches: firstly, non-monetary techniques, wherein the landscape assessment is accomplished through esthetic preferences in which one scenario is compared with another, and, secondly, monetary techniques, wherein a valuation exercise is implemented to estimate use values and/or nonuse values. In any case, numerous assessment techniques can be identified in literature. For the purposes of this review, the focus will be on the second approach.

As is widely recognized, landscape is a non-excludable and non-rival good; these characteristics make it, in general, a public good. The application of economic nonmarket techniques is in effect required when competitive markets do not exist or fail to estimate policy impacts and environmental changes. However, single elements or parts of landscapes could be owned by private parties, rather than public bodies. This is an important condition for the valuation and makes defining a credible scenario for future policies and regulations increasingly complex.

Landscape economics can be described as a branch of environmental economics with a specific focus on landscape attributes. One of the first books about the economic valuation of landscape was Price's book (1978). In this text, Price outlined how the esthetic characteristics and holistic perception of landscapes lead to partial inconsistency in the economic results, which are dependent on rationality and the completeness of individual preferences. Nevertheless, from the 1980s to the current day, many scholars have attempted to estimate the economic value of landscape through indirect or direct methodologies. The first case (*revealed preferences*) involved an investigation of observed consumer behavior (*Travel Cost Method* and *Hedonic Pricing Method*), while in the second case (*Contingent Valuation Method* and *Choice Experiments*) individuals were questioned about the values directly (*stated preferences*).

Table 41.1 summarizes a number of experiments and compares the various methods and valuation goals adopted during the last 23 years.[3] The first consideration relates to the fact that CVM and CE cover about 67 % of all applications, with the latter being prevalent more recently, while HPM represents approximately 26 % and only a few studies deal with TCM (7 %). The CVM[4] was in effect one of the first approaches to be employed to value esthetic preferences. In an initial application, involving an analysis of the visual impacts of a power plant near Lake Powell (USA), Brookshire et al. (1976) argued that various types of bias were identified during the bidding game elicitation process. Since its inception, the method has, in an objective and complete manner, denounced its own limitations in representing individual preferences in relation to landscape values.

Individual perceptions of landscapes are clearly difficult to measure empirically and to test into the model. Economists have employed various measurable characteristics as proxies of landscape qualities. Table 41.1 highlights the fact that, in many applications, the identification of attributes represents the valuation goal. Some complementary techniques, such as factorial, principal components or multi-criteria analyses, are implemented to create factors, or indices, to be used in a second step.

The complexity and holistic nature of landscapes have been clearly revealed by HPM experiments. A key argument for using this method is that the proximity of a scenic landscape improves residential living conditions and increases home values. There are three main types of landscape under investigation: rural, forest (natural), and urban. The application of HPM essentially concerns the latter type. As is well known, the method is devoted to identifying externality values through revealed preferences and indirect utility function. Indeed, the limits of the applicability of the hedonic price function and its possible interpretation which willingness to pay have long been discussed. This has generated a substantial amount of literature on the topic. Recently, Rouwendal and Weijschede-van der Straaten (2012) observed that the benefits generated by a landscape depend on the number of people living in a region, yet, in the absence of protective measures, a negative impact can result when the population grows. It is no coincidence that some studies have dealt with the valuation of urban open spaces as an incentive for their regulation and protection.

By contrast, the lack of applications of TCM documents reveals a low level of interest in assessing the recreational benefits of landscape. There have only been a few attempts to estimate recreational demand. In many cases, this method is instead used to compare different estimates obtained using CVM and CE. Since TCM also refers to the indirect utility function, the predominance of the use of CVM and CE is more obvious. However, it should be noted that, although mention is made of conditions such as biodiversity at risk, the fragility of the landscape, future

---

[3] The sample is based only on articles published by reviews and journals that are mainly, but not exclusively, indexed by SCOPUS. It intentionally employed a sample with a focus on landscape rather than other forms of environmental services.

[4] As is known, the individuals express their willingness to pay (WTP) for a benefit or their willingness to accept (WTA) a damage through a contingent market, where supply and demand are simulated.

Table 41.1 Sample of case studies on economic landscape valuation and related methodologies (1990–2013)

Years	Methods	Articles number	Evaluation purposes
1990–1995	CVM	4	Rural landscape
			Agricultural landscape
			Landscape conservation
	HPM	2	Countryside characteristics
			Forest and urban landscapes
	TCM	1	Forest and urban landscapes
1996–2000	CVM	6	Landscape (wildlife habitat, natural parks, woodlands, boreal forest)
			Landscape attributes and changes
	HPM	5	Landscape attributes
			Urban landscape
			Land uses
	CE	2	Landscape attributes
			Forest landscapes
2001–2005	CVM	2	Landscape indicators
			Forest protection
	HPM	2	Urban landscapes: open spaces and forest parks
2006–2010	CE	12	Landscape quality
			Rural landscape
			Landscape policies
			Evocative landscape
			Cultural landscape
	HPM	7	Urban landscape
			Landscape attributes
			Green spaces
			Forest view
	CVM	5	Landscape
			Biodiversity
			Protection policies
	TCM	1	Rural landscape
2011–2013	CE	12	Landscape attributes and effects
			Rural landscape
			Landscape externalities
			Landscape conservation and improvement
	CVM	7	Urban landscape
			Forest management
			Landscape attributes
			Cultural landscape
			Cultural services
	HPM	4	Urban landscape
			Landscape attributes
			Green spaces
	TCM	3 (together with CVM)	Landscape
			Cultural services

landscape patterns, alternative strategies of management, and the components of risk and uncertainty do not hold a central place in the valuation.

Finally, the recent dominance of CE over the traditional CVM, which is well documented in the sample, is indicative of the success that this technique has enjoyed among applied economists. Its flexibility, combined with a robust theoretical framework, headed by the Random Utility Theory (RUT), has been universally recognized (Hoyos 2010). Moreover, CE is specifically designed for valuing attributes and treating the heterogeneity of preferences in the choice process. What about risk and uncertainty, though? Although choices under risk were generally considered in CE, only one case in the sample indicates a specific interest in this direction. It is therefore worth giving a more thorough consideration to this topic.

41.3 Choices Valuation Under Risk and Uncertainty

The social utility theory was applied to uncertain scenarios by weighting the social utilities about the possible states of the world by the subjective probability of those states. This represents the factor commonly known as "expected utility" (von Neumann and Morgenstern 1947). Various risk communication methods deal essentially with subjective expectations, which are also termed "perceived risk," or subjective risk perceptions. In fact, the perception of environmental risks is affected by multiple factors, some relating to rational thinking, others to unconscious, irrational thinking. The latter was not initially taken into account in the theory of decision making under risk that dominated applied economics studies for at least 50 years.

During the 1950s and 1960s, behavioral decision theory (Wright 1984) drew new strength from cognitive psychology and the behavioral sciences. However, of the various theories of human behavior, expected utility theory has proven to be the most fundamental model, at least until the contribution of prospect theory (Kahneman et al. 1982) and its more advanced version, cumulative prospect theory (Tversky and Wakker 1995). Tversky and Kahneman (1974), have already argued that people base their uncertain decisions on a limited number of rules. This heuristic of choice is based on likelihood, on the available information, and, finally, on the anchoring effect.[5] As Daniel McFadden (1999) has illustrated, choice behavior can be regarded as a decision-making process that is influenced by the perception of subjective convictions, which are themselves based on the information, as well as prejudices, attitudes, motivations, and preferences. Only under

[5] During the decision-making process, anchoring occurs when individuals use an initial piece of information to make subsequent judgments. This implies a psychological phenomenon according to which the utility of individuals is determined by changes in the level of wellness in relation to an initial state (*reference point*), rather than in absolute terms. Behavior of this kind is reproduced from a utility function that is termed the *value function* and a system of weights, which transforms the objective probabilities of the outcomes of the decisions into subjective utility.

certain technical conditions, which include completeness and transitivity, can the latter be represented by a numerical scale that is defined in terms of utility.

Standard economic models require consumers to be able to process information using strict "Bayesian rules." Preferences are primitive, consistent, and immutable, while the cognitive process is simply looking for a maximum point, given the market constraints. It should be noted that the introduction of a new framework based on RUT did not resolve all of the problems. It is not immune to experimental disproof, and therefore, it is not surprising that the valuation of nonmarket goods, based on hypothetical scenarios, has often fudged the issue of choice valuation under risk.

CE on environmental resources have typically assumed no uncertainty in the hypothetical markets, even though the outcomes of environmental policies were uncertain. In fact, uncertainty reduces the value of the benefits derived from the environment (Arrow and Fisher 1974), and the inclusion of information relating to probabilities in the future scenarios may not be straightforward. Applied studies have often found that respondents were risk averse (Macmillan et al. 1996), and in some cases, the economic results were not fully consistent. In any event, omitting information on risks may contribute to a hypothetical bias and impair the validity of the stated-preference valuation. Viscusi and Zeckhauser (2006) have analyzed environmental risk perception in relation to global climate change and citizens' attitudes to related policies. The results highlight the fact that the valuation process is influenced by a mixture of decisions based both on rationality and on significant framing effects. At the same time, individuals provide impressive probability distributions in their responses and make choices influenced by an evident anchoring effect.

41.4 Final Considerations About Risk Affecting the Landscape

Landscape valuation has been used primarily for two purposes: to demonstrate that its economic value was previously underestimated and to push protective measures into public policies. Placing a value on possible landscape changes was not, at least until the present day, the main goal of the valuation because of the elusive characters of changes (local/global) and the indefinable nature of individual perception.

How, therefore, can risk and uncertainty be taken into account in the landscape valuation? It may be useful to identify different sources of risk and to try to distinguish those concerning the local scale from those relating to global changes.

The valuation of risks affecting the landscape and the ability to prevent damage and reduce degradation depend on a thorough understanding of the levels of danger and vulnerability. Human actions that cause progressive or sudden changes in the landscape represent potential risks and hazards in addition to the lack of actions and governance in place. From the point of view of vulnerability, landscapes, which are less well protected, where the tendency toward the urban development is greater, can be considered to be more at risk. The more a landscape is intact, the higher the level of social attention to its preservation and the lower the level of vulnerability in relation to human pressure will be. Considering the level of danger, we can identify three groups of external risk factors: environmental (air, soil, water pollution, and climate change), physical-structural (earthquakes, landslides, and floods), and human. The third factor represents the real issue to be faced in identifying the risk components and the elements of vulnerability of local landscapes.

The same area is indeed exposed to various human risk factors. However, the hazard factors are not mutually exclusive and usually act synergistically. Population growth, changes in land uses, the presence/absence of planning regulations, and illegal constructions often overlap in describing different local configurations of risk. They can only be assessed using a bottom-up approach, starting from an identification of landscape values and the effects of human hazards encountered in the study area, before then refocusing on elements of a more general or global character.

In light of these considerations, it is perhaps necessary to reverse the point of view on the economic valuation of landscape. Despite its holistic nature and the analytical difficulties involved in describing the possible changes, efforts could be focused on monetizing the risk of irreversible loss of value, rather than costs and benefits referring to actual scenarios. This goal would seem to have a greater strategic importance because it could make it possible to accord a value to preservation and regulation policies.

References

Antrop M, Sevenant M, Tagliafierro C, Van Eetvelde V, Witlox F (2012) Setting a framework for valuing the multifunctional landscape and its multiple perception. In: Van der Heide CM, Heijman W (eds) The economic value of landscape. Routledge, New York

Arrow K, Fisher A (1974) Environmental preservation, uncertainty, and irreversibility. Q J Econ 88:312–319

Barker T, Qureshi MS, Köhler J (2006) The costs of greenhouse-gas mitigation with induced technological change: a meta-analysis of estimates in the literature. 4CMR, Cambridge Centre for Climate Change Mitigation Research, University of Cambridge, Cambridge

Brookshire DS, Ives BC, Schulze WD (1976) The valuation of aesthetic preferences. J Environ Econ Manage 3:325–346

European Landscape Convention (2000) http://www.coe.int/t/dg4/cultureheritage/heritage/Landscape/Publications/Convention-Txt-Ref_en.pdf

Hoyos D (2010) The state of the art of environmental valuation with discrete choice experiments. Ecol Econ 69:1595–1603

Kahneman D, Slovic P, Tversky A (eds) (1982) Judgment under uncertainty: heuristics and biases. Cambridge University Press, Cambridge

Macmillan D, Hanley N, Buckland S (1996) A contingent valuation study of uncertain environmental gains. Scot J Polit Econ 43(5):519–533

McFadden D (1999) Rationality for economists? J Risk Uncertainty 19:5–275

Price C (1978) Landscape economics. Macmillan, London
Price C (2012) Subjectivity and objectivity in landscape evaluation. An old topic revisited. In: Van der Heide CM, Heijman W (eds) The economic value of landscape. Routledge, New York
Rouwendal J, Weijschede-van der Straaten JW (2012) What can hedonic analysis tell us about the value of landscape? In: Van der Heide CM, Heijman W (eds) The economic value of landscape. Routledge, New York
Schaeffer PV (2008) Thoughts concerning the economic valuation of landscapes. J Environ Manag 89:146–154
Stern NH (2007) The economics of climate change: the stern review. Cambridge University Press, Cambridge
Tversky A, Kahneman D (1974) Judgment under uncertainty: heuristics and biases. Science 185:1124–1131
Tversky A, Wakker P (1995) Risk attitudes and decision weights. Econometrica 63:1255–1280
Van der Heide CM, Heijman W (2012) Landscape and economics. In: Van der Heide CM, Heijman W (eds) The economic value of landscape. Routledge, New York
Viscusi WK, Zeckhauser R (2006) The perception and valuation of the risks of climate change: a rational and behavioral blend. Clim Change 77:151–177
von Neumann J, Morgenstern O (1947) Theory of games and economic behavior. Princeton University Press, Princeton
Wright G (1984) Behavioral decision theory. Sage, Beverly Hills

Chapter 42
Towards an Integrated Economic Assessment of Landscape

Marta Bottero, Valentina Ferretti, and Giulio Mondini

Abstract The economic assessment of landscape represents a complex task, and integrated approaches are therefore needed in order to properly consider all the relevant aspects. A valuable support in such a context is offered by the Conjoint Analysis (CA) methodology which refers to a variegated set of mainly statistical methodologies that aim at studying individual choices using preference expressed about various profiles, i.e. several versions of a product or service. Starting from a real case concerning the vineyard landscape of the Langhe area (Italy), the present contribution investigates the use of the Conjoint Analysis for the estimation of the economic value of the landscape.

Keywords Conjoint Analysis • Choice experiments • Landscape • Vineyards

42.1 Introduction[1]

According to the European Landscape Convention (2000), landscape planning and management need specific decision processes that are based on scientific analysis and evaluation. Among the various available approaches for landscape analysis, a very important role is played by economic evaluation (Bottero 2011).

There are many economic aspects associated with landscape. The economic activities related to the use and transformation of landscape have various effects and repercussions on the same; according to the literature in the field of economic

[1] The contribution is the result of the joint effort of the authors. Despite the global responsibility for the work being equally shared among the three authors, it should be noted that Marta Bottero was responsible for the research outlined in paragraphs 42.1, in section "State of the art" in paragraph 42.2 and in section "Definition of attributes and levels" in paragraph 42.3 while Valentina Ferretti undertook the research described in section "Theoretical background" in paragraph 42.2, in section "Presentation of the area" in paragraph 42.3 and in section "Questionnaire" in paragraph 42.3. The abstract and paragraph 42.4 are the result of the joint work of all the authors.

M. Bottero (✉) • V. Ferretti • G. Mondini
Interuniversity Department of Regional and Urban Studies and Planning (DIST), Politecnico and Università di Torino, Viale Mattioli 39, 10125, Turin, Italy
e-mail: marta.bottero@polito.it; valentina.ferretti@polito.it; giulio.mondini@polito.it

R. Gambino, A. Peano (eds.), *Nature Policies and Landscape Policies*,
Urban and Landscape Perspectives 18, DOI 10.1007/978-3-319-05410-0_42

analysis, this is tantamount to saying that landscape is an (positive or negative) externality. In general terms, externalities are defined on the basis of the effects (favorable or unfavorable) on the production or consumption of one person by the production or consumption of another, without there being any kind of monetary transaction between the two to balance the costs or benefits of these effects. Furthermore, landscape, especially in modern society, is seen all the more as a limited resource. From the point of view of economic analysis, this is the same as saying that landscape can be considered an "economic good," in other words a good available in an insufficient quantity to meet requirements for the same, and for which there is a problem of efficient allocation of resources, guaranteed or not, as the case may be by the spontaneous actions of the market (Marangon and Tempesta 2008).

Conjoint Analysis (CA) is an evaluation method that can be applied for estimating the economic value of landscape. CA is a stated preference technique that asks respondents to state their preferences and opinions towards hypothetical scenarios. CA has the ability to obtain values for various elements of a good and to estimate multidimensional changes regardless of whether an entry fee is paid or not. This ability to investigate the "part-worths" of a good is well suited to the nature of environmental goods and landscape. Conducting a CA will allow the quantification of value for specific features of a site and the observation of trade-offs people are prepared to make (Kinghorn and Willis 2008).

Starting from a real case concerning the vineyard landscape of the Langhe area (Italy), the present study investigates the use of the Conjoint Analysis and proposes a methodological approach for the estimation of the economic value of the landscape.

42.2 Conjoint Analysis and Landscape Evaluation

Theoretical background. The Conjoint Analysis (CA), developed by Luce and Tukey (1964) and Krantz and Tversky (1971), was initially applied in commercial psychology and marketing literature by Green and Rao (1971). As CA subsequently adapted to new research trends, its practical application increased in environmental research (Wittink and Cattin 1989; Ness and Gerhardy 1994; Green et al 1999; Alvarez-Farizo and Hanley 2002; Sayadi et al. 2009).

The term Conjoint Analysis refers to a variegated set of mainly statistical methodologies which aim to study individual choices using preferences expressed about various profiles, i.e. several versions of a product or service (Gustafsson et al. 2001; Bravi and Giaccaria 2006), which have in common a number of features (Alvarez-Farizo and Hanley 2002):

1. They are based on a set of attributes or features describing the good, service, project, or policy, each taking a number of prespecified levels.
2. These levels and attributes are combined to build up descriptions of hypothetical bundles, using experimental design techniques.
3. Individuals are asked to state their preferences over these alternatives, using a number of different protocols.

4. During the decision-making process, individuals appraise the worth of each combination, and their choice demonstrates prioritization among the different combinations of features. It is assumed that the total worth of a particular product choice is determined by the different part utilities (part-worths) of each feature level (Sayadi et al 2005). Responses are then analyzed using statistical models.

A central feature of this approach is that the utility derived from a good or service can be decomposed into part-worths relating to different attributes of that good or service (Lancaster 1966). The approach can equally be applied to project or policy evaluation.

CA is a technique that is especially suitable for analyzing decisions, and in particular, when it comes to understanding the process by which consumer/individuals develop their preferences for products or services (Sayadi et al. 2005).

Many variants of the Conjoint Analysis exist, and the present contribution focuses on the Choice Experiment one, that seemed more suitable to be developed in the context of the economic evaluation of the landscape.

In Choice Experiments, individuals are faced with a series of choices over pairs or three-way combinations of alternatives. From each choice set, respondents must choose their preferred option, taking into account that the *status quo* is typically included in the choice set (Alvarez-Farizo and Hanley 2002).

State of the art. Conjoint Analysis has not been widely applied to estimate the value held for landscape, being most commonly used to estimate the value of environmental goods where it was first applied. Table **42.1** summarizes the main scientific works that have been developed considering the application of CA for economic landscape evaluation. In the table, the area of the application, the aim of the research, the size of the sample, the set of considered attributes, and the definition of the payment vehicle are shown. As it is possible to see, the domain of the applications is vast and the CA has been applied in the field of cultural heritage and archaeological sites, forest areas, and rural landscapes. The main objective of the applications is the definition of conservation and management strategies for the landscape; mention should be made to some researches aiming at supporting the design of tourism policies.

Table 42.1 Main works available in the literature concerning the application of CA for the estimation of the economic value of landscape

Authors	Region	Aim	Sample	Attributes	Payment vehicle
Campbell et al. (2007)	Ireland	Improvement in landscape	Residents (796)	Mountain land, stonewalls, farmyard, tidiness, cultural heritage, cost	Cost for the implementation of the alternative (annual tax)
Horne et al. (2007)	Finland	Management of recreational forest areas	431	Site-specific species richness, forest scenery, costs of management	Cost for the implementation of the alternative (annual tax)
Hanley et al. (1998)	Scotland	Conservation of protected areas	Residents and tourists	Woods, archeology, heather moors, wet grasslands, dry stone walls	Cost for the implementation of the alternative (tax)
Alvarez-Farizo and Hanley (2002)	Spain (Zaragoza Plain)	Evaluation of the landscape impacts of wind farm	488	Cliffs, habitat and flora, landscape	Cost for the implementation of the alternative (tax)
Bottazzi and Mondini (2006)	Italy (UNESCO Cinque Terre National Park)	Analysis of the tourist demand	Tourists and residents	Landscape, accessibility, services, food, cost	Cost for spending a day enjoying the alternative under examination
Bullock and Collier 2011	Ireland	Conservation of peat lands cultural landscape	1,000	Activities, trails, time, peat, wildlife, rarity, wildlife viewing, price	Cost for supporting the creation of the Peat lands National Park
Meyerhoff et al. (2009)	Germany (Saxony)	Biodiversity conservation and forests		Habitat for endangered species, biodiversity, forest and structure, landscape diversity, cost	Contribution to forest conversion fund
Kinghorn and Willis (2008)	UK (UNESCO Hadrian's Wall)	Conservation and management of the archeological site	Tourists (149)	Research and excavation, interpretation, artifacts, reconstruction, facilities, price	Cost for the entry fee
Rambonilaza and Dachary-Bernard (2007)	France (Brittany)	Conservation and management of rural landscape	Residents (284) and tourists (230)	Scrubland, hedgerows, farm buildings, cost	For tourists: resort tax increase. For residents: local tax increase

42.3 Methodological Proposal

Presentation of the area. The Piedmont Region, located in the North West of Italy, represents an exceptional testimony of the long-term winemaking tradition in Italy. Different combinations of climate and ecosystems have determined a suitable environment for the development of numerous grape varieties. This has been the fundamental basis for the establishment of viticulture as a productive activity of significant importance since ancient times.

This cultivar-cultural tradition is particularly evident in the territories of Langhe-Roero and Monferrato (Fig. 42.1), not only for the strong and unquestioned economic and social role of modern wine production but also and above all for the structure of the stunning landscape of this area that modeled through centuries on this wealth of knowledge, expertise, and meanings connected to the vineyard, to the production and the daily consumption of wine. The current landscape is the result of a strong attachment to the land by countless generations of winemakers and centuries of constant hard work, necessary for the implementation of an agrarian transformation of exceptional size (SiTI 2013).

Given the aforementioned characteristics, the landscape of Langhe-Roero and Monferrato has recently been nominated part of the UNESCO World Heritage List. In particular, the property forming the UNESCO site consists of a series of six components (named "Langa del Barolo," "Grinzane Cavour Castle," "Hills of Barbaresco," "Nizza Monferrato and Barbera," "Canelli and Asti Spumanti," "Monferrato of the Infernot") that represent specific natural, anthropic, and perceptive characters which, in their essence and reciprocal relations, help represent the numerous aspects of the millenary "culture wine," on which the landscape has been remarkably molded via an ongoing relationship between man and nature. In order to investigate the importance of the different characteristics of the vineyard landscape of the Langhe area in Italy, the authors developed a methodological proposal using choice-based Conjoint Analysis and focusing on the area of Langa del Barolo (Fig. 42.2).

Definition of attributes and levels. The first step of the method consists in defining the attributes and the levels of the evaluation model. In this case, six

Fig. 42.1 Vineyards in the Langhe area (Source: SiTI 2013)

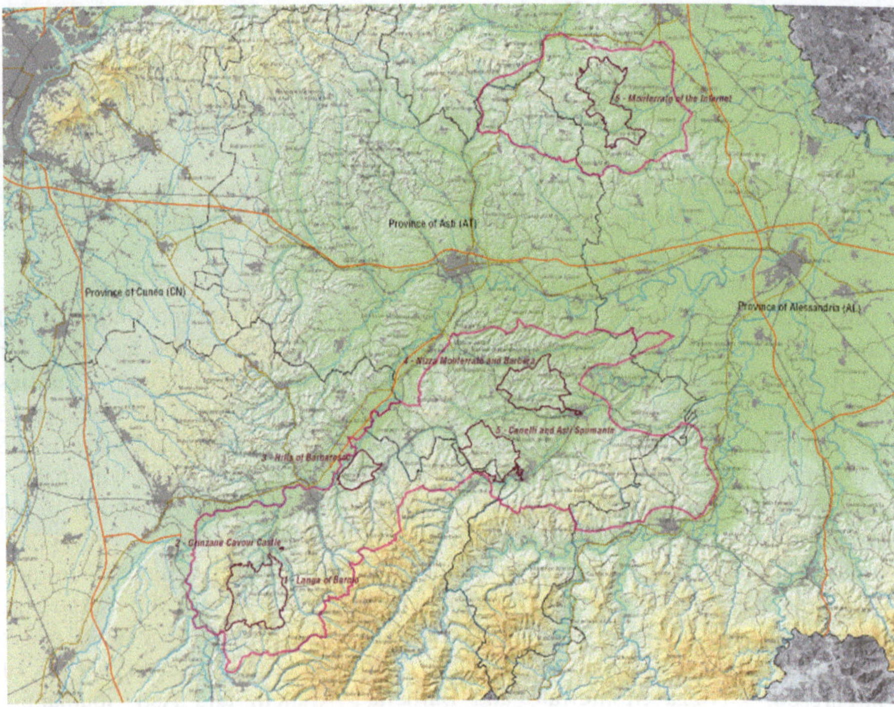

Fig. 42.2 The six components included in the serial property recently nominated as part of the UNESCO World Heritage List (source: SiTI 2013)

attributes have been identified, namely, (i) nature and protected areas, (ii) vineyards, (iii) landscape quality, (iv) intangible heritage, (v) cultural accessibility, and (vi) cost. Each attribute has been associated to three levels where one level refers to the *status quo*, while the other levels represent possible options for landscape management. Mention has to be made to the fact that the aforementioned attributes consider the specific characters and values of the landscape under investigation as they were defined in the dossier for the UNESCO nomination. Moreover, the structuring of the evaluation model has been designed with the help of a specific focus group where the proposed attributes and levels were discussed by many experts in the field of environmental assessment and landscape management in order to reach a common vision. In details, the attributes considered for the evaluation are defined in Table **42.2**.

Questionnaire. The second step of the model consists in the development of the questionnaire for the investigation of the individual preferences of the respondents with reference to different landscape profiles. In particular, the questionnaire comprised three parts. The first asked a series of questions regarding peoples' attitudes towards the particular landscape under analysis, the second contained the choice experiment questions, while the third collected socioeconomic information (age, gender, municipality of residence, job, income, etc.).

Table 42.2 Attributes and levels of the evaluation model

Attributes	Description
Nature and protected areas	Ecological-natural value of the area, considering the integrity of the vegetation and the presence of animals. The attribute varies among very high, high and medium/low
Vineyards	Extension of the surface destined to vineyards. It can vary among very large amount of vineyards, medium amount of vineyards, and small amount of vineyards
Landscape quality	It takes into account the perceptive-visual quality of the landscape, that can vary among high, medium and low quality
Intangible heritage	It considers the cultural heritage of the area, such as the presence of village feasts, centers for the diffusion of the wine culture, museums, etc. The presence of these elements can be high, medium, or low
Cultural accessibility	It concerns the possibility of directly experiencing the territory, getting into contact with the local producers, visiting the cellars, tasting wine and food, etc. This experience can be extremely easy, medium easy, or extremely difficult
Cost	The attribute varies according to the typology of the interviewed. (Mention has to be made to the fact that the levels for the cost attribute were defined via a pre-test. In this pre-test a small sample of tourists and residents was asked about the Willingness To Pay for an hypothetical landscape scenario where the attributes were set at the maximum level. The values thus obtained were further elaborated in order to produce the three levels considered in the final model). In case of tourists, the cost considers the global expenditure for spending a day in the Langhe area, including the cost for internal transports, meals, entrance fees to museums/cellars/farms; this cost varies among 100 €, 60 €, and 20 €. In case of residents, the cost refers to an extra-tax to be collected by the Regional Authority in order to afford the investments for the improvement of landscape management; this tax varies among 50 €, 25 €, and 0 €

Having decided to adopt the choice-based CA approach, the next step in the protocol design was to decide on (i) the number of choice tasks, (ii) the number of profiles for each task, and (iii) the selection of suitable profiles to be included in each task. For the latter, the authors have generated an orthogonal design using the SPSS software (www.spss.com), where nine combinations of attributes were selected. In order to keep the cognitive burden on the respondents reasonably low, we decided to develop two versions of the questionnaire, a version A with five tasks and a version B with four tasks, each with three profiles (including the *status quo*). To better explain the experimental design, Fig. 42.3 shows the graphical representation of the first task in our final designed CA questionnaire (version A). In particular, each combination of attribute levels represents a specific landscape alternative (profile), and the last profile refers to the *status quo* of the area under investigation.

Mention has to be made to the fact that one of the most critical aspects of every CA application is the importance of bearing an appropriate segmentation of the reference target population in mind (Hagerty 2008).

Which alternative do you prefer?		
Alternative 1	Alternative 2	*Status quo*

Nature and protected areas: **medium/low** Vineyards: **medium amount of vineyards** Landscape quality: **high** Intangible heritage: **high** Cultural accessibility: **very difficult** Cost: **100 €**	Nature and protected areas: **very high** Vineyards: **large amount of vineyards** Landscape quality: **low** Intangible heritage: **high** Cultural accessibility: **very difficult** Cost: **60 €**	Nature and protected areas: **medium/low** Vineyards: **large amount of vineyards** Landscape quality: **medium** Intangible heritage: **medium** Cultural accessibility: **medium easy** Cost: **20 €**

Fig. 42.3 Graphical representation of the first choice task

In our case, it was decided to interview both residents and tourists in order to consider all the relevant interests associated to the particular landscape under analysis and to better support the preparation of future plans and programs for the area.

42.4 Conclusions and Further Developments of the Work

The implementation of the evaluation according to the proposed methodology will allow the importance of the different attributes constituting the landscape to be defined as well as the economic value of the landscape to be assessed, on the basis of the preferences expressed by residents and tourists in the area. The application will satisfy one of the basic principles of sustainability, according to which it is necessary to move from eco-centric values towards an anthropocentric approach where the quality of life has a fundamental importance. The attributes and levels that have been considered in the study are suitable for representing the complexity of the landscape system under investigation, and they grant an adequate quality level of the collected information and the reliability of the numerical results. The outputs of the evaluation model will support the definition of the future policies for the site. First of all, the analysis will provide information on residents and tourists preferences about landscape management scenarios. Given the successful process of the UNESCO inscription, this information will have a crucial importance in the design of the actions that are a fundamental part of the management plan.

References

Alvarez-Farizo B, Hanley N (2002) Using conjoint analysis to quantify public preferences over the environmental impacts of wind farms. An example from Spain. Energy Policy 30:107–116

Bottazzi C, Mondini G (2006) L'analisi della Domanda Turistica nei processi di gestione dei paesaggi culturali. Aestimum 49:15–29

Bottero M (2011) Assessing the economic aspects of landscape. In: Cassatella C, Peano A (eds) Landscape indicators. Assessing and monitoring landscape quality. Springer, Dordrecht

Bravi M, Giaccaria S (2006) La Conjoint Analysis (CA) nelle valutazioni immobiliari. Aestimum 48:39–59

Bullock CH, Collier M (2011) When the public good conflicts with an apparent preference for unsustainable behaviour. Ecol Econ 70:971–977

Campbell D, Hutchinson WG, Scarpa R (2007) Using choice experiments to explore the spatial distribution of willingness to pay for rural landscape improvements. Working paper in economics 6/07, Department of Economics, University of Waikato, Hamilton

European Landscape Convention (2000) Available at: http://www.coe.int/t/dg4/cultureheritage/heritage/Landscape/default_en.asp. Accessed 22 July 2014

Green PE, Rao VR (1971) Conjoint measurement for quantifying judgmental data. J Mark Res VIII:355–363

Green PE, Wind J, Rao VR (1999) Conjoint analysis: methods and applications. In: Dorf RC (ed) The technology management handbook. CRC Press, Boca Raton

Gustafsson A, Herrmann A, Huber F (2001) Conjoint measurement: methods and applications. Springer, Berlin

Hagerty MR (2008) Can segmentation improve predictive accuracy in conjoint analysis? J Acad Mark Sci 2(4):353–355

Hanley N, MacMillan D, Wright RE, Bullock C, Simpson I, Parsisson D, Crabtree B (1998) Contingent valuation versus choice experiments: estimating the benefits of environmentally sensitive areas in Scotland. J Agric Econ 49(1):1–15

Horne P, Boxall PC, Adamowic WL (2007) Multiple-use management of forest recreation sites: a spatially explicit choice experiment. For Ecol Manage 207:189–199

Kinghorn N, Willis K (2008) Valuing the components of an archaeological site: an application of choice experiment to Vindolanda, Hadrian's Wall. J Cult Herit 9:117–124

Krantz H, Tversky A (1971) Conjoint measurement analysis of composition rules in psychology. Psychol Rev 78:151–169

Lancaster KJ (1966) A new approach to consumer theory. J Polit Econ 2:132–157

Luce RD, Tukey JW (1964) Simultaneous conjoint measurement: a new type of fundamental measurements. J Math Psychol 1:1–27

Marangon F, Tempesta T (2008) Proposta di indicatori economici per la valutazione del paesaggio. Estimo e Terrritorio 5:40–55

Meyerhoff J, Liebe U, Hartje V (2009) Benefits of biodiversity enhancement to nature-oriented silviculture: evidence from two choice experiments in Germany. J For Econ 15:37–58

Ness M, Gerhardy H (1994) Consumers preferences for quality and freshness attributes of eggs. Br Food J 96:26–34

Rambonilaza M, Dachary-Bernard J (2007) Land-use planning and public preferences: what can we learn from choice experiment method? Landsc Urban Plan 83:318–326

Sayadi ST, Gonzalez Roa MC, Calatrava Requena J (2005) Ranking versus scale rating in conjoint analysis: evaluating landscapes in mountainous regions in southeastern Spain. Ecol Econ 55: 539–550

Sayadi ST, Gonzàlez-Roa MC, Calatrava-Requena J (2009) Public preferences for landscape features: the case of agricultural landscape in mountainous Mediterranean areas. Land Use Policy 26:334–344

SiTI, Istituto Superiore sui Sistemi Territoriali per l'Innovazione (2013) Dossier di Candidatura Unesco per il sito piemontese I Paesaggi Vitivinicoli di Langhe-Roero e Monferrato, Turin

Wittink DR, Cattin PH (1989) Commercial use of conjoint analysis: an update. J Mark 53:91–96

Chapter 43
Protected Areas: Opportunities for Decentralized Financial Mechanisms?

Luca Cetara

Abstract Nature conservation policies suffer from financial restrictions worldwide. The level of understanding of financial needs of and funding sources for protected areas (PAs) is low everywhere. Landscape areas that are often buffer zones of other conservation areas could greatly benefit from even small investments in PAs due to the benefits for some economic sectors, ecosystems, and human well-being. Global expenditure for PAs would amount to 76.1 billion US $ annually and is expected to increase. Environmental protection expenditure for biodiversity conservation in EU did not increase over the last decade. Tight financial procedures, cost-effective management, and financial planning of PAs do not seem to be enough. Increasing public funds for conservation is unlikely to deliver satisfactory results. Alternative mechanisms to raise funds at the site level are needed to complement traditional ones. PA finance qualifies as an ideal field where mobilizing dispersed knowledge to fund conservation. This is a portfolio, or investment problem, suitable for the application of the "decentralization argument": many decentralized sources can build a diverse, stable, and secure portfolio, by addressing direct users or beneficiaries of PAs, their goods and services. PA networks are governance models relying on the benefits they deliver to their members that should support decentralization and provide the lacking incentives to dispersed creation and testing of novel financial mechanisms. Decentralized institutions enjoying greater freedom in action are desirable in uncertain situations and when creativity and innovation are needed. PA networks as a governance model may support decentralization.

Keywords Financial mechanisms • Finance • Protected areas • Decentralization • Funding • Nature conservation

---

L. Cetara (✉)
European Academy of Bolzano, Representing Office, Via Ludovisi, 45 00187 Rome, Italy
e-mail: luca.cetara@eurac.edu

R. Gambino, A. Peano (eds.), *Nature Policies and Landscape Policies*,
Urban and Landscape Perspectives 18, DOI 10.1007/978-3-319-05410-0_43

43.1 Enhancing Landscape Function Through Well-Functioning Protected Area Systems

A systematic lack of understanding about the financial management, needs, and funding sources of protected areas (PA) was detected in different regions (Bovarnick et al. 2010). Weaknesses in PA governance are known to undermine their operational efficiency and cost-effectiveness, in a situation of structural underfunding that can bring about a failure to deliver conservation targets.

Any PA governance model should focus not only on areas subject to strict nature conservation policies but also on those presenting actual opportunities for local economic development. Landscape areas[1] are often buffer zones, representing unique test sites for novel revenue-generating mechanisms, where development activities can be incorporated that complement and support classical conservation targets – also by collecting financial resources.

Financing PAs and ensuring the provision of their services in landscape areas, where biodiversity protection matches regulated production activities often in the presence of resident communities, means pursuing jointly multiple environmental, social, and economic objectives and testing novel landscape policies. Properly managed national PA systems in particular help avoid excessive conversion to other land uses and protect ecosystems and the services they provide to development and human well-being (UNEP 2009).

Underestimating the potential of investing in PAs could result in a failure to seize benefits for whole national economies from relatively small financial allocations (TEEB 2010), in particular in economic sectors such as agriculture, fisheries, forestry, nature-based tourism, and energy.

Thus, financial mechanisms aiming at ensuring a proper management of PAs play both landscape conservation and economic development function.

43.2 A Growing Need for Financial Resources in Protected Areas

Prudential estimates of the global expenditure for managing PAs and financing nature conservation worldwide, on the basis of the 11 targets set by the Convention on Biological Diversity (CBD),[2] indicate a need for 76.1 billion US $ annually to manage conservation sites of global significance and "particular importance for

[1] IUCN Category V Protected Landscape and Protected Seascape – area covers entire bodies of land or ocean with a more explicit management plan in the interest of nature conservation but is more likely to include a range of for-profit activities (see Dudley in this book).

[2] Convention on Biological Diversity (CBD). Strategic Plan for Biodiversity 2011–2020. COP X Decision X/2. Tenth meeting of the Conference of the Parties to the Convention on Biological Diversity Nagoya, Japan 18–29 October 2010.

biodiversity" (Target 11) and reduce the extinction risk of globally threatened species (12). PA financial needs would amount to 1.09 billion US $ for lower-income countries and 2.82 billion US $ for higher-income countries – to cover, respectively, 69 % and 50 % of the needs (McCarthy et al. 2012). These figures might increase further, since threats to biodiversity are likely to grow steadily and the surface of territory subject to conservation has increased: in 1990–2010, global PA coverage widened from 8.8 % to 12.7 % in terrestrial and from 0.9 % to 4 % in marine areas under national jurisdiction (Bertzky et al. 2012) and is set to further augment (CBD SPB 2011).

On the other hand, in Europe the available figures on environmental protection expenditure (EPE) for biodiversity conservation (United Nations et al. 2003)[3] at the country level show some volatility, a small dimension in absolute terms, and often shrink over time (European Commission 2011). Public expenditure seems to have followed a similar trend over the last few years: coherently, the whole current and historical expenditure for nature conservation has been covered by public funds. The small funds earmarked to protect biodiversity by the private sector do not seem – in contrast – to have been significantly oscillating as a share of total industry expenditure (EC 2011). Private investment on biodiversity conservation remains low everywhere, even though private funds are quite unanimously considered a promising source for PAs (Mitchell 2007). Private investment could play an essential role in diversifying financial sources, by complementing traditional ones and reducing the risks associated with income fluctuation (Flores et al. 2008).

43.3 Is Efficient Management Enough?

Tight financial procedures and criteria to enhance cost-effective management and financial planning in PAs do not seem to be enough. Rationalizing the expenditure and introducing a professional approach to financial management at the site level is a conventional claim (IUCN 2000; Gutman and Davidson 2007b), but a higher cost-effectiveness would not suffice to meet the financial needs of most PAs. The level of spending on PAs is insufficient to cover the costs for managing the existing sites and to create a global PA system (Emerton et al. 2006). Many national parks, especially in developing countries, lack funds to pay for staff salaries, patrol vehicles, or wildlife conservation programs (Saporiti 2006). Shortfalls are experienced for different countries and ecosystem types, more intensively in low-income countries (Balmford et al. 2002).

---

[3] According to the Classification of Environmental Protection Activities (CEPA), the relevant category of expenditure (protection of biodiversity and landscape) is defined as follows: "Protection of biodiversity and landscape refers to measures and activities aimed at the protection and rehabilitation of fauna and flora species, ecosystems and habitats as well as the protection and rehabilitation of natural and semi-natural landscapes" (EUROSTAT 2007).

In theory, a more efficient governance of PAs is independent from new financial mechanisms. Nonetheless, a claim for simply increasing public funds spent on biodiversity conservation in national accounts is unlikely to deliver a satisfactory result, if the financial sources and mechanisms in use are not altered. Alternative mechanisms to raise funds at the site level are needed to complement traditional ones.[4]

43.4 A Classification of Financing Mechanisms for PAs and Nature Conservation

Financial mechanisms are designed to raise, generate, or mobilize funds to cover the costs for the implementation of conservation programs (Flores et al. 2008). Classical sources of PA finance include domestic government budgets, international assistance, multilateral funds, bilateral donors, and private and community funds. A classification of these standard categories is presented in Table 43.1 (Emerton et al. 2006).

Table 43.1 Financing mechanisms for PAs

Category and aim of the financial mechanism	Possible tools	Sources of funding
1. Attracting and administering external flows	Government budgets, bilateral funds, multilateral funds, donor budgets, NGO grants, private and voluntary donations (philanthropic foundations, corporate funding, personal donations)	International, domestic
2. Encouraging conservation activities among the groups using or impacting on PAs	Cost-sharing, benefit-sharing, investment funds, enterprise funds, fiscal instruments, arrangements for private or community management of PA land, resources and facilities	Users or beneficiaries of PA goods and services
3. Introducing market-based charges for PA goods and services, attempting to capture the willingness-to-pay of PA beneficiaries	Resource-use fees, tourism charges, PES schemes	Users or beneficiaries of PA goods and services

[4] Traditional sources of PA funding include: (i) national and international public budgets, (ii) national and international NGOs, (iii) tourism, and (iv) emerging green markets (Gutman and Davidson 2007a).

The first category refers to external funds, the last two ones on resources that can be collected at the site level. External funds depend limitedly on the users of a PA, suffer from competition for public funding and development assistance, and are increasingly difficult to access for PAs; the last two categories can be more easily activated through independent, site-level management.[5]

43.5 The Decentralization Argument as a Governance Option?

Often PA reliance on centralized public funding is subject to an increasing competition and fiscal reform and changes in government-managed tools are unlikely in the short run. Thus, feasible financial mechanisms are increasingly likely to depend on decentralized knowledge of the territory and be implemented at the site level, information on their efficiency and effectiveness is limited, and innovation is required in the field.

Thus, PA funding qualifies as an ideal field where mobilizing dispersed knowledge in order to collect funds. Many decentralized sources can build a diverse, stable, and secure funding portfolio supporting PA financial sustainability. Such an instrument, in line with the "decentralization argument" presented above, is expected to promote cost and management effectiveness, allow for long-term planning and security, and incentivize PA managers to generate and retain funds. Thus, support should be given to the search for diverse sources of funding, and information spread about the mechanisms tested, by setting up appropriate incentives or removing barriers to their adoption. On the contrary, an undersupply of financial tools may pressure on the traditional mechanisms and produce a loss in the effectiveness of conservation policies.

As observed in fields with significant decentralization in production (Cowen 2006), decentralizing implies a more independent governance system, influenced by existing institutional arrangements allowing for specific solutions and policy mixes. For securing funds to PAs, site-level mechanisms qualify public and private users of the PA or its goods and services as "customers"(recipients of the benefits flowing from a PA) to be selected aiming at assuring a sustainable income flow to the PA, compatible with its goals and political, institutional, and geographic context (IUCN 2000). Nevertheless, private-oriented solutions, still scarcely explored (Mitchell 2007), present the advantage to trigger novel mechanisms and mobilize funds over which competition is lower than over public funds.

---

[5] An important exception is the fiscal instruments (e.g., taxes and subsidies) that require a government intervention to be changed. Their receipts can be directly reinvested in conservation. Indirectly, fiscal measures can encourage businesses to adopt biodiversity-friendly actions. Fiscal policy cannot be considered manageable at the site level.

43.6 PA Financial Mechanisms and Decentralization

The adoption of sound financial planning in PAs is considered as a preliminary condition for testing novel mechanisms (Flores et al. 2008) that should secure stable and long-term access to funds, timely and appropriate allocation, full costs coverage, and assurance of an effective and efficient management of PAs on conservation and other targets (Emerton et al. 2006). The most suitable mechanisms for a decentralized adoption address the direct users or beneficiaries of PAs, their goods and services,[6] and are usually managed at the site level – when the legal framework allows for some flexibility in decision-making.[7] Surveys on the economic values of PAs, stakeholder analyses for identifying "PA customers," and an economic valuation for setting up a coherent business strategy for the site may be required (IUCN 1998).

43.7 The Potential of Private Finance to Support Conservation Policies

Recently, new governance models have been launched aiming at channeling private funds to biodiversity conservation and complement the shrinking traditional sources. Here, some are recalled:

1. "Privately owned PAs" are not reported in official PA statistics for their surface nor recognized as conservation sites (Mitchell 2007), but contribute to achieve conservation targets. There are four sub-categories: individual, cooperative, NGO, and corporate IUCN Vth World Congress on Protected Areas (2003), also known as World Park Congress. They can be found both in developed and developing countries, should focus on the quality of protection rather than on the extension of protected land, aim at making profit, and are supported by their owners' intrinsic motivation and recognition of a bequest value (Langholz et al. 2000). They can use a mix of financial mechanisms to secure their sustainability and set up compensation policies.[8]
2. "Public-private partnerships" (PPP) for nature conservation are diffused in countries where legal regulations supporting independent management of PAs coexist with structurally limited government funds for conservation. PPPs can

[6] Categories 2 and 3 of Table 43.1.

[7] In Romania, for instance, the national law provides for the annual budget of state-managed parks being predetermined, and redistribution of funds takes place if further funds are collected on the initiative of a PA. This centralized mechanism clearly disincentives the adoption of decentralized sources of finance in Romanian PAs.

[8] As it happens with NGOs managing more PAs (Mitchell 2007). Quite interesting is also the recent experience in South Australia with PAs established on private lands (see Leaman and Nicolson 2012 and for the innovative conservation policies in Australia Figgis et al. 2012).

take different forms, e.g., autonomous park agencies, tourism partnerships, and biodiversity management practices (Saporiti 2006). The parties involved in the transaction gain mutual benefits. Sound management is required in order not to sacrifice the sustainability of conservation to the provision of goods and services.

3. "Corporate Ecosystem Valuation" (CEV) is a promising approach for business depending on, or taking advantage from, typical PA services. The basic assumption is that often business production processes, especially for some industries, strongly rely on long-term flows of goods and services supplied by ecosystems, which are inputs in the production function of the industry (e.g., green markets), while others affect the risk management system of the firm (e.g., regulating services). Thus, investing in biodiversity conservation makes economic sense for companies whose production, resource security, and risk safety depend on the availability and quality of ecosystem services (WBCSD 2007).
4. "Pay-per-nature-view" or "use" mechanisms exist especially in sites suitable to attract visitors (Font et al. 2004). Mechanized or administered by PA staff, they present a varying incidence of marginal revenues on the marginal cost of managing the access. Usually they depend on local natural and landscape resources. Among the existing funding methods are entrance and user fees, concessions and leases, and direct operation of commercial activities (e.g., "mushrooms collection permit" in Italy).
5. "Payments for ecosystem services" (PES) schemes aim at defining mutually beneficial exchanges between suppliers and users of ecosystem services, trying to realize net money transfers to the suppliers. Suppliers are landowners (private or public) for whom sustainable management of their land (resulting in the provision of ecosystem services) is an option with an opportunity cost – the minimum amount of money they would accept to leave aside the alternative use of their land and provide the services. The recipients are the beneficiaries of the services, e.g., businesses or private citizens, through public institutions[9] (Vatn et al. 2011). Interestingly, PES can be applied to establish real markets where a sufficient amount of information and knowledge is available (e.g., pharmaceuticals).

43.8 Can PA Networks Support Decentralized Finance for Conservation?

PAs are diverse and decentralized by nature. Diversity is required both in PAs and finance, especially due to the opportunities that different locations may offer, based on management category, natural resources, economic values, and presence of landscapes (IUCN 1998).

[9] It is worth noticing that in the latter case, the difference between a PES and a public subsidy fades (as discussed in Vatn et al. 2011).

The economic rationale for the existence of a PA network relies on the quality and value of the benefits it delivers to its members. The payment of a "network entrance fee" has to be justified, especially if financial innovation is achieved in a decentralized and independent manner by single PAs. Thus, a network should supply services that support decentralization and provide the lacking incentives to dispersed creation and application of financial mechanisms that, without the proper incentives, are unlikely to be developed and tested, since failures in trials are not offset and may result in a too low level of investment in innovation.

As a governance model alternative to central planning – where power, discretion, and finances concentrate in political institutions – networks should collect and share experiences and promote consultations rather than issuing directives and allocating funds, leaving more autonomy to their members. A network acts as an independent authority. Not holding any strict legal power, it is likely to adopt economic and other instruments (often voluntary) instead of "command and control" legislation. Decisions ultimately depend on the willingness of its members to support them. A network can set up voluntary rules for its members, through internal consultations and by involving qualified stakeholders (e.g., governments).

The decentralization argument assumes that greater freedom in action and independent choice are especially desirable in situations of uncertainty (Knight 1921). If little information is available on the effectiveness of financial mechanisms, diversity and innovation are highly positive outcomes.

A "venture capital metaphor" quite properly describes the case. Venture capital (VC) is a suitable technique to address uncertain situations, where the actual degree of success for initiatives is unknown. A VC investment foresees, after an assessment of a business plan for a novel undertaking, the destination of funds to support the start-up phase and eventually reap the resulting benefits. The likely failure of the project and the resulting loss of money are factored in the investment calculation. From the investor's standpoint, diversification of investment over more initiatives mitigates the risk and works as a bet on better ideas, so that the peaks should compensate the losses suffered on less-effective projects. Positive spillovers of VC include support to innovation, opportunity to undertake activities that would not have been started, and exploitation of decentralized knowledge bringing diversity in the outcomes.

"Nature" could require less creativity than other fields (e.g., innovation, arts creation), since the ends to be achieved are clear enough (Cowen 2006). This does not cancel the structural problem of declining availability of funds for conservation and the inefficiency of the most used options. A diverse and wide portfolio as well as a rationalization in spending might help.

PA networks can participate in leveling the plain field for innovation by providing incentives to dispersed creation of funding sources needed to maximize innovation and trials: training on financial management to create a background on which newer mechanisms can be built (seeding phase), support to spreading of knowledge, testing and falsification of new mechanisms at the site level (start-up and testing phases), and insurance or compensatory services to cover the costs of conservation impossible to finance with autonomously collected funds, but deemed important to implement from a regional or national standpoint.

References

Balmford A, Bruner A, Cooper P, Costanza R, Farber S, Green RE, Jenkins M, Jefferiss P, Jessamy V, Madden J, Munro K, Myers N, Naeem S, Pavola J, Rayment M, Trumper K, Turner RK (2002) Economic reasons for conserving wild nature. Science 297:950–953

Bertzky B, Corrigan C, Kemsey J, Siobhan K, Ravilious C, Besançon C, Burgess N (2012) Protected planet report 2012: tracking progress towards global targets for protected areas. IUCN/UNEP-WCMC, Gland/Cambridge

Bovarnick A, Fernandez Baca J, Galindo J, Negret H (2010) Financial sustainability of protected areas in Latin America and the Caribbean: investment policy guidance. United Nations Development Programme (UNDP) and The Nature Conservancy (TNC)

Convention on Biological Diversity (CBD) (2010) Strategic plan for biodiversity 2011–2020. COP X Decision X/2. In: Tenth meeting of the conference of the parties to the convention on biological diversity Nagoya, Japan, 18–29 Oct 2010

Cowen T (2006) Good & plenty: the creative successes of American arts funding. Princeton University Press, Princeton

Emerton L, Bishop J, Thomas L (2006) Sustainable financing of protected areas: a global review of challenges and options. IUCN, Gland/Cambridge, x + 97 pp

European Commission (2011) Environmental protection expenditure in Europe – data 1995–2009. Publications Office of the European Union, Luxembourg

EUROSTAT (2007) Environmental expenditure statistics 2007 edition. General government and specialized producers data collection handbook. Office for Official Publications of the European Communities, Luxemburg

Figgis P, Fitzsimons J, Irving J (eds) (2012) Innovation for 21st century conservation. Australian Committee for IUCN, Sydney

Flores M, Rivero G, León F, Chan G et al (2008) Financial planning for national systems of protected area conservancy: guidelines and early lessons. The Nature Conservancy, Arlington

Font X, Cochrane J, Tapper R (2004) Pay per nature view: understanding tourism revenues for effective management plans. Leeds Metropolitan University, Leeds

Gutman P, Davidson S (2007a) A review of innovative international financial mechanisms for biodiversity conservation with a special focus on the international financing of developing countries' protected areas. WWF Macroeconomics Program Office, Washington, DC

Gutman P, Davidson S (2007b) The global environmental facility and payments for ecosystem services – a review of current initiatives and recommendations for future PES support by GEF and FAO programs. UN Food and Agricultural Organization – FAO, PESAL Papers Series No. 1, Rome

IUCN (1998) Economic values of protected areas: guidelines for protected area managers. IUCN, Gland/Cambridge

IUCN Financing Protected Areas Task Force of the World Commission on Protected Areas (WCPA), in collaboration with the Economics Unit of IUCN (2000) Financing protected areas. IUCN, Gland/Cambridge, viii + 58 pp

Knight FH (1921) Risk, uncertainty, and profit. Hart, Schaffner & Marx/Houghton Mifflin Company, Boston

Langholz JA, Lassoie JP, Lee D, Chapman D (2000) Economic considerations of privately owned parks. Ecol Econ 33:173–183

Leaman G, Nicolson C (2012) Innovative measures for establishing protected areas on private lands in South Australia. In: Figgis P, Fitzsimons J, Irving J (eds) Innovation for 21st century conservation. Australian Committee for IUCN, Sydney

McCarthy et al (2012) Financial costs of meeting global biodiversity conservation targets: current spending and unmet needs. Science 338(6109):946–949

Mitchell BA (2007) "Who's doing the protecting in protected areas?" A global perspective on protected area governance. Georg Wright Forum 24(3):81–91

Saporiti N (2006) Managing national parks. How public-private partnerships can aid conservation. The World Bank, Washington, DC. pp 1–4

TEEB (2010) In: Kumar P (ed) The economics of ecosystems and biodiversity ecological and economic foundations. Earthscan, London/Washington, DC

UNEP (2009) Ecosystem management program. A new approach to sustainability, UNEP, Nairobi

Vatn A, Barton D, Lindhjem H, Movik S, Ring I, Santos R (2011) Can markets protect biodiversity? An evaluation of different financial mechanisms. Noragric report no 60. University of Life Sciences, Norway

WBCSD, IUCN (2007) Business and ecosystems. Markets for ecosystem services. New challenges and opportunities for business and the environment. A perspective IUCN, World Business Council for Sustainable Dvelopment. IUCN/World Business Council for Sustainable Development, Gland/Geneva, 17p.:ill

Part III
Experiences and Practices

Chapter 44
The Langhe Landscape Changes

Danilo Godone, Matteo Garbarino, Emanuele Sibona, Gabriele Garnero, and Franco Godone

Abstract Agricultural policies and socioeconomic constraints are strong drivers of change in human-dominated rural landscapes of Mediterranean Europe. Changes in rural landscapes can have a strong influence on the perspectives of protection and improvement of the natural and cultural heritage. A shift towards quality production, favoured by institutional financial support, has been recently observed in hilly productive Mediterranean sites. An example of this situation is the Langhe region (NW Italy), where woody plantations such as vineyards and orchards have been cultivated on hillslopes for centuries. In this chapter, we assess the landscape evolution occurred in this study site. Land use changes in the 1954–2000 period were assessed by object-oriented analysis of aerial photographs and quantified by spatial statistics capturing and measuring different elements of landscape change. The expansion of orchards from 1954 to 2000 caused an increase of landscape heterogeneity and the fragmentation of field crops. Orchards expansion has reduced other land uses occupying up to 55 % of former field crops, 24 % of vineyards and 15 % of forests. Changes in rural landscapes, traditionally dominated by vineyards, field crops and forests, were so observed in the Langhe region.

Keywords EU-CAP • Historical aerial photographs • Land use change

D. Godone (✉) • E. Sibona
Department of DISAFA and NatRisk, Università degli Studi di Torino, Via L. Da Vinci 44, I-10095, Grugliasco, TO, Italy
e-mail: danilo.godone@polito.it; emanuelemarco.sibona@unito.it

M. Garbarino
Department of D3A, Università Politecnica delle Marche, Via Brecce Bianche, I-60131, Ancona, Italy
e-mail: m.garbarino@univpm.it

G. Garnero
Interuniversity Department of Regional and Urban Studies and Planning (DIST), Politecnico and Università di Torino, Viale Mattioli 39, 10125, Turin, Italy
e-mail: gabriele.garnero@unito.it

F. Godone
CNR – IRPI, Str. delle Cacce, 73, 10135 Torino, TO, Italy
e-mail: franco.godone@irpi.cnr.it

R. Gambino, A. Peano (eds.), *Nature Policies and Landscape Policies*, Urban and Landscape Perspectives 18, DOI 10.1007/978-3-319-05410-0_44

44.1 Introduction

In this contribution, land use change in a hilly region of southern Piemonte (northwestern Italy), called Langhe, is studied. This landscape is a traditional complex mosaic of forests, vineyards, orchards and other cultivated fields. The entire Langhe region is also famous for the quality of its wine and food and is candidate to be included in the UNESCO World Heritage List.

Land use changes observed in the area are particularly strong and seem influenced by local and European rural policies. A rigorous photogrammetric processing of historical aerial images was performed thanks to the availability of rare historical camera calibration certificates. The goal of the contribution is the quantification of the landscape changes occurred in the area between 1954 and 2000 and, in particular, the determination of transition paths and transition rates of the main land use categories of this ecosystem. Finally, the effects of the EU-CAP and local policies on the rural landscape of Langhe region are discussed (Godone et al. 2014).

44.2 Materials and Methods

Study area. The study area is part of Langhe hilly region (NW Italy – 44°40′ N, 07°59′ E), which is characterised by strong agricultural tradition and is widely renowned for its high-quality wine production like Barolo and Barbaresco (Delmastro 2005). Study site elevation ranges from 190 to 634 m a.s.l. and the climate is characterised by humid summer and dry winter seasons; annual precipitation ranges from 800 to 1,100 mm with a main minimum in July, a secondary one in winter, and a peak in autumn. Total annual rainfall reaches 730.4 mm while mean annual temperature averages 11.9 °C.

A set of four historical aerial photographs, taken by the Gruppo Aeronautico Italiano (GAI) flight, which represents the first available flight covering almost the entire Italian territory after the Second World War, was employed in the analyses. They were stored in the archive of Italian National Research Council (CNR-IRPI, Torino), where historical and recent aerial images concerning hydrogeological phenomena are stored. GAI flight was carried out in 1954–1955 with a flight height ranging from 10,000 to 5,000 m a.s.l. having a medium scale of 1:33,000 (Acosta et al. 2005).

In the Italian Military Geographic Institute (IGMI) historical archive, the camera calibration certificates of the investigated flight were retrieved, thus allowing a rigorous image orientation. Photograms were oriented, in Z-Map software, by Automatic Aerial Triangulation assuring an overall accuracy of 2.22 m. The oriented images were then orthorectified and mosaicked at 1-m resolution. A recent, 1:1,000 RGB, orthoimage (Terraitaly – IT2000™, Blom C.G.R. S.p.A) having a ground resolution of 1 m, was employed in the diachronic analysis.

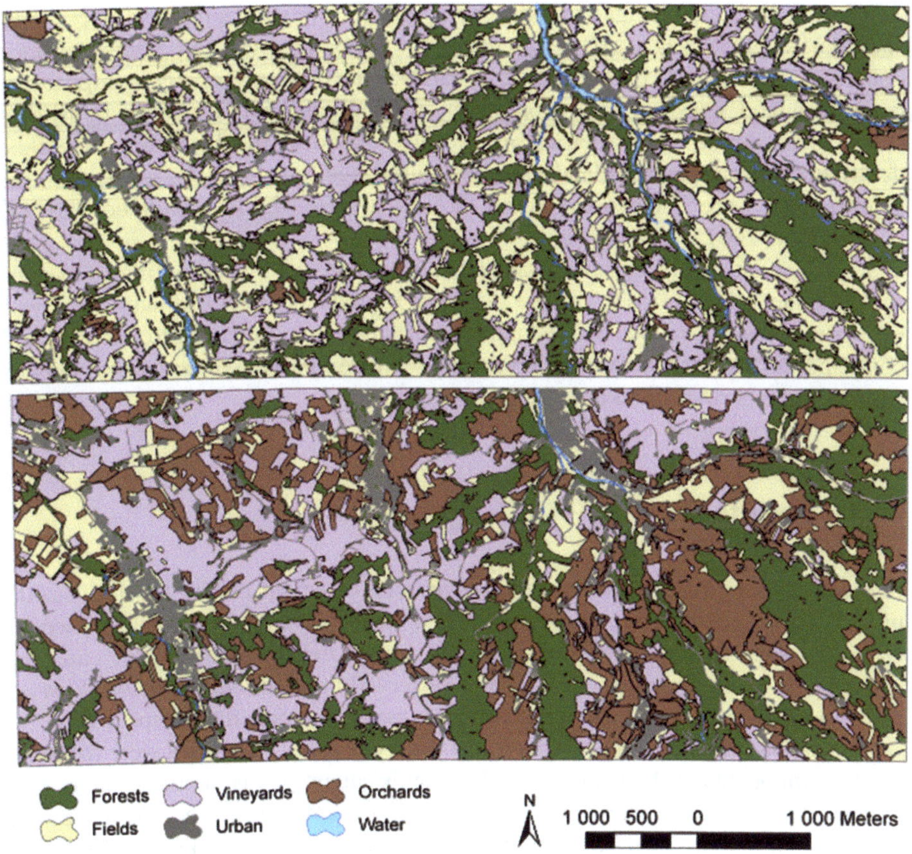

Fig. 44.1 Land use classifications of 1954 (*upper* map) and 2000 (*lower* map) orthoimages

In order to delineate land use classes in the investigated orthoimages (1954–2000), automated segmentation with the eCognition software (scale parameter, 100; shape factor, 0.5) was used to draw polygons in the test area. The segmented images were manually classified into six categories of land cover (Fig. 44.1). An accuracy assessment was performed on each map resulting in the K statistic ranging from 0.86 (90.2 % overall accuracy) for the 1954 image to 0.87 (90 % overall accuracy) for the 2000 image.

Landscape analysis. The analysis was performed by using the "change detection" free extension in ArcView environment (Chandrasekhar 1999), which allowed the computation of a matrix describing the transition between each pair of land cover categories by quantifying its extent per unit time. The permanence of a certain land cover category in the studied period was also calculated by the kappa (K) statistic (Romero-Calcerrada and Perry 2004) that ranges from 0 (category's surface unchanged) to 1 (all categories' surface changed).

Then, to analyse changes in landscape pattern, Fragstats software (McGarigal and Marks 1995) was used to calculate, by applying an 8-cell neighbourhood definition, key metrics representative for landscape configuration and composition: patch size and density, edge, contagion, connectivity and diversity (Cushman et al. 2008). Since many metrics are closely related at the landscape level and describe similar aspects of landscape structure (Neel et al. 2004), ten landscape-level metrics were selected excluding those that were highly correlated ($r > 0.8$) (Tischendorf 2001).

Landscape structure was also analysed at the class level by computing 13 metrics for the 6 land cover classes, for the two time periods. Indirect ordination analysis (PCA) was used to reduce the redundancy of landscape metrics into uncorrelated components (McCune and Grace 2002), allowing comparison of land cover classes from all time periods (Tinker et al. 1998).

44.3 Results

In the study site, remarkable changes have been observed during the 1954–2000 period, e.g. a general increase in landscape heterogeneity. The total surface of *Orchards* showed an increase of 24.6 %, to the detriment of *Fields* category (−26.9 %). A slight increase (3.6 %) was observed in the *Urban* class too. *Forests* and *Vineyard* categories experienced an increase and a reduction, respectively.

The kappa statistic indicated that forests and vineyards categories were the most stable components of the landscape, while fields category was the least stable. The noteworthy variation experienced by the *Orchards* category pushed us to deepen our analysis on class level transitions. Only 3 % of the total surface of *Orchards* category remained unchanged, while 55 % of it was former fields, 24 % vineyards, 15 % forests and 3 % other categories.

Principal component analysis provided an ordination scatterplot of land cover changes. The first principal component accounted for 35 % of the total variation and reflected variations of aggregation, core area, mean patch area and shape complexity. The second component explained an additional 30 % of the total variation and was correlated to connectivity, Euclidean nearest neighbour and patch density. The scatterplot confirmed that *Orchards* category experienced the strongest change in the landscape (Fig. 44.2). This category increased both in patch density and patch size mainly at the expenses of the *Fields* category that resulted more fragmented and reduced its mean patch size. *Forests* and *Vineyards* categories showed a similar pattern of change where they increased the aggregation, the connectivity and the size of their patches. A different pattern was observed for *Urban* areas that increased in patch/edge density and shape complexity.

Land use change study requires careful approaches and should be carried out by employing trustworthy data sources in order to correctly reconstruct historical landscape dynamics (Burgi and Russell 2001; Garnero 2013; Minucciani and Garnero 2013). For this reason, the adopted rigorous photogrammetric approach

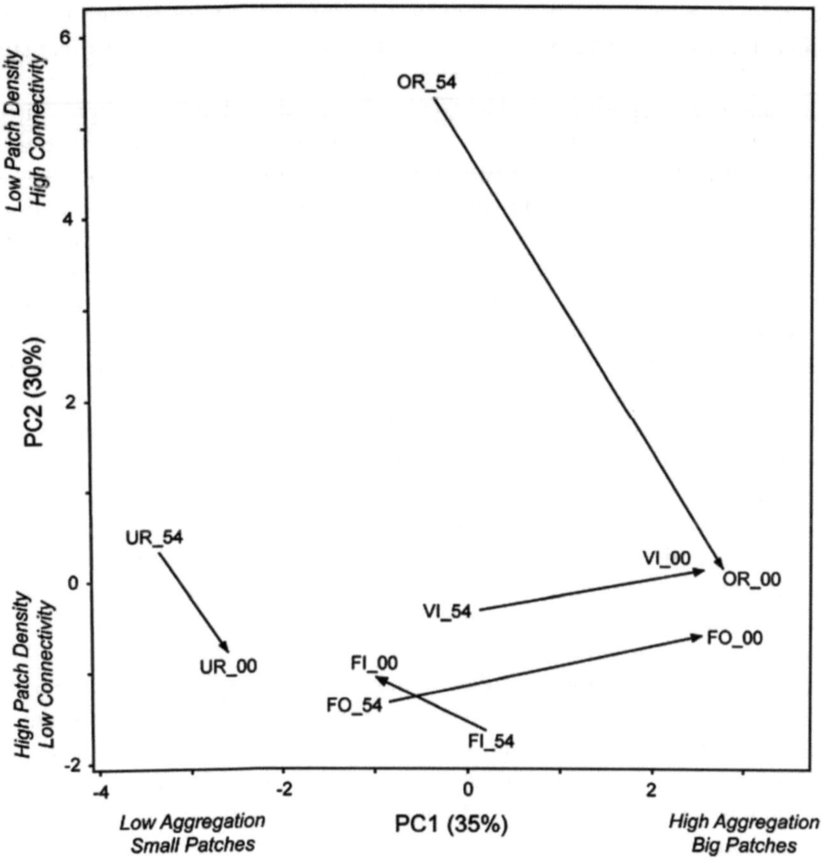

Fig. 44.2 Principal component analysis of five land cover classes (*FO* forests, *OR* orchards, *VI* vineyards, *FI* fields, *UR* urban areas) for 1954 and 2000 periods

assures accurate processing results (Rocchini et al. 2012). The described image processing together with the high classification accuracy assured a reliable land use change investigation.

Among the six land use categories, *Orchards* increased from 1954 to 2000, replacing other agricultural areas, like *Fields* and *Vineyards* and *Forests*. The expansion of *Orchards*, coupled with the increase of *Urban* areas transformed the traditional landscape featured by vineyards, crops and forests in a more fragmented mosaic of various land use categories.

Forests remained almost unchanged; the reforestation pattern observed in the study area is in agreement with other Italian and European (Sitzia et al. 2010; Cocca et al. 2012) mountainous and hilly areas. The non-marginal, productive, vocation of the investigated area is also proved by a notable increase of inhabitants recorded in the last decades, in contrast with Italy and other European countries (Peroni et al. 2000). Vineyards slightly expanded in the most accessible and productive sites, limiting orchards expansion. However, a general reduction of their surfaces in marginal and less accessible sites in favour of orchards is confirmed by other studies in the Mediterranean territory (Corti et al. 2011).

The land use category defined as *Orchards* was almost entirely represented by hazelnut (*Corylus avellana* L.) orchards. In accordance to the explained results, in Piemonte region the hazelnut cultivated surface increased by 20 % in the last 1990–2000, with the highest increment peak during the 1990–1995 period (Valentini and Me 2002). Furthermore, in the 1981–2000 period, the surface expansion triggered an increase of hazelnut production, nearly doubling the output tons, and of its price from 1.66 to 1.96 €/Kg. A cultivation shift from vineyard towards hazelnut is detectable from the historical records, concerning Cuneo province, on cultivated surfaces (ISTAT 1971–2001). In the last twenty years of the study period (1980–2000), hazelnut orchards have nearly doubled their surfaces, while vineyards have shown a remarkable decrease. These records confirmed the results observed in our study area through change detection analysis.

The strong increase of hazelnut surfaces occurred in Piedmont at the beginning of the 1990s was probably favoured by several European Council regulations and national rural policies. A key policy measure concerning hazelnut cultivation in Italy was a decree (DM 2/12/93) of the Italian Ministry of Agricultural and Forestry Policies which recognised hazelnut's Protected Geographical Indication (PGI) under the appellation *Nocciola Piemonte*. Likewise, in 1996, the European Union registered, with a regulation, the "Piedmont hazelnut" as a PGI (EC 1107/96), and, more recently, in 2007 the Langhe hazelnut was registered within the Community Plant Variety Office (CPVO) with a new name: *Tonda Gentile Trilobata*, for an efficient preservation. The rapid expansion of hazelnut plantations is radically transforming the rural landscape of Langhe region questioning the role of the EU common policy as a driver of land use change.

References

Acosta A, Carranza ML, Giancola M (2005) Landscape change and ecosystem classification in a municipal district of a small city (Isernia, Central Italy). Environ Monit Assess 108:323–335

Burgi M, Russell EWB (2001) Integrative methods to study landscape changes. Land Use Policy 18:9–16

Chandrasekhar T (1999) Change detection extension for ArcView. ESRI, Redlands

Cocca G, Sturaro E, Gallo L, Ramanzin M (2012) Is the abandonment of traditional livestock farming systems the main driver of mountain landscape change in Alpine areas? Land Use Policy 29:878–886

Corti G, Cavallo E, Cocco S, Biddoccu M, Brecciaroli G, Agnelli A (2011) Evaluation of erosion intensity and some of its consequences in vineyards from two hilly environments under a Mediterranean type of climate, Italy. In: Godone D, Stanchi S (eds) Soil erosion issues in agriculture. InTech, Rijeka, pp 113–160

Cushman SA, McGarigal K, Neel MC (2008) Parsimony in landscape metrics: strength, universality, and consistency. Ecol Indic 8:691–703

Delmastro M (2005) An investigation into the quality of wine: evidence from piedmont. J Wine Res 16:1–17

Garnero G (2013) Le nuove basi geografiche di supporto alla pianificazione ed alla gestione territoriale. Mondo Digitale XII(47):216–225

Godone D, Garbarino M, Sibona E, Garnero G, Godone F (2014) Progressive fragmentation of a traditional Mediterranean landscape by hazelnut plantations: the impact of CAP over time in the Langhe region (NW Italy). Land Use Policy 36:259–266

McCune B, Grace JB (2002) Analysis of ecological communities. MjM Software Design, Gleneden Beach

McGarigal K, Marks BJ (1995) FRAGSTATS: spatial pattern analysis program for quantifying landscape structure. U.S. Department of Agriculture, Forest Service, Pacific Northwest, Portland

Minucciani V, Garnero G (2013) Available and implementable technologies for virtual tourism: a prototypal station project. In: Murgante B et al (eds) ICCSA 2013, Part IV, published in lecture notes in computer science LNCS 7974. Springer, Heidelberg, pp 193–204. ISSN: 0302-9743. doi:10.1007/978-3-642-39649-6-14; Scopus:2-s2.0-84880708338

Neel M, McGarigal K, Cushman S (2004) Behavior of class-level landscape metrics across gradients of class aggregation and area. Landsc Ecol 19:435–455

Peroni P, Ferri F, Avena GC (2000) Temporal and spatial changes in a mountainous area of central Italy. J Veg Sci 11:505–514

Rocchini D, Metz M, Frigeri M, Delucchi L, Marcantonio M, Neteler M (2012) Robust rectification of aerial photographs in an open source environment. Comput Geosci 39:145–151

Romero-Calcerrada R, Perry GLW (2004) The role of land abandonment in landscape dynamics in the SPA Encinares del rio Alberche y Cofio, Central Spain, 1984–1999. Landsc Urban Plan 66: 217–232

Sitzia T, Semenzato P, Trentanovi G (2010) Natural reforestation is changing spatial patterns of rural mountain and hill landscapes: a global overview. Forest Ecol Manag 259:1354–1362

Tinker D, Resor C, Beauvais G, Kipfmueller K, Fernandes C, Baker W (1998) Watershed analysis of forest fragmentation by clearcuts and roads in a Wyoming forest. Landsc Ecol 13:149–165

Tischendorf L (2001) Can landscape indices predict ecological processes consistently? Landsc Ecol 16:235–254

Valentini N, Me G (2002) Attualità e problematiche della coltura del nocciolo in Italia: la situazione piemontese. In: Proceedings of 2nd National Congress on "Nocciolo", Giffoni Valle Piana (SA), pp 133–140

Chapter 45
Cultural Landscape and Royal Historical System in Piedmont Region

Maria Grazia Vinardi

Abstract The Conference of Kraków clarified the meaning of cultural landscape as the result of "the prolonged interaction in different societies between man, nature and the physical environment. It is testimony to the evolving relationship of communities, individuals and their environment. Its conservation, preservation and development focus on human and natural features, integrating material and intangible values". The contribution develops the relationship between conservation and sustainability, seen as the exploitation of the heritage, virtuous propeller for the expansion of these environments, which have been influenced by a historical continuity of functions and have been the economic source for noble complexes, such as holiday destinations or temporary residences of the owners. The land, crops and historical sediments represent a complex heritage of artistic and cultural as well as economic values. The idea is to identify, in addition to the recognition of certain places and objects as heritage of interest (world, European and national), the historical systems connected to it, to propose strategies destined to pursue continuity in this virtuous historical relationship or new programmes that take part in this aim. Yet, these presences and their systems are currently besieged by politically and economically driven local transformation processes and by plans and legislation that are often indifferent to their values and have little involvement in the communities concerned.

Keywords Historical systems • Conservation • Innovation

The 2000 Kraków Charter (International Conference on Conservation 2000) defines the cultural landscape as the result of the "prolonged interaction in different societies between man, nature and the physical environment (. . .). Its conservation, preservation and development focus on human and natural features, integrating material and intangible values". The land, crops and settlements represent, in addition to their value as historical documents and within a development process, a resource for the conservation of the identities of populations and also an economic heritage in support of the same, nobler testimonies. Investigating certain situations, which provide examples of the intersection between local resources and the

M.G. Vinardi (✉)
Interuniversity Department of Regional and Urban Studies and Planning (DIST), Politecnico and Università di Torino, Viale Mattioli 39, 10125, Turin, Italy

R. Gambino, A. Peano (eds.), *Nature Policies and Landscape Policies*, Urban and Landscape Perspectives 18, DOI 10.1007/978-3-319-05410-0_45

presence of residential and institutional architectures, helps shed light on development strategies, which, in Piedmont, have been inseparable from the awareness of the relationship between the historical uses and architectural consistencies of the infrastructures. The exploitation of the heritage also derives from the relationship between conservation and sustainability, seen as the virtuous "engine" for strengthening those environments that are testimony to how the historical continuity of functions has constantly promoted a technological and economic upgrade joined with the persistence of cultural identities.

The contribution aims to explain the relationship between protection and enhancement of heritage, taking into account the universal relevance, constantly supported by specific historical and artistic values. In these contexts, such specific values consist not only the individual building but also the territorial roots, as a result of a complex strategies of development. In the case of the royal estates, which are included in the list of UNESCO WHS, this interaction suggests the existence of peculiar territorial strategies extending largely beyond the site's boundaries.

The idea is to identify, in addition to the recognition of the cultural interest, certain local historical systems, in order to recognise which strategies can guarantee the continuity of the said virtuous historical relationships or define programmes that are compatible with this aim.

In the past, certain places were ennobled by the distinguished presence of royal or aristocratic residence with grounds and gardens and were also occupied by productive enterprises. Emblematic is the case of Venaria Reale (1659–1679), a leisure site (*delitie*) of royal court, with its farms. After the establishment of the complex as a royal hunting estate, in the second half of the eighteenth century, the new Mandria della Venaria was opened, joined the Apertole, a short distance southeast, thus expending the activities of the royal estate of Venaria Reale. This royal palace represents emblematically the close interaction between territorial context, urban settlement, the palace, park and gardens. This interaction is also visible in other royal sites: in Agliè, ancestrally ruled by the San Martino family, at the time of acquisition of the Feud as part of the estate of the Royal House of Savoy (in the second quarter of the eighteenth century), control of the local territory was reorganised, with the setting up of the new La Mandria farm and the expansion of others. The possessions of the Church and those of the Knightly Orders of Saints Maurizio and Lazzaro and of the Order of Malta also adopted, each in their own area, farming strategies, which supplied Abbotts and Knights with resources to reinvest in the maintenance and expansion of civil and rural real estate and in the construction of new architectures, canals and routes.

The virtuous intersection between land governance, irrigation system and crops, observing agreements and customs, and infrastructural programmes for the adjustment of roads as of the second half of the eighteenth century, which created direct links between the most important towns, bypassing the farm system, enabled big investments in noble architectures and buildings for productive efficiency on one hand and absolute control of the local territory on the other.

The integration between conservation of the cultural landscape and development became an action to increase value, widespread but specific to each situation, in view of the particular morphologies of the "regions" and of the locations, distinguished by farming activities and different natural features.

Today these systems are besieged by politically and economically driven local transformation processes and by plans and legislation that are often indifferent to their values and have little involvement in the communities concerned.

The case study regards a system of residential and farming settlements around Pinerolo and the fact that, in the nineteenth century, they were lucky enough to be progressively purchased by the upper middle classes who preserved their landscape and architectural features and values. Their identity has been conserved to this day, thanks to the decisions made during works for the 2006 Turin Winter Olympics, when the area was suddenly involved due to the need for a new and extensive use of infrastructure and the consequent expansion programmes, which often resulted inevitably in the separation of the environmental and historical continuity of the properties. In addition to this, the layout of the area was transformed for intrinsic, productive and economic reasons, such as changes in crops and technologies, the plans and fortunes of the property owners and the people that worked in the area, resulting in prosperity or abandon.

The rural complexes in the area are the following farms: Le Peschiere, Ai Nana, Colombretto, Galetta, Losetta and Pezzia. However, the local landscape conservation and transformation processes also involve the Il Gruato, Pavia, Ciabot Bas, Butal, San Luigi, Biscorno and Biscornetto farms and the Bersano and Torrione Villas, recorded in the Cadastral Register of the Royal House of Savoy in 1783.

The first documents found related to the area date back to 1434 and refer to a share cropping deed of Cascina Gruat, in the Chiamboni region, now belonging to the Peschiera and Colombretto estates. The description is confirmed by the persistence on the territory of the imprint of a very historical use of the land and the cultural landscape, which is still configured in relation to habit and to the use of water from the River Lemina, with its derivations, recorded in detail in the "Valbe di Pinerolo" Register. The eighteenth-century survey identifies the region and precisely indicates the building complex, grape cultivations, routes and irrigation channels using expressive symbols, outlining a heritage which, apart from the change in some crops, seems to be more or less the same today. The six farms with civil grounds and gardens were all owned by the Bertea family in the second half of the nineteenth century. Ernesto Bertea, a lawyer and art lover, as well as being a talented artist himself, brought together a circle of artists at the Villa, whose members included Antonio Fontanesi, Alfredo d'Andrade, Vittorio Avondo, Federico Pastoris, Leonardo Bistolfi, Davide Calandra, Carlo Follini, Marco Calderini and others. Thanks to the sequence of maps (Valbe di Pinerolo Register of the eighteenth century, the French Cadastral Register of 1808–1811, the Rabbini Cadastral Register of 1866) and the sources kept in the family's private archives, it is possible to reconstruct the transformations of the sites and factories, along with the expansions, reorganisations and architectural restoration of the main residence and its surroundings (Fig. 45.1).

The Ai Nana farm complex is also of local importance and the civil building and chapel still retain their original identity (Castagnova 2006). Ownership of the

Fig. 45.1 G.B. Reale, Mappa della Città di Pinerolo, 1783, ASCP, Fondo urbanistica 2G, Particolare del Quadro d'unione

complex was transferred from Count Luigi Mattone to Ernesto Bertea in 1867. The first construction phase seems to have been the tower at the entrance, possible a dovecote or a gatehouse. The chapel, devoted to Saint Giovanni Battista, built in the eighteenth century, and the wing of the farm building are two perpendicular bodies joined by an awning that houses the outside staircase leading to the house.

Another Bertea property, deriving from successive inheritance divisions among the Nana brothers, is Colombretto, not far from Le Peschiere. The events surrounding this property involved changes in ownership that testify to the importance of vineyards and grape cultivations and the consistency of the constructions, comprising farmhouses, vat rooms, cellars and chapel. The same document also defines the name of the region, which refers to the presence of the most important farm in the middle of the eighteenth century, the Losa farm.

The Galetta walled complex also retains evidence of the late sixteenth-century construction style, with productive farm, civil buildings, "casi da terra" (lean-tos and open sheds) and enclosure, made up of a linear structure with vaulted ceilings on the ground floor, flanked by rooms with wooden coffered ceilings. The importance of the site of the Losetta farm, in addition to the connection within the Galetta, Colombretto and Losa farm system, derives from the maintenance of certain elements that qualify the territory in question, also in that, next to the farm, which has rural features of documented interest, there is an area occupied by barriers affording protection against the River Lemina, built in 1754 by hydraulic engineer, Michele Buniva. In 8 April 1829, the architect G. Camusi of the

Fig. 45.2 Pinerolo, Torino. Tenuta Bertea

Engineering Corps, as requested by Mr Gio. Battista Filippa, visited the Losetta farm, south of the Lemina, to see whether the works carried out by the Ospedale di Carità, which owned Losa farm on the other bank of the river, had not prejudiced the Filippa property (Fig. 45.2).

Once again, the eighteenth-century maps identify the region and, thanks to clever symbols, the construction, the grape cultivations, the roads and paths and the irrigations channels, registering a heritage which seems to be confirmed today, apart from changes to some cultivations. The farms that have been detached from the old system in the wake of road works include the Chiabotto Pezzia farm, which was part of this system. It became part of the Bertea estate as a result of the sale, as documented in the deed dated 26 January 1841, by Ms Teresa Artero, widow of Giovanni Enrico Pezzia, of the farm known as Chiabotto, consisting of 12 "giornate" of land, 47 "tavole" (Piedmontese unit of land measurement equivalent to about 30 m^2) and 4 ft of house, barn and vegetable garden, lawn meadow and grape cultivations, for the sum of 32 250 lira. The farm is of particular interest for the purposes of the conservation of the landscape, subject to defence in that it is a limit element of the historical farming content of the town of Pinerolo, characterised by the existence of the farms and their appurtenances acquired in the nineteenth century by the Bertea family, even though the cultivation of grapes that characterised the area for so long has been abandoned and replaced by the production of maize, wheat or left wild.

The new Cultural Heritage and Landscape Code (Repubblica Italiana 2004) has made it possible to progressively bind the farms and buildings of interest with annexed appurtenances, grounds, gardens and limited pertinent spaces, but up to now, it has not been possible to extend the environmental limitation to the entire system, and this procedure is only being launched now. The difficulties that we are faced with are not those related to the retracing of the property deeds and the cadastral references, but those regarding the definition of a boundary which is not linked solely to a single and prevalent property, extending defence also to other properties, so that its limits are historically and structurally defined.

The condition of the farmland subject to restrictions should create a bridge (or a virtuous relationship) between the need for conservation and those of growth, sustainability and innovation. These choices require a vision expanded as much as possible to the complete surroundings, recognising the specifics and the strong continuity with the past in order to ensure productivity in the correct use of resources: an aim that has nothing to do with the exultation of tourism and the food and wine sector alone.

Farming landscapes on the plane, expanding suburbs and abandoned complexes in the middle and high valleys, hill landscapes with high plateaus, hills, clearings and woods were the historical requirements for the construction of productive settlements but also for prestige and promotion in the technical and artistic fields. The active defence of this heritage does not, however, mean that the consistencies cannot change, but implies an appropriate reorganisation to properly exploit them, with promotional, adaptive and productive strategies. The results of the resources recently lavished on limited sector of heritage must be verified with a view to sustainability and continuity in time. The separation of architectural and landscape assets from their historical structural conditions and characteristics has been unavoidable, but in future, flows of tourism cannot be the only source of income. Plans must be made for the independent management of these assets, detached from heritage-related contexts, which no longer exist. Just think of the royal farms of Racconigi now split between numerous private owners who have independently and uncontrollably changed the consistency and types of use within each building, land parcel or construction.

There is not even any point in imagining the utopia of inflexible legislation that standardises a sort of manual of uses, restorations and reuses. For those that take care of conservation, while relating to a general plan, the intervention has to take into consideration the specifics of architecture so that a heritage of knowledge, even when it has still to be officially sanctioned, is not lost.

Recognition through knowledge is the first form of defence and exploitation of those elements that critical historical judgement has overlooked up to now. A perfect example is the existence of that intersected system of presences and uses built up over time by a consolidated habit, of which the simplicity of the elements and techniques has promoted not only functional efficiency but permanence and adaptation in time, proposing them today as a cultural limitation, but mainly as possible resources when there is a desire to appreciate them. The farm construction and landscape heritage is rich in such presences. The system of farms, for instance, ennobled by aristocratic and middle-class progression into civil residences and villas, enriched by grounds and gardens, testimonies to culture and events.

In the territory, an age-old process of conservation and innovation of layouts and transport infrastructures (roads, railways, irrigation and navigation channels) has created presences that we now recognise as authentic works of art, like those to overcome rivers and valleys, with the engineering of bridges (in stone, brick and iron). The conservation of the territory also comprises that of the complex network of farmland and irrigation connections, testimonies to knowledge and technologies of the past but also of the flexible attitude towards innovation and projects sustained by a cultured and meditated planning.

The conservation of these repertories does not mean generalising and opening up a vast catalogue of works throughout the whole territory without any form of discrimination, or selecting them because they are more well known or analysed, but creating responsibility towards a programme that filters the sustainable components that are susceptible to comparison in the innovation processes. Artificial canals and locks, derivations for irrigating the land, which have had to adapt in their technologies and management, often make use of historical tracks and arrangements, not necessarily due to a sensitivity towards appreciation and defence, but to the fact that this more cost-effective and makes good use of resources. These include those for the setting up of new sources of energy, intended to gradually replace those of a traditional nature and also the recycling of waste (in the territory to which reference has been made, this is the case of the Losa farm, part of the surroundings of which has been reduced to the status of dumping areas). The virtuous relationship between conservation and innovation also proposes the need to avoid exceeding in the exploitation of the cultural heritage from the point of view of tourism without worrying about the consumption that this involves. Lastly, that of the need not to break attention down into sectors, prioritising the aim of a virtuous intersection between the assets at different levels and in the variety of consistencies. Monuments, architectures, sites, infrastructures, environment and territory make up a combination of events, presences and overwriting in the design of the Italian landscape, to which to devote commitment in terms of recognition and provisions, as irreplaceable expressions of culture and beauty.

References

Castagnova S (2006) Il Palazzo Ai Nana a Pinerolo: tra conoscenza e conservazione. Master Degree, tutor Maria Grazia Vinardi, Politecnico di Torino, Turin

International Conference on Conservation (2000) The Charter of Krakow. Principles for conservation and restoration of built heritage. International Conference on Conservation, Krakow

Repubblica Italiana (2004) Cultural heritage and landscape code. Legislative Decree no. 42/2004. Rome

Chapter 46
Regional Management Tools at Local Level: The Po and Orba Regional River Park

Franca Deambrogio and Dario Zocco

Abstract According to the ELC definition of "landscape," its management and protection are to be based on the use of land and natural resources for farming and other activities. In this context, farms are both "partners" as consumers of the territory and at the same time "competitors" with other productive activities. The Authority of the Po and Orba River Park launched an Information Desk (INFOFIUME) to promote the joint planning of such activities within and beyond the boundaries of the protected area, to increase and protect biodiversity, and promote landscape. The centrality of farms on the discussed issues very often collides with policies that should take into account biodiversity and landscape while supporting the structural and organizational choices oriented to renewal agriculture.

Keywords Trademark FQA • Information Desk (INFOFIUME) • Collective Project • Landscape • Biodiversity

46.1 The Context

The Authority of the *Aree Protette del Po vercellese-alessandrino e del Bosco delle Sorti della Partecipanza di Trino* (below as Po and Orba River Park) protects an area of 19,000 ha. At the same, around the Park gravitates an area with very different characteristics: on one side the hills of Monferrato and on the other the rice fields of Vercelli and the Alessandria planes. In the area, cities such as Casale Monferrato and Valenza can be found and industrial areas of considerable size. There are also important industrial activities such as mining and cement industry. Agriculture, particularly in the lowland area, is mainly organized in one-crop models, rice growing or cereal-horticultural. Breeding farms are now scarce,

F. Deambrogio (✉)
Sportello INFOFIUME, Parco Fluviale del Po e dell'Orba, Viale Lungo Po Gramsci, 8, Casale Monferrato, Italy
e-mail: sportello@parcodelpo-vcal.it; deafranca@alice.it

D. Zocco
Parco Fluviale del Po e dell'Orba, Viale Lungo Po Gramsci, 8, Casale Monferrato, Italy
e-mail: dario.zocco@parcodelpo-vcal.it

R. Gambino, A. Peano (eds.), *Nature Policies and Landscape Policies*,
Urban and Landscape Perspectives 18, DOI 10.1007/978-3-319-05410-0_46

while on the hillside area, there are quite a few agricultural enterprises (especially high-quality viticulture) and significant wooded areas, after the low agricultural profitability resulted in the abandonment of the land (Regione Piemonte, 1995). In this context, the Park Autority has reasoned out of an action, in addition to the protection, enhancement and interaction with the territory outside the Protected Area to reduce the enviromental and landscape impact.

46.2 Park and Territory

The Po (for the Vercelli and Alessandria Tracts) and Orba River Park. The special nature reserve of Garzaia[1] di Valenza, first nucleus of Po River Park, was established by Regional Law no. 51/1979. The current denomination of Park Authority was established later on in 1990. Following the reorganization of the Protected Areas (Piedmont Region R.L. No. 19, June 29, 2009) the Park Authority has joined the "Bosco delle Sorti della Partecipanza di Trino, in important heritage of lowland forest.

Since 1995, the Park Authority initiated researches on the Socioeconomic Plan for the promotion and development of compatible activities, as requested by Law no. 394/1991 and the Regional Law no. 36/1992. On the basis of the outcomes, a homogeneous territory called "Tourist Area of the Po River Park" was identified, consisting in all the municipalities of the River Po Park of Vercelli/Alessandria and other neighboring municipalities that are directly or indirectly included in the Po belt or in the vicinities of the Park, which covers an area of 100,000 ha (Gaido and Bollati, 1996). The analysis underlying the drafting of the Socioeconomic Plan considered the Park as a territorial context where the River Po plays a leading role. Although its morphology and width do not extensively involve urban and production activities, the aim of the Plan was to make the Park Authority a driving force for the development of the territory. The Park Authority promotes the development of compatible activities beyond the area of close relevance, to avoid the creation of a large protected territory whose surroundings are environmentally problematic.

Among the areas of economic interest, the main focus is on tourism, boosted by strong historical and architectural attractions, the naturalistic features of the landscape, and also those characteristics of the territory arising from productive activities and in the first place from a model of agriculture that combines the protection of the territory with the enhancement of its productions (Deambrogio, 2003).

The Information Desk (INFOFIUME). After setting the target, the Park Authority identified the subjects to involve in order to obtain a better quality/value of the territory. The sectors to promote a sustainable development are agriculture and tourism. The local authorities involved (the municipalities identified by the Socioeconomic Plan) consist of 51 towns (24 of which in the Park area

[1] *Garzaia* is a nesting site for red herons.

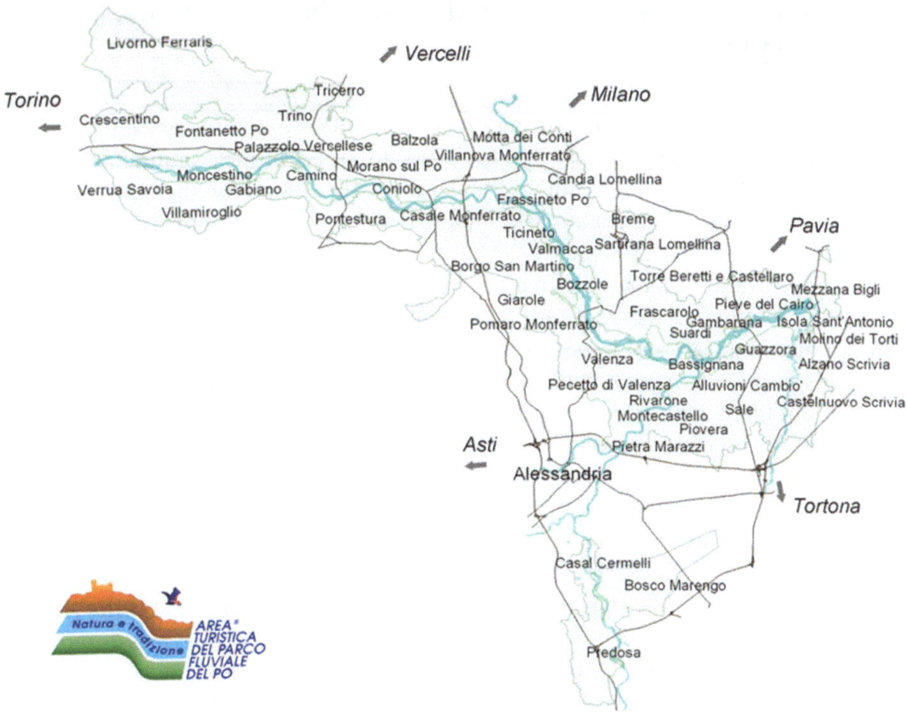

Fig. 46.1 Territory of the tourist area of Po and Orba River Park. Over 100,000 inhabitants, 51 municipalities, and 4 provinces

and 24 in its surroundings), in addition to the 3 municipalities of the Natural Reserve of the Orba River (Fig. 46.1).

The possibility of involving these areas using ordinary tools appeared difficult to achieve, but within the project "Promotion of sustainable multifunctional management of river areas" (AGENDA 21-2001), an Information Desk (INFOFIUME) to activate sustainability policies has been started up (Bergoglio et al., 2000).

The INFORMAFIUME desk promotes projects, organises briefings and reports funding targeted towards a sustainable tecniques and reducing impacts on the environment. The Info Desk has an office in Casale Monferrato but is also itinerant as for farmers it is essential the presence of technicians of the "INFOFIUME Sportello" in the farm in order to identify the main issues and the concrete feasibility of sustainability techniques applicable to business characteristics.

The Information Desk manages the trademark FQA[2] and promotes the Collective Projects under the Rural Development Plan (below, *Piano di Sviluppo Rurale*).

[2] *Fornitore di Qualità Ambientale* (Environmental Quality Provider).

46.3 The "Environmental Quality Supplier" Trademark (FQA)

FQA is a trademark of manufacturing process, based on the voluntary adherence of farms and tourist accommodation facilities to a procedural guideline (www.parcodelpovcal.it) that takes into consideration all aspects of the production of goods and services. The trademark aims to reduce the environmental impact on biodiversity and landscape.

In conjunction with the trademark, a Stakeholder Forum (representatives of the farms, environmental associations, local authorities, etc.) was established, taking into account requests and issues of the territory.

Farms required organic and/or integrated farming for every crop and bred animal, but there are of course also business guidelines for the increase of natural resources, the preservation of the landscape, and the production of traditional local products.

The procedural guideline also calls for the construction of new buildings and the refurbishment of existing ones, to maintain the landscape value, with particular attention for "historical farmsteads."

The assignment of the trademark goes through the definition of a specific Action Plan that identifies the main issues and the concrete feasibility of environmental improvements according to the characteristics of the business, thus creating a "system" where the peculiarities of every subject are integrated.

46.4 The Collective Project

Also the Park Authority aims to increase the extension of natural areas, linking and connecting them. In particular, the need to raise awareness for the quality of the territory and the safeguard of the protected area has arisen, where outlays are seen as investments, the constraints are seen as new opportunities, and the Park is seen as an engine of economic flows and a promoting tool for the whole area. In this perspective, it is necessary to use up-to-date and relevant financial means to make the area at issue more cohesive and environmentally more valuable.

The Collective Project, as requested by the agri-environmental measures of the Rural Development Program for Piedmont Region (RDP) (Regione Piemonte, 2006), concerning the "Protection and enhancement of agri-environmental features for landscape and environmental purposes," has served as a great tool for the wider spread of natural features in the territory. This action actively funds the creation of hedges, rows, woods, wetlands, and ponds and the retrieve of dimple springs and retting ponds.[3] The Collective Projects accommodate the individual applications of farms enclosed in the area at issue.

[3] Tanks formerly used for maceration, relinquished for a long time. They host unique ecosystems.

It is crucial to extend the preexisting natural fragments or to create them, given the constant widespread of abandoned natural habitats. Consequently, the planning aims at increasing the spread of ecological network (Deambrogio and Pavese, 2012).

Moreover, the project cannot forget the sustainability of the area, obtained with a newfound compatibility between agricultural production and environmental conservation and with the increase of landscape diversification and tourist presence. The Park Authority has submitted two Collective Projects:

1. The first in 2003 (PSR:2000–2006 – action F7), involving companies for a total of 52 ha of refurbished and newly created areas of hedges, rows, and wetlands in the municipalities of Valenza, Sale, Moncestino, Villamiroglio, Valmacca, Bozzole, and Frassineto Po (all under the jurisdiction of the Alessandria Province). The project has accounted for 36 % of the regional areas erected with this type of financing. The Collective Project welcomed new participants until the completion of the Rural Development Plan (PSR). As for the companies involved, for example, it has been possible to use hedges to serve different purposes. In various business projects, hedges have been used to protect organic crops. In another intervention, it was possible to recover a "historical" black locust hedge (from the 1920s) that completely encloses the participant company. The hedge of an industrial plant for the production of elderberry jam was enhanced. Moreover, many other interventions on hedges and wetlands in floodplains were realized within the Project, with the recovery of declining riparian vegetation.
2. The second in 2012 (RDP 2007–2013 – action 216, the only opening of the planning). Six rice farms in the towns of Livorno Ferraris, Ronsecco, Crova, and Santhià are participating. The interventions include the construction of a large wetland, hedges, and woods and the recovery of two springs. The close connection with the project Eco-Rice LIFE09/NAT/IT/00093 (Environmental Department of the Province of Vercelli) allowed to intervene in the rice-growing area, where there is great need to create/increase the number of natural areas and their connections, especially since plain areas have been declared preferential, with particular regard to the plains identified by the Directive 79 /409 (Birds Directive) and Directive 92/43 (Habitat Directive). The contacts with farmers have been numerous, but at the same time, there have been several difficulties in adhering to a guideline which requests some constraints to the ordinary activity and introduces a new "vision": the return to procedures that are often seen as obstacles to cultivation. One of the limitations of such interventions is a ten years commitment at the end of which there is no constraint of preservation measures.

46.5 Conclusions

The Po and Orba River Park is located in a heavily man-made area: it is therefore necessary to aim for a balance between environment, social structure, and economic activities and to spread a long-lost natural network. In this context, it is essential to involve farmers, ensuring their profitability, both for primary production and land preservation and management.

Moreover, the Park Authority intervenes in regional planning, especially in the agricultural sector, to raise awareness of the importance of the land as a resource largely underestimated on the basis of supported residual role of agriculture.

The sensitivity expressed by the farms participating in the FQA trademark and in the Collective Projects, are concrete premise for a greater involvement of farmers in a management that can lead to positive benefits for the environment and can meet the needs for ecosystemic services and socioeconomic diversification, in the frame of agricultural policies.

References

Bergoglio MT, Cristakdi L, Minciardi MR, Rossi G, Zocco D (2000) Progetto INFOFIUME: Promozione della gestione sostenibile multifunzionali delle fasce fluviali

Deambrogio F (2003) Il mondo agricolo in prima fila. Informafiume 14:7

Deambrogio F, Pavese B (2003) Progetto Collettivo, Parco Fluviale del Po tratto vercellese-alessandrino e dell'Orba

Deambrogio F, Pavese B (eds) (2012) Aziende agricole & ambienti naturali. Parco Fluviale del Po e dell'Orba

Gaido L, Bollati C (1996) Piano di sviluppo economico e sociale per la promozione delle attività compatibili

Parco del Po, Enea (2003) Analisi Ambientale, Progetto Parchi in qualità. 2001–2003

Regione Piemonte (1995) Piano d'Area delle Aree Protette del Po. D.C.R. no. 982-4328, 8 March 1995. Turin

Regione Piemonte (2006) Piano di Sviluppo Rurale 2000–2006. Turin

Chapter 47
The Landscapes of the Portofino Nature Regional Park

Franca Balletti and Silvia Soppa

Abstract The uniqueness of the landscapes of the Portofino Park lies in significant surviving features from different stages of its "territorialisation" (from its historic landscape, through an intermediate landscape, to its present-day landscape). Different societies have shaped the physical and morphological characteristics of its landscape and conferred it a functional and symbolic value upon the space within it. However, even though formal aspects of its landscape have been more or less preserved (turning it into an "emblematic" place, where the historical relationship between nature and culture is still intact), its environmental values, uses and functions have changed. For many, the area of the Monte is simply a backdrop for the transformation of old fishermen's houses and farm buildings into luxury seasonal residences, for the use of terraces as in English landscape gardens, and where the evident abandonment and marginalisation of industrial architecture and the "natural" landscape cause hydrogeological instability.

Keywords Landscape transformation • Government • Governance • Insiders • Outsiders • Planners • Affected population

47.1 The Historic Landscape

Over time the promontory of Portofino and surrounding areas have assumed the form of a *rural landscape*. It represents the complex interaction between human activities, nature, settlements, architecture, the organisation of trade and communication routes and the farming system.

Most land-use patterns on the promontory were shaped by the land management systems established by the Abbey of San Fruttuoso, which, by concentrating holdings and emphyteusis, mark it with the alternation of olive groves, vineyards and orchards close to the oldest built-up areas (Mortola, Poggio, Galletti, San Fruttuoso di Capodimonte, Olmi) and chestnut groves in the inland areas.

F. Balletti (✉) • S. Soppa
Department of Sciences for Architecture, Polytechnic School, University of Genoa, Stradone S. Agostino, 37, Genoa 16123, Italy
e-mail: francaballetti@libero.it; s.soppa@libero.it

R. Gambino, A. Peano (eds.), *Nature Policies and Landscape Policies*, Urban and Landscape Perspectives 18, DOI 10.1007/978-3-319-05410-0_47

The woodland heritage was the centre of interest and attention for local communities: it helped prevent flooding in the area and provided wood for shipbuilding, on which the other part of the local economy was based (fishing and transport of foodstuffs). Moreover, it supplied thc *lisca* used to make fishing nets and in the manufacture of farming tools as well as everyday items, which were traded in the nearby coastal settlements.

Most of the population lived on the produce of Mediterranean subsistence multi-cropping.[1] The need to exploit particularly difficult land led to the introduction of the "a four-crop farming" method. The belts of olive groves provided the area's main crop; below them there were grapevines; below the rows of grapevines wheat and rye were produced. Fruit trees – especially cherry, almond, peach and fig trees, whose fruits represented the main part of the rural diet – were also grown.

Some grapevines were planted beside the houses and climbed right up onto the roof – known as the "*solario*" – sustained by special stones with holes in them which protruded from the wall. The need to make picking grapes as easy as possible often led farmers to modify the slope, which, in some cases, was even replaced by a slightly sloping roof.

The area of Portofino was thus taken up by vegetable gardens and meadows, both in level zones and valley bottoms, where cattle were bred, while the hill slopes were planted with vines, olive trees, and figs and covered by large extensions of chestnut groves, where charcoal pits were widespread.

The process of "territorialisation" of the promontory, which occurred over eight centuries, has and has led to complex relationships between buildings, land use, physical impacts and the general recognition of the value of the landscape's natural structure. This gradual, strong evolution has created a *man-made landscape* and has been shaped according to social and economic needs strictly connected with sustenance.

47.2 The Landscape of the *Bourgeois*

The agricultural and maritime organisation of the area remained unchanged until the nineteenth century, when romantic travellers, mainly from England and Germany, discovered these places and appreciated their climate and landscape. The period was characterised by an interest in the area, particularly certain parts of it, for example, the coastal villages and the scenic spots situated on the two gulfs, seen from the perspective of a typically "urban" perception of natural spaces that was able to profoundly modify its peculiar historical features in a very short space of time.

[1] This activity is combined with lucrative coral, tuna and greater amberjack fishing, a historical remnant of which is the *tonnarella* (a netting fence for catching tuna) near Punta Chiappa.

Fig. 47.1 Borgo Portofino view from Monte

Transformation of the area began with the construction of the road connecting Santa Margherita Ligure to Portofino (1864), which until then had been reachable only by sea or along difficult mule tracks, and the subsequent construction of new residential buildings, especially villas with large gardens.

From then on, tourism – hotels and bathing and boating establishments – gradually became the main economic resource of the local population, sparking a transformation of the area, taking in the coastal area of the promontory of Portofino. However, above all it mainly brought a cultural transformation in the population itself, which had to adjust to a very different pace and way of life, in contrast with preceding generations.

The increasing use of land transport, in particular the construction of the railway line, brought about a sudden reversal in priorities: the main economic interests were no longer connected to trade with hinterland areas but depended on the new centrality of coastal settlements and caused the gradual yet steady abandonment of the higher areas of land.

The promontory and adjacent coastal areas underwent an unprecedented process of colonisation, which would bring about the evolution of the historical cultural landscapes, completely changing their spatial and functional organisation and attributing new, exclusive values to the context. This was the rise of the *hedonistic landscape*, a starting place for the discovery of unexplored horizons, connected with aesthetic pleasure, harmony, the perception of beauty and hidden rationales. It was a period of homage in the descriptions of George Gordon Byron (1822), Friedrich Nietzsche (1877) and Guy de Maupassant (1889).

The danger of the distortion and loss of the identity of these places which lay in these alterations had already been sensed in the first three decades of the twentieth century by the same elite who had been largely responsible for these very changes.

It was at this time that the intellectual and scientific communities lent their support both to protecting the area through the establishment of a national park and at the same time to increasing its tourism potential by promoting the construction, no less, of a coastal road linking Camogli to Portofino.[2] Realisation of the huge tourism potential of the Portofino area meant that it was deemed opportune to place the promontory under the control of a body established specifically for the purpose of protecting its natural and artistic heritage. In 1935 a protection law not only established the Ente Autonomo del Monte di Portofino but also made it possible to conduct a study of the promontory's natural vegetation, thus enabling its high scientific value to be discovered and its protected botanical status to be declared. Thus, was born an interest in the Monte, which went beyond mere aesthetic beauty and took the form of recognition of the profound value of its natural resources.[3]

47.3 The Protected Landscape

The notion of protected landscape, in which preserving and safeguarding the natural environment takes precedence over the local needs of the local population, began to gain traction.

During this stage, the structure of the built environment and respective infrastructure, with its historical origins, remained unchanged. However, the historical relationship between man and nature had been broken. The area ceased to be the daily world of rural life – which had descended directly from the tenth century – and underwent significant fragmentation in terms of functions and meanings. Following the major transformations that took place at the beginning of the century, the park area was "frozen" by the Ente Autonomo del Monte di Portofino; however, part of the adjacent territory was "overrun" by buildings, often in a chaotic, haphazard manner, while areas further away and less desirable from the tourism point of view were simply abandoned.

From this moment on, a clash of interests between social classes with different perceptions of the area arose: between those who wished to protect the integrity of the landscape for scientific or aesthetic reasons, those firmly in the "elite camp"

[2] The proposals regarding this area may appear to be contrasting. However, at that time it was thought that its value would be enhanced by the building of major public works, which would create employment in a period when the economic recession connected to the Wall Street Crash was particularly deep. As it was, this aspect came to nothing, in mainly as a result of the outbreak of World War II.

[3] The study of the flora had been assigned to Professor Augusto Béguinot, holder of the Chair of Botany at the University of Genoa and director of the "Hanbury" Botanical Institute. His studies would lay the groundwork for the articles of botanical interest contained in Law no. 1251/35.

who wished to restrict collective enjoyment of the good in order to exploit it for their own (business and leisure) interests, and the inhabitants themselves who asserted their right to use the area without any restrictions or prohibitions and according to the same models which were becoming established outside the protected area.

For their part, local communities, whose lives were already far removed from the customs and traditions that had governed the use of the area in centuries past, despite not subscribing to the new values proposed and means of protecting them, were unable to express their own positions and laid themselves open to political instrumentalisation and the individual interests which ultimately exerted a strong influence on the area. Thus, in spite of protection measures,[4] the progressive transformation of these places was consolidated still further: Camogli, San Fruttuoso, Portofino and Santa Margherita Ligure emerged as centres of excellence for tourism. Areas close to the built-up ones surrounding the park were considered for their high market value and transformed into new zones of building expansion. The woodland and farming area of the inland hills were abandoned or neglected, merely fulfilling the role of "enhancing backdrop" for the residential coastal zones, thus losing part of the identifying values which had stratified over the course of history.

Fig. 47.2 Monte Portofino

[4] See the Regulation of 1937 which drew up the urban planning regulations for Monte di Portofino, the application of Law no. 1497/39 to the entire promontory, and drafting of the Piano Territoriale Paesistico (large-scale land planning), which would subsequently be approved in 1958.

47.4 The Birth of the Portofino Park

In the 1980s the processes of land transformation and usage which during the postwar period had often taken the form of incongruous actions regarding the value of the area became manifest and deeply rooted: the dividing up of ownership of the most desirable pieces of land, the presence of large property developers who took possession of extensive areas of the promontory and the inaccessibility of private properties of major public interest, speculative financial transactions and constant attacks on protection the area, evidenced by several deliberate fires (Fig. 47.3).

This situation was denounced by a number of high-profile intellectuals (such as Salvator Gotta and Indro Montanelli) who brought the strong conflicts and interests involving the Portofino landscape to the public's attention, while on the institutional front, the region of Liguria confirmed its willingness to establish a park in order to protect the Portofino landscape and legislate in favour of a more active and more focused protection on development needs.[5]

Once again the local population was unable to play an important role in events which involved it directly. On the contrary, not believing in the possibility of action on the part of institutions which would take their interests into account (including

Fig. 47.3 The view from Monte Portofino towards Genoa

[5] Regional Law no. 40/77, governing the preservation of the natural value of parks and the promotion of parks and wildlife reserves in Liguria – including the Portofino area – and the more recent legislation contained in Regional Law no. 32/86 were also enacted during this period, establishing the Ente Regionale del Monte di Portofino and regulating the system of areas of naturalistic and environmental interest in Monte di Portofino.

economic and financial incentives), the few remaining farmers chose to leave the area. Thus, the process of abandonment of farming areas and deterioration of the natural heritage, which had begun during the period in which the Ente Autonomo del Monte di Portofino was active, between 1935 and 1978, accelerated and continued to accelerate.

At the time, the policy of passive conservation of the area had been unable to gain the support of the local population, despite being an effective instrument against speculation. On the contrary, limited powers of the authority made its inhabitants hostile to the suggested "park solution", seeing it as the ultimate threat to the survival of all of their activities on this land.

Against this background, the clash between the different perceptions of the landscape held by a large number of increasingly varied stakeholders clearly emerged, highlighting the lack of shared values between those called upon to participate in decision making, each one of them represented a different position about the management of this landscape and assigned a different role to the park establishment, to the idea of nature and to the concept of protection. The stakeholders involved in the debate (business owners, intellectuals, external users, local associations, institutions and political parties) shifted the means of debate, the weight, the cross-cutting nature and dynamicity of, positions and the complexity and stability of, the rules of the game. It was an increasingly fragmented group, lacking a shared perception of the landscape and unable to bring about effective programming and planning actions.

In reconstructing this *landscape process*, the dissonance which most clearly emerges is the one between the rapidity of the most recent transformations due to human impact which have been capable of consuming the land at varying speeds and producing severe imbalances and the slow, meaningful rhythm of historical "stratification", in which the relationship between man and the environment is consolidated and a landscape characterised by a high degree of complexity and richness takes shape.

Similarly, the rift in the perception of this area in symbolic terms has become more and more evident. On the one hand, its social and cultural values have become universal and the park takes on the role of a metaphor representing an environment where the relationship between man and nature is respected. On the other hand, within the area itself, the values which led to its protection are weakened: the uses, functions and relationships which formed its specific identity are changing, without them being replaced by others which are equally respectful of the places concerned. In the absence of appeals made on an ethical basis or people being made aware of their shared responsibility, even relations between the stakeholders concerned change, fragmenting into many different forms of behaviour, which are increasingly difficult to steer back to an outlook inclined towards a scenario of development, transformation and adequately shared management.

References

Anonymous (1876) Saggio storico di Portofino. Genova. New edition (1984) ATESA, Bologna
Banti G (1987) Boschi e macchie del Monte di Portofino. Agricoltura Liguria 1 (Jan–Feb)
Bariletti AN, Medri G (1972) Il Monte di Portofino. Sagep ed, Genova
Bertollo A (1995) Nozarego, la Cervara e Paraggi: profile di paesaggi, di storia e di folklore civile e religioso dell'area orientale del Monte di Portofino. Tigullio
Bocciardo E (1933) Paraggi. Genova
Brandolini P (1991) Osservazioni sul promontorio di Portofino nell'ambito delle aree di interesse naturalistico-ambientale in Liguria. Genova
Celesia E (1874) Petrarca in Liguria. Genova
Christen E (1952) Il promontorio di Portofino e la protezione del paesaggio (cronache urbanistiche). Urban 9:96–92
Cimaschi L (1955) La tutela del Monte di Portofino in uno statuto medieval della Comunità di Camogli. Bollettino Linguistico 1–4
Degli Uberti F (2007) Il Dittamondo. Third book, Chapter XI, CISVA
Ente Autonomo Monte di Portofino (1943) Raccolta di legislazioni, di documenti e di studi per gli anni 1940/41. Genova
Gnecco G (1770) Riflessioni sopra l'agricoltura del Genovesato cò mezzi propri a migliorarla e a togliere gli arbusti e i suoi vizi inveterate. Genova
Goffis CF (1950) Portofino in una prosa d'arte del 1534. Bollettino Ligustico 2
Gotta S (1964) Portofino. Cenni storici di Salvator Gotta. Portofino
Gravina L (1921) Camogli ed i suoi cari dintorni di Rita, Portofino Vetta, e S. fruttuoso. Guida artistica illustrata. Livorno
Luxardo F (1876) Storia del Comune di Santa Margherita Ligure. Tipografia della Gioventù, Genova
Magnasco A (1876) Saggio storico civile religioso del Comune di Portofino nella Riviera Ligure di Levante. Tipografia delle Letture Cattoliche, Genova
Mercenaro G (1992) Viaggio in Liguria. Genova
Mazzino E (1959) La tutela del Promontorio di Portofino. Chiavari
Quaini M (1973) Per la storia del paesaggio agrario in Liguria. Note di geografia storica delle strutture agrarie della Liguria medioevale e moderna. Camera di Commercio, Industria, Artigianato e Agricoltura, Savona
Scarin E (1953) Alcune osservazioni sul piano paesistico del promontorio di Portofino. Novara
Scarsella AR (1914) Annali di Santa Margherita Ligure dai suoi primordi sino all'anno 1914 scritti per uso dei Sammargheritesi colti dal Signor Attilio Regolo Scarsella. Rapallo

Chapter 48
The Alpi Liguri Nature Regional Park

Adriana Ghersi

Abstract The recently established (2007) Alpi Liguri Regional Park is naturally and culturally related to Piedmont (Maritime Alps and Marguareis protected areas) and France (Mercantour protected area). The park's uniqueness is due to the coexistence of diverse habitats, which intermingle, as they lie on the boundary between the Mediterranean and the Alpine areas. Here, natural and rural landscapes are profoundly interconnected. Human presence can guarantee the conservation not only of traditional activities (pastures, transhumance, chestnut woods) but also of the natural protected heritage. The establishment of a "protected landscape" to connect the various "natural park areas" would have represented a valid approach to conserving biodiversity and involving communities in park policies.

Keywords Ligurian Alps • Biodiversity • Communities • Protected landscapes

48.1 Mediterranean Landscape and Alpine Landscape: Extraordinary Biodiversity Values

The Ligurian Alps are one of the richest areas in terms of biodiversity in the whole Alpine chain, hosting a variety of habitats in a very rapid sequence.

The Alps, which are very close to the sea, interfere with Mediterranean habitats, giving birth to an unusual mix of flora and fauna. In the lower part of the valleys, the transition from olives and vineyards to dense woods and open pastures is very sudden. A wide number of species can live here, including also glacial relicts and rare endemisms. On the Toraggio (1,971 m) and Pietravecchia (2,040 m), only 20 km from Sanremo, the Mediterranean influence can be observed in the species on the southern rocky falaises, yet extremely nearby, we find rhododendron prairies, with orchids and lilies, and fresh woods of larch and white fir with marmot and chamois, black woodpecker, eagle-owl and black grouse. The narrow covered passages inside

A. Ghersi (✉)
Department of Sciences for Architecture, Polytechnic School, University of Genoa, Genoa, Italy
e-mail: aghersi@arch.unige.it

R. Gambino, A. Peano (eds.), *Nature Policies and Landscape Policies*, Urban and Landscape Perspectives 18, DOI 10.1007/978-3-319-05410-0_48

Fig. 48.1 The crest of Piancavallo (1,896 m) with the Mongioie ridge (2,630 m) in the background, from S. Bernardino di Mendatica

small ancient villages anticipate the *chiaroscuro* of the breathtaking stone-cut Alpine paths.[1]

This particularly biodiverse border area became a protected area in 2007[2] through the institution of the Alpi Liguri Regional Natural Park, involving 7 municipalities (Rocchetta Nervina, Pigna, Triora, Montegrosso Pian Latte, Rezzo, Mendatica and Cosio d'Arroscia) in three valleys (Nervia, Argentina, Arroscia at the watershed with Roya and Tanaro) in the Province of Imperia. It is characterized by four separate "natural park" areas,[3] along the ridge at the border between France and Piedmont.

[1] 200 km of itineraries: among them, the Alpine military path (carved out in the rocks between France and Italy), the geological Flysh itinerary, the Marenca Road (an old pastural path from Imperia to Tende saddle, towards the high grazing areas of Saccarello) and the Salt Road (a Roman Liguria-Piedmont trade route for salt, oil, wine, vinegar and cereals, from S. Lorenzo al Mare to the Ellero Valley, via the Saline pass) (Brancucci and Paliaga 2008).

[2] Instituted by the Regional Law no. 34 of 15.11.2007; after a long period of conflicts between local communities, the park has been in full operations since 2011, with three dedicated staff members (legal/administrative department, Pigna; scientific/technical department, Rezzo; and information points at Pigna, Rocchetta Nervina, Triora; Environmental Education Centre, Mendatica), see the park's Sustainability Assessment 2011.

[3] The 4 protected natural areas host numerous species of plants and animals and European, national or regional protected habitats. Inside these areas, there are 7 Sites of Community Importance (SCIs) (IT1313712, IT1314609, IT1314610, IT1314611, IT1315421, IT1315313, IT1315504) and 4 Special Protected Areas (SPAs) (IT1313776, IT1314677, IT1315380, IT1314679, IT 1315504).

The Piancavallo zone, one of Liguria's most important natural areas, for its high-quality woods of larches, pines, beeches and mixed mesophilous formations, is also one of Europe's major European karstic zones, with grottoes, hollows and syphons (see, e.g., the *Gorge Gola delle Fascette*).

The Saccarello (2,203 m)–Frontè–Monega crest is the most elevated Ligurian group, characterized by species-poor grasslands and meadows and by the magnificent beechwood of Rezzo (De Moro 1988). Pastures maintain a high level of biodiversity. In the northwestern portion of the Argentina Valley, a number of carsic forms and subvertical cliffs (the Realdo and Loreto falaises) are found.

The Gerbonte, Toraggio and Pietravecchia mountains host different habitats and a large number of endemic species. The Gerbonte forest (622 ha) consists of an evolving mix of beeches, maples, larches, pines and firs. Mounts Toraggio and Pietravecchia are considered some of the most interesting in the Alps: the geological condition, selective erosion on different lithotypes, proximity to the sea of heights above 2,000 m and alternation of glacial and interglacial periods have created varied microenvironments of biogeographical interest.

The state-owned Testa d'Alpe Forest is practically unique in Liguria for its predominance of white firs, mountain maples and sylvan pines, with herbaceous ridges and many fascinating little lakes and waterfalls of the Barbaira stream.

48.2 An Attempt to Establish a Protected Landscape in Order to Conserve Biodiversity and to Involve Local Communities

In 2003, the Liguria Region made provision for the institution of Protected Landscape[4] status as an extension of protected areas,[5] in order to strengthen the protection of natural areas and to foster socioeconomic development in wider areas of the region.

---

[4] IUCN category V, "Protected Landscape", is conceived as a functional and practical mechanism for the protection of biodiversity, cultural diversity and the sustainable use of natural resources (see Vols. 1–3 on Values of Protected Landscapes and Seascapes, by IUCN's World Commission on Protected Areas).

[5] Law no. 13/2003, art. 1-bis. The actions of active conservation can integrate the development of compatible activities and services, within a framework of planning and management in common synergy with the protection of regional natural protected areas. The law underlines the need to define and support activities and measures to maintain and improve landscape identity and environmental quality, focusing on the prohibition of activities and action that can have a negative impact on landscape identity and environmental quality (opening of quarries, waste disposal and introduction of genetically modified organisms, particularly with regard to agriculture and breeding).

In the Alpi Liguri Regional Park, the Protected Landscape was conceived to connect the separated protected natural park areas and to involve local communities in park policies.[6]

Unfortunately, following a heated debate, in 2009, Decision no. 272 of the Constitutional Court revoked the status of Protected Landscape from the Alpi Liguri Regional Park. This led to a lack of effectiveness of the main policies promoted by the park.

The IUCN "new conservation paradigm" (Phillips 2002; Dudley 2008) underlines the need to link protecting biodiversity with landscape conservation: successful management of protected landscapes depends on both the effective conservation of the environment and the safeguarding of the social and economic vitality of the people who live within them (Bissonette and Storch 2003). For this reason, in the Alpi Liguri Regional Park, the Protected Landscape may represent a strategy (Ferrara 1994; Gambino 1994; Gambino et al. 2008) for conserving nature in a highly complex context.

Although it has not been possible to establish its status of Protected Landscape, the role of the Park should remain that of maintaining silviculture, agriculture and sheep farming in order to preserve its various natural areas of high ecological value, the diversity of landscapes and the area's identity. Eligio Bertone, the park's director, reports an increase of biodiversity (Teofili and Clarino 2008) as a result of the felling of a large area of the beechwood of Rezzo, thus introducing a new approach and providing an opportunity for many different species in an ancient wood where only beeches and a small number of animals could live.

Moreover, the status of protected landscape provides a strategy for associating natural protected areas and their landscape, involving the local community in the park management (Migliorini et al.1999; Ingegnoli and Giglio 2005), through different actions aimed at a wider enhancement of the local resources (Farina 2009) and at enhancing the quality of life of local people and tourism products and services.

The Ligurian-Provençal alpine culture characterizes the Ligurian Alps, linking France, Liguria and Piedmont: the pastures and the transhumance[7] of cattle represent the main cultural elements connecting the different local communities of the Maritime Alps. The history of the Brigasca culture has left its mark on many aspects of this landscape. The Brigasca transfrontier community, speaking a Ligurian-

[6] Under Liguria's regional Law no. 34/2007, the Alpi Liguri Regional Natural Park is defined as being characterized by "natural park" areas and a "Protected Landscape". The Natural Park is divided into four separate areas situated at the border between France and Piedmont (6,041.21 ha). Surrounding and connecting these areas is the "Protected Landscape" (6,771.79 ha).

[7] Transhumance moved flocks and shepherds, using different parts of the area in different seasons: "*arp*" or higher alpine pastures (July–August); "*autunas*" or middle alpine pasture, alpine villages (September–November); "*primaglie*" or lower alpine grazing areas (May–June); "*bandite*" along the ridges towards the coast, near Sanremo and Capo Mele (November–April); this movement created social, economical and cultural relations between different populations (Massajoli P 1984). "*Chaque année, vous entendez, tous les ans depuis des millénaires, il a fallu conduire les bêtes aux pâturages d'été (…). Ce va et vient régulier, ce rythme, cette respiration n'ont jamais é té interrompus*" Jean Blanc in Repetto A, Campora M (2007).

Provençal language, lived on hay and grain cultivation on the dry-wall terraces on the slopes of the mountains, reaching high altitudes.[8] These people bred Brigasca sheep, and the shepherds were free to cross the French Italian border,[9] long before the European Union. These people were legendary for the inhabitants of the low valley: they spent their time on the crests, nomads like their flocks, keeping the mountain coast relationship alive.

The park should be a means of stimulating the local community to take care of their landscape and recognize the value of their natural resources and cultural identity (Colantonio Venturelli and Muller 2012), incentivizing human presence in the alpine areas by building a cooperative network between the various stakeholders and improving their living conditions (Giacomini and Romani 1982) by the exploitation of a high-quality tourism products and services.[10] In recent years, the Alpi Liguri Park has built relations with the neighbouring parks[11] in Piedmont and France, attending a European Group of Territorial Management at the local and regional level, with the aim of drawing up and implementing joint landscape programs, providing for an initial pilot programme of an international system of local government authorities.[12]

---

[8] The valley of French Briga Alta (1,310 m) was considered Genoa's bread basket in 1600 and 1700.

[9] The continue conflicts about boards and their changes are well represented in the legend of the "Framargal" to determine the boarders of Briga and Roccaforte di Mondovì communities' pastures, two chosen team had to start walking from their village towards the other village at the singing of the cock, but the Briga's men watered their cock to make it sign early (Parodi A, Pockaj R, Costa A (2012) Nel cuore delle alpi liguri. A. Parodi, Genova, p. 160) There is a similar story about cocks for the boarder of Chianti between Florence and Siena.

[10] Main Traditional fairs/events: Cosio d'Arroscia: herbs fair, living crib, Mendatica: transhumance fair, goats palio, "white cooking" show, Monterosso Pian Latte: chestnut fair. Typical products: Pigna beans (slowfood protection), Triora bread, cheese and diary products from cattle, goats and sheep, herdsmen's "white cooking" (mountain potatoes and vegetables, garlic, grass-pea, scorzonera, spontaneous herbs and fruits, chestnuts and mushrooms), Ormeasco wine, honey and lavender (lavender was a sheep's by-product, hundreds of quintals were picked up and Realdo had 3 distillers until 1958).

[11] An example of the cooperation of the Alpi Liguri Regional Park and the Natural Park of Marguareis is a tourist informative brochure about 7 trekking itineraries to discover Ligurian Alps, from Nava to Tende. Mercantour and Maritime Alps were twinned in 1986 and received prestigious awards, such as the European Diploma from the European Council and the European Chart of Sustainable Tourism.

[12] Various projects and research programmes have been implemented in order to share knowledge and experiences and cooperate with a view to obtaining financial assistance (Alcotra, OCOVA AlpMedNet), and these constitute the common basis for an ambitious project concerning an international system of local and regional government authorities from Mount Argentera (3,300 m) to the sea. The project has been submitted to UNESCO World's Heritage List. The project's partners are the Mercantour National Park, the Maritime Alp and Marguareis Park, the Alpi Liguri Natural Park, Hanbury Botanical Gardens and several other Natural Reserves, the Liguria Region, the Municipalities of Tende and Limone, the Chamber of Commerce of Cuneo and many other stakeholders.

Fig. 48.2 Chestnut dryer in the wood of Triora

In this respect, transfrontier cooperation represents a real opportunity to explore landscape conservation strategies, by rewarding the economy of the local community,[13] with a view to application of the ELC (article 9, Transfrontier Landscapes), giving an added value to traditional culture (Luginbühl 2012) and solving conflicts (e.g. concerning wolf predation of cattle (Boitani 2000; Farina 2003). Farina 2003), in order to show people that the park can help and sustain their activities.[14]

[13] Resident population (in the 7 municipalities of the park): 2,590 (15 people work in the protected park areas in summer pastures).

[14] A common action plan will focus on several objectives, from the maintenance of pathways to the communication of the image of this system, linking environmental education to sustainable development. The aim is to create a transfrontier identity by sharing know-how and through experimental activities (such as maintenance of paths and refuges, educational programmes for schools on biodiversity and recycling, a compensation fund for damage caused by wolf predation of cattle, increased use of sustainable energy, saffron or helices cultivation, short supply chains, park branding, a "preserved nature and high quality products" label) and scientific projects (such as the creation of an international Biosphere Reserve or the drawing of a cartography of the whole area, with a species-specific monitoring and the first biological transfrontier inventory), fostering bilingualism, social, productive and cultural relationships and exchanges to raise awareness among young French Italian people of their shared natural and cultural heritage.

References

Bissonette JA, Storch I (eds) (2003) Landscape ecology and resource management. Linking theory with practice. Island Press, Washington, DC

Boitani L (2000) Action plan for the conservation of wolves in Europe, vol 113, Nature and environment. Council of Europe, Strasbourg

Brancucci G, Paliaga G (2008) Atlante dei geositi della Liguria. Guida alla lettura del paesaggio geomorfologico ligure. Servizio Parchi e Aree Protette della Regione Liguria. Grafiche Amadeo, Imperia

Colantonio Venturelle R, Muller F (eds) (2012) Paesaggio culturale e biodiversità. Principi generali, metodi, proposte operative. L. Olschki, Florence

De Moro G (1988) La valle di Rezzo. Dominici, Imperia

Dudley N (ed) (2008) Guidelines for applying protected area management categories. IUCN, Gland

Farina N (2003) La presenza del lupo nella storia e nella cultura delle comunità alpine liguri. R Nì d'Áigüra, 39: I–IV

Farina A (2009) Ecology, cognition and landscape: linking natural and social systems. Springer, Dordrecht

Ferrara G (1994) Parchi naturali e cultura dell'uomo. Maggioli, Naples

Gambino R (1994) I parchi naturali europei dal piano alla gestione. NIS, Rome

Gambino R, Talamo D, Thomasset F (2008) Parchi d'Europa. ETS, Pisa

Giacomini V, Romani V (1982) Uomini e parchi. Franco Angeli, Milan

Ingegnoli V, Giglio E (2005) Ecologia del paesaggio: manuale per conservare, gestire e pianificare l'ambiente. Esse Libri, Naples

Luginbühl Y (2012) La mise en scène du monde. La construction du paysage éuropéen. CNRS, Paris

Migliorini F, Moriani G, Vallerini L (1999) Parchi naturali. Guida alla pianificazione e alla gestione. Franco Muzzio, Padua

Phillips A (2002) Management guidelines for IUCN category V protected areas: protected landscapes/seascapes. IUCN, Gland/Cambridge

Repetto A, Campora M (2007) Pastres Pastori Bergers. Regione Piemonte, Ecomuseo della pastorizia di Pontebernardo (CN), Cuneo

Teofili C, Clarino R (2008) Riconquistare il paesaggio. La Convenzione Europea del Paesaggio e la Conservazione della Biodiversità in Italia. WWF Italia/MIUR, Rome

Chapter 49
Towards the Park of Florence Hills

Gabriele Corsani and Emanuela Morelli

Abstract Florence's landscape structure has a significant homogeneous status that traces back to medieval times. The introduction of the sharecropping system gave a major impetus to the transformation of such landscape pattern, and the eighteenth and nineteenth centuries saw its completion, from both a technical and aesthetical viewpoint, with the establishment of villas and farmsteads, churches, scattered houses and hamlets crossed by an ample road network and paths traversed by numerous brooks. This economical and social system rapidly imploded after the second half of the twentieth century while the flatland area, which had been governed by the same rules up till then, had seen a wave of disorderly increasing urbanization; considerable parts of ancient settlements in the hilly areas were preserved; thanks to government zoning restrictions aimed at safeguarding natural heritage and local town planning policies that embraced and enhanced such objectives. Now, about 50 years later from these changes, the hilly landscape is starting to show further signs of deterioration, which, if not duly counteracted, could bring about adverse effects in the next few years. In this respect, it is important to raise a new awareness towards this territory, in the light of previous and subsequent land use constraints that entailed a wide range of solutions for the Florence hills parks project, to no great avail however. Besides political and administrative intentions, the involvement of the population is fundamental, given the vital role it shall play in environmental education programs especially for children, in line with the European Landscape Convention.

Keywords Florence hills • Rural parks • Landscape fragmentation • Ecological connectivity • Environmental education

G. Corsani (✉) • E. Morelli
Department of Architecture, University of Florence, Via P.A. Micheli, 2, 50121 Florence, Italy
e-mail: gabriele.corsani@unifi.it; emanuelamorelli@tin.it

R. Gambino, A. Peano (eds.), *Nature Policies and Landscape Policies*, Urban and Landscape Perspectives 18, DOI 10.1007/978-3-319-05410-0_49

49.1 Florence and the Historical Formation of Its Hilly Region

Florence virtually represents the focus of a parabola formed by the hills that encircle it to the north, east and south. The city can be regarded as the harbour of a gulf of land that faces the plains to the west and is crossed by the river Arno. Moving clockwise from north to west, the hilly territory comprises the municipalities of Calenzano, Sesto Fiorentino, Vaglia, Fiesole, Bagno a Ripoli, Impruneta, Scandicci, Lastra a Signa and Signa (Barbieri 1977).

The geological and morphological composition of these hills comprises the peaks of the Chianti Mountains to the south and is dominated to the north by the mountain range of Monte Morello.

Thus, we have a harmonious setting of hills of moderate height, with a prevalence of argillite and clayey – sandy silts, alongside hills characterized by a more massive structure composed of marly-calcareous Flysch, in particular, alberese (Monte Morello, Monte Acuto) and sandstone (Fig. 49.1).

The environmental equilibrium of the territory that surrounds Florence, characterized by continuity that is still evident today, traces its roots back to the eleventh and twelfth centuries, when agriculture became widespread and castles, towers and hamlets traversed by extensive road networks and paths running alongside or intersecting streams and brooks began to appear. Major changes in this settlement pattern occurred around the beginning of the fourteenth century when the land began to be cultivated extensively. At the same time, the sharecropping system had come to dominate agriculture, and fortified houses were transformed into villas.

Fig. 49.1 View of Florence from the eastern side, from the road between Candeli and Villamagna (Source: Rodolico (1959))

Sharecropping is the factor determining the physical shift in the religious and social system that underlay the rural society of that time. The main unit of cultivation, that is, the farm, entailed intensive land cultivation while ensuring the preservation of biodiversity through crop rotation, demarcation of farm roads with dense hedges and boscage planted in arid or steep grounds.

The eighteenth and the nineteenth centuries saw the apotheosis of this evolutionary process with the development of new farming techniques, such as the strategic location of trees along field boundaries to stabilize terraces that gave the landscape a 'combed' appearance. Both landscape types have retained some of these picturesque traits, as we can see from innumerable descriptions and iconographic depictions.

This landscape pattern remained more or less unchanged until the Second World War, which brought about epochal changes. The Agrarian Reform introduced in 1950 and, most of all, the development of manufacturing workshops in the plains to where the young people from the hills migrated resulted in the severe depopulation of many rural villages in just over a decade, with a consequent detrimental impact on the landscape and the social structure that had become consolidated over the past centuries.

Historically speaking, one of the most appreciated features of the landscape character described herein is represented by the 'Apparita' viewpoints that overlook the plains in the direction of Florence and offer an enchanting view of the city as you travel along road (Fig. 49.2).

Fig. 49.2 A glimpse of Florence from the Viale dei Colli, on the southern side, with the upper part of the cathedral dome looming beyond the trees (Source: Rodolico (1959))

49.2 Land Use Ordinances

Between the 1950s and the 1970s of the twentieth century, a set of landscape restrictions, in particular, provisions pursuant to law 1497/39 with a view to safeguarding the hilly areas in the municipalities referenced above, were imposed by the Ministry of Public Education. The reasons behind the first restriction for the preservation of the Monte Morello area in the municipality of Sesto Fiorentino (Ministerial Decree 22/12/1952) are a clear expression of the bi-univocal relationship between Florence and its surrounding countryside, given the need to

> (...) retain the benefits bestowed on the city of Florence amongst which the possibility of easily accessing a site dominated by a magnificent forest - type vegetation richly punctuated by breath-taking panoramic views over the city and its surrounding countryside, by keeping a careful watch over urban planning policies that tend to focus on mere commercial purposes that could deprive this site of its natural beauty. (Va.Aa 1952)

In 1953, the same restriction was extended to the Municipalities of Florence and Vaglia. Monte Morello in particular, stands out for its uncommon beauty reflected by its massive dimensions, its dark green forest vegetation and the roughness of its rocky areas.

Within such traditional farming context, the hill ranges, viewed as 'natural balconies that offer a stunning view over the underlying plains and the city of Florence (Regione Toscana 2013), represent a *unicum* of considerable naturalistic and environmental value'.

49.3 Hilly Area Development Plans

In continuation with architect Giuseppe Poggi's plan for the urban extension of the city of Florence (1865) and his proposal (never implemented) to extend the stretch of road known as Viale dei Colli that would travel straight through the hills that run south-westwards up to the Arno river (Corsani 2011), the criteria adopted in 1934 for a new urban development plan envisage the integration of the neighbourhoods that surround the urban core, besides the enhancement of the green belt, with the explicit advice that no urban development of whatever type should ever

> jeopardize the supreme treasure of Florence that consists of its hilly landscape as it appears today. (*Il verde nella città* 1934)

Edoardo Detti's urban development plan for Florence (1962) is the first formal deed recognizing in the surrounding hills such a value of sedimentation that they can be regarded as equivalent to a historical centre. Such protection gave rise to a virtuous emulation effect on the urban development plans for the Florentine hilly area – still under elaboration at the time – and resulted in the subsequent encompassment of the directives at an inter-municipal level, the so-called safeguard plan (Di Pietro et al. 1966).

In the mid-1960s, the entire hill range around Florence was subjected to urban planning restrictions. The result is of great significance in that it establishes for each area of concern specific safeguard measures and outlines, at the same time, a convergence of opinions on the matter of landscape management rarely found among neighbouring municipalities, due to the lack of a prescriptive approach. In relation to the subject matter of this paper, the Inter-municipal Plan of Florence, first drafted in 1965, envisages

> The setting aside of specific areas of Monte Morello and Monte Pilli – Poggio di Firenze for the future parkland development, and the linking of both sites by means of the Colli Alti road that would wind through the hilly territory. (Di Pietro et al. 1966)

This linkage was eventually built along the stretch that runs through the Municipality of Sesto Fiorentino.

The first proposal, clearly aimed at the creation of territorial parks in the Florentine areas, was formalized at the end of the 1970s by a multidisciplinary group coordinated by Gian Franco Di Pietro and outlines the prototype of the Monte Morello Park in terms of environmental protection, landscape preservation and social objectives (Di Pietro et al. 1979).

The urban development plan for Florence drawn up by Comune di Firenze (1992) with the aim of reviving reciprocity in the relationship between urban and rural areas envisages on the one hand, a green wall (*murazione verde*) for defining a boundary in the urban framework with the creation of rows of trees and new green gates and on the other hand, the establishment of the historical park of the Florentine hills and the Park of the Arno River and its tributaries, to be extended to other neighbouring municipalities (Comune di Firenze 1992; Valentini 2005). Such proposals never came into force and the idea of creating two territorial parks as if they were two distinct entities does not seem to be sufficiently motivated.

Subsequent studies promoted by the municipal administration of Florence are orientated towards an overall, systemic vision of the different components of the Florentine rural and urban landscape. In particular, we wish to point out the following:

- The research on the development of an urban park in the Florentine hilly area commissioned in 1999 to the University of Florence (coordinator: Prof. Augusto Boggiano). The objectives entail, among the other things

> The installation of a network system of trails that constitute a functional framework for tourist and recreational fruition of the hilly land where the urban park is to be located. The project guidelines envisage the valorisation of historical-cultural characters and the safeguard and development of natural characters besides the environmental connectivity between northern and southern hill ranges, viewed as elements that form a single system and accomplished by means of connecting corridors that cut through the areas of the Hill Park and the Arno River park. (Boggiano 2004)

- The Proposal for the creation on an urban park along the Arno river banks (Progetto Interreg IIIC Sud RiverLinks), that is, a network of parks and open spaces closely linked to each other. The Arno is intended here not only as a simple river but also as a receptor of a network system which constitutes, together with its tributaries, the bearing frame of the landscape structure of which the hills form an integral part.

The attention towards the hilly landscape is also confirmed by regional planning tools, such as the current Regional Design Plan and Regional Landscape Plan (*PIT, Piano di Indirizzo Territoriale* and *PPR, Piano Paesaggistico Regionale*) and the Provincial Coordination Plan (*PTCP, Piano Territoriale di Coordinamento Provinciale*). After taking a census of the areas of ecological connectivity throughout the provincial territory, the hilly rural area that encircles the Florentine urban core became the structural invariant of the *PTCP*, i.e. the framework for the reclamation of potential park areas and Natural Protected Areas of Local Interest (*ANPIL, Aree Naturali Protette di Interesse Locale*). As a matter of fact, this framework is marked by geological, floristic, faunal, ecological, morphological, environmental and agricultural peculiarities, regarded as forms of anthropization of considerable value in the light of their historical, formal and cultural meaning and civil values (*PTCP* 2013).

49.4 Current Status and Project Proposals

The synthetic reconstruction of the safeguard processes and the diagnosis of the fragmentation status should be viewed as design objectives of the proposals related to the development of an urban park in the hilly area. These are aimed at the orientation of positive trends already underway as well as the promotion of necessary trends that can enable the formation of a unitary organism, as the identified hilly region can be actually defined. Urbanization processes only had a marginal impact on the hilly areas, whereas, with respect to the plains, the relationships mentioned above were modified, trivialized, fragmented or wiped out altogether.

Besides more evident urbanization phenomena, other types of transformations with equally profound effects are affecting this characteristic landscape pattern. In this respect, we wish to point out the following:

- Abandonment of cultivated areas with the consequent uncontrolled growth of wooded areas and the diffusion of infesting species
- Privatization and modification of paths and private roads due to the fragmentation of farmlands or facilitated by the indifference of municipal authorities
- Delimitation and fragmentation of private areas pertaining to historical buildings with the introduction of ornamental vegetation that is inconsistent with the natural context and lastly, the privatization of vantage points
- Abandonment of historical quarries which, if redeveloped, could acquire a strategic role in landscape valorisation policies

The agrarian hilly landscape has taken on a strong and well-defined identitary connotation by virtue of its historical relationship with Florence. Such relationship has undergone many modifications, and the city has grown to be the conveyor of an anthropic pressure that becomes more effective according to the degree of affinity between these two distinct worlds.

New ecological contaminations between territory and society are likely to emerge if the city is able to regain its positive role within its surroundings (Sargolini 2013), while ensuring the permanence of historical *vacua* as genuine 'full spaces', generous receptacles of resources.

Also for this reason, we believe first and foremost that the municipalities involved in the hill park project should launch a landscape survey campaign in primary and junior high schools through exploratory researches and meetings; botanical, ecology and history courses; and drafting of thematic maps, pamphlets and the like.

References

Barbieri G (ed) (1977) Aree verdi e tutela del paesaggio. Guaraldi, Rimini

Boggiano A (ed) (2004) Passeggiare Firenze: studio di fattibilità di percorsi pedonali all'interno delle aree extraurbane. Comune di Firenze, Florence

Comune di Firenze (1992) Variante generale del P.R.G.'62 – P.R.G.'92, Relazione generale. Litografia I.P, Florence

Confederazione Nazionale Sindacati Fascisti Professionisti e Artisti (1934) Il verde nella città. In: Comitato per lo studio dei criteri da seguire per un piano regolatore di Firenze (ed) Per la Firenze future. Stianti, San Casciano, pp 231–234

Corsani G (2011) Il Viale dei Colli di Firenze. Storia dell'urbanistica/Toscana XIII–XXI:61–76

Di Pietro GF, Greppi C, Pettini P, Sica P, Dallerba GF (1966) Piano Intercomunale del Comprensorio Fiorentino, Studi, ricerche, documenti. Stab. Grafico Tipocolor, Florence

Di Pietro GF, Errera G, Omodei Zorini L, Piussi P (1979) Il parco territoriale di Monte Morello. Analisi delle risorse e metodologia di intervento per la formazione di parchi territoriali nell'area fiorentina. Provincia di Firenze, Florence

Provincia di Firenze (2013) PTCP, Variante adottata con Delibera del Consiglio Provinciale di Firenze 1/2013. Available via http://www.provincia.fi.it/territorio. Accessed 20 Nov 2013

Rodolico F (1959) Il paesaggio fiorentino. Le Monnier, Florence

Sargolini M (2013) Urban landscapes. Environmental networks and quality of life. Springer, New York

Regione Toscana (2013) Piano paesaggistico regionale della Toscana. Florence

Valentini A (2005) Progettare paesaggi di limite. Firenze University Press, Florence

Vv.Aa (1952) Monte Morello. Available via http://www.comune.sesto-fiorentino.fi.it/Engine/RAServeFile.php/f/D.M._23_dicembre_1952_Monte_Morello.pdf. Accessed 15 Nov 2013

Chapter 50
Protected Area Planning, Institution and Management in Apulia Region

The Lama San Giorgio Nature Regional Park

Nicola Martinelli and Marianna Simone

Abstract The main objective of the preliminary plan for the Institution of the *Lama San Giorgio* Park in Apulia was to integrate the karst channel, locally known as *Lama*, into park planning processes. Indeed, of the possible areas of study, the *Lama* catchment area appeared to be most consistent with the functions of karst valleys, which is not only of a hydraulic nature but also concerns the landscape and settlement systems. Furthermore, the *Lama* catchment basin preserves the physical connection and ecological connectivity of the park to the southeast of the *Bari Valley*. This was the base to develop both the area knowledge framework and the institutional context to be implemented in the governance of the local communities within the basin and the institutionalised phases of the Regional Conferences at the Parks Office (2005–2007). However, some 5 years have passed since that stage and the definitive institution of the park, which has in the meantime been extended to include the nearby *Lama Giotta* area, is still awaited.

Keywords Planning • Ecological connectivity • Participation • Governance

Although the 1970s and 1980s were a time of intense regional activity for the planning of wildlife parks in Italy, Apulia took little part in this trend; indeed, before the Framework Law no. 394/1991 (Gambino 1991) and Regional Law no. 19/1997 (regulations for the institution and management of protected areas in

---

This contribution is the fruit of the joint reflection of both authors. Paragraphs 49.2 and 49.4 are the work of M. Simone, whilst 49.1 and 49.3 are of N. Martinelli.

N. Martinelli (✉)
A6 Lama San Giorgio Protected Area Plan, Politecnico di Bari, Dicar DPT, via Orabona, 4, 70125 Bari, Italy
e-mail: nicola.martinelli@libero.it

M. Simone
Master IUAV, Tolentini, Venice, Italy
e-mail: marianna.simone78@gmail.com

R. Gambino, A. Peano (eds.), *Nature Policies and Landscape Policies*,
Urban and Landscape Perspectives 18, DOI 10.1007/978-3-319-05410-0_50

the region of Apulia), it was listed as one of the regions with the least parks and nature reserves in Italy, with just 425 ha of protected territory.

Now, however, the region has 2 national parks; 3 marine reserves; and 18 state nature reserves, wetlands and protected areas covering a total of 221,000 ha of land that is 11.3 % of the region's territory. Even more importantly, the legislative process is under way for the approval of many of the 33 wildlife parks, which are included in the list annexed to the regional law. Among these are many areas of natural interest within the hydrography networks, which are geographically structured by the small karst valleys locally known as *Lame* and *Gravine* (ephemeral streams and gorges).

50.1 Study Area

The study concentrates on the Conca di Bari where an ample network of *Lame* converges. This network allows the torrential regime flow of surface water from Murge Plateau towards the Adriatic. It shapes the karst landscape of the Murge (Ripiani di Terra di Bari, Sestini 1963). The erosional channels also constitute important eco-landscape structures, "corridors" that are built into the surrounding landscape. These are characterised by the mobility of the landscape elements of the eco-mosaic they cut through.

One of the vastest erosional valleys, *Lama San Giorgio*, is located in the southeast of the above-mentioned Valley which begins in *Murge* uplands, some 385 m a.s.l., and flows into the Adriatic Sea in the municipality of Bari. It runs alongside five settlements during its 42 km. Therefore, as the *Lama San Giorgio* Park cuts across a significant part of the urban region of Bari, it establishes important relationships between natural and anthropic elements connecting different landscapes: broad-leaved heliophilous vegetation (sub-Mediterranean vegetation) with sclerophyll evergreens (Mediterranean vegetation), arable crops and olive groves with *tendone* vine training systems and irrigated coastal and subcoastal vegetable gardens, coastal towns with Murge "farmers' villages" and *masserie*.

For this reason, no pre-established framework regarding the physical organisation of the park, its unequivocal morphology and, at the same time, its complex close proximity (the park in the urban region), medium proximity (*Murgia* southeast) or wider relations (regional park system) seemed equipped to represent the unique nature of the *San Giorgio* area.

Furthermore, if we consider the institutional and administrative aspects of the project, those involved in the drafting of the plan for the park were faced with a lack of systematic comprehensive frameworks for regional territory, which could be used as a basis for the creation and strengthening of ecological networks. Consequently, reference was made to the following: the national soil protection policy (Law no. 189/1983), current environmental policy (Law no. 431/1985, Law no. 394/1991 and Regional Law no. 19/1997), community projects (CFS 2000–2006, Natura 2000 Network) and national projects (BioItaly, REN). In terms of socioeconomic planning, reference was made to the Apulian Regional Operational Programme.

Fig. 50.1 Aerial view of *Lama San Giorgio* at the watershed of the rural Annunziata church (eighteenth century) in the Bari urban region

In the meantime, although it was publicly announced in the final drafting stage of the plan in question (2000), a regional initiative for the preparation of a feasibility study regarding the Regional System of protected interconnecting areas in the environmental system provided the confirmation of a chosen planning approach which aimed at overcoming the vision of "natural sanctuaries" (Fig. 50.1).

50.2 Methodology Adopted

A number of guidelines can be extrapolated from the interpretation of the physical and institutional context in which the drafting of the plan evolved. They relate to:

- The choice of study area and protected area management
- The strengthening of the connectivity and connection functions of the *Lama San Giorgio* corridor
- The creation of shared and sustainable local development policies

Choices regarding study area and Park management The *Lama* engages with the territorial areas and administrative and management contexts and territorial relationships which cross the proximity scales. This needs a precise definition of the study area and implies a *scaling-up* process. This process gradually adjusted the hierarchy of the environmental and institutional reference park systems. In this way, it was possible to obtain its most coherent definition, which would then, in turn, be carefully examined by local communities (Fusco Girard and Nijkamp

1997) during the official meetings provided for by the new regional law (the Regional Conferences). Furthermore, the protection of the "basin system" (the hydrography network and the surrounding areas) has proved, although in disciplinary terms, to be a matter of particular importance. A global defence system has been realised in the *Conca di Bari* territory since the disastrous floods of 1929. However, in recent years there have been extreme and uncontrolled land-use transformations along the relevant areas of the riverbed, the riparian strips of the *Lame* (Milillo and Trisorio Liuzzi 1996) and, in particular, the *Lama*. It is, therefore, vital not to underestimate the effects these transformations could have on hydraulic risk. Among the possible areas of study, the *Lama* hydrography basin – which spans some 30,000 ha – appeared to be the one which most closely reflected the functions of the karst channel. By functions we refer not only to hydraulic functions but also to eco-landscape functions and integration with the settlement system in the wider context (Forman, 1995). Firstly, the catchment basin became the basis for the creation of a knowledge framework for the protected area, which is articulated into systems (geology, forestry, settlements, etc.). It assumed the role of an institutional context for the consultation/listening stage involving the local communities in the 13 municipalities within the area.

Strengthening the functions of connectivity The "corridor" form and function of *Lama San Giorgio* guided the interpretative and planning stage in an attempt to conciliate two demands – not to be considered as alternatives – one relating to the unitary nature of the system and the other to the diverse nature of the system components. Analysed longitudinally the erosional channel presents three primary functional attributes (Mininni 2001):

- The function of a longitudinal conduit which runs from the hills to the sea and is subject to flows and movements within the corridor: for the most part water and natural elements. Allowing and facilitating the correct flow along the *Lama* is the first requirement relating to deep valley maintenance and management provided for in legislation.
- The transversal filter-barrier function, or rather increased or decreased resistance to permeability and the crossing of natural elements from the margins of the erosional channel to ground level, is influenced by local level management and is, therefore, closely associated with existing local planning;
- The habitat function, in the cross-scale approach of the plan, becomes increasingly linked to the hierarchical relationship of the *Lama* with the surrounding territory. It is, therefore, dependent on the existence of the other two functions, and it progressively forces systems for the management of the park to shift their focus from the macro-natural to the micro-natural scale of the ground level.

The functions of connection (form) and connectivity (function) become priority factors in terms of the Land Use Designation of the Park. It could be said that the plan for the park represents a *strategic plan* in relation to the connective function, which is capable of mobilising the actions and intentions of the communities (Fig. 50.2).

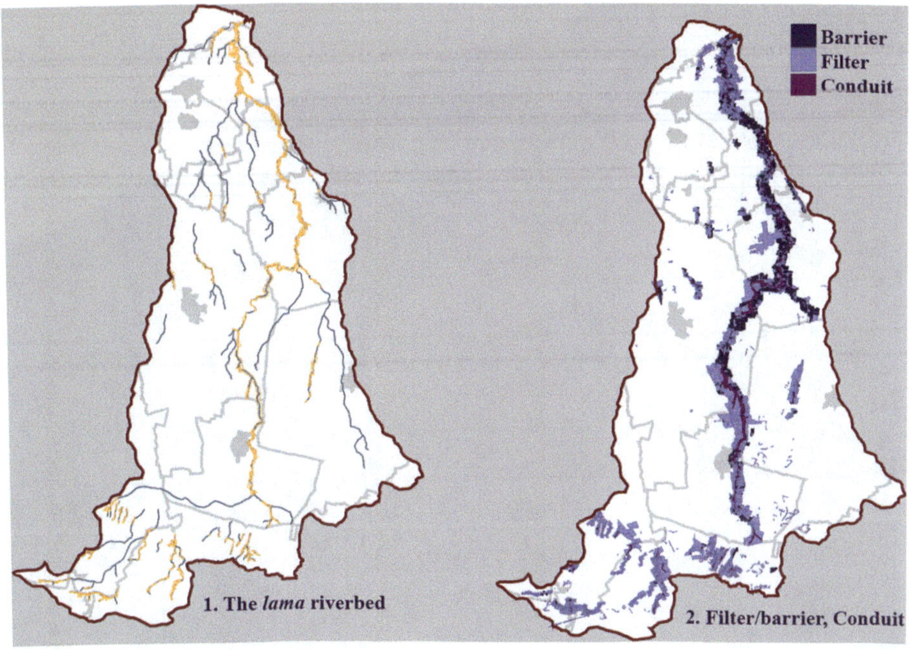

Fig. 50.2 Drawings of the preliminary plan for the park including the perimeter of the catchment basin of the lama, the hydrographic networks and the three ecological functions as well as the early stages of the zoning

50.2.1 The Start-Up of Shared and Sustainable Local Development

> The main elements of difficulty which emerge from the planning of parks in Italy refer to the relationships between the managing bodies of the parks and the local communities. These difficult relationships are frequently aggravated by the absence of a territorial-landscape planning approach which extends the concept of safeguarding territory to vast areas of territory and integrates it with development processes. (Peano 1998)

In the experience of the plan, there was an attempt to establish an initial consultation phase with the local population, which allowed for the progressive gathering of knowledge relating to the *Lama* area. This process was followed by an open discussion of the choices involved in the plan in an attempt to actively involve the local population in the creation of the park (Fig. 50.3).

All this was made possible through the organisation of a number of in itinere public forums. These were opportunities for structured listening and were developed according to research/planning processes. The timing was intentionally different from the institutional timing provided for in regional law (Regional Conferences). In this way, the Province of Bari – the commissioning body of the work – acknowledged the mediation role assumed by this particular planning team

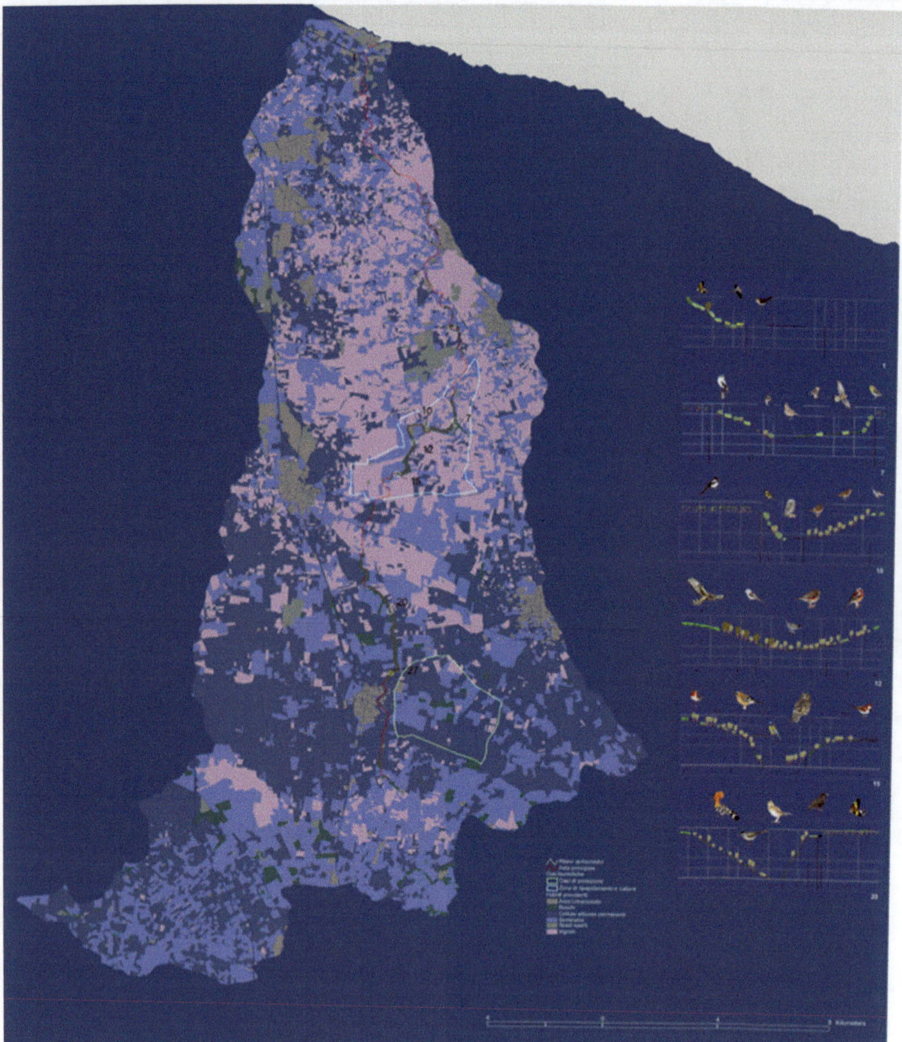

Fig. 50.3 The system of habitat

activity. This was viewed as being a necessary and pressing part of the involvement process for communities within the park.

50.3 Results and Discussion

The knowledge framework, which has been created, is arranged in systems and implemented in the Territorial Information System of the Protected Area. The knowledge framework presents some features, which could be employed in future research approaches regarding different elements of the hydrography network in the *Conca di Bari*. This is supported by a similar knowledge gathering process, which has recently gotten under way (Besio and Monti, 1999). It regards the *Lama Giotta* catchment basin, which is bordered to the W with that of *Lama San Giorgio*. The process was activated by the Province of Bari due to the pressing demands local populations made on the Regional Council of Apulia for the inclusion of *Lama Giotta* in the list of regional parks.

In order to evaluate the efficiency of the plan, it was necessary to await the conclusive stage of the Regional Conferences (held in 2005 and 2007). They broached the subject of the proposed perimeter and land-use designation of the park. These proposals had already been approved at a provincial level (2001) and were again discussed with the competent territorial bodies and the relevant associations for definitive approval.

Waiting for the definitive institution of the *Lame San Giorgio and Giotta* Park, a regional decision, which was strongly opposed by the local communities directly overlooking the *Lama* course, was introduced.

Based on an old wastewater management project, the Regional Government plans to dump wastewater from the purification plant mains of the Municipality of Casamassima in the *Lama San Giorgio* riverbed. Although the regional council guarantees the respect of the Merli Law tables for the dumping of treated wastewater, and the presence of water in the valley is not in contrast with its function as an ecological corridor, local communities do not trust the quality of the treated wastewater and strongly oppose this decision. Therefore, whilst recognising the plan as a basic reference for safeguarding and valorisation policies for the *Lama*, the municipalities have formed a Consortium and, in synergy with Environmental Associations and the third sector, they rally to prevent the Apulian water authorities from emitting treated wastewater into the *Lama* riverbed. The Parks Office, within the Regional Government, does not seem to have respected the processes for the involvement of the communities. These processes were implemented during the drafting of the Plan (2000–2001) and the Regional Conferences (2005–2007) and, as we have already mentioned, they fostered unprecedented bottom up initiatives in local communities.

References

Besio M, Monti C (1999) Dal cannocchiale alle stelle, strumenti per il nuovo piano. Franco Angeli, Milan

Forman RTT (1995) Land mosaics. Cambridge University, Cambridge

Fusco Girard L, Nijkamp P (1997) Le valutazioni per lo sviluppo sostenibile nella città e nel territorio. Franco Angeli, Milan

Gambino R (1991) I Parchi Naturali. Nis, Rome

Milillo F, Trisorio Liuzzi G (1996) Effetti della sistemazione congiunta idraulica e idraulico-forestale del Torrente Picone (Puglia). In: Proceedings of Conference La difesa dalle alluvioni, Florence, 4–5 Nov 1996

Mininni MV (2001) Relazione del Preliminare di Piano dell'Area Protetta Regionale 1A6 Lama San Giorgio

Peano A (1998) Un percorso nei parchi italiani. Urbanistica Dossier 8

Chapter 51
The Environmental Issue in Sicily

Ignazia Pinzello

Abstract In Italy, despite the number of protected natural areas and the abundance of scientific research upon the landscape, the level of control of human pressure on the environmental system is still seriously inadequate. Among the main factors is the disconnection between urban/regional and sectorial planning, in the frame of the detachment of the nature conservation policies from the landscape and territorial policies, which is the focus of the present research. The regional situation is rendered even more serious by the lack of an up-to-date planning law (the Sicilian planning law was passed in 1978), the poor diffusion of territorial and vast area planning (the Regional Master Plan has never been endorsed and just one out of nine provincial capitals has endorsed the Territorial Provincial Plan) and the difficulty of integration of territorial policies and specialized planning tools, falling within the competence of different agencies (regional and provincial councillorship, *Soprintendenze*, Port authorities, free associations of municipalities, etc.). The difficulty of a complete and efficient application of the environmental assessment tools regarding projects, master plans and programmes (such as EIA, SEA). All of these questions today impose a reflection about the existing tools to address such territories. These tools have the limit to be disconnected among them.

Keywords Protected natural areas • Park (national, regional, natural) • Urban planning • Regional planning • Natural reserve • Landscape plan

51.1 Introduction

The attention to the environmental issue started in the 1960s, a period that coincided with the birth of the first environmental organizations (World Wildlife Fund, Friends of the Earth and Greenpeace). The first conference that addressed the issue on "Sustainable Development" was held in Stockholm in 1972. On that occasion, the principles of social equity and also man's responsibility towards the environment were enshrined. Following the numerous statements and documents being

I. Pinzello (✉)
University of Palermo, Piazza Marina, 61, 90133 Palermo, Italy
e-mail: ignazia.pinzello@unipa.it

R. Gambino, A. Peano (eds.), *Nature Policies and Landscape Policies*,
Urban and Landscape Perspectives 18, DOI 10.1007/978-3-319-05410-0_51

signed by all the member states, facing each other in an integrated vision towards the problems relating to environmental, social and economic issues.

At first, the European Union (EU) was worried about struggling with punctual interventions of the damages caused by polluting sources. Such policies all revealed insufficient, recognizing that the environmental degradation phenomena have a dimension beyond that of the local one.

The spread of environmental culture is a very important affirmation of nature conservation and the way to consider the protection of the environment not only as an aesthetic demand but also as a social requirement, induced by the international, national and regional legislator to make laws.

It is from the Belgrade Charter (1975) that a new relationship is considered between human beings and their society. A new ethical relationship is going to be faced regarding the differences, which are continually growing among people. If we consider addressing this problem in a sectorial manner, it will make us lose sight of the relationship systems, which concretely put together those different causes, and in other words, it would not permit us to individuate in the right way as to obtain social equity.

The Charter, confirming the new ethic request, confers a leading role to the environmental education system, being convinced that it would contribute to provide a new process of knowledge, growth and awareness of the role that the individual plays in society.

In 1982, the General Assembly of the United Nations, trying to reach this purpose, adopted the World Charter for Nature. This document pointed out the importance of international cooperation concerning social, economic and cultural problems, including security and peace themes, in the knowledge that "Humanity takes part in Nature and life depends on the continuous running of natural systems; energy and nutriment sources".

Man, who has considered nature and its resources as inexhaustible and "subservient" to satisfy his needs and not just the primary, is gradually entrenching in the principles that he is not in opposition to nature, but is part of a living system. Man must establish an ethical relationship with the resources, by becoming aware of their degree of reproducibility and the "waste" that is made through irrational use. The irresponsible behaviour causes damage to the community both in terms of economic and in health. Among these, an example is given of the process of unplanned urbanization. This phenomenon, besides producing agricultural soil erosion, heavily weighs on the loss or reduction of the identity.

The urban process invades territories rich in history, where landscape is the outcome of the relationship through man, nature and society. The diversity reduction has led to the destruction of entire ecosystems and it is the cause of ecological disasters. For this reason, the biodiversity conservation is going to become a greater and more urgent issue and that is why it seems necessary to provide initiatives and territorial policies in order to mark a change of direction and control the indiscriminate use of resources. The environmental conservation and particularly the protection of some specific sites by means of normative measures is only the first step towards ensuring the ecosystem safety; a system of relationships between natural, seminatural and anthropic environment is to be built to avoid their isolation promoting a genetic exchange.

The environmental and landscape conservation policies in Sicily seem to be disarticulated and not coordinated. The causes lie in an outdated legislative framework that is often contradictory, with a considerable number of uncoordinated plans and the allocation of similar matters to different administrative bodies. The case of the Regional Land and Environment council who has the expertise in the field of environmental and protected areas, with the Regional Department of Cultural and Environmental Heritage and Education responsible for the planning and management of cultural heritage and landscape.

Unfortunately, a law, dating from 1978, still today regulates Sicily's regional planning. From that time and especially in the last few years, many laws have been presented on the territory government, but they have almost completely ground to a standstill due to the alternation of regional governments.

This makes it difficult to implement strategies for the planning and management of innovative methods in the participation, in the training and in the management of planning tools, consistent with the objectives set by the international documents signed by all the member states.

51.2 The Protected Natural Areas System

The protected natural areas are intended as a "privileged" part of the territory, where environmental aspects are integrated with historical and cultural ones. Where the signs of the relation between man and nature are still recognizable. These areas are considered as museums *en plain air* thanks to their environmental and landscape-related aspects, especially for their exclusivity.

In 1981, Sicily, regarding the environmental issue, marked a change of direction as to the regional urban policy with the passage of the Regional Law no 98, integrated and modified by the Regional Law no 14/88, for the establishment of parks and reserves within the regional territory. If on one hand some important results have been obtained (preserving over 12 % of the territory of the island), on the other hand, those areas of particular natural and cultural interest, which play a crucial role in the sense of defining the local landscape characteristics next to the towns, have been excluded.

In particular, we are referring to the peri-urban agricultural areas taking on the character of historical significance like the olives of Castelvetrano, the citrus plantations of Palermo and Catania, the terraced vineyards of the Etna Valley, the rural (farms and mills) and religious (monasteries and hermitages) architectures, the rural villages, the archaeological areas and, in conclusion, the network of connections formed by historical infrastructures like paths and sheep tracks.

The Sicilian protected areas, taken as one body, consisting of 4 regional parks (Etna, Nebrodi, Madonie and Alcantara) and together with 89 natural reserves, 205 Sites of Community Importance (SCI) and 29 Special Protection Areas (SPA), give us a “patchy” system. It is characterized by rare but representative geological and botanical emergencies, monumental trees, faunal biodiversity and habitat of great scientific and cultural interest.

These sites are configured as fundamental points of a more complex system and of a new developing and planning processes.

The protected area planning is more than other planning systems, as it takes on a knowledge process and an interdisciplinary and trans-disciplinary interpretation. Since the diversity, which is a peculiar trait of the single areas and the territories they refer to, it is not possible to deal with a plan as a model, but integration between plans and regional planning is required. Nevertheless, some reflections have to be made on the management and planning tools of the natural reserves, the natural parks and the landscape.

The natural reserves[1] are provided by the plan of arrangement for the A zone (drawn up by the scientific provincial Board/Council) and the plan of use for the B zone or pre-reserve zone[2] (drawn up by the town council or the municipalities where the reserve is located). The plans are as effective as the detailed plans and according to their formation; adoption and publication have to keep the arrangements in force concerning the detailed plans themselves.

The lack of these tools, especially concerning the plan of use, can compromise the reserve favouring those illegal phenomena, which can assume such proportions as to force a pre-reserve perimeter revision, whose role is to “integrate the surrounding territory inside the environmental protection system”. More important, it must be noticed that the drawing up of the two plans is not contemporaneous and even its responsibility is committed to two different institutions.

Since the reserve is understood as the whole of A and B zones, it would be better if the two plans were drawn up at the same time and have dealt with one administrator. It does not mean that municipal planning does not work or interact with the natural reserve planning; on the contrary, it departs from the conviction that their boundary is not synonymous with isolation, but it gives the occasion to involve more agents in the managing and planning processes.

What is going to be shown is the interaction among protected area planning, urban planning and the economic programme, as to consider their relations not hierarchical but integrated. The natural parks regulating tools are the territorial plan of the park and the socioeconomic multiyear programme. The territorial plan is a special one, integrated, drawn up by the Park Authority concerning naturalistic and landscaping contents. It is understood as a regional plan. With regard to the territorial contiguity of the park in relation to the municipal territory, there are

---

[1] Natural reserve typology (Regional Law no 14/1988, Art. 7): (1) integral natural reserve, which protects the whole natural environment and where only scientific interventions are allowed; (2) oriented natural reserve, which protects the natural environment where only proper interventions are allowed (cultural, agricultural and stock-raising); (3) special natural reserve, particularly located to the biological, biological-forest, geological and ethno/anthropological conservation; and (4) genetic natural reserve, which protects the gene pool of animals and plants of the region.

[2] Regional Law no 14/1988, art. 22.

two types of plans to compare: the territorial plan of the park and the master plan; both refer to environmental, economic and social policies, especially in those regions provided with modern urban laws. This is not the situation in Sicily. The D zone of the park (intended as control zone or pre-park zone) is viewed as the most problematic since the indefiniteness as by laws enacted. In fact, it keeps up the expectations of the C zones (expansion zone) and the municipal master plan, asking for some housing needs.

It is in conflict both with the environmental planning principles – apart from the administrative contest – and with the contents of the national framework law upon the protected areas (Framework Law no 394/91) that recognize the D zones as contiguous areas "to ensure the values conservation of the protected areas".

The annual socioeconomic programme, drawn up by the Park Authority, is the reference tool upon which the municipalities and their programmes and plans are based. This programme underlines the importance of the synergy between municipalities and Park Authorities, first dealing with the relation between the park and its surroundings, in order to combine protection and development policies. Nevertheless, not being contextual, the drawing up of the territorial plan and also the socioeconomic programme connected to the complexity of the planning approval procedures may lead them into conflict or make difficult any reconciliation as in the intended use. The knowledge that the modifications or even the destruction of the environmental resources are as fast as the irreversible damages they produce in a very short lapse of time.

51.3 Landscape Planning

The division of competences between two departments as to the landscape and environmental protection makes it difficult to find effective solutions to the institutional requests as determined by the National Law no 431/1985 which, alternatively to the real drawing up of a landscape plan, provides for a "territorial urban plan, specifically applied to the environmental and landscape values". Regarding the landscape plan formulation, the Region of Sicily does not follow the procedures according to the law quoted above, but another one: National Law no1497/1939.

This situation obviously suffers the lack of an organic regional legislation upon landscape; there are also legal provisions that have taken place for many years and have finally led to the attribution to the *Assessorato ai Beni Culturali e alla Pubblica Istruzione* (Regional Department of Cultural Heritage and Public Instruction)[3], an exclusive competence in matters of landscape. If the transfer of such

[3] The department shall exercise the power granted to the central and peripheral body of the State in matters of environmental and cultural heritage through the BB.CC Department. The law no. 80/77 has laid down rules for the protection, valorization and social uses of the environmental and cultural heritage within the regional territory and also set up the *Sovrintendenza* (Provincial

competences, concerning all the ordinary administrative regional statutes, is defined by a decree of 1977 (on the grounds of which Law no 431/85 would interfere only in a second time, giving the regions a chance to draw up urban territorial plans having a landscape value or definitely, real landscape plans. In Sicily, it had just happened according to the President of Republic Decree (DPR) no 637 of 1975.

In fact, in the implementation of the regional statute for the protection of the Sicilian countryside, the legislature demands that the regional administration have all the powers of the central and peripheral administrations of the State relating to the protection of the countryside ex Law no1497/39.

Indeed, landscape protection is a department's competence, pertaining to its peripheral organisms: the *Sovrintendenza* which provide their proper functions related to cultural and environmental heritage,[4] as they are expected to, and, specifically, to those sections connected to the architectural and landscape heritage as to DPR no 805/75.

The Landscape Territorial Plan (LIP) itself is intended as a department's[5] competence, although revised in its contents by Law no 431/85. This can be applied to those of the *Sovrintendenze*[6] pertaining to the cultural and environmental heritage. It is a faculty, not an obligation, as shown by the regulation. It is the demonstration that the central authority may confer the drawing up of the plan to an external or peripheral department which will elaborate on it, on the grounds of the central planning principles.

Plans drawn up by the *Sovrintendenza* affecting the provincial level are contradictory to the provisions of Law no 431/85, which refer to the regional territorial plan, both the characters of environmental plans that effect morphological and nonadministrative areas.

With regard to the PTP formation procedures, on the one hand, the Sicilian region has to meet the indications of Law no 1497/85, and on the other hand, the plan cannot leave its contents out of the landscape patrimony consideration as granted by Law no 431/85.

In 1996, on the grounds of this context, the *Assessorato BB.CC.AA e PI* issued the Guideline for the Regional Landscape and Territorial Plan[7] which subdivides the regional territory into 18 ambits; for each one of them, a drawing up of a landscape plan is expected. These ambits have been specified in relation to the morphological traits, not the historical or geographical ones, of the natural systems.

The Cultural Heritage and Landscape Code and the Agreement signed between the State and Sicily's region have entered into force, now making the guidelines

---

Department of Cultural and Environmental Heritage) on the grounds of a provincial system, organized in technical and scientific sections.

[4] Regional Law no 116/1980, art. 2.

[5] Regional Law no 80/1977, art. 3.

[6] Royal Decree no 1357/1940, art. 23; President of Republic Decree no 805/1975, art. 31; Regional Law no 116/1980, art. 2.

[7] Italian National Law no 1497/1939.

clear. These guidelines do not have a decisional but an addressing character, whose measures need to be conformed to.

In relation to the points expressed above, the absence in regional laws in matters of landscape and landscape planning, the Cultural Heritage and Landscape Code, which entered into force in Italy in May 2004, can be applied in Sicily exclusively as to concern its contents, not its procedures.

In relation to landscape delegations, for example, while in all the other regions of Italy, which coherently with the Cultural Heritage and Landscape Code have made laws and planned in matters of landscape, the building and landscape concessions are issued by the pertaining municipality by means of the *Sovrintendenza,* which control action, in Sicily. The building concession is issued by the municipality, whereas the landscape concession is issued by the *Sovrintendenza* as the competent regional and external authority pursuant to the law of 1939. Landscape planning in Sicily has discovered a further level of complexity, because of the department's decision to commit to the *Sovrintendenze* by the drawing up of the landscape plans, decentralizing the unitary planning action, promoted by the guidelines in 1996 and committed to the proper department of the region. The *Sovrintendenze* have elaborated their own documents (real plans, in absence of the normative part) in relation with their own provincial contexts obviously incompatible with the guidelines ambits, inconsistent and incoherent with a wide and unitary vision in order to preserve the Sicilian landscape.

After all, starting from the general framework of the main legislative measures in matters of environmental and landscape issues, it seems that the pursued policies have been declared by a state of emergency. It means a lack of coordination and integration among the different plans, looking at the territory as a complex and articulate unit, and not as the total of unsolved elements.

References

Bevilacqua P (2011) La Terra è finita. Laterza, Bari

Brunetta G, Moroni S (eds) (2011) La città intraprendente. Carocci, Rome

Caracciolo A (1988) L'ambiente come storia. Il Mulino, Bologna

Daly HE (2001) Oltre la crescita. L'economia dello sviluppo sostenibile. Edizioni di Comunità, Turin

Quarta C (ed) (2006) Una nuova etica per l'ambiente. Dedalo, Bari

Querini G (2007) La tutela dell'ambiente nell'Unione Europea. Un'analisi critica. Franco Angeli, Milan

Settis S (2010) Paesaggio Costituzione Cemento. Einaudi, Turin

Settis S (2012) Azione popolare. Einaudi, Turin

Tacchi EM (ed) (2004) Sostenibilità ambientale e partecipazione. Franco Angeli, Milan

Vv.Aa (2010) Per un'altra campagna. Maggioli, Santarcangelo di Romagna

Chapter 52
Revitalising the Historical Landscape: The *Grange* in Southern Europe

Claudia Matoda

Abstract Landscape can be seen as a layering of different historical traces; this coexistence of several witnesses implies the possibility of several ways in which landscape policies can be oriented. This contribution aims to explore and analyze the relationship between the presence of historical sediments (namely, the system of monastic *grange*) and landscape policies. *Granges* are particularly interesting as they represent a medieval strategy for territorial control and economic resource exploitation. Through a systematic territorial design, *grange* became satellites of network hierarchical organization, modifying the existent territorial structure and generating new territorial systems. This contribution, treating *grange* as a paradigm, is articulated into two sections. The first part aims to propose an exam of European experiences, providing a synthetic census of contemporary strategic attitude and ideas for valorization towards *grange*, within legislative systems. The second section regards an overview on the effectiveness of contemporary strategies. Through this approach, we would like to provide a comparative view on different local approaches towards a transnational territorial phenomenon.

Keywords Historical landscape • Cistercian architecture • Rural valorization

52.1 Grange as Pivots in the Historical Landscape

This contribution aims to analyze a particular intersection between historical landscape and contemporary territorial strategies, through the analysis of the landscape policies concerning *grange*. *Grange* may be considered as significant examples of medieval planning: this network of farms, depending from a monastery or an ecclesiastic institution, implied the creation of a system of rural exploitation and an occasion for modifying country landscape. This contribution aims to focus

C. Matoda (✉)
Interuniversity Department of Regional and Urban Studies and Planning (DIST), Politecnico and Università di Torino, Viale Mattioli 39, 10125, Turin, Italy
e-mail: claudia.matoda@polito.it

R. Gambino, A. Peano (eds.), *Nature Policies and Landscape Policies*, Urban and Landscape Perspectives 18, DOI 10.1007/978-3-319-05410-0_52

on the relation between this historical grid and nature-landscape strategies, verifying the validity of this example of history and landscape policies integration. We also want to discuss the coherence of the strategies focused on the *grange* with the wider regulatory framework regarding landscape.

The productive use of territory (and its consequent need of a consistent strategy that could encompass landscape, economy and rulers) during the Middle Ages is a still disregarded topic in literature (Corner 1999). The lack of attention to medieval structures for work implies not only an incomplete view on historical productive capacity but (consequently) an undeniable difficulty in encompassing these ancient entities in contemporary strategies. However, there is a particular type of medieval structures for work that have captured scientific attention, namely, *grange*.

Granges are, firstly, a practical solution to a medieval logistic problem. The purchase of *latifundia* by Cluniac houses during the eleventh century (Migne 1844–1855, PL CXLIX, coll. 738–740) and the almost contemporary difficulties of Cistercians monks in administrating their lands (Righetti Tosti-Croce 1983) implied the need for a system of points for territorial control. These pivotal elements were the *grange*. A *grange* was a satellite farm, belonging to a monastery and used for several productive activities (stocking goods, milling corn, stable).

Cistercian *grange* had a stricter dependence from the abbey, from territorial and administrative point of view (Higounet 1983). Historical sources attest the importance of agriculture for the survival of the institution (Waha 1978; Bouton-Van Damme 1974) and demonstrate the indisputable role played by rural landscape in the material sustenance of ecclesiastical community. The evolution and the permeation in the economic panorama through the acquisition of landed estates and farms brought also a territorial and hierarchical organization and implied a transformation of the agriculture landscape (Higounet 1980). Moreover, it is attested the significant role played by Cistercians as actors in rural history, through reclaims of swampy lands and deforestations (Righetti Tosti-Croce 1983). The role played in water control is also well known, stating from the deviation of the river Aube in order to serve the Abbey of Clairvaux (*Desciptio monasterii Claraevallensis; PL,* CLXXXV, coll. 596–573).

Granges are not architectural emergences in a uninhabited territory, but material traces of a process of anthropization. Thus, they can be considered as crucial elements that should be inserted in a regressive study on the phases of utilization of a site (Moreno 1990). An effective landscape strategy should encompass these two aspects: architectural evidence and rural dimension.

52.2 The European Charter of the Cistercian Abbeys and Sites: A Normative Overview

The most relevant instrument regarding the system of Cistercian *grange* is the European Charter of the Cistercian Abbeys and Sites. This agreement has been developed in a particular strategic context.

In 1987, the Council of Europe launched the "Cultural Routes programme" to emphasize the transnational contribution to the creation of a shared cultural patrimony. It is necessary to wait until 1993 for the draft of a charter devoted to Cistercian heritage, based on the principle of highlighting the cultural (and also touristic) potential of the network of *grange*. The initiative is born outside the frame of Cultural Routes, but it has been incorporated in the programme in 2010. Countries included are Belgium, Czech Republic, Denmark, France, Germany, Italy, Poland, Portugal, Spain, Sweden and Switzerland.

Which is the role played by landscape in this policy? The concept of "Cultural Route" doesn't include landscape. As specified in Sect. 1, paragraph 1 of the resolution CM/Res (2010)53, the definition of a Cultural Route is

> An itinerary or a series of itineraries based on a historic route, a cultural concept, figure or phenomenon with a transnational importance and significance for the understanding and respect of common European values.

The previous CM/Res (2010)52 shows a marginal role for the landscape approach. Landscape valorization is not included within the eligibility criteria of a new itinerary within the frame, although it can be glimpsed in the "field of cultural tourism and sustainable cultural development" (Sect. 1, paragraph 5). A signal of attention to the landscape question is clear in the part II, "List of priority fields of action": Sect. 2, paragraph 2 underlines the necessity of promoting the work with other authorities (Council of Europe, UNESCO and ICOMOS) connected with "heritage restoration, protection and enhancement, landscape and spatial planning". The following subsection emphasizes the importance of enhancing European heritage sites generally not exploited by tourism, with a particular attention in rural (and industrial) areas. This point of emphasizing rural culture is underlined also in Sect. 2, paragraph 5. It seems particularly significant that as declared further (Sect. 4, paragraph 1), the certification "Cultural Route of the Council of Europe" is awarded by the Governing Board of the Enlarged Partial Agreement on Cultural Routes (EPA) only after consultation with the Steering Committee on Culture (CDCULT) and the Steering Committee for Cultural Heritage and Landscape (CDPATEP). Thus, even if the resolution doesn't include the theme of landscape, there is a part of the awarding process that should include a particular attention to landscape.

The point which we would like to focus on is the apparent lack of connection between these tools.

52.3 The Effectiveness of Cultural Routes Policies in the Valorization of Rural Landscape

Cultural Routes are seen as social passages. It seems appropriate to borrow a consideration concerning the impact of circuits on the surrounding territories. The creation of touristic itineraries is seen as a starting point in face of a process of decline of landscape that can be avoided only through a consistent coordination between different administrative subjects (Español Echániz 2012; Balchin and Sýkora 1999).

In the case of *grange*, the consequences of the marginalization of the importance of landscape seem attested by an important fact. The evidence comes from a particularly interesting tool, namely, a monitoring study on the "Cultural Routes" appeared in 2010. The research, devoted to observing results on innovation, competitiveness and clustering of the European Cultural Routes programme, examines the sustainability and quality of the programme, analyzing the effectiveness in promoting less known cultural sites and heritage. Between the monitored itineraries, it seems that only the "Routes of the Olive Tree" and the "*Iter Vitis* Route" have a consistent component of valorization and management of historical landscape, seen as an economic element (Khovanova-Rubicondo 2010). In this case, the safeguard of rural areas is a fundamental aspect, while in the experience of *grange*, it can be easily disregarded.

Consequently, the landscape, as in European Landscape Convention, article 1, defined as "an area, as perceived by people, whose character is the result of the action and interaction of natural and/or human factors" (Council of Europe 2000), seems slightly disregarded in "European Charter of the Cistercian Abbeys and Sites" and the connected regulatory framework.

52.4 Conclusions

The promotion of *grange* should include different aspects of landscape: connections and routes, cultivated lands, deforestations and millruns. All these aspects were at the basis of an effective architectural and cultural quality and should be explicitly promoted in the regulatory framework (Hewlett 1973; Bishop and Lange 2005). This integration is still missing and the consequences have been demonstrated through the monitoring; only in the cases in which landscape plays an explicit role in the definition of the "Cultural Route" that it seems to be worth preserving. It would be necessary to go beyond the great historical picture, in order to recover also the underlying economic structure.

References

Balchin P, Sýkora L (1999) Regional policy and planning in Europe. Routledge, London

Bishop I, Lange E (eds) (2005) Visualization in landscape and environmental planning. Technology and applications. Taylor & Francis, London/New York

Corner J (ed) (1999) Recovering landscape: essays in contemporary landscape architecture. Princeton Architectural Press, New York

Council of Europe (2000) European Landscape Convention. Treaty series no. 176, Florence

De La Croix Bouton J, Van Damme JB (1974) Les plus anciens textes de Cîteaux (Cîteaux. Commentarii Cistercienses, studia et documenta, 2). Abbaye Cistercienne, Achel

De Waha M (1978) Aux origines de Cîteaux. Rapports entre l'Exordium Cistercii et l'Exordium Parvum. In: Combier G, Deroux C, Préaux J (eds) Lettres latines du moyen âge et de la Re-naissance. Latomus, Bruxelles, pp 152–182

Español Echániz I (2012) Paysage et infrastructures de transport: les routes. In: Facettes du pay-sage. Réflexions et propositions pour la mise en œuvre de la Convention européenne du pay-sage. Editions du Conseil de l'Europe, Strasbourg, pp 55–116

Hewlett G (1973) Reconstructing a historical landscape from field and documentary evidence: Otford in Kent. Agric Hist Rev 21(2):94–110

Higounet C (1980) Le premier siècle de l'économie rurale cistercienne. In: Ronzani M (ed) Isti-tuzioni monastiche e istituzioni canonicali in Occidente (1123–1215), Atti della Settimana internazionale di studio, Mendola 1977. Milano, pp 343–368

Higounet C (1983), Essai sur les granges cisterciennes. In: L'économie cistercienne. Géographie-mutations du Moyen Age aux Temps modernes, Troisièmes Journées internationales d'his-toire, Abbaye de Flaran 1981. Auch, pp 157–180

Khovanova-Rubicondo K (2010) Impact of European cultural routes on SMEs' innovation and competitiveness. Available via http://www.coe.int/t/dg4/cultureheritage/culture/routes/StudyCR_en.pdf. Accessed 10 July 2013

Migne JP (1844–1855) Patrologia Latina

Moreno D (1990) Dal documento al terreno: storia e archeologia dei sistemi agro-silvo-pastorali. Il Mulino, Bologna

Resolution CM/Res(2010)52 on the rules for the award of the "Cultural route of the council of Europe" certification (Adopted by the committee of ministers on 8 Dec 2010) at the 1101st meeting of the Ministers' deputies). Available via https://wcd.coe.int/wcd/com.instranet.InstraServlet?Index=no&command=com.instranet.CmdBlobGet&InstranetImage=1734454&SecMode=1&DocId=1671428&Usage=2. Accessed 5 June 2013

Resolution CM/Res(2010)53 establishing an enlarged partial agreement on cultural routes (Adopted by the committee of ministers on 8 Dec 2010 at its 1101st meeting of the Ministers' deputies). Available via http://conventions.coe.int/Treaty/EN/PartialAgr/Html/CulturalRoutes201053.htm. Accessed 5 June 2013

Righetti Tosti-Croce M (1983) Architettura e economia: "strutture di produzione cistercensi". AM 1:109–134

Chapter 53
Nature, Landscape and Energy: The Energy Masterplan of Emilia-Romagna Po Delta Regional Park

Anna Natali and Francesco Silvestri

Abstract In the last years, natural parks seem to be affected by a growing interest in energy planning. This attitude is easily explicable with a willingness to contribute to CO_2 reduction, to struggle against climate change, to promote virtuous patterns in energy efficiency, to preserve territory from landscape impacts that energy production equipment could generate (solar fields, wind poles, obliged modification in local agriculture to feed bioenergy plants). This contribution depicts how energy planning is handled by natural parks. After a brief review of the still few existing projects throughout Europe, we focus on the experience run by the Emilia-Romagna Regional Park Po Delta, the first one in Italy to design a real strategic plan for sustainable energy, with guidelines and prescriptions on landscape and nature conservation. The emerging picture offers insights on different issues – why should a natural park deal with energy planning, how to design this kind of actions in a protected area, and which role is assigned to planning and regulation – to incentive schemes and public goods provision.

Keywords Energy • Climate change • Natural parks • Po River

53.1 Introduction: Energy, Climate Change, and Nature Conservation

Protected areas (National and Regional Parks and Natura 2000 Sites) are institutions established to deal with flora and fauna conservation and sustainable development. At the same time, nature conservation is deeply challenged by climate change, a problem that is directly connected with energy.

In the last years, many protected areas have implemented efficient energy and CO_2 abatement projects, often through pilot projects. Although the energy consumption of parks and the communities who live in is difficult to label as problematic,

A. Natali (✉) • F. Silvestri
eco&eco, Economia ed Ecologia, Ltd, Via G. Oberdan 11, 40126 Bologna, Italy
e-mail: anna.natali21@libero.it; silvestri@eco-eco.it

R. Gambino, A. Peano (eds.), *Nature Policies and Landscape Policies*,
Urban and Landscape Perspectives 18, DOI 10.1007/978-3-319-05410-0_53

protected areas would have the opportunity to act as relevant nodes of an energy network, both by emphasizing their nature of "social laboratories", where to experience good practices in the field of renewable and saving energy, and by behaving as centres of attraction and addressing skills for the benefit of the rural zones (Silvestri 2007, Osti and Silvestri 2009; Natali and Silvestri 2010).

Many reasons may justify a role for natural parks in the (renewable) energy policy at the local scale:

1. First of all biodiversity, whose protection is the primary mission of a natural park, is strictly influenced by climate change, and the latter is strongly connected with consumption and production of energy.
2. A natural park, as a territory and as a local authority, is also expected to contribute to the objectives of energy policy defined by the higher levels of government in the "burden sharing" framework and in the "think global and act local" rationale.
3. A protected area is a place where to test righteous patterns with regard to a variety of environmental issues, including energy (Setti 2012).

Moreover, in Italy the deep uncertainty concerning the future of the intermediate levels of government (Provinces, Municipalities' Consortia, Mountain Municipality Communities) raises the real possibility that in the future natural parks will be asked to play a crucial role in coordinating and supporting environmental and energy programmes at the subregional level.

53.2 Parks and Energy Planning

An increasing number of natural parks throughout Europe parks are progressively involved in the energy issue, both with occasional actions (such as renewable energy production in their own buildings, coordination of joint-purchase groups to install PV modules in residential buildings inside the park, reforestations to broaden carbon sinks capacity according to Tokyo Protocol, purchase of electrical cars, bicycles and vehicles to transport visitors and staff workers, addressing inhabitants to energy saving), and in a more systematic way through a strategic vision (Bruckmeier and Tovey 2009).

At a regional level, many projects are carried on by French Regional Parks, such as Vercors (Energy and Territory Programme, aimed at supporting local municipalities and inhabitants in reducing CO_2 emissions and committing the park to cover at least 55 % of its energy requirements with renewable energy); Baronnies Provençales (promoting Energy sobriety to contrast climate change); and Morvan (improving energy performance at the local level, with special emphasis on wooden biomass and building techniques).

At a national level, we must point out the Sustainable Energy project in the Italian Abruzzo, Lazio and Molise National Park (with 400 photovoltaic modules and two wind rotors producing a part of the energy requested by the park administration and a bike sharing service for tourists and inhabitants, financed by Italian

Ministry of Environment, Land and Sea); Fossil Free Park, supported by Italian Ministries of Economic Development and of Environment and involving Belluno Dolomites National Park, Pollino National Park and Adamello Brenta Regional Park, to run pilot projects on renewable energies; and the Maramures Mountains National Park (Romania) providing four villages with modern heating systems fed with firewood from local certified forests.

A number of projects funded by European Union programmes involve partners developing specific actions in their own territory, at the same time pursuing common objectives in a unitary framework. Some examples of this are (1) the Intelligent Energy Europe project "Wise-Plans – Co-operation between Communities for Energy Action Plans", involving Ticino Regional Park (Italy) and Doñana National Park (Spain); every park committed to draw the Sustainable Energy Action Plan (SEAP) of the park community, with a target of reducing CO_2 emissions by 5 %; (2) the Climaparks project (Interreg Italy-Slovenia) that promotes research, studies and awareness campaigns on climate change issues related with biodiversity conservation and energy saving in nine Regional and National Parks: four in Slovenia (Triglav, Strugnano, Sicciole, Škocjanske Jame) and five in Italy (Friuli Dolomites, Julian Prealps, Romagna Chalk Seam, Veneto Po Delta and Emilia-Romagna Po Delta).

Inside Climaparks, the Emilia-Romagna Regional Park Po Delta decided to draw its own energy master plan. The next section of this chapter illustrates this document that is currently the only strategic energy plan being implemented in Europe by a natural park.

53.3 The Po Delta Energy Master Plan

The master plan focuses on energy-related policies and activities, with a particular emphasis on biodiversity conservation. The document's approach is strategic and focuses on the definition of objectives, lines and pilot projects to be implemented in the area. Actions aiming at energy efficiency of the park as a local organization (office activities, machineries, vehicles) are left aside and postponed to a later stage.

The master plan is divided into two macro-sections:

(1) An analytical one, where data, figures and qualitative information on plans and activities related to energy efficiency are collected, following the *fil rouge* of economic activities and problematic energy consumption in the different zones that compose the park territory: the coastal lagoon, with fishery and seashell farming; the wetlands, characterized by former fishery recovers and huts (the so-called *capanni*); the seacoast, with bathhouses and seaside services; and the outback areas, with agriculture, and country-house accommodations (eco&eco 2006). Moreover, in this section a list of the most interesting energy projects implemented in the whole protected area, by different economic and institutional operators, is reported.

(2) A strategic one, dealing with objectives, lines and projects to be addressed in the area, thanks to the Regional park support.

Table 53.1 A synopsis of the energy master plan of Po Delta Regional Park

Strategic objectives	Specific objectives	Lines
Protecting biodiversity through climate change	1. Promoting the rational use of energy and reduction of CO_2 emissions in the Regional Park	1. Implementing a Regional Park SEAP
		2. Implementing energy efficiency in buildings
		3. Supporting local planning on energy and CO_2 abatement
	2. Implementing intelligent energy schemes in productive sectors	4. Facilitating efficient and noninvasive equipment and devices
		5. Promoting energy-efficient solutions adaptable to local productions
	3. Reducing emissions in transports	6. Renewing the transport fleet of park
		7. Implementing visitors sustainable mobility projects
		8. Promoting CO_2 abatement with respect to boats for seashell farming
Contributing to the overall objectives of energy efficiency according to burden sharing	4. Supporting local production of energy from renewable sources	9. Promoting low-impact plants and micro-systems
		10. Promoting brownfield recovering to produce renewable energy
	5. Supporting energy efficiency in agriculture and forest companies	11. Encouraging energy efficiency in agricultural firms
		12. Sustaining short supply chains and closed cycles in productions
Acting as an experimental and coordination player on energy matters	6. Supporting research and information on climate change and energy matters	13. Financing R&D on energy and climate change
		14. Publishing popular and widespread circulation documents on energy and climate change
		15. Favouring good practices dissemination
		16. Cooperating with other local and public bodies
	7. Involving local municipalities in efficiency energy policies	17. Creating a network of actors on energy and climate change
		18. Supporting weaker municipalities in implementing efficient energy policies

Table 53.2 Energy and CO_2 saving and pay back allowed by the energy master plan implementation

Project	Energy saving	Pay back	Total emissions reduction
	kWh/year	years	kg CO_2/year
(a) Regional Park SEAP	–	–	–
(b) Efficient Lagoon hut (grid connected)	4,000	8	78,000
(b) Efficient Lagoon hut (stand alone)	1,300	24	56,550
(c) Eliceo Wood reforestation		Not relevant	173,400,000
(d) Goro Solar village	3,312	10	24,292
(e) Covenant of Mayors participation	–	–	–
(f) Efficient Camping	44,500	7	243,000
(g) Efficient Bathhouse	10,950	8	254,560
(h) Efficient Agritourism accommodation	41,000	8	108,000
(i) Ecoprofit in the Delta	395,000	3	348,803
(l) Sustainable Mobility in Park Stations	25,000	Not relevant	6,700
(m) Mussels farming boats (kit)	400	When consumption of 8,000 l of fuel is reached	240,000
(m) Mussels farming (replacement)	256		61,440
(n) Zero Emission Saltpans	675,000	8	70,000
(o) Efficient Fishery Station	47,656	13	96,000
(p) Converting Ex SIVALCO brownfield	800,000	7	540,000
Total		**8 (average)**	**175,527,344**

The contemplated projects are composed of both intangible (studies, communication and information campaign) and tangible actions, all measurable in terms of produced/saved energy, CO_2 abatement contribution and monetary costs and benefits. Moreover, to address the viability of each action, a survey of the available financial funds is provided. The synopsis of the master plan is represented in Table 53.1.

The master plan designs 15 pilot actions, gathered in a portfolio that allows on one hand to quantify the needed investments and on the other to appreciate the results obtainable in terms of monetary value of energy efficiency and of saved CO_2. The final balance of the master plan claims that a total investment of 12 millions of euro will produce an annual economic value of 1.6 millions of euro, with an Average Global Pay Back of about 7 years, even considering the projects without direct profitability. But the most impressive result is the annual CO_2 abatement, estimated in 175 t of CO_2 (Table 53.2).

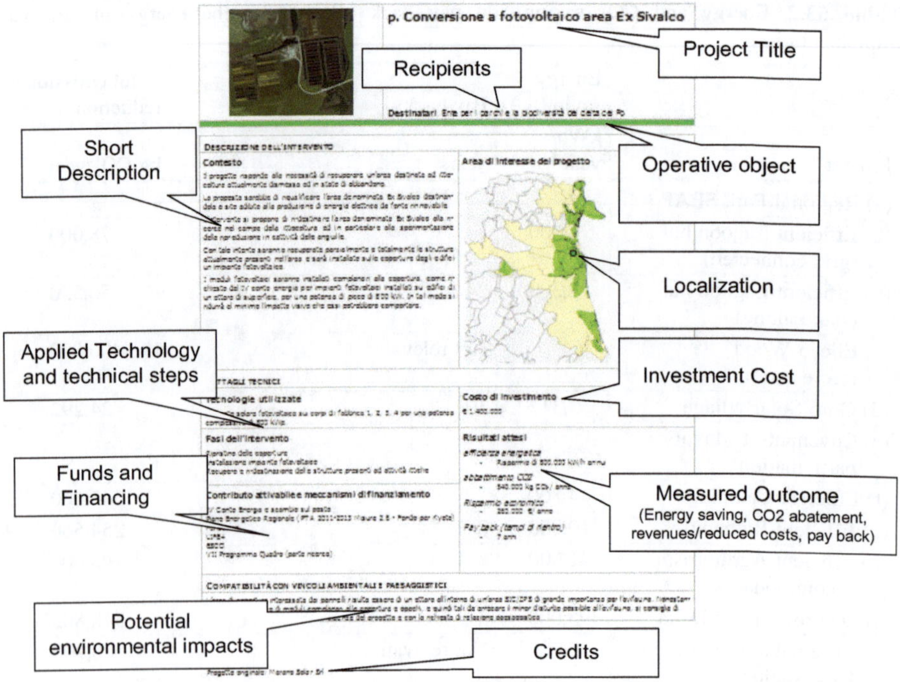

Fig. 53.1 A representative summary sheet of the projects constituting the energy master plan

The most interesting character of the energy master plan conceived by the park is the privileged attention payed to the potential impact on biodiversity conservation, in particular to the effects of the implementation of each project on the Natura 2000 sites. This assessment leads to a series of practical recommendations regarding the technological solution (PV or biomass), the size of the plants (with a threshold in 100 kWp) and their localization (internal or external to a Natura 2000 site).

Each project is illustrated both by means of a narrative description and through a specific summary sheet form, with a related more exhaustive sheet in the annexes, when the project requires a further technical explication. An example of the summary sheet is reported in Fig. 53.1.

53.4 A New Role for Protected Areas in Energy Planning

The Po Delta energy master plan aims at providing a guideline to local communities, to face the energy and climate change challenges from a biodiversity protection and sustainable development perspective. In this way, the park intends to claim its competence in this topic in connection with other administrative levels and in cooperation with other local municipalities and public bodies.

To comply to this general objective, the master plan has been drawn as a strategic tool, typical of widen areas, and not as a SEAP, more fitting for municipalities that need to implement practical actions to abate CO_2 and to cut the energy bill. Notwithstanding its strategic orientation, the master plan is able to point out feasible targets in terms of monetary saving and CO_2 abatement, thanks to the conception of a 15 pilot projects portfolio.

As mentioned above, the master plan puts the Po Delta in the "avant-garde" of parks throughout Europe, facing the complex relationship between nature conservation and energy planning. This master plan is a first step, to be followed on one side by the implementation of a SEAP, addressed at introducing the energy efficiency topic inside the administrative organization, and on the other side by the prospected adhesion of the park to the Covenant of Mayors (CoM), playing the role of coordinator of the whole area.

The case of the Po River Delta, as many other experiences mentioned in the Sect. 1 of this chapter, attests that parks begin to be aware of their potentiality in dealing with the energy topic. The pioneers that have just started to go along this new path are organizations of a quite large size, no matter if regional or national, with a strong commitment with their territory. They seem to look at energy planning as a means, rather than as an objective itself: a tool to provide a better performance in biodiversity conservation and to pursue sustainable development, in the climate change scenario and in the EU 20-20-20 strategy framework.

References

Bruckmeier K, Tovey H (eds) (2009) Rural sustainable development in the knowledge society. Ashgate, Farnham

Eco&eco (2006) Il programma di sviluppo socio-economico del Parco regionale del Delta del Po Emilia-Romagna. Coop Albatro, Ferrara

Natali A, Silvestri F (2010) Sortirne insieme. Un programma di ricerca sui beni collettivi nei parchi. Parchi 59:47–53

Osti G, Silvestri F (2009) Practical knowledge and institutional mediation in a controversial case of clam farming in Goro. In: Bruckmeier K, Tovey H (eds) Rural sustainable development in the knowledge society. Ashgate, Farnham

Setti L (2012) Una programmazione puntuale per recuperare il ritardo. Il Divulgatore 3/4. Available via http://www.ildivulgatore.it/pdf/2012/02-art2.pdf. Accessed 15 Nov 2013

Silvestri F (2007) Il ruolo della conoscenza per lo sviluppo delle aree fragili: il caso di Goro. Sviluppo locale, XI(27). Available via http://www.eco-eco.it/download/pubblicazioni/art_goro.pdf. Accessed 15 Nov 2013

Chapter 54
How to Manage Conflicts Between Resources' Exploitation and Identity Values

The Agri Valley Amid Oil Supply and the Lucanian Apennines National Park

Mariavaleria Mininni

Abstract In the mid-1990s, the valley landscape of Basilicata found itself at a crossroads between two contrasting development models. One was based on the supply of hydrocarbons, with a view to starting up a heavily industrialised process, whilst the other drew its inspiration from local development policies, based on nature tourism and high-quality agriculture. Basilicata attempted to find a middle ground, which could respect the principles of ecological, economic and social sustainability in agreement with oil company activity. Twenty years later, the Agri Valley is a far cry from achieving the following two goals for the future, putting itself forward as a model for a sustainable economy and becoming a large-scale centre for energy production. The construction of the biggest onshore extraction plant in mainland Europe in the heart of the Lucanian Apennines National Park is producing a serious environmental impact. If we consider the current political, economic and financial climate, the now well-established cultural concepts of *smart growth* and the OECD *Better Life Index* forecast, which considers the well-being industry and the *green economy* to be the main driving forces for the global economy over the next 20 years, does it still make sense to obsessively consider oil as the area's main driving force for growth? The aim of this contribution is to highlight and discuss the conflicts and paradoxes which stem both from different ways of interpreting the term resource and from the ambiguous nature of identity values in an area where resources' exploitation implies deeply contrasting notions of landscapes.

Keywords Oil • Royalties • Local development • Basilicata

R. Rizzi is the author of all thematic maps

M. Mininni (✉)
DiCEM UNIBAS, Viale dell'Ateneo Lucano, 10, 85100 Potenza, Italy
e-mail: mariavaleria.mininni@unibas.it

R. Gambino, A. Peano (eds.), *Nature Policies and Landscape Policies*,
Urban and Landscape Perspectives 18, DOI 10.1007/978-3-319-05410-0_54

54.1 The Identity Features of the Area. Oil?

According to some local legends, which have been passed down through the generations, sightings of "tongues of fire" rising from the Apennine Mountain Range go as far back as the fifteenth century: we are here talking about the so-called torches which were actually methane burning as it leaked out from the earth's surface.

The discovery of oil in Basilicata dates back to 1902 when the Military Mining Inspectorate of that time commissioned a report on the oil reserves in the municipality of Tramutola in the province of Potenza.

During the Second World War, the oil products of the Upper Agri Valley were mostly used to cope with the international embargo; in 1942, it produced at most 700 tonnes of oil. That is 7 % of national production at that time which reached a total of 10,000 tonnes a year.

It was, however, only through Enrico Mattei in 1958 that the search for reserves intensified, despite the rugged nature of the land, and the following year, significant oil and gas reserves were discovered in Grottole, Ferrandina, Rotondella and Pomarico in Basilicata.

The 1970s oil shocks led Europe to once again seek energy resources in safe territory.

A vast seismic prospecting campaign and perforation activity did not, however, bring important results until the 1990s and led to the discovery of five oil reserves in the Agri Valley and the Camastra-Alto Sauro area: Monte Alpi, Monte Enoc/ Volturino, Cerro Falcone, Costa Molina and Tempa Rossa.

The Viggiano Oil Centre was built in 1996 in order to allow for the processing of crude oil and it has the capacity to process some 104,000 barrels a day. In 2007 some 32.35 million barrels were brought to the centre of Viggiano with some 95 million barrels being processed a day (Regional Government of Basilicata).

The production cycle consists of a gathering network; that is a series of wells from which crude oil is extracted which are connected to the Centro Olio Val d'Agri COVA (Agri Valley Oil Centre) by way of underground pipes. Today there are 39 wells in the "Agri Valley" plant of which 26 are currently in use.

After an initial treatment process, the oil is transported via a pipeline, which runs down the River Agri in the Agri Valley to the Taranto Oil Centre.

54.2 The Park, the Valley and the Extraction Basin

After a lengthy and much debated process lasting almost 20 years, in 2007 the Lucanian Apennines National Park-Agri Valley-Lagonegrese became the 24th and latest addition to Italy's national parks (Presidential Decree 8th December 2007). The area, which stretches across some 45,000 h, is just one-third of the originally planned area. The lengthy, problematic and much debated institutional process,

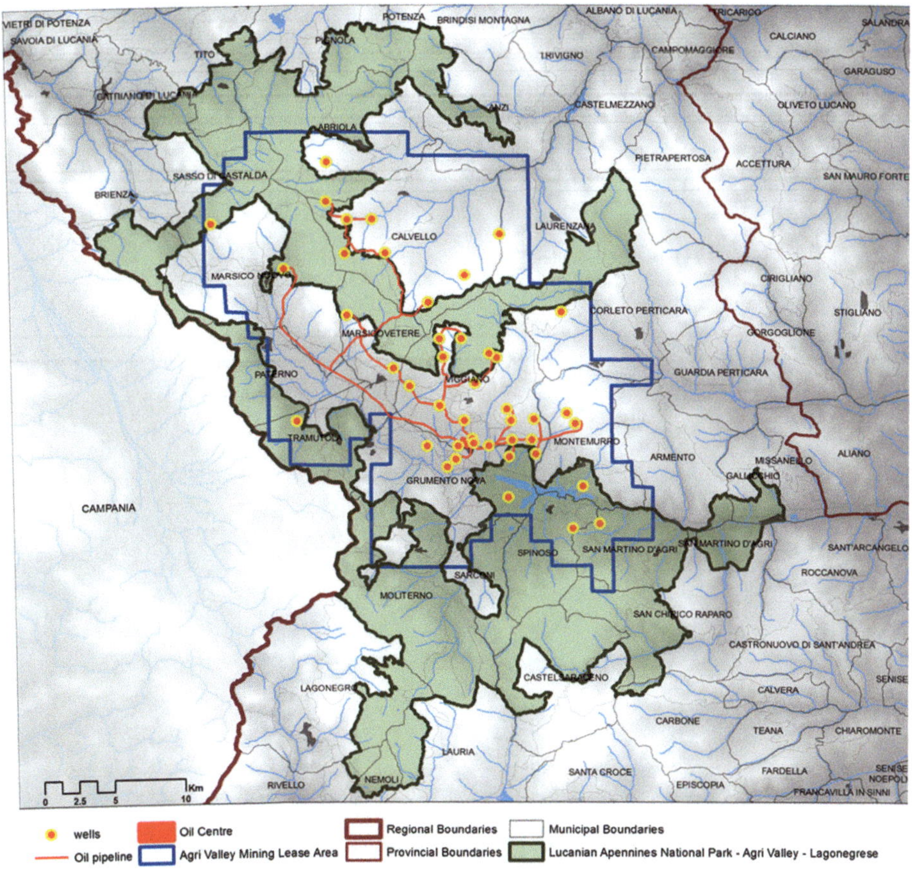

Fig. 54.1 Case-study area

based on the binding opinion of the municipalities involved, led to an irregular delimitation of the park due to requests for the exclusion of valley areas made by the municipalities of Sarconi, Marsico Nuovo and Marsicovetere and of much of the municipality of Viggiano where oil company activity already takes place (Fig. 54.1).

Following the creation of the park, seven of these sites now lie within the perimeter of the protected area.

A number of Sites of Community Importance and Special Protection Areas have not been included in the proposed delimitation of the park. A total of 29 municipalities[1] have a part of their territory within the park: seven sites are part of the Natura 2000 network (four sites of community interest and three special protection areas),

[1] The municipalities are Satriano, Lagonegro, Moliterno, Sarconi, Anzi, Paterno, Nemoli, Spinoso, Marsico Nuovo, Marsicovetere, Grumento Nova, San Martino d'Agri, Montemurro, Viggiano,

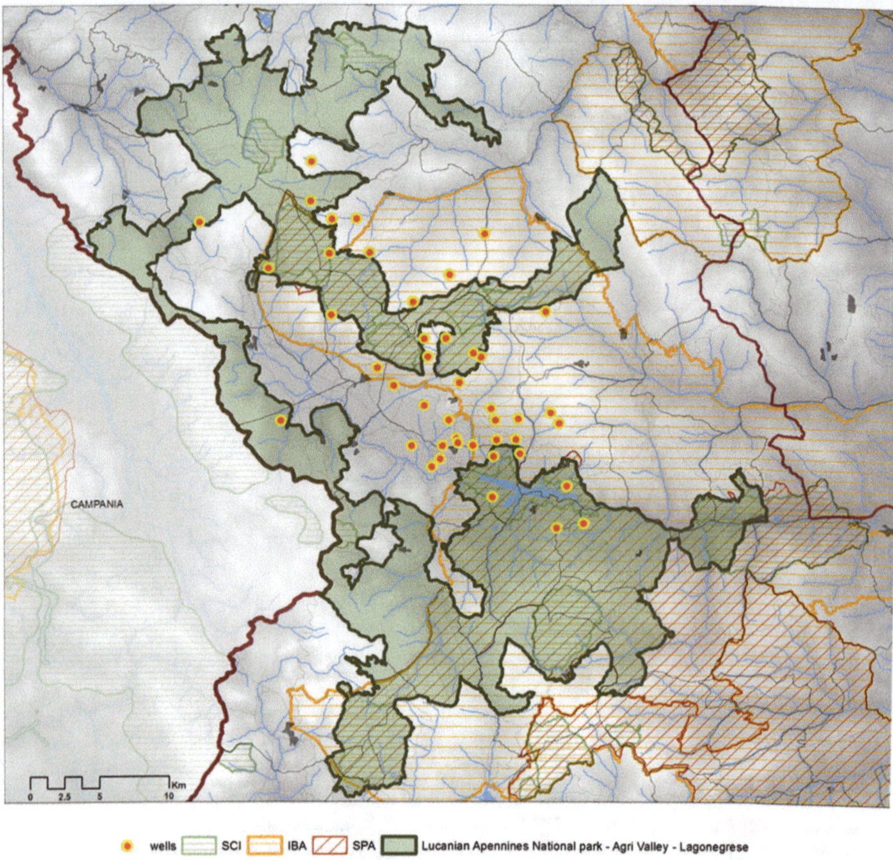

Fig. 54.2 Sites of community importance and special protection areas in the case-study area

and some areas which are included in landscape plans such as Mount Volturino and Mount Madonna di Viggiano, to the north, are part of the Sellata-Volturino-Madonna di Viggiano Landscape Plan (Fig. 54.2).

The original delimitation of the park took the natural conformation of the area into account. It spanned across three river valleys, the Basento, the Agri and the Sinni, which are characterised by a unique wealth of biodiversity. Indeed, the area is home to a wealth of flora with numerous plants endemic to the Balkan Peninsula and the Mediterranean. It also stands out for its geological features, for its fauna (*Lepre Italica,* the Italian hare), for the dense vegetation which covers its mountainsides (beech forests mostly *Aquifolium fagetum*), not to mention its immaterial, historic and cultural heritage, which is kept alive through its traditions.

Tramutola, Sasso di Castalda, Gallicchio, Castelsaraceno, Pignola, Abriola, Armento, Carbone, Rivello, Calvello, Lauria, Tito, Brienza, San Chirico Raparo and Laurenzana.

The central part of the Agri Valley, which is also the most anthropised area where the extraction and processing of oil takes place, remains excluded. The delimitation of the park is the fruit of numerous compromises, omissions and negotiations: in the end the northern part will be linked to the southern part by the ecological corridor between Paterno and Tramutola.

The park is located at the centre of a complex system of parks and protected areas of important environmental and nature-related value: to the south the Pollino National Park, to the west the Cilento and Vallo di Diano National Park and the Gallipoli Cognato Regional Park to the northeast.

The park area is, therefore, influenced by the flora and fauna of surrounding parks, and this ensures genetic exchange between the populations hosted in this vast system of protected areas.

At a national and European level, thanks to its strategic position, the park plays a vital role in strengthening the network of parks and protected areas in the APE Apennines Park of Europe. Indeed, it acts like a bridge between the southern ridge of the Apennines overlooking the Tyrrhenian Sea and the one facing the Ionian Sea. It serves as an ecological corridor between the parks in the Apennines of Basilicata, the Cilento Park to the north and the Pollino Park to the south (Stanzione 2007).

What is more, the area must deal with difficult issues associated with the risks it faces. Indeed, the area is subject to environmental risks, hydrogeological risks and water risks and the entire park is characterised by severe instability. Some mountain villages, which act as morphological divides between water catchment areas, and urban settlements and infrastructures in the valley are consequently at risk of flooding.

The area is also considered to be extremely dangerous due to its elevated seismic risk values.[2] Furthermore, there are numerous uninhabited homes, and in some municipalities, these can amount to as much as 50 % of all housing. This figure is not only an indicator of depopulation but also the result of financial incentives provided for rebuilding after the 1980 earthquake.

54.3 A Conflict of Interests Within the Same Territory: Protected Areas and Oil Extraction

The nature conservation strategies within the framework of regional park policies are inextricably linked to territorial development policies and to the coordination of other sectoral policies. This is necessary if we are to attempt to reach some form of coherence and compatibility with the vision of Basilicata as a Green Region, which regional authorities have always striven to adopt.

[2] Regarding the Agri Valley, the ordinance places almost all municipalities in the 1st risk category, whilst only municipalities in the southeast belong to the second risk category.

Fig. 54.3 The area of *Grumentum*

In other words, also the archaeological area of *Grumentum* (Fig. 54.3), a Roman colony which was home to one of the most important cities in Lucania and located in the most southerly part of the Agri valley, there is also another story to be told; that is one in which the territory is working toward the definition of the identity values and assets which characterise this landscape.

For decades this area has been a battlefield for conflicting territorial and development plans. On the one hand, you have the outstanding natural beauty of its unspoilt natural environment. On the other hand, however, the area has oil reserves, which are also a natural resource. This story is not only linked to the relationship local inhabitants have with the representation of oil in the social imaginary but also to that of economies which would be capable of supporting a *place-based policy* and the taking on the challenge regarding the future of *rural inland areas* if conflicts were managed appropriately (Barca 2012).

Through two distinct memorandums of understanding[3] (MoU) signed in 1998, the regional government of Basilicata, the State and the owners of the oil companies agreed to a detailed programme guaranteeing that the exploitation of oil reserves would go hand in hand with investments aimed at improving the natural and cultural heritage of the area. The agreement also included the recognition of royalties and the environmental cleanup of disused extraction sites: on the one hand, multi-sectoral measures were implemented for compensation, monitoring and environmental safeguarding; on the other hand, with a view to sharing the resources made available from ENI funds with the entire region, infrastructural projects were

[3] Protocolli di Intesa (PdI).

implemented for the 30 municipalities identified by Regional Law 40/1995 as "areas or (basins) for oil extraction".

It is not easy to deduce to what extent the exploitation of the natural resource petroleum can generate comparative advantages and economic growth and, thus far, unanimous agreement regarding this matter has not been reached (Percoco 2007).

Both awareness of the effect that petroleum extraction has on ecological, economic and social systems and public awareness of and participation in decision-making processes are of vital importance if trust and collaboration are to be established between territorial administration and the local population. This would not only facilitate social process but also the enhancement of the area in all potential directions.

The *AgriBioDiversity* study (2007) carried out research along these lines. This was followed by the project Action Plan for "*Biodiversità in Val d'Agri*" developed by the University of Basilicata in 2009 in collaboration with ENI, *Fauna and Flora International* and IUCN: the *World Conservation Union.* The aim of the study was to better understand the biodiversity of the area with a particular focus on the effect of petroleum extraction and processing on territorial transformation and ecosystem processes. The study attempted to analyse the effectiveness of the three main points considered in terms of environmental impact:

1. *Environmental compensation* (activities associated with supporting forestry policies: for the most part reforestation and forest viability)
2. *Environmental monitoring* (the implementation of an environmental monitoring system, set out in the MoU in 2000 but still incomplete[4])
3. Programmes for the sustainable development of the area (planning the allocation of royalties)

If we consider the third point, we can see that over time objectives, procedures and planning strategies as well as the allocation of funds amongst those interested have been characterised by three different approaches.[5]

- *1999–2000*. In compliance with the provisions of Regional Law 40/1995, which establishes an annual fund amounting to 3 % of the quantity of liquid and gas hydrocarbons extracted (royalties), the *Plan for the use of the funds to develop economic activities and increase production and industry in the Agri Valley*[6] 1999 defines their use in two yearly plans managed by the Department of Industry planning office. The 1999 Plan collects the applications made by the municipalities in the oilfield during an earlier consultation phase and then adopts a six-pronged approach.
- *2001–2002*. In the second half of the 1990s, following an extremely positive trend, the regional industrialization process which was based on the

[4] ENI in Basilicata Local Report 2012.

[5] Favia (2007).

[6] Piano di utilizzo del fondo per lo sviluppo delle attività economiche e l'incremento produttivo ed industriale della Val d'Agri.

establishment of large nonlocal businesses suffered a crisis. As a result, the way royalties were spent had to change and this led to a rethinking of the regional development model which, in turn, led to the consideration of sectors which had, until that time, been neglected (agriculture and tourism). Through these sectors natural, entrepreneurial and local identity resources could be enhanced.

- *2003–2006*. In recent years, the area as a whole has been subject to a special planning system, which refers to the use of financial resources from various sources. The *Agri Valley, Melandro, Sauro, Camastra Operational Programme*[7] (OP) and the *Agri Valley Integrated Territorial Project*[8] (ITP) are both currently under implementation. The use of royalties falls within a more complex planning framework: the Operational Programme encourages high-quality territorial development with a view to keeping and attracting human capital: company incentives are accompanied by a series of "context" initiatives which are aimed at the inhabitants and the territory.

In addition to ENI funds, the area of petroleum extraction currently benefits from further resources which come from the *Regional Programme for Innovation "Territory of excellence"*, cofinanced by the ERDF.

Of late, the future of the area has been something of a concern. ENI has started up the *Agri Valley Sustainable Development Project* (ASD) in collaboration with the regional government of Basilicata in order to reach a new shared agreement. On the one hand, this would allow for further mitigation of environmental impact and the development of territory through solid initiatives focused on higher levels of local employment and enterprise. On the other hand, it would help maintain production levels and allow for further development of the oilfield, thus preventing a 10 % yearly drop in production levels.

From this brief examination, we can attempt to deduce that (i) the culture of sustainability and the attention dedicated to places is effectively permeating the local production system, (ii) compatibility between safeguarding and development and their relaunch in terms of smart growth and sustainable growth represents the real challenge for the future of this territory, and (iii) greening strategies (relating to the economy and the social dimension) must become the driving force and guide criteria for the next regional planning process where a specific focus must be placed on cohesion policies in rural inland areas.

References

Barca F (ed) (2012) Ministero delle Politiche di Coesione Territoriale: Metodi e obiettivi per un uso efficace dei fondi comunitari 2014–20. Rome

Eni (2012) Eni in Basilicata local report. Available via https://www.eni.com/files/documenti/eni-in-basilicata.pdf. Accessed 01 Jan 2014

[7] Programma Operativo (PO) Val D'Agri, Melandro, Sauro, Camastra.

[8] Progetto Integrato Territoriale (PIT) Val D'Agri.

Favia F (2007) AgriBioDiversity Project Action 6 Monitoring of socio-economic forces, from Protocol 1 background modification. A socio-economic group contribution.
Percoco M (2007) Petrolio e sostenibilità. Analisi economiche e politiche di sviluppo locale per la Val d'Agri. Consiglio Regionale della Basilicata, Potenza
Stanzione L (ed) (2007) In Basilicata. Guida alle escursioni. Edizioni di pagina, p 159

Chapter 55
Planning and Management in the Otranto-Leuca Nature Park

Integration of Approaches for an Active Conservation

Annalisa Calcagno Maniglio and Marianna Simone

Abstract The Natural Regional Park of Costa d'Otranto, Santa Maria di Leuca, Bosco di Tricase, is a unique case in the Apulia region because the entire park is stretched along the coast. Along its 57 km there are many natural and historical values. In these areas the increased tourist settlements have recently created serious problems in the landscape, such as the fragmentation of continuity in the protection of the park, the reduction of ecological functions and the loss of landscape identity. The park now has a great opportunity because the territorial plan is currently being prepared: this can enable the positive integration between different planning policies, not only with the protection of natural and cultural values and the environment but also by playing a key role in achieving new sustainable economic development for the entire territory.

Keywords Regional park • Planning • Conservation • Sustainable tourism

55.1 Natural Regional Park of Costa d'Otranto, Santa Maria di Leuca, Bosco di Tricase

The Natural Regional Park of *Costa d'Otranto*, *Santa Maria di Leuca*, *Bosco di Tricase*,[1] established by the Regional Law of 26 October 2006, covers 3,227 ha at the extreme east of the Salento Peninsula, has a coastline of 57 km and comprises

A. Calcagno Maniglio is the author of paragraph 54.1. M. Simone is the author of paragraphs 54.2 and 54.3.

[1] http://www.parcootrantoleuca.it/. Accessed 13 Jan 2014.

A.C. Maniglio (✉)
Department of Sciences for Architecture, Polytechnic School, University of Genoa,
Stradone S. Agostino, 37, 16123, Genoa, Italy
e-mail: maniglio@polis.unige.it

M. Simone
Master IUAV, Tolentini, Venice, Italy
e-mail: marianna.simone78@gmail.com

R. Gambino, A. Peano (eds.), *Nature Policies and Landscape Policies*,
Urban and Landscape Perspectives 18, DOI 10.1007/978-3-319-05410-0_55

Fig. 55.1 A view of a coastal tower near Otranto (Source: Andrano 2013)

12 municipalities located on the sea side: Alessano, Andrano, Castrignano del Capo, Castro, Corsano, Diso, Gagliano del Capo, Ortelle, Otranto, Santa Cesarea Terme, Tiggiano and Tricase. The coastline is characterised by high coastal cliffs (up to 130 m above sea level), deep canyons and numerous famous grottos. The rocky emergence of the Capo d'Otranto, with the Punta Palascìa lighthouse, is shown in nautical maps as the most eastern part of Italy, a watershed between the Adriatic Sea and the Ionian Sea.

The founding document of the park outlines the will to:

> Conserve and safeguard the natural resources, recuperate the historical-architectural values, improve the ecological functioning of the natural environment, promote sustainable recreational activities and develop compatible economic activities, with the aim of improving the quality of life of the resident populations.

The recognition of the whole coastline of the Salento Peninsula as a marine protected area (Framework Law no. 394/91) is being discussed by the Ministry of the Environment. The whole park has, in fact, a landscape and environmental development along the coast, a geomorphic structure and an uninterrupted, visual and perceptive land-sea relationship (Fig. 55.1).

This unique mix combines historical architectural heritage (ancient sites, farms, towers built between the fifteenth and sixteenth centuries to defend the coast of Salento) and many archaeological ruins, including marine archaeology: important memories of an ancient human presence in this strategic crossroads of civilizations. Of considerable interest for their rarity and ecological significance are some of the woodlands and a wide array of plant species that colonise the rocky soils. Along the slopes down towards the sea, terraced strips built with dry stone walls bear witness to ancient agricultural activity.

Numerous noble villas in various eclectic styles located along the coast denote the residential vocation of this coast, which began in the nineteenth century.

Between the wealth of natural and cultural heritage, in recent years, the area has undergone significant tourist-induced anthropization, which conflicts in many ways with the objectives of protection and enhancement of the park: it creates fragmentation and reduces ecological functionality and landscape identity.

In the drafting of the plan, the park's boundaries between the coastline and the immediate hinterland are interrupted in their continuity by the extent of tourism in the coastal centres of S. Cesarea Terme, Castro, Tricase and S. Maria di Leuca; it is a territorial transformation identified in the planning instruments resulting in the fragmentation of vast park zones and requires an integration between urban, territorial and landscape planning; it requires the identification of planning and project strategies capable of integrating active conservation of a protected area with new sustainable development strategies to avoid damage to the landscape, local environmental and cultural identity.

In the IUCN Durban Congress in 2003, the need for integration between conservation and development, and between park and territory, was established. It was – as stated by Attilia Peano – a "conceptual revolution" which recognised not only the role that protected areas play in nature protection and the designation of reserves but also as

> Territorial areas of coordinated planning between various competent institutional subjects, inside and outside boundaries, in which to develop converging actions towards an integrated strategy for protection, enhancement and sustainable development of the territory. (Peano 2013)

The park is intended therefore as a "laboratory" for managing environmental, cultural, social and economic values of an area while also managing to contribute to conservation, "becoming spring boards for economic development". Programmes and guidelines have also suggested how to experiment with new forms of enhancing the cultural heritage and natural resources through sustainable studies, consistent with the development of the local economy; it is suggested, with examples of best practice, how to promote new business activities compatible with resources to be protected, for example, forms of ecotourism, agribusiness or cultural routes for sustainable tourism.

The plan of the park, in the process of being drafted, has amongst its objectives "a design geared to transforming existing goods into an engine for local economic and social development, capable of integrating resources and services within an area defined by a strong territorial identity". It is an instrument with complex aims: in addition to the multiple values and characteristics to protect, and rules to apply, it must tackle the many problems of rehabilitation and enhancement of the natural and cultural heritage and policies reconnecting systems of bio-cultural relationships interrupted by recent settlement in the territory; it must seek to harmonise local planning policies, including the natural protected area, and carry out a careful study of sustainability for the park and innovative and suitable management of its resources.

The debate that preceded the establishment of the park was marked by successive actions and phases: from studies carried out on the main resources and identities to protect to the analysis of links between nature, culture and society and to the measures to implement in the park with a view to identifying answers capable of accommodating the needs of new users (e.g. tourists) or of facilitating the wishes of local communities of a new rural and territorial development.

55.2 The Problem of the Protected Area's Boundaries: A Necessary Ecological and Cultural Reconnection

The park's boundaries have been adjusted in the plan currently being drafted. They are defined according to natural resources, excluding urbanised ones.

Observing the boundaries and the development of the built-up areas (Fig. 55.2) shows a systematic interruption in relation to the coastal tourist settlements that generate a fragmentation in the continuity of the system of coastline protection. The PTCP of Lecce,[2] which envisages Salento as a *large park*,[3] and has amongst its objectives the creation of a spatial organisation of settlements in which concentration and dispersion are present together and integrated, introduces a new model of tourist accommodation dispersed across the territory in order to decompress the coastal stretches. Yet it has failed to deal with the fragmentation of the cultural and natural networks of the coast.

The will to protect, which attributed an important environmental value, is often insufficient for requalifying places that represent important areas of life for the population, as well as tourist attractions. It is an approach that contrasts with the proposals of the PPTR,[4] which extends its protection policies to rural and urban landscapes of everyday life and which addresses fragmentation in its *territorial projects* through suitable ecological networks, integrated enhancement and requalification of coastal landscapes.

Organising the parks of Salento with various forms of protection for the areas involved[5] means the protection system will assume a continuity that affects the coastal areas around Salento.

There exists a correspondence, with short breaks, between the SCIs affecting the water surface and the cover of the coast, the exception being the fracture corresponding to the stretch of sea facing the park. Thus, the ongoing measure is important, which provides for the establishment of a protected marine area for the peninsula of Salento by the Ministry of the Environment (Figs. 55.2 and 55.3).

---

[2] Provincial Territorial Coordination Plan, approved in 2008. http://www3.provincia.le.it/ptcp/ptcp/index.htm

[3] "(. . .) A complex habitat that takes the form of a park (. . .) not just referring to a place of leisure, but to a set of situations in which environmental values contribute in an essential way to building (. . .) the main activities and social practices" (Viganò 2006).

[4] The Regional Territorial Landscape Plan of Puglia, adopted on 02 August 2013.

[5] Sites of Community Importance (SCI), Special Protected Areas (SPA), protection of the hydrogeomorphic structure in the range of the coastal areas to the depth of 300 m (Law no. 431/1985) and Protected Landscapes of the Puglia Region.

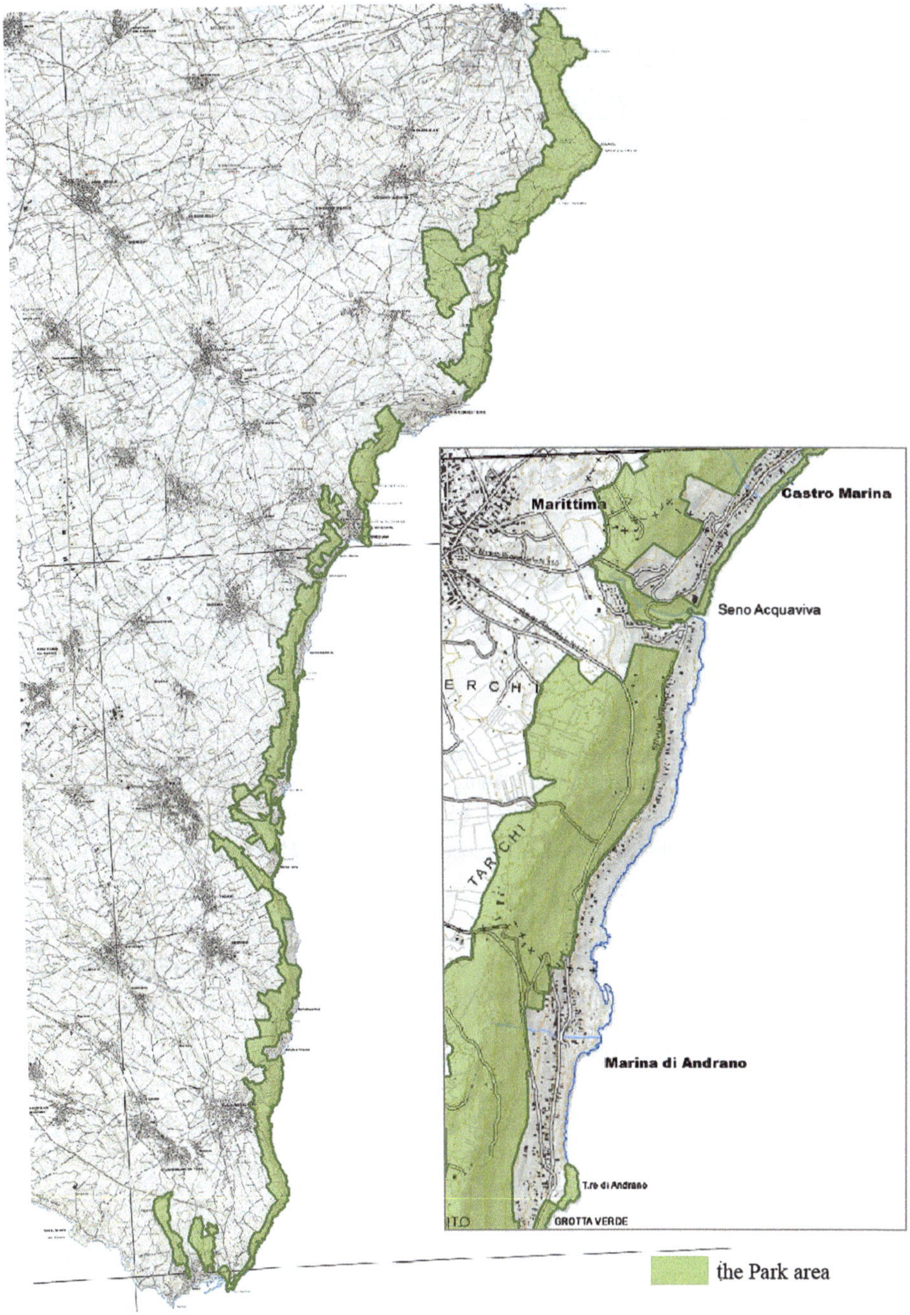

Fig. 55.2 The overlap between the park area and the development of the urbanised areas illustrates the interruption in relation to the main settlements along the coast

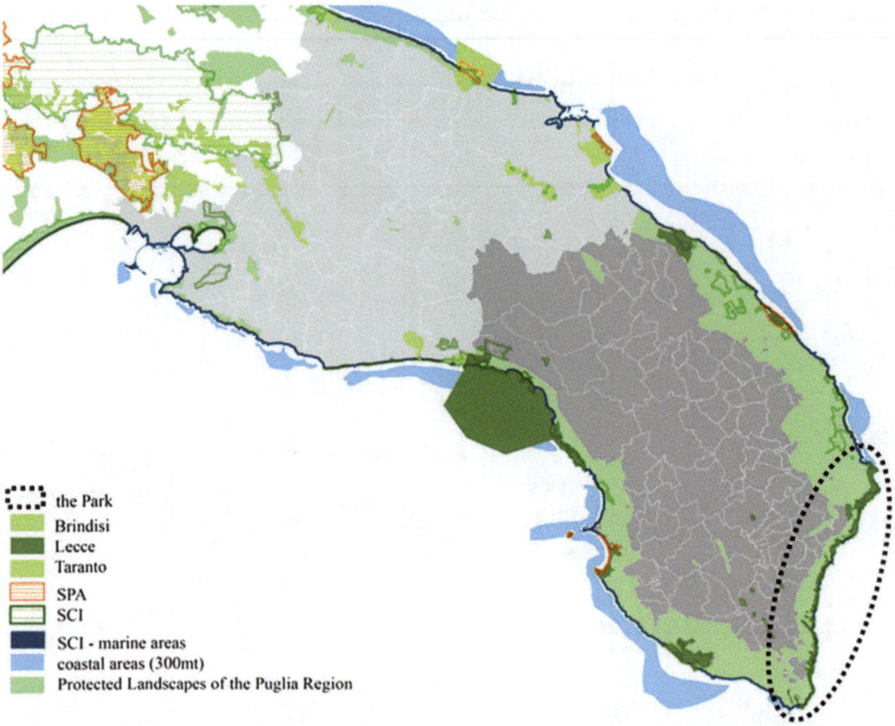

Fig. 55.3 Continuity of protection along the coastline between the park system, SCI, SPA, the side of the coastal territories and protected landscapes

55.3 Environmental Protection, Land-Use Planning and Landscape Policies: Questions of Governance

Following the adoption of the PPTR and the proposal of the Territorial Plan of the Park[6] and with some of the municipalities committed to updating their planning instruments, it is important to reflect on the interaction between urban, territorial and landscape planning. The PPTR pays attention to integrated enhancement and requalification of coastal landscape, the problem of the loss of identifiable characteristics and the banalisation of the landscape and ecology of coastal areas. The objective of fighting the tendency towards an organisation of the coastline made up of residences and tourist facilities implies taking the concept of "coastal zone" as a transition zone between sea coast and inland: an area that comprises territories of considerable variability according to their geomorphic and environmental features and societal history. The European strategy for integrated management of coastal zones shows how coastal areas, characterised by elevated environmental fragility

[6] Delivered on 11 July 2013. http://paesaggio.regione.puglia.it/

and ecological diversity and with a high level of anthropic pressure, need integrated development strategies in order to balance active protection and enhancement of territories while involving local communities.

The PPTR has not, however, seized the opportunity to define guidelines specifically geared towards interventions in coastal landscapes. The proposal of the park plan entails a zoning of the park, which seems to approach the demand for reconnection between environmental and cultural values of the area. It develops project strategies to extend the boundaries (as spaces of integration) in the surrounding landscapes including diversified contexts, with the aim of recuperating and enhancing anthropic, historical-cultural and agro-silvo-pastoral values. At the same time, the municipalities upgrade their territorial plans and manage the external areas adjacent to the park, identifying areas of compensation in relation to the limitations placed on the internal territories of the park, in apparent contrast to the requirement of a synergic relationship between park and context, for reasons of ecological and local development (Peano 2002a). It can only occur with the interaction between the three levels of planning.

References

Bushell R, Eagles PFJ (eds) (2007) Tourism and protected areas: benefits beyond boundaries. IUCN, Gland

Cannas R, Solinas M (2006) Il turismo dei parchi: tendenze in atto. Rapporto sul Turismo Italiano 2005–6. Mercury, Florence

Eagles PFJ, McCool SF, Haynes C (eds) (2002) Sustainable tourism in protected areas: guidelines for planning and management. Best practice protected areas guidelines. IUCN, Gland

Gambino R (2001) Terra e mare nella pianificazione territoriale-ambientale. In: Moschini R (ed) La gestione integrate delle coste ed il ruolo delle aree protette marino-costiere. ComunicAzione, Forlì

Peano A (2002a) Piano per il parco e piano regolatore comunale: la ricerca di un dialogo. National Congress INU Marche. Rivista della Federazione Italiana Parchi e Riserve Naturali 36. http://www.parks.it/federparchi/rivista/P36/78.html

Peano A (2002b) Parchi e turismo: dal conflitto alla cooperazione. Parchi naturali e turismo. Cerabona, Turin

Peano A, CED PPN (eds) (2000) Intégrer le tourisme dans les stratégies du territoire. Formation continue en TPE et PME touristiques. Féd. Parcs Naturels Régionaux de France, Paris

Viganò P (ed) (2006) Territori della nuova modernità. Piano Territoriale di Coordinamento Provinciale, Provincia di Lecce, Assessorato alla Gestione Territorile. Electa, Naples

Vv.Aa (2003) Parchi, natura, turismo. La situazione in Italia. Pirmo Rapporto sul Turismo Natura. AGRA, Rome

Chapter 56
A Regional Planning for Protected Areas of Sustainable Development in the Mercantour and Maritime Alps

Marco Valle and Maria Giovanna Dongiovanni

Abstract The strategic planning has a growing relevance in the national and international context of territorial development and has been for a long time the center of a lively technical-professional, scientific, and cultural debate. In this context, the design of the future of the territories is the path that all local authorities should set out, using the valuable contribution of scholars, social workers, professionals, and all those who feel they can contribute to territorial development. The experience presented here is a management plan of a Site of Community Importance (SCI), located in the Mercantour and Maritime Alps protected area. The management plan of this SCI is a super-local tool called to meet the needs of an area as a whole, crossing the concept of the administrative border, even national, to direct the management decisions to the pursuit of a common goal. Its drafting has been made possible thanks to the contribution of many experts in different fields and to the involvement of local stakeholders, with the goal to ensure the preservation of the structure and function of the habitats and the long-term conservation of the species, while taking into due consideration the socioeconomic factors and values.

Keywords Management • Strategic planning • Natura 2000 network

56.1 Strategic Planning in Italy

Since the 1980s, and especially in the 1990s, various European cities and regions tested new procedures for planning, going beyond the traditional urban planning mainly based on restraints and control of land use. After a long period in which the regulation of land uses had been entrusted to laws for development which did not take into account the environmental aspects, finally the establishment of the Ministry of the Environment (July 8 1986, Law no. 349), provided the opportunity

M. Valle (✉) • M.G. Dongiovanni
SiTI Istituto Superiore sui Sistemi Territoriali per l'Innovazione, Via Pier Carlo Boggio, 61, 10138 Turin, Italy
e-mail: valle@siti.polito.it; mariagiovanna.dongiovanni@siti.polito.it

R. Gambino, A. Peano (eds.), *Nature Policies and Landscape Policies*,
Urban and Landscape Perspectives 18, DOI 10.1007/978-3-319-05410-0_56

and set the stage to give an important role to environmental issues in the process of territorial planning, it began to become "strategic" – where strategic means aimed at building a future which offers a high quality of life consistently with a high level of protection of natural resources (which are unique and non-renewable). From that date, in fact, some European directives were incorporated into legal system, and so in Italy too, the procedure of Environmental Impact Assessment became compulsory for some kind of projects: a first step toward the integration of all the components of the territorial system in the evaluation and planning process.

In 1986, therefore, a slow process of integration with the community model starts: Italy begins to adapt to a system of planning rules aimed at integrating the country into a more complex reality, in which it is necessary to weave networks of synergistic relationships aimed at the exchange of experiences and good practices. So, the awareness that membership of the European Union is not only a matter of geographical location or of involvement in economic agreements with other states but a status requiring a systemic approach to the problems of the territory as a whole (as a human-environment system) begins to spread. The goal is to achieve optimum utilization of resources and to improve the economy of the country through the enhancement of local systems in synergic with their own territories.

This enlargement of cultural boundaries, which brings a brand new way to interpret and design the development of man-nature systems, lays down some indispensable conditions, such as the testing of methodologies and development processes inspired by a new system of governance. This governance model is, ultimately, what we here call "strategic spatial planning."

The scheme most widely used for planning tools starts with the definition of the goal to be reached that is the future overall scenario that the territory wants to achieve. The scenario is the aggregation of a series of macro objectives, which should thus be clearly delineated in order to identify, in the next step, all the actions which contribute to build the vision of the area projected in the future. At the same time, projects should be identified, specific material or immaterial initiatives, that will help to actually implement the strategic actions outlined.

The focal point of this process is the "consciousness of common interests," being able to achieve a transformation that, in the long term, ensures a comprehensive improvement of the life quality of citizens.

The strategic aspect of planning does not rely only on time but also on space. Development, and therefore planning, does not dwell anymore on the localism of administrative boundaries. On the contrary, it is necessary to strive to focus on the territorial point of view (more than on the local one), being inspired by a more efficient and effective interrelationship between decision-making, participation, and concrete actions, in order to meet the need to plan interpreting the needs of the territory, so that the plan is a tool in which the whole community can identify itself. Thus the plan, despite of being valid and effective within the intangible boundaries designed by the administrative limits, manages to frame the broader strategic vision of development of the entire territory, allowing it to fit within the flow of transformation of the vast territory, sharing vision and objectives to put together the mosaic of sustainable development.

In this regard, it would be useful to quote the words of Daniel Burnham (1846–1912), an American architect and urban planner who was one of the leading exponents of American City Beautiful movement and author, in 1909, of the urban plan of Chicago. Burnham gives some indications, still very topical, on the need to plan the development looking at the territory, rather than to dwell in the local community, and keeping in mind that this plan should be the instrument able to lead the country into the future:

> Make no little plans. They have no magic to stir men's blood and probably themselves will not be realized. Make big plans; aim high in hope and work, remembering that a noble, logical diagram once recorded will never die, but long after we are gone will be a living thing, asserting itself with ever-growing insistency. Remember that our sons and grandsons are going to do things that would stagger us. Let your watchword be order and your beacon beauty. Think big. (Burnham 1909)

56.2 The Territory and the Case Study: The Management Plan for a Site of Community Importance in the Mercantour and Maritime Alps Protected Area

The management plan of a Site of Community Importance (SCI) fits perfectly in the context outlined here, since it is a super-local tool called to meet the needs of an area as a whole, crossing the concept of the administrative border, even national, to direct the management decisions to the pursuit of a common goal.

Based on the existing regulations and guidelines, the management plan of a SCI plays a more operational than strategic role. In this case, however, considering the characteristics of the area (transnational area, different national laws, a high number of different actors to be involved in the process), it is preferred to give the plan a more strategic tone, so that the document can serve as a "collector" of all the strategic and programmatic courses of the two protected areas, without prejudice to the conservation objectives peculiar to a SIC management plan.

The SCI management plan is an instrument of planning designed to ensure the preservation of the structure and function of the habitats and the long-term conservation of the species, while taking into due consideration the socioeconomic factors insisting on the area. This is the operational aspect of the Natura 2000 network, "a coordinated and coherent system of areas for the conservation of biodiversity," introduced by the European Union and regulated by the Habitats Directive (92/43/EEC) and by the Birds Directive (79/409/EEC).

The Maritime Alps Natural Park and the Natura 2000 site called IT1160056 "Maritime Alps" are characterized by certain environmental and territorial features, such as the extraordinary biological richness in species of fauna and flora and the complexity of natural or seminatural habitats and geological and geomorphological peculiarities, which together constitute an environmental and landscape mosaic rich and varied, in terms of both nature and perception. These outstanding elements stress out the uniqueness of the climatic characteristics and of geological and biogeographic processes, which shaped the landscape and get the biological component to evolve, sometimes giving rise to adjustments and specializations unique to this area of the Alps. These extraordinary features allow the Maritime Alps Natural Park to be inscribed, along with the Ligurian Alps, in the list of the ten "hot spots" of Mediterranean floristic biodiversity (Médail and Verlaque 1997), thanks to the presence of numerous endemic and subendemic species of this Alp sector (Casazza et al. 2005). Although the settlements and urban areas are representative of a very small percentage of the planning area (less than 1 %) and although the human presence is subject to seasonal fluctuations related to tourism and excursions, the interactions with productive activities exploiting the natural resources are significant (hydropower industry, mining, tourism, livestock, productive use of forests, etc..) and detectable both within the protected area, both in the immediate vicinity.

In a territorial framework showing both an outstanding natural value and the usability of certain economic resources, it is easy to understand how the conservation of natural values and the development of productive activities can often be complex to manage; thus, it becomes necessary to outline effective and decisive compromises.

Based on these considerations, the approach for the management plan has been marked with the integration planning tools provided by law for the territory of the Maritime Alps Natural Park, to which the SCI IT1160056 "Maritime Alps" overlap almost completely and to ensure coherence with instruments provided by the Mercantour National Park – the Projet de Charte du Mercantour National Park – and the Plan du Parc and Document d'Objectif (DOCOB) for the Natura 2000 site "The Mercantour."

With respect to the guidelines at the international level and the relevant legislation at the national and regional level, orienting the draft of the management tools for protected areas, which include a variety of instruments, the management plan is characterized by its interdisciplinary nature and its character of coordination with other planning and management tools (area plan, naturalistic plan, socioeconomic development plan, forest business plan, pastoral management plan – implemented in the context of activity 3, ecological water plan) and sectorial studies developed within the Integrated Cross-Border Program Alcotra: "Cross-Border Space Maritime Mercantour: the natural and cultural diversity at the heart of sustainable and integrated development."

This methodological approach has provided, among other things, the establishment of a working group characterized by a high degree of inter- and multidisciplinary approaches, both from the point of view of entities involved and from the point of view of professionalism and skills.

In a more enlarged view, the SCI management plan has also laid the bases for the actual realization of an ecological cross-border network, thanks to the creation of a strategy for the protection of biological diversity and landscape.

The elaboration of the SCI management plan "Maritime Alps" has therefore prepared the strategic vision of the ecological network, outlining the operational path to be implemented and actively contributing to the process of regional interconnection.

References

Casazza G, Barberis G, Minuto L (2005) Ecological characteristics and rarity of endemic plants of the Italian Maritime Alps. Biol Conserv 123:361–371

Médail F, Verlaque P (1997) Ecological characteristics and rarity of endemic plants from southeast France and Corsica: implications for biodiversity conservation. Biol Conserv 80:269–281

Erratum to Chapter 23: Participation and Regional Governance. A Crucial Research Perspective on Protected Areas Policies in Austria and Switzerland

Norbert Weixlbaumer, Dominik Siegrist, Ingo Mose, and Thomas Hammer

R. Gambino, A. Peano (eds.), Nature Policies and Landscape Policies,
Urban and Landscape Perspectives 18, DOI 10.1007/978-3-319-05410-0_23

Prof. Thomas Hammer's email ID was incorrect. The correct email ID should read as: thomas.hammer@cde.unibe.ch

The online version of the original chapter can be found under
DOI 10.1007/978-3-319-05410-0_23

N. Weixlbaumer (✉)
Department of Geography and Regional Research, University of Vienna, Universitätsstraße 7, 1010 Vienna, Austria
e-mail: norbert.weixlbaumer@univie.ac.at

D. Siegrist
Institute for Landscape and Open Space, HSR University of Applied Sciences, CH-8640 Rapperswil, Switzerland
e-mail: dominik.siegrist@hsr.ch

I. Mose
ZENARiO – Center for Sustainable Spatial Development, Applied Geography and Environmental Planning Research Group, Carl von Ossietzky University Oldenburg, D-26111, Oldenburg, Germany
e-mail: ingo.mose@uni-oldenburg.de

T. Hammer
Centre for Development and Environment CDE, University of Bern/Switzerland, Bern, Switzerland
e-mail: thomas.hammer@cde.unibe.ch

R. Gambino, A. Peano (eds.), *Nature Policies and Landscape Policies*,
Urban and Landscape Perspectives 18, DOI 10.1007/978-3-319-05410-0_57

Index

R. Gambino, A. Peano (eds.), *Nature Policies and Landscape Policies*, Urban and Landscape Perspectives 18, DOI 10.1007/978-3-319-05410-0

MIX
Papier aus verantwortungsvollen Quellen
Paper from responsible sources
FSC® C105338

If you have any concerns about our products,
you can contact us on
ProductSafety@springernature.com

In case Publisher is established outside the EU,
the EU authorized representative is:
Springer Nature Customer Service Center GmbH
Europaplatz 3, 69115 Heidelberg, Germany

Printed by Libri Plureos GmbH
in Hamburg, Germany